Neurotoxicology

Neurotoxicology

Edited by

Mohamed B. Abou-Donia

Professor of Pharmacology and Neurobiology
Deputy Director of Toxicology
Duke University Medical Center
Durham, North Carolina

CRC Press
Taylor & Francis Group
Boca Raton London New York

CRC Press is an imprint of the
Taylor & Francis Group, an **informa** business

CRC Press
Taylor & Francis Group
6000 Broken Sound Parkway NW, Suite 300
Boca Raton, FL 33487-2742

© 1992 by Taylor & Francis Group, LLC
CRC Press is an imprint of Taylor & Francis Group, an Informa business

First issued in paperback 2019

No claim to original U.S. Government works

ISBN-13: 978-0-367-45028-1 (pbk)
ISBN-13: 978-0-8493-8895-8 (hbk)

Visit the Taylor & Francis Web site at
http://www.taylorandfrancis.com

and the CRC Press Web site at
http://www.crcpress.com

Library of Congress Cataloging-in-Publication Data

Neurotoxicology/editor, Mohamed B. Abou-Donia.
  p. cm.
  Includes bibliographical references and index.
  ISBN 0-8493-8895-3
  1. Neurotoxicology. 2. Nervous System—drug effects. I. Abou-Donia, Mohamed B.
[DNLM: QV 76.5 N49453]
RC347.5.N484 1992
616.8—dc20
DNLM/DLC
for Library of Congress   91-43467
                                                                                CIP

Library of Congress Number 91-43467

PREFACE

Neurotoxicology is concerned with the adverse effects of toxic agents on the nervous system. It deals with the effect of these agents *in vivo* in the intact animal or *in vitro* in tissues or cell cultures to evaluate their risk and hazard to human health.

Neurotoxicants may cause serious damage to the nervous system, i.e., brain, spinal cord and peripheral nerves. Although many of these disorders are reversible when promptly treated, irreversible changes may also occur, some of which are fatal. Long-term neurotoxic effects may involve succeeding generations. Although the major visible consequence of many neurotoxic agents is behavioral dysfunction, there are usually other significant alterations from molecule to organism that precede or accompany clinical manifestations at various levels. Neurotoxic alterations occur at the following targets: molecular changes at the DNA or RNA level resulting in changes in protein synthesis; cellular interaction leading to neurophysiological alterations; tissue lesions resulting in neuropathological changes; organism changes with consequent neurobehavioral or neurological abnormalities.

Neurotoxicological studies are carried out in animals. No single animal species is adequate for all neurotoxicity studies for extrapolation to humans. Animal models used in toxicologic investigations are those that closely resemble humans in most aspects of neurotoxicity, e.g., clinical signs, behavorial condition, physiological changes, pathological lesions, and biochemical alterations. Neurotoxic agents bind to specific molecular and cellular targets in the nervous system. Molecular and cellular changes are limited by the repair process. All of these processes are dependent on the chemical, as well as on the strain and species of experimental animals.

Modern neurotoxicology has been an active area of research since the latter part of the 19th century. Many neurotoxic effects first became known through incidences of human poisoning. As early as 1898 massive poisoning with tri-*o*-cresyl phosphate (TOCP) resulted in delayed neurotoxicity in as many as 40,000 persons in the United States. Other major neurotoxic episodes include TOCP poisoning in Morocco in 1958, mercury poisoning in Iraq in the 1950s, collective suicide with cyanide in Guyana in the 1970s, and methyl isocyanate poisoning at Bhopal, India in the 1980s.

Advances made in synthetic chemistry in recent years and consequent rapid expansion in research have resulted in the development of many useful therapeutic drugs, industrial chemicals, and agricultural products. These advances, however, have not been without risk to human health. The pharmaceutical industry has developed new and effective drugs that may cause nervous system damage resulting from overdosing due to mistakes or misuse. Chemicals of drug abuse are of great concern since they affect persons of all socioeconomic levels irrespective of age, sex, race, or culture.

Since World War II thousands of organic chemicals belonging to many chemical groups, e.g., chlorinated hydrocarbons, organophosphorus compounds, carbamates, and pyrethroids, have been synthesized and used as pesticides. The use of these highly neurotoxic chemicals, which are intentionally contaminating the environment to kill some form of life, has led to injury in nontarget organisms including humans. Some of these chemicals or their effects are long lasting and persist in the environment, resulting in the contamination of the atmosphere and pollution of the air and water resources. Ground and surface waters in many locations are permeated by neurotoxic organic and inorganic chemicals and their breakdown products.

Advances in the chemical industry have resulted in the introduction of many products for household use. Other chemicals such as tobacco and alcohol that may be consumed socially are also abused and cause many neurologic diseases.

Accidental exposure to chemicals represents the most common cause of acute medical illness in developed countries, while they are the second most common cause of death, after infectious diseases, in many developing countries. This is of particular significance, since developing countries that have been traditionally plagued with three traditional maladies—poverty, illiteracy, and disease—have been greatly affected with the adverse effects of chemicals resulting from rapid and environmentally unsound industrialization and urbanization.

Episodes and neurotoxic diseases and concerns of long-term adverse effects of neurotoxicants have increased the awareness of people worldwide and initiated research programs for careful studies of neurotoxic effects of chemicals and the development of appropriate antidotes and treatments. This has also led to the establishment of public policy for the protection of health and environment from the neurotoxic effects of chemicals. Development of antidotes or treatments for neurotoxicities induced by chemicals is sometimes handicapped by an inadequate understanding of their mechanisms of action. This lack of understanding is further complicated by other factors such as the identity of the offending chemicals, whether the exposure is limited to a single chemical or a mixture of chemicals, the amount of intake of the chemical, exposure to other drugs or chemicals, the elapsed time after exposure, idiosyncrasies, pregnancy, the metabolic bioactivation or inactivation of the chemical, and the extent of its distribution, storage, and excretion.

Several neurotoxicology books are general references or cover specific topics. These books were designed to provide either a rapid overview of neurotoxicity or an in-depth understanding of a specific subject. *Neurotoxicology* presents a multidisciplinary volume on neurotoxicity. It derives from the Editor's experience in teaching a course about neurotoxicology and provides a coherent understanding of the field resulting from the integration of multilateral ideas covering several disciplines: chemistry, biochemistry, anatomy, physiology, and pathology relating to the action of neurotoxic agents on the nervous system. It accurately details the neurotoxic agents, their mechanisms of action on the nervous system, and the diagnosis and treatment of their adverse effects. It describes the methods for studying the mechanisms of action of neurotoxic agents: biochemical, electrophysiological, and pathological. Other factors affecting the neurotoxic action are also covered, i.e., developing nervous system, nutrition, alcohol abuse, and metabolism and toxicokinetics. Diagnosis and detection of neurotoxic diseases are also included for humans and experimental animals. These topics have been supplemented with adequate references to allow follow-up and in-depth reading.

Neurotoxicology is a reference for graduate and medical students, individuals engaged in research in neurotoxicology, regulatory toxicologists, industrial hygienists, as well as individuals desiring to understand the adverse effects of neurotoxic agents on the nervous system that are manifested as behavioral or neurological disease. This book should be of value to individuals interested in studying injuries to the nervous system caused by neurotoxic agents in the diet, environment, the field, or the workplace. It should stimulate further research in screening drugs and chemicals for neurotoxic action and in investigating the mechanisms of action of neurotoxicity to develop antidotes and treatments for them. This should lead to a better understanding of neurotoxic diseases and result ultimately in improving human health. It is hoped that this multidisciplinary textbook will provide information that will benefit and safeguard society against neurotoxic diseases.

Mohamed B. Abou-Donia

THE EDITOR

Mohamed B. Abou-Donia is a professor of Pharmacology and of Neurobiology, and Deputy Director of the Toxicology program at Duke University Medical Center. Dr. Abou-Donia graduated in 1960 from Alexandria University, Alexandria, Egypt, with a B.S. degree (with honor) and received his Ph.D. degree in 1967 from the University of California, Berkeley. He obtained training at Texas A&M University from 1967 to 1970. He served as an Assistant Professor at Alexandria University from 1970 to 1973, and as an Assistant and Associate Professor of Pharmacology at Duke University Medical Center in 1977 and 1979, respectively. He became a Deputy Director of the Duke University Toxicology program in 1981, a Professor of pharmacology in 1984 and a Professor of neurobiology in 1988 at Duke University Medical Center.

Dr. Abou-Donia is certified by the American Board of Toxicology (1981, 1986, 1991) and is a Fellow in the Academy of Toxicological Sciences (1982, 1987, 1992). He is a member of the American Association for the Advancement of Sciences, the American Association of Pathologists, the American Chemical Society, the American College of Toxicology, the American Institute of Chemistry, the American Society of Neurochemistry, the American Society for Pharmacology and Experimental Therapeutics, the Entomological Society for Pharmacology and Experimental Therapeutics, the Entomological Society of America, the International Society of Xenobiotics, the New York Academy of Sciences, Sigma Xi, Society of Environmental Toxicology and Chemistry, the Society for Neuroscience, and the Society of Toxicology. Dr. Abou-Donia is the author or co-author of approximately 200 research papers. He has presented approximately 130 papers at national and international meetings, and 30 invited lectures at universities and institutes. His current major research interests include the molecular pathogenesis of chemically induced neurodegenerative disorders.

CONTRIBUTORS

Mohamed B. Abou-Donia, Ph.D.
Professor
Departments of Pharmacology and Neurobiology
Deputy Director of Toxicology
Duke University Medical Center
Durham, North Carolina

Martha M. Abou-Donia, Ph.D.
Senior Clinical Research Scientist
Department of Clinical Neurosciences
Burroughs Wellcome Company
Research Triangle Park, North Carolina

Peter G. Aitken, Ph.D.
Associate Research Professor
Departments of Cell Biology and Neurobiology
Duke University Medical Center
Durham, North Carolina

Syed F. Ali, Ph.D.
Neurotoxicologist
Adjunct Professor of Toxicology
Division of Neurotoxicology
National Center for Toxicological Research
Jefferson, Arkansas

Rebecca J. Anderson, Ph.D.
Manager, R&D Projects
Scientific Affairs Administration
Boehringer Ingelheim Pharmaceuticals, Inc.
Ridgefield, Connecticut

Stephen Bondy, Ph.D.
Professor
Department of Community and Environmental Medicine
University of California
Irvine, California

Nell B. Cant, Ph.D.
Associate Professor
Department of Neurobiology
Duke University Medical Center
Durham, North Carolina

Louis W. Chang, Ph.D.
Professor
Department of Pathology
University of Arkansas for Medical Sciences
Little Rock, Arkansas

Jeffry F. Goodrum, Ph.D.
Research Scientist
Brain and Development Research Center and
Department of Pathology
University of North Carolina
Chapel Hill, North Carolina

G. Jean Harry
Systems Toxicity Branch
National Institute of Environmental Health Sciences
Research Triangle Park, North Carolina

Cynthia M. Kuhn, Ph.D.
Associate Professor
Department of Pharmacology
Duke University Medical Center
Durham, North Carolina

Daniel M. Lapadula, Ph.D.
Principal Scientist
Department of Toxicology/Safety Evaluation Center
Schering-Plough Research
Lafayette, New Jersey

Richard B. Mailman, Ph.D.
Professor
Departments of Psychiatry and
Pharmacology
University of North Carolina
Chapel Hill, North Carolina

Pierre Morell, Ph.D.
Professor
Brain and Development Research
Center and Department of
Biochemistry
University of North Carolina
Chapel Hill, North Carolina

Toshio Narahashi, Chairman
Department of Pharmacology
Northwestern University Medical
School
Chicago, Illinois

James P. O'Callaghan, Ph.D.
Senior Neuroscientist
Neurotoxicology Division
U.S. Environmental Protection
Agency
Research Triangle Park, North
Carolina

Cynthia S. Payne, M.D.
Associate
Department of Radiology
Duke University Medical Center
Durham, North Carolina

Hugh A. Tilson, Ph.D.
Director
Neurotoxicology Division
Health Effects Research
Laboratory
U.S. Environmental Protection
Agency
Research Triangle Park, North
Carolina

CONTENTS

PART VI. EVALUATION AND DIAGNOSIS OF NEUROTOXICITY

*To my wife Martha
and my children
Rick, Steve, and Suzanne*

Part

I

Introduction

Introduction
M.B. Abou-Donia

Chapter

1

Introduction

Mohamed B. Abou-Donia
*Departments of Pharmacology and Neurobiology
Duke University Medical Center
Durham, North Carolina*

This book is about neurotoxicology, the area of toxicology that deals with the adverse effects of neurotoxic chemicals on the nervous system. It deals with neurotoxic agents and the nervous system, as well as the methods to study neurotoxicities and the factors affecting neurotoxic action. There are more than 1000 known neurological disorders in humans. More than 50 million Americans are affected annually by these diseases that cost more than $120 billion. These illnesses affect people of all ages including the elderly and children.

Some neurologic diseases result from bacterial or viral infections, others are related to the immune system. Many nervous system disorders are of unknown etiology. The extent that neurotoxic chemicals may contribute to nervous system diseases is not clear. Although these chemicals produce their own specific neurologic deficits, they may also affect other neurologic diseases through potentiation or synergism. Interaction between neurologic diseases and chemically induced neurotoxicities may result from several mechanisms, e.g., increased accessibility of neural tissues to neurotoxic agents by changing membrane permeability, changes in the activity of drug metabolizing enzymes, or a joint neurotoxic action at the target. This is of particular importance in the case of exposure to chemicals that interfere with the cholinergic system such as organophosphorus esters and diseases that involve the cholinergic system such as Parkinson's and Alzheimer's diseases.

It is of great interest that the Congress of the United States has designated the decade which began in 1990 as the Decade of the Brain, with an annual budget for federal neuroscience research of approximately $1.2 billion. Such congressional interest in neurosciences reflects the public concern regarding neurologic diseases including those induced chemically. Research into the mechanisms of neurologic illnesses should contribute to the understanding of how they affect the nervous system and improve ways for prevention and provide sensitive methods for diagnosis and effective treatment of these diseases. A brief introduction to the nervous system follows.

THE NERVOUS SYSTEM

The nervous system is the organ that controls all processes of life. It is divided into two parts (Figure 1):

1. The central nervous system (CNS), composed of the brain and spinal cord.
2. The peripheral nervous system, consisting of all other nervous system tissue: ganglia and a network of nerve fibers that connect the CNS to the body.

Nervous tissue is composed of functional units known as neurons or nerve cells (Figure 2). These cells do not have the ability to reproduce and have limited power to repair. Neurons consist of

1. a cell body or perikaryon
2. one or more branching afferent processes known as dendrites, which transmit impulses toward perikaryon and
3. an efferent process known as the axon, which conducts impulses away from the perikaryon.

Dendrites and axons are collectively called nerve fibers. Branches present along the length of nerve fibers are called collaterals; those at the ends are called terminal arborizations. The synapse is a region in which the terminal arborizations of the axon of one neuron come in proximity to the dendrites or perikarya of succeeding neurons.

The Brain

The brain (encephalon), which constitutes the major tissue of the nervous system, contains approximately 200 billion nerve cells and a trillion supporting cells. Brain neurons are all formed before birth and no new neuronal cells are produced thereafter. It is divided into three sections (Figure 3):

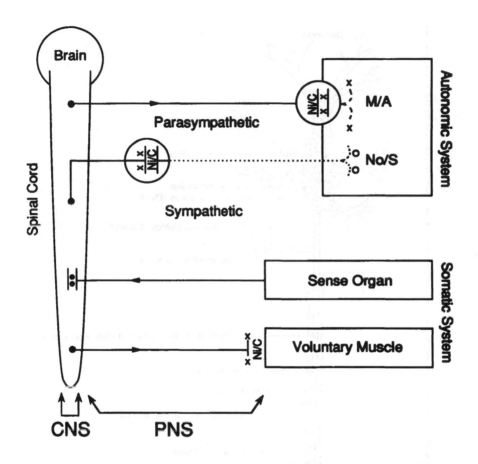

Figure 1. Schematic presentation of the nervous system depicting the central and peripheral nervous systems. The peripheral nervous system is further subdivided to autonomic and somatic nervous system.

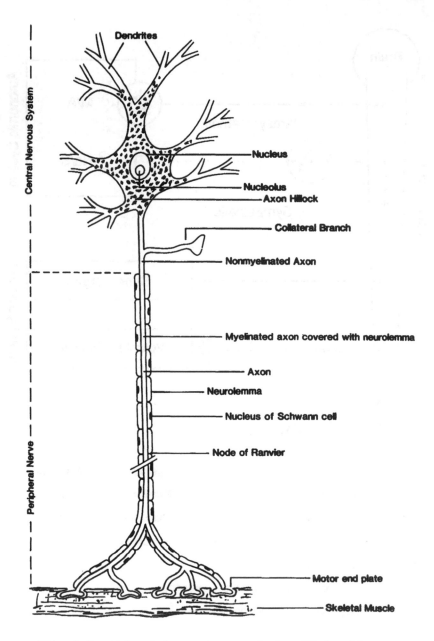

Figure 2. Presentation of a lower motor neuron showing cell body and the axon.

1. Forebrain
    a. Cerebrum
    b. Limpic system: amygdala, hippocampus, and septum
    c. Thalamus
    d. Hypothalamus

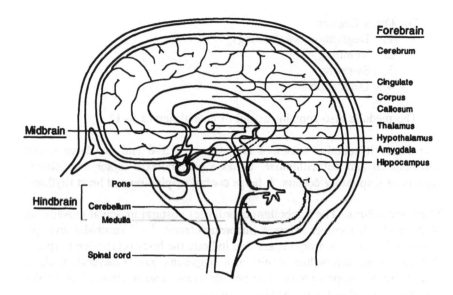

Figure 3. Sagittal view of the brain showing its three subdivisions: forebrain, midbrain, and hindbrain and their parts.

2. Midbrain
3. Hindbrain
   a. Pons
   b. Medulla oblongata
   c. Cerebellum

Brain Functions

Medulla Oblongata. The medulla is the lowest part of the brain and contains most of the ascending, i.e., the cuneate and gracilis tracts, and descending, i.e., corticospinal (pyramidal) tracts. It also contains the following centers that regulate many functions of the body:

A. Respiratory Centers
   1. Inspiratory
   2. Expiratory
   3. Others: coughing, sneezing, etc.
B. Cardiac Center
   1. Accelerator
   2. Depressor
C. Vasomotor Center
   1. Vasopressor
   2. Vasodilator

D. Other Centers
   1. Deglutition (swallowing)
   2. Vomiting
   3. Sweating
   4. Defecation

The medulla also contains nuclei of cranial nerves 9–12.

The Pons. It contains nuclei of cranial nerves 5–8 and two large tracts: corticospinal and corticolumbar. It has a pneumotaxic center that inhibits activity of respiratory centers. It helps control respiration and heart rhythms.

The Cerebellum. The cerebellum coordinates postural muscular movements of the body. It receives afferent (incoming) tracts, i.e., vestibular and spinocerebellar, from sensory organs that indicate the body orientation in space. Damage to the cerebellum results in a staggering gait. Chemicals such as neurotoxic organophosphorus compounds cause degeneration of vestibular and spinocerebellar tracts resulting in ataxia.

The Midbrain. The midbrain contains the nuclei of the third and fourth and part of the nucleus of the fifth cranial nerve. It also contains the red nucleus that is the origin of the rubrospinal tract. The superior and inferior colliculi are present on the dorsal surface of the midbrain beneath the cerebellum. The superior colliculi contain reflex centers for visual, auditory, and tactile stimuli. The inferior colliculi are reflex centers for auditory stimuli.

The Limpic System. The amygdala, hippocampus, and septum are parts of the cortex. They help regulate emotion, memory, and certain aspects of movement.

The hippocampus and medial region, parts of the temporal lobe, which are connected with the cerebral cortex, play an important role in the process of memory consolidation, i.e., consolidation of new learning. On the other hand, cortical areas are more important for the storage of knowledge, e.g., vocabulary and its everyday use. Working or transient memory, e.g., during conversation, is dependent upon the prefrontal cortex. Neurotransmitters such as dopamine, norepinephrine, and ACh influence certain neurons in this area. In Alzheimer's disease, these systems exhibit dysfunction. Studies on the mechanisms of memory, indicate that it involves a persistent change in the relationship between neurons. Such changes take place through stable biochemical alterations in the neurons.

The Thalamus. The thalamus is the major relay for visual and auditory information going to the cortex. Pain and proprioceptive pathways also relay through the thalamus. It is also responsible for subjective feeling.

The Hypothalamus. It has the following functions:

1. Regulation of body temperature: sweating and shivering
2. Regulation of metabolism, satiety (satisfaction, e.g., full gratification of appetite or thirst), water, and fat
3. Sleep
4. Sexual activity
5. Control of emotions: anger, fear, and pleasure. Furthermore, the hypothalamus neurons secrete hormone-releasing factors that travel via blood to the anterior pituitary glands where they trigger the release of pituitary hormones into blood stream:
   a. Somatotrophin (STH): growth hormone
   b. Thyroid-Stimulating Hormone (TSH): stimulates thyroid's thyroxin production
   c. Adrenocorticotropic Hormone (ACTH): stimulates release of adrenal cortical steroids
   d. Lactogenic Hormone (prolactin): stimulates growth of mammary glands
   e. Follicle-Stimulating Hormone (FSH): stimulates ovarian follicle maturation and spermatogenesis
   f. Luteinizing Hormone (LH): stimulates ovulation, progesterone release, and corpus luteal development in females, and testosterone production in males.

The hypothalamus releases two other hormones: antidiuretic hormone (ADH, vasopressin), which stimulates water reabsorption by the kidney, and oxytocin, which stimulates smooth muscle (e.g., uterus) contraction.

Cerebrum. The cerebrum is composed of the corpus striatum and cerebral cortex. The corpus striatum is part of the voluntary motor system and is modulated by the extrapyramidal motor system. Damage to this motor system is noted in Parkinson's disease and results in

1. paralysis characterized by tremor and rigidity
2. interruption of sensory inputs resulting in the "perception" of the sense of smell, sight, hearing, taste, touch or orientaton in space with sensory tracts and
3. processing of intellectual or emotional information such as reasoning, judgment, memory or integrated language function with associated neurons.

The voluntary motor system is divided structurally and functionally into two subdivisions: the pyramidal system and the extrapyramidal system.

The nerve cell bodies of the pyramidal system are present in the motor area of the cerebral cortex. Its axons pass down the internal capsule, cerebral peduncles, to the ventral portion of the pons and medulla. At the pyramidal decussation (the end of the medulla) they cross to the opposite side. Because of this crossing, the left cerebral hemisphere controls the right side of the body and vice versa. Then they enter the spinal cord on the lateral part. This system controls the movements of the body, primarily the fingers and muscles of the head and neck, e.g., eye, voice, etc. Neurotoxicants may affect the voluntary motor system by damaging any of the following tissues: cortex, internal capsule, cerebral peduncles, or medulla.

An upper motor neuron is a neuron in the pyramidal system, since it terminates on a nerve cell in the brainstem or spinal cord. Conversely, a lower motor neuron is a fiber that originates from the brainstem or spinal cord and terminates in the muscle (Figure 2). Pyramidal lesions result in dysfunctions in fine motor control movements, e.g., fingers, speech (laryngeal muscles), facial expression, etc. Generally, upper motor neuron lesions produce motor deficiencies ranging from tremor to paralysis of increased tone (spastic), and lesions of lower (spinal) motor neurons produces paralysis of decreased tone (flaccid).

The extrapyramidal system controls movements such as posture and walking. This system is present in all five levels of the brain as follows:

1. the cortex and corpus striatum in the cerebrum
2. the thalamus and subthalamic nucleus of the diencephalon
3. the substantia nigra and red nucleus of the midbrain
4. the positive nuclei and the purkinje cells and dentate nucleus of the cerebellum
5. the reticular formation of the neuraxis from the spinal cord to the diencephalon.

Cranial Nerves. The functions of the twelve cranial nerves that emerge from the brain are shown in Table 1.

The Spinal Cord

The spinal cord has eight cervical, twelve thoracic, five lumbar, five sacral, and one coccygeal pair of dorsal-ventral nerve roots. The sensory (afferent or incoming) nerves attach to the dorsal (back) side of the cord. The motor (efferent or outgoing) nerves leave the ventral (front) side of the cord. The spinal cord conducts sensory information to the brain and conducts motor commands from the brain to the effector organs.

Anatomy of the Spinal Cord

The spinal cord is symmetric about its midline divisions, the ventral median fissure and the dorsal median sulcus. Each half has a central portion

TABLE 1.
Functions of the Cranial Nerves

Nerve Number	Name	Sensory Function	Motor Function
I	Olfactory	Smell	None
II	Optic	Vision	None
III	Oculomotor	Eye muscles	Eye movement, pupil contraction
IV	Trochlear	Eye muscles	Eye movement
V	Trigeminal	Chewing, skin of face and head	Chewing movement
VI	Abducens	Eye muscle, lateral rectus	Eye movement
VII	Facial	Taste buds of the tongue, skin of ear	Facial expression, secretion from nasal mucose and salivary glands
VIII	Vestibulocochlear	Hearing, equilibrium	None
IX	Glossopharyngeal	Taste, visceral, e.g., blood pressure	Swallowing, glandular secretion
X	Vagus	Visceral, taste, skin of ear	Visceral muscles, e.g., heart, respiratory tract, gastrointestinal tract, glands; swallowing, speech
XI	Accessory	None	Speech and swallowing; movement of head and shoulder
XII	Hypoglossal	None	Movement of tongue

of gray matter surrounded by white matter. The gray matter is composed of cell bodies and nuclei, whereas the white matter is mostly myelinated axons. The white matter is divided into three areas:

1. dorsal columns (funiculi)
2. lateral columns
3. ventral columns

Similarly, the gray matter is divided into three divisions:

1. dorsal horns: contain integrative interneurons
2. ventral horns: contain large motor cells
3. lateral horns: contain preganglionic cells for autonomic nervous system

Ascending and Descending Tracts (Table 2)
The white matter is comprised of numerous ascending and descending tracts (Figure 4):

1. long ascending tracts that contain the majority of the dorsal columns

TABLE 2.
Functions of Ascending and Descending Tracts

Tract	Function
A. Ascending (sensory, affectors)	
1. Fasciculus gracillis	Conscious proprioception
2. Fasciculus cuneatus	Touch and pressure, vibration, sense and two-point discrimination
3. Posterior spinocerebellar tract	Reflex proprioception
4. Anterior spinocerebellar tract	Reflex proprioception
5. Lateral spinothalamic tract	Pain and temperature
6. Ventral spinothalamic tract	Light touch
B. Descending (motor, effectors)	
1. Lateral corticospinal tract	Voluntary motor control (pyramidal)
2. Rubrospinal tract	Muscle tone and coordination
3. Tectospinal tract	Visual and auditory reflexes
4. Vestibulospinal tract	Balance reflexes
5. Ventral corticospinal tract	Voluntary motor control

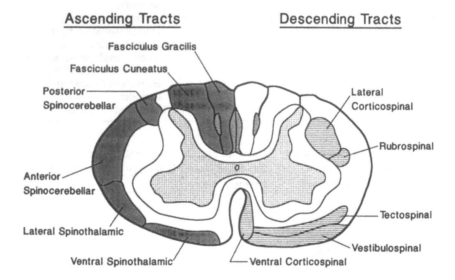

Figure 4. The main ascending and descending tracts in their spinal cord. Dotted central area is the gray matter.

2. short intersegmental tracts comprised of mixed ascending and descending tracts that surround the gray matter and

3. long descending tracts that are present between the border ascending tracts of the ventral and lateral columns and the central intrasegmental tracts.

Peripheral Nerves

Peripheral nerves are bundles of neurons that connect the body to the central nervous system as follows: cervical region to the neck and arm, the thoracic region to the trunk, the lumbar region to the legs, and the sacral region to the bowels and bladder.

The peripheral nervous system consists of the two divisions: the somatic and autonomic nervous systems (Figure 1). The somatic nervous system connects the spinal cord to voluntary skeletal muscles. The autonomic nervous system connects the spinal cord with the internal organs. It consists of two subdivisions: the sympathetic and the parasympathetic nervous systems. The sympathic nervous system functions during states of stress and excitation by mobilizing energy and resources. On the other hand, the parasympathetic nervous system functions by conserving energy and resources during resting state. There are two types of neurons: myelinated and unmyelinated.

Myelinated Axons

Myelinated axons are covered with a segmented myelin sheath. These fibers contain a specialized Schwann cell, which encircles the axon with several layers of myelin. The space between the segments is called the node of Ranvier where the axon contacts extracellular fluid. The myelin is high in lipid content and insulates as well as protects the axon membrane.

Unmyelinated Axons

These axons also have Schwann cells that are not wrapped around the axon. In addition to the protection provided by Schwann cells, neurons get further protection by forming bundles that are protected by connective tissue to form fascicle. A nerve is made up of a number of fascicles.

Spinal Peripheral Nerve

A spinal peripheral nerve is typically called a mixed nerve because it contains both motor and sensory components. Its motor neurons are somatic, which connect to skeletal muscle and postganglionic sympathetic autonomic motor, which connect to visceral structures. Sensory neurons are somatic sensory, e.g., pain, touch, temperature, position, sense, etc. Some of the sacral peripheral nerves carry parasympathetic fibers.

Peripheral nerves may be classified into four types:

1. large myelinated, somatic afferent and efferent fibers
2. smaller myelinated, preganglionic sympathetic efferents
3. unmyelinated postganglionic sympathetic efferents
4. small unmyelinated non-proprioceptive afferent fibers.

Nerve Conduction Velocity

Conduction velocity of the axon is determined by its diameter and myelination state. Axons with large diameter conduct faster than small-diameter ones. Also, a myelinated axon conducts faster than an unmyelinated axon.

Axonal Conduction

Neurons or nerve cells move signals by transmitting impulses along the axons. Axonal conduction depends on the permeability of its membrane to Na^+ and K^+ and upon the sodium pump. This process involves the opening and closing of ion channels. These are pores or molecule tunnels in the membrane that allow Na^+ and K^+ to cross the membrane. Ion pump may be blocked by specific chemicals. Na^+ channels are blocked by tetrodotoxin (from the puffer fish), tarichotoxin (from certain newts), and saxitoxin (from marine dinoflagellates). In contrast, tetraethylammonium ions block the K^+ channels that prolong the time course of action potentials by delaying depolarization, while neuronal excitability remains unchanged. The difference in ion concentration between the inside and outside of the axon produces an electrical current resulting in voltage changes across the membrane. At the initiation of nerve impulse, the polarity of the electric current reverses at the site of stimulation on the cell's membrane. This reverse is called an action potential. The action potential then travels along the membrane to the axon at a speed of up to several hundred miles an hour. When the action potential reaches the end of the axon, it triggers the release of a chemical neurotransmitter, which transmits the impulse across the synapse to the next neuron.

Following its release, the neurotransmitter binds to receptor molecules present on the surface of the next neuron at the postsynaptic membrane. This leads to changes in the outer membrane of the neuron and triggers an action such as the contraction of a muscle or increased activity of an enzyme-releasing cell.

Neurotransmitters

Acetylcholine (ACh)

ACh is synthesized in the cell body of the neuron and transported to the axon terminal. Following nerve stimulation, an action potential travels down the axon and when it reaches the terminal Ca^{2+} rushes in and ACh is released from the presynaptic membrane into the synapse and attaches itself to the ACh receptor. When this occurs at the neuromuscular junction, sodium ion channels are opened in skeletal muscle, causing its contraction. ACh is immediately broken down by acetylcholinesterase (AChE).

This system is very important in neurotoxicology, since AChE is the target for a large number of neurotoxic chemicals, e.g., organophosphorus compounds and carbamate insecticides. These chemicals inhibit AChE resulting in the accumulation of ACh at the receptor site and leading to overstimulation of the receptor and eventual breakdown of the cholinergic system, which causes paralysis and, ultimately, death.

Amino Acids

Glycine. Glycine inhibits the firing of neurons.

Glutamate and Aspartate. These amino acids act as excitatory signals. They also stimulate *N*-methyl-D-aspartate (NMDA) receptors that are involved in learning and memory. Benefit usually results from the stimulation of NMDA; overstimulation, however, may cause nerve cell damage or death.

Gamma-Amino Butyric Acid (GABA). GABA inhibits the firing of neurons. Its activity is increased by benzodiazepine (Valium) and by anticonvulsant drugs.

Catecholamines

Dopamine. Dopamine in the brain is involved in three different roles.

1. Regulation of endocrine system. It initiates the hypothalamus to make hormones and store them in the pituitary gland; it then triggers their release into blood stream.
2. Regulation of movement. Dopamine is lacking in the brains of Parkinson's disease patients who exhibit muscle tremors, rigidity, and difficulty in moving. Levodopa is used to treat these patients.
3. Cognition and emotions. Drugs that block dopamine receptors in the brain are helpful in treatment of schizophrenia.

Norepinephrine. This neurotransmitter is present in many nerve fibers and is secreted by the adrenal gland in response to stress or events producing excitation. Norepinephrine is deficient in patients with Alzheimer's disease and in chronic alcoholics that develop Korsakoff's syndrome.

Serotonin. This neurotransmitter is present in the brain, the lining of the digestive tract, and blood platelets (as well as many other tissues). It induces strong contraction of smooth muscle and is involved in states of consciousness, mood, depression, and anxiety.

Peptides. The brain contains specific peptides, called opioids, that function to kill pain (like opium) or cause sleepiness. Enkephalin and endorphins are endogenous opiates produced by the brain. Recently substance P has been discovered in C fibers.

Second Messengers

Neurotransmitters are considered to be first messengers. A second messenger transmits the chemical message of a neurotransmitter from the cell membrane to the biochemical apparatus of the cell. The time required for the second messenger to convey the message may be a few milliseconds or minutes. Long-term changes in the nervous system may be mediated by second messengers.

CLASSIFICATION OF NEUROTOXIC ACTIONS

Injuries to the nervous system by neurotoxic chemicals may be divided into two classes depending on the nature of the effect: nonselective and selective. In nonselective neurotoxicity, damage results from hypoxia (low oxygen content; apoxia is lack of oxygen). Selective neurotoxicants produce damage that is limited to localized target structures.

Blood-Brain Barrier

Selectivity of nerve tissues to injuries by neurotoxic chemicals is related to the accessibility of the nervous system to neurotoxicants. The central nervous system is protected by the blood-brain barrier, which is not permeable to polar, water-soluble compounds. Chemicals that are nonpolar and lipid soluble can penetrate the blood-brain barrier. In the immature brain, this barrier is not effective, which may allow the accumulation of neurotoxic chemicals, such as lead, in the brain of children. The blood-neural barrier in the peripheral nervous system is present in some places and absent in others.

Non-Selective Neurotoxic Action

The structural unit of the nervous system is the neuron or cell body. It conducts nerve impulses throughout the body. The neuron has the following characteristics: it cannot reproduce itself, has little ability for anaerobic metabolism, has a high metabolic rate, and cannot survive for more than a few minutes without oxygen or energy.

The brain requires one fifth of the oxygen and energy (glucose) consumed by the body to maintain its constant activity. The hourly flow of blood through the brain is approximately thirteen gallons, which accounts for one fifth of the blood pumped by the heart. Blood is pumped to the brain by two pairs of arterial trunks: the vertebral arteries and its internal carotid arteries. The veins of the brain drain into dural sinuses and superficial venous plexuses and eventually to the internal jugular veins.

Although the central and peripheral nervous systems are highly vascularized, hypoxia is confined to the central nervous system. Also, not all parts of the brain are equally supplied with blood, which results in selective damage to brain areas under hypoxic condition. Brain white matter requires less oxygen than gray matter, which makes it less sensitive to lack of oxygen. Brain cells with high metabolic activity, such as cortical or hippocampal pyramidal cells and cerebellar purkinje cells, are particularly sensitive to hypoxia. The order of sensitivity to hypoxia in decreasing order is neurons, oligodendrocytes, astrocytes, microglia, and cells of the capillary epithelium.

There are three types of anoxia: anoxic, ischemic, and cytotoxic.

Anoxic Hypoxia

Anoxic hypoxia develops from inadequate oxygen supply in the presence of adequate blood flow. There are two principal causes of this type of hypoxia.

Respiratory Paralysis. Some neurotoxicants interfere directly with respiration. For example, α-tubocurarine chloride can cause respiratory paralysis by interference with the action of acetylcholine. α-Tubocurarine, the active neurotoxicant of curare, is an ACh antagonist at nicotinic receptor sites. It competes with ACh in the skeletal muscle end plate. The major effect of the compound is a prolonged flaccid paralysis of striated muscles.

The antidotal treatment for this neurotoxicity is the administration of reversible acetylcholine esterase inhibitors such as neostigmine and physostigmine.

Interference With Oxygen-Carrying Capacity of Blood. Examples are the production of carboxyhemoglobin by CO, or methomoglobin by nitrites.

$$\text{Hemoglobin (Fe}^{2+}) \xrightarrow{\text{CO}} \text{CO (Fe}^{2+}) \text{ hemoglobin (carboxyhemoglobin)}$$

$$\text{Hemoglobin (Fe}^{2+}) \xrightarrow{\text{nitrite}} \text{hemoglobin (Fe}^{3+}) \text{ (methemoglobin)}$$

These neurotoxic actions are discussed in detail in Part V, *Neurotoxic Agents*.

Ischemic Hypoxia

Ischemic hypoxia develops from deficiency of blood supply in the brain. This can result from

1. cardiac arrest from toxic substances which lead to decrease in arterial blood pressure
2. extreme hypotension from vasodilation can cause brain ischema

3. hemorrhage or thrombosis of cerebral vessels causes local ischemic hypoxia of the brain areas supplied by these vessels or
4. carbon monoxide, which initially combines with hemoglobin and causes anoxic hypoxia without primary loss of blood circulation. However, prolonged exposure to CO results in ischemic hypoxia or stagnant hypoxia.

Cytotoxic Hypoxia

Cytotoxic hypoxia develops from interference with "cell metabolism" in the presence of an adequate supply of both blood and oxygen. This damage may result from severe hypotension but is most common with chemicals that inhibit cell respiration. Many neurotoxic chemicals that damage the cell body produce neurotoxicity via cytotoxic hypoxia.

Cytochrome Oxidase Inhibitors. Cyanide and azide inhibit cytochrome oxidase and produces cytotoxic hypoxia. Cyanide also causes hypotension through its effect on the heart.

Metabolic Inhibitors. Examples are dinitrophenol, malonitrile, and methionine sulfoximine.

These effects are discussed in detail in Part V, *Neurotoxic Agents*.

Repeated or Prolonged Severe Hypoxia

Repeated hypoxia episodes from any mechanism cause damage to the blood-brain and lead to the development of a pattern of diffuse sclerosis (hardening inflammation) of the white matter, or leukoencephalopathy, as a consequence. Cytotoxic hypoxia and ischemia from metabolic inhibitors, which also depress cardiovascular functions, are particularly likely to damage the white matter of the brain.

Selective Neurotoxic Action

In this type, neurotoxicants may cause peripheral neuropathy in addition to CNS effects. This group may be divided into the following categories.

Demyelinating Substances

These chemicals affect myelin-forming cells, i.e., oligodendroglia in CNS or Schwann cells in peripheral nervous system. The damage results in "encephalopathy" if central white matter is involved, or "polyneuritis" if peripheral cells are damaged. When myelin loss is predominant, with relative sparing of the axon, the condition is called "segmental degeneration." Examples of neurotoxicants that are specific demyelinating agents are triethyltin, lead and hexochlorophene, all of which are discussed in detail in Part V, *Neurotoxic Agents*.

Axonopathy

Central-Peripheral Axonopathy. The type of axonopathy that produces primary axonal degeneration followed by secondary degeneration of the myelin is known as Wallerian-type degeneration. It is also known as central-peripheral axonopathy.

1. Central-Peripheral Proximal Axonopathy. This type of axonopathy results in degeneration of the axon in areas proximal to the spinal cord. An example is β,β'-iminodipropionitrile (IDPN) which is discussed in Part V, *Neurotoxic Agents*.
2. Central-Peripheral Distal Axonopathy. This type of axonopathy is also known as dying back neuropathy, since early neuropathologic alterations occur in the most distal portions of the peripheral nerve. This type of neuropathy may be classified into two groups according to the morphological changes produced by neurotoxic chemicals.
   a. Neurofilamentous Neuropathies. In these neuropathies, neurotoxic changes are characterized by an accumulation of the 10-nm neurofilaments in axonal swellings above nodes of Ranvier in the central and peripheral nervous system, which begins at the distal portions of the axon. Examples of neurotoxicants producing this type of neurotoxicity are carbon disulfide, acrylamide, and *n*-hexane and its related chemicals.
   b. Tubulovesicular Neuropathy. Some organophosphorus esters such as tri-*o*-cresyl phosphate produce this type of neurotoxicity. This effect is discussed in detail in the Pesticides chapter in Part V, *Neurotoxic Agents*.

Effect on Axonal Conduction. Some chemicals, e.g., chlorinated hydrocarbon insecticides and pyrethroids, cause neurotoxicity by interfering with the normal flow of ions in the axon. Ion movement in the axon involves the opening and closing of ion channels. Ion channels are water-filled molecular tunnels present in the cell membrane that allow the passage of ions or small molcules in and out of the cell. These effects are discussed in detail in Part V, *Neurotoxic Agents*.

Effect on the Synapse

After receiving a stimulus, an action potential passes across the axonal membrane. When it reaches the nerve ending, it releases a neurotransmitter at the synapse. The neurotransmitter, e.g., ACh, binds to an ACh receptor on the postsynaptic membrane, which triggers an action, e.g., muscle contraction. ACh is then hydrolyzed by AChE. Chemicals that inhibit this enzyme, e.g., organophosphorus and carbamate insecticides result in the accumulation of ACh at the receptor site leading to paralysis and, eventually death.

Cell Body

Some neurotoxic chemicals exert their toxicity by directly affecting the cell body resulting in cell death. An example is triphenyl phosphite, which produces type II organophosphorus compound-induced delayed neurotoxicity (OPIDN) which is characterized by cell body necrosis and axonal degeneration. This effect is discussed in detail in Part V, *Neurotoxic Agents*.

The text is divided into six major sections: Introduction; Cellular Neuroanatomy; Mechanisms of Neurotoxicity including Biochemical, Electrophysiological, and Pathological Studies; Factors Affecting Neurotoxicity encompassing Disposition, Metabolism, and Toxicokinetics, Developmental Neurotoxicology, Nutritional Neurotoxicology, and Alcohol and Neurotoxicology; Neurotoxic Agents; and Evaluation and Diagnosis of Neurotoxicity including Principles and Methods for Evaluating Neurotoxicity in Test Animals, Neurobehavioral Toxicology, and Neurological Examination.

The *Introduction* includes general information relating to neurotoxicity, a brief overview of the nervous system, and classification of neurotoxic actions. *Cellular Neuroanatomy* discusses the anatomy, biochemistry, and physiology of neural cells.

The biochemical studies subsection encompasses four chapters. *Cytoskeletal Proteins* describes the cytoskeletal elements and their involvement in the mechanisms of some neurotoxicities. *Neurotypic and Gliotypic Proteins as Biochemical Indicators of Neurotoxicity* discusses the selective vulnerability of nervous tissues to neurotoxicity and the use of neurotypic and gliotypic proteins as biochemical indicators of chemically induced neurotoxicity. *Axonal Transport* discusses property of and techniques for studying axonal transport and alterations in axonal transport induced by neurotoxicants. *Neurotransmitter Receptors* describes the utility and assay of receptor study of neurotoxicants and gives an account of the action of neurotoxic agents on neural receptors.

The electrophysiological studies subsection includes three chapters. *Cellular Electrophysiology* discusses the basic mechanism of nerve conduction, principle and technique of voltage clamp studies, and the action of neurotoxicants on ion channels. *Electromyographic Methods* describes the anatomy and physiology of the neuromuscular junction, electromyographic recording techniques, and the use of electromyography in the studies and diagnosis of neurotoxic diseases. *Brain Slices* includes discussion on the blood-brain barrier, brain slice techniques, and the use of brain slices in neurotoxicity studies.

The pathological studies subsection contains one chapter, *Basic Histopathological Alterations in the Central and Peripheral Nervous Systems*, which describes the pathology of various parts of the neuron and the classification of various neuropathies. It presents the techniques for neuropatho-logical studies and neuropathological alterations induced by neurotoxic chemicals.

Disposition, Metabolism, and Toxicokinetics describes the behavior of the toxicant in the body: absorption, distribution, elimination, and biotrans-formation. *Developmental Neurotoxicology* discusses the development of the nervous system and the mechanism of action of neurotoxicants on the de-veloping nervous system. *Nutrition and Neurotoxicology* gives an overview of basic nutritional elements and their effect on the function of the nervous system. *Alcohol and Neurotoxicology* is discussed in relation to alcoholism and the mechanism of its adverse effects on the nervous system. *Neurotoxic Agents* encompasses all classes of chemicals that produce adverse effects on the nervous system: metals, solvents, gases and vapors, other industrial neurotoxic chemicals, pesticides, drugs, and naturally occurring toxins. It describes the neurotoxic action and its mechanism for each neurotoxic agent. The metabolic behavior of neurotoxic chemicals is discussed and treatment and antidotal medication for each neurotoxicity is described.

Principles and Methods for Evaluating Neurotoxicity in Test Animals describes the experimental design and factors affecting neurotoxicity studies. *Neurobehavioral Toxicology* classifies behavioral approaches and describes tests for behavioral studies. *The Neurological Examination* presents the neu-rological examination for humans exhibiting neurologic dysfunctions includ-ing the motor and sensory systems as well as reflexes and coordination.

REFERENCES

Carey, J. (1990). Brain Facts. *A Primer on the Brain and Nervous System.* Society of Neurosciences, Washington, D.C.

Kandel, E.R. and J.H. Schwartz (1985). *Principles of Neural Science,* 2nd ed. Elsevier, New York.

Walton, J. (1985). *Brain's Diseases of the Nervous System,* 9th ed. Oxford University Press, New York.

Part

II

Cellular Neuroanatomy

Cellular Neuroanatomy
N. Cant

Cellular Neuroanatomy

Nell B. Cant
Department of Neurobiology
Duke University Medical Center
Durham, North Carolina

THE STRUCTURE AND FUNCTION OF THE NERVOUS SYSTEM

The brain and spinal cord consist almost exclusively of cells. There is very little interstitial substance so that the cells lie close together, their external membranes separated only by narrow extracellular spaces. There are two major classes of cells. The first is the class of excitable cells or neurons, cells that possess the properties of irritability and conductivity that allow the nervous system to receive, process, and convey information. All neurons are reactive to appropriate stimuli, the activity taking the form of changes in the electrical properties of the cell membrane. Nerve cells contact each other at specialized sites known as synapses and thereby form continuous functional networks, which are highly organized. The second class of cells in the nervous system, the nonexcitable or supporting cells, are several times more numerous than the neurons, but their functions are less well understood. In the central nervous system, these include the ependymal and neuroglial cells. In the peripheral nervous system, they include the Schwann cells.

Neuronal Morphology

A fundamental characteristic of the nervous system is morphological diversity. Neurons differ dramatically in their structure from one part of the nervous system to another. There are some useful generalizations that apply to most or all neurons, and these will be emphasized in the following discussion. It should be remembered that the nervous system does not hesitate to break the normal rules when a particular function may be better served thereby.

The soma or cell body is the part of the nerve cell (and of the supporting cell) that is visible with most routine stains of nervous tissue (Figure 1A). The region between the cell bodies, an area that appears empty in the routine preparations, is actually packed with processes arising from both the neurons and the glial cells associated with them. This tangle of processes, or neuropil, is not stained unless special procedures are used (e.g., Figure 1B), and can be appreciated fully only in the electron microscope (Figures 2 and 3).

All neurons contain a nucleus, endoplasmic reticulum, a Golgi apparatus, mitochondria, microtubules, and other organelles typical of all cells (Figures 2 and 4). In addition, however, neurons are specialized for integration and communication of activity. They integrate inputs from diverse sources, generate electrical signals and transmit them to sites distant from the site of activation. In this way, they affect the multiple events involved in sensory, motor and mental activities. These functional aspects of neural activity call for structural specializations.

Nerve Cell Processes

The most striking of the structural specializations is in the elaboration and arrangement of the many branches or processes that arise from the cell bodies of neurons. Some of the seemingly endless diversity in neuronal structure is indicated in Figure 5, a drawing of some characteristic neuronal types from different subdivisions of the nervous system. All neurons have a cell body in which the nucleus is housed and, consequently, the machinery for protein synthesis. The cell body itself varies greatly in size from one population of neurons to the next. Compare, for example, the sizes of the cell bodies of different neurons shown in Figure 1A. From the cell bodies radiate processes that may be one of two kinds: the dendrites, of which there are usually several; and the axon, of which there is only one.

Dendritic morphology varies with respect to the numbers and lengths of the dendrites, their branching patterns, and the types of appendages that adorn them. Dendrites may branch many times, as in the example from the cerebellum illustrated in Figure 5A, or they may be relatively simple and unbranched (Figure 5F). Some have protrusions from their surface known as dendritic spines (Figure 5G). Others terminate in sprays of appendages of characteristic shapes (e.g., Figure 5B). Sometimes, as in the example from the olfactory bulb (Figure 5D), one of the dendrites may form specializations not present on the other dendrites of the same cell. These neuronal types and the other examples illustrated in Figure 5 represent only a few of the many possibilities; in every part of the brain and spinal cord the dendrites of neurons

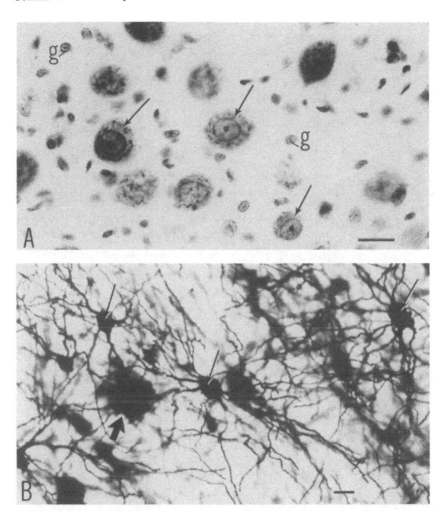

Figure 1. Photomicrographs of nervous tissue prepared in two different ways. A. Neurons (arrows) and glial cells (g) in a section through the cochlear nucleus of the cat stained with a Nissl stain. This stain reveals the cell bodies of neurons and glial cells but does not stain the many processes characteristic of these cell types, because the staining agents have a particular affinity for the ribonucleic acids, which are located almost exclusively in the cell bodies. The cell body of every neuron in the section is stained. The spaces between the cell bodies that appear empty in a preparation such as this are actually occupied by dendrites, axons and axonal terminals as well as by glial processes. B. Neurons (small arrows) in a section through the lateral geniculate body of the cat prepared with the Golgi method. (This micrograph was made at a lower magnification than that shown in Figure 1A; therefore, the cell bodies appear smaller). The cell bodies of only a few neurons in the section are impregnated by this technique, but all the many processes arising from them are visible as well. If all of the neurons were impregnated it would be impossible to study them, since the entire section would be filled with branching processes. A few glial cells (large arrow) are also impregnated. They have so many processes that the image appears as a blurry spot in this photograph. The Nissl method and the Golgi method complement each other, and they are often used together in studies of the nervous system. Each scale bar represents approximately 25 μm.

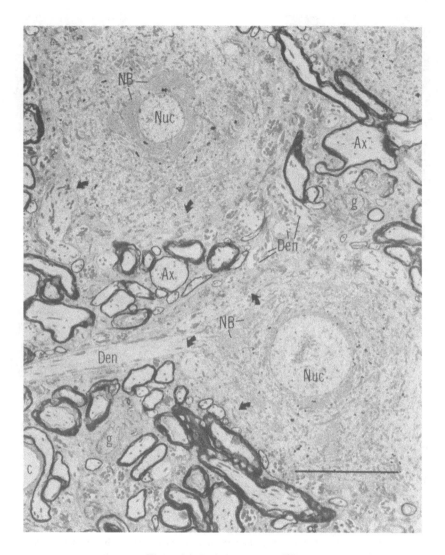

Figure 2. *(Explanation on p. 29)*

exhibit a characteristic structure, often different from that found anywhere else. Presumably, the structural differences reflect differences in the functional demands placed upon the neurons.

The Axon

The axon of a neuron consists of a thin, relatively unbranched process that may extend for very long distances and a terminal portion, often highly branched, that contacts another neuron or an effector organ. Branching patterns of axons, like those of dendrites vary widely and are characteristic

Figure 2. Electron micrograph of a section through two neurons in the cochlear nucleus of the cat. Both cell bodies contain a nucleus (Nuc) surrounded by the part of the neuron known as the perikaryon, which contains the same constituents as the cytoplasm of other cell types, such as mitochondria, lysosomes, a Golgi apparatus, and smooth and granular endoplasmic reticulum. The granular endoplasmic reticulum forms stacks that have become known in neurons as Nissl bodies (NB). Between the neurons is the neuropil containing dendrites (Den), axons (Ax), and the processes of glial cells. The cell bodies of two glial cells (g) are also included in the section. A part of a capillary (c) is seen at the lower left. In this low magnification micrograph, axonal terminals forming synapses with the cell bodies and dendrites can just barely be seen, but the surfaces of both of these cells are covered almost entirely by synaptic terminals. Some of the larger ones are indicated by arrows. Compare this micrograph to the light micrograph of Figure 1A. The same type of cell is shown in each, but in this figure all of the other constituents of the nervous tissue are seen as well. A disadvantage of electron microscopy is that the sections must be cut extremely thin so that it is rare to be able to follow the attachment of a dendrite or axon to its parent cell body. Thus the branching patterns of a neuron cannot be appreciated in the electron microscope. The neuroanatomist who wishes to understand the structure of nervous tissue therefore uses both light and electron microscopic techniques to piece together a thorough description. The scale bar represents approximately 10 μm. (Modified from N.B. Cant and D.K. Morest, Pergamon Press Ltd., 1979. With permission.)

of particular neuronal types and particular parts of the nervous system (e.g., Figure 5C and D). The shapes, sizes, and patterns of arborization of the terminal endings may differ even on different branches of the same axon. For example, branches of axons in some parts of the auditory system terminate in very large swellings, in fact, the largest in the brain (Figure 4II). The same axon also gives rise to branches that form smaller terminals in other areas. The axons of some neurons travel for long distances before terminating; others arborize in the vicinity of their parent cell body. Most terminal arborizations are characterized by multiple swellings, the terminal boutons; the boutons form the synaptic connections made by the neurons with other cells.

Neuronal Membrane: Structure, Potential, and Channels

The ability of nerve cells to produce signals for communication with one another and with effector cells depends on the separation of electrical charges across the cell membrane. In all cells, including neurons, electrical charge is carried by positive and negative ions, including sodium, potassium and chloride ions. Concentration gradients across the relatively impermeable cell membrane are maintained by metabolic ion pumps, which use energy to move ions. In the resting or steady state, the separation of ionic charge leads to a steady potential difference across the membrane, with the inside of the cell relatively more negative due to a very slight excess of negatively charged ions inside the cell. Such a membrane is referred to as polarized. The membranes of neurons, like those of other cells, are made up of lipids, which are highly impermeable to ions, and of proteins and glycoproteins some of which are inserted into the membrane to form pores or channels through which ions may move easily if the channel happens to be open. Some

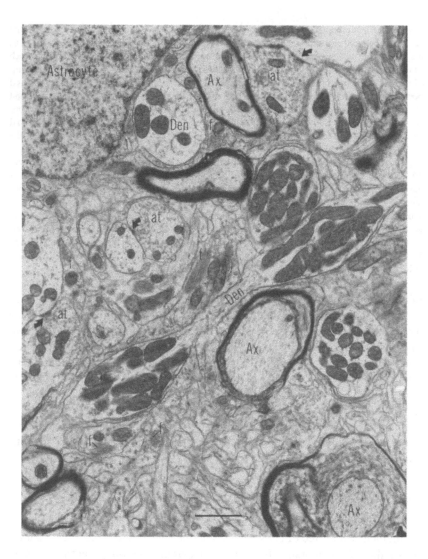

Figure 3. Electron micrograph of a section through the neuropil of the cochlear nucleus of the cat. Myelinated axons (Ax) give rise to axonal terminals (at) which form synaptic contacts (arrows) with dendritic profiles. Several dendritic profiles (Den) that do not receive synapses within the plane of the section are also seen. Interspersed among these neuronal elements are the multiple, sinuous processes that arise from astrocytes, the cell body of one of which is seen in the upper left of the figure. The glial processes can often be recognized by their characteristic complement of filament bundles (f). Other small processes visible in the section may be glial processes, unmyelinated axons, or small dendrites. Scale bar represents approximately 1 μm.

membrane channels are always open; others may be open or closed. There are many different types of channels, and one neuron may contain up to ten or more types. Membrane channels in different neurons and in different parts of the same neuron vary with respect both to the ions that can enter them

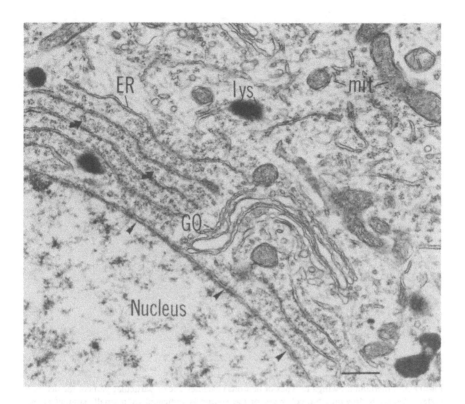

Figure 4. Electron micrograph of a section through part of a neuronal cell body. The nucleus is surrounded by a double membrane known as the nuclear envelope (arrowheads). Stacks of endoplasmic reticulum (ER) are associated with numerous ribosomes (arrows) which, in nervous tissue, often lie between the layers of membrane rather than attached directly to them. Also illustrated are a portion of the Golgi apparatus (GO), lysosomes (lys) and mitochondria (mit). Scale bar represents approximately 0.5 μm.

and also to the stimulus that will cause them to open. Many of the ion channels in neurons are closed during the resting state. If the channels are caused to open by an appropriate stimulus, ions move through them and, consequently, the potential difference across the membrane changes temporarily. When the stimulus is removed, metabolic pumps return the ionic concentrations on each side of the membrane to their resting values.

Electrical Signal

Neurons generate two types of electrical signals (Figure 6). Different parts of the neurons and different kinds of membrane channels are involved in each type. One of the forms of electrical signal is a localized, graded change in the potential difference across the membrane. The graded potential changes are effective over short distances and play an essential role at the

Figure 5. Freehand drawings of types of neurons characteristic of particular parts of the nervous system. On each drawing, the cell body of the neuron is indicated by a small, straight arrow and the axon by an a. A. Purkinje cell of the cerebellum. One or a few main dendritic trunks branch repeatedly into a fine network of dendrites that are covered with spinous processes. B. Granule cells, also found in the cerebellum. The small cell body gives rise to short dendrites that terminate in claw-like structures. The axons form branches that run parallel to one another through the cerebellum. The swellings along the axon contain vesicles and make synaptic contacts with the spines of the Purkinje cells. C. Spiral ganglion cell of the auditory system. The dendrite terminates in the cochlea where it is contacted by the receptor, the cochlear hair cell. The elaborate axon enters the brain where it branches to supply several different regions. In each region that it innervates, its patterns of terminal branching are different. D. Mitral cell of the olfactory bulb. One of the dendrites of the cell terminates in a bushy cluster of appendages (arrow) not seen on the other dendrites of the same cell. The axon terminates in a series of tufts of terminals. E. Small stellate cell characteristic of many parts of the nervous system, often serving as an interneuron. The dendrites may be smooth or spinous. The axon branches in the vicinity of the cell body. F. Large multipolar neuron characteristic of the reticular formation of the brain stem. The dendrites are very long and relatively unbranched. The axon may travel for several millimeters and terminate in many different target zones. G. Pyramidal cell of the cerebral cortex. There are two sets of dendrites: the apical dendrite, which is long and extends toward the cortical surface to end in a tuft of branches; and the basal dendrites, which branch in the vicinity of the cell body. Both are covered with spines. The axon may branch locally as well as send terminals to distant targets. H. Globular cell of the cochlear nucleus. The dendrites form a bushy tangle. The relatively large axon terminates in a structure known as the calyx of Held, the largest synaptic structure found in the mammalian brain (curved arrow). I. A cartoon of a neuron, like those often used to portray neurons in illustrations of connectivity in the nervous system. It is well to remember that most of the relevant structure of the neuron is missing from such a depiction.

points of contact of two cells, but any one, by itself, may have little effect on the overall activity of the neuron. At peripheral sensory receptors, the graded potentials, which arise in response to environmental stimuli, are known as generator or receptor potentials. Between neurons, they are known as postsynaptic potentials and may be either excitatory or inhibitory. Between nerve cells and muscle cells, the graded potentials are known as motor end plate potentials. The localized membrane potential changes allow individual cells to perform an integrative function in that the neurons effectively add together the inhibitory and excitatory inputs from many different sources. If the excitatory input reaches a level known as threshold, the cell will produce an action potential, which is the second form of electrical signal generated by neurons. The action potential is an explosive, all-or-none change in the membrane potential difference and its purpose is to conduct a signal from one part of a neuron to another, often over long distances.

Many different types of membrane channels are involved in the production of graded potentials. Some are sensitive to changes in the level of chemical neurotransmitters, while others react to stimuli such as changes in light levels, temperature, or osmotic pressure, or to mechanical distortion. The channels involved in the production of an action potential are sensitive to changes in the membrane potential itself. The graded membrane potential changes can thus lead to the production of action potentials when the potential changes in the excitatory direction are great enough to cause opening of sufficient numbers of the electrically excitable channels.

It is useful to classify the individual parts of nerve cells according to their function and the types of potential changes that they generate. Thus, we consider the receiving, conducting, and transmitting portions of the nerve cell. Most neurons are polarized, with receptive ends and transmitting ends. These are usually, although not always, joined by a conducting segment.

Signal Processing and Synaptic Transmission

The dendrites and, in some cases, the soma of the neuron are specialized for receiving and integrating information. One very striking feature of nerve cells is the enormous surface area afforded by the many dendrites and their protrusions (Figure 5). This receptive surface of neurons is specialized for the generation of graded potentials. The dendritic surface receives synaptic inputs from other nerve cells (Figures 3 and 8) or, in the peripheral nervous system, from receptors. The balance of activity in excitatory and inhibitory inputs determines whether the cell is brought to theshold. Ion channels located in the membranes of dendrites in the central nervous system are specialized for activation by specific chemical transmitters. Upon stimulation by the appropriate transmitter, the normally closed membrane channels open, specific ions move through them from one side of the membrane to the other and, as a consequence, the electrical potential across the membrane changes at that point, giving rise to an excitatory or inhibitory postsynaptic potential

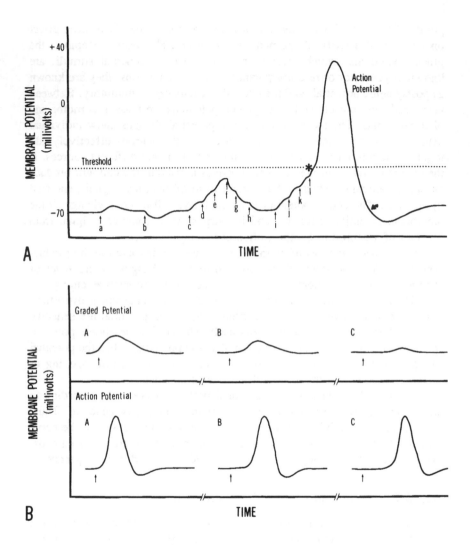

Figure 6. *(Explanation on p. 35)*

(abbreviated EPSP and IPSP, respectively). Channels located in the receptive surface of nerve cells or receptor cells in the peripheral nervous system are specialized to respond to specific environmental stimuli. Mechanisms for the rapid reuptake or destruction of the transmitter ensures that the event is transient and that the membrane is rapidly repolarized in the absence of continued stimulation.

Axonal Conduction

The multiple dendrites funnel information toward the single axon, which has the function of conducting information in the form of action potentials

Figure 6. Drawings to illustrate distinguishing features of the two types of electrical signals generated by nerve cells. Both illustrate the membrane potential across a neuronal membrane plotted against time. (A potential difference arises from a separation of charge. It is measured in volts, which is a measure of the amount of work required to move a unit charge from one side of the membrane to the other). A. At the beginning of the trace, the membrane is at its resting potential of −70 mV. At the time indicated by the arrow labeled a, an excitatory input to the membrane is stimulated and the membrane is slightly depolarized. This graded, positive-going potential is known as an excitatory postsynaptic potential (EPSP). The EPSP is shortlived and in the absence of continued stimulation, the membrane returns to its resting potential. At the time indicated by the arrow labeled b, an inhibitory input to the membrane is stimulated and the membrane potential becomes even more negative than the resting value. The membrane is said to be hyperpolarized. This graded potential is known as an inhibitory postsynaptic potential (IPSP). Like the EPSP, it is shortlived and the membrane returns to its resting value if the stimulus is not maintained. At the arrow labeled c, another EPSP is elicited, but before the membrane returns to its resting potential, more excitatory inputs are stimulated (arrows d and e). As a result, the membrane becomes even more depolarized. The increase in depolarization with repeated stimulation of excitatory inputs over time is known as temporal summation. Simultaneous activation of several excitatory inputs would also give rise to a larger depolarization, a phenomenon known as spatial summation (not illustrated). At the times indicated by arrows f, g, and h, inhibitory inputs are activated and the membrane potential is brought back to resting levels. At the times indicated by the arrows labeled i-l, another series of EPSPs are elicited. In this case, the depolarization reaches a level known as threshold, at which point an action potential is evoked (at the point indicated by the asterisk). The action potential is an explosive, all-or-none change in the membrane potential. During the action potential, the membrane polarity is actually reversed, the inside of the membrane becoming slightly more positive relative to the outside. The action potential is shortlived, and the membrane potential falls rapidly back toward resting levels. Sometimes, as it falls, the potential actually becomes even more negative than resting levels (curved arrow), a phenomenon known as after-hyperpolarization. Action potentials cannot be summated. In fact, for a short time after the action potential is elicited, the membrane is incapable of generating a second one, a period known as the absolute refractory period. After this, there is a short period of time (approximately coinciding with the time of the after-hyperpolarization) during which only a greater than normal stimulus will elicit another action potential, a period of time known as the relative refractory period. B. A two-part drawing to illustrate differences between graded potentials and action potentials as they spread from one point to other points along the membrane. In both traces, three consecutive locations (A, B, and C) some distance apart along the nerve cell membranes are illustrated. The upper traces might represent locations along a dendritic surface, whereas the lower trace might represent locations along an axon. In the illustrations in the upper trace, the arrows represent the time of elicitation of a graded potential at location A. At point B, after a delay, a potential change is seen that is of a smaller amplitude and a slower rise time than that at point A. At point C, further along the membrane, the latency of the change is even longer and the amplitude is less than that at point B. If the membrane potential were recorded even further along the dendrite, there would be a point at which no change in responsiveness to the stimulus at A would be detectable. For this reason, these graded potentials are referred to as local potentials. In the lower trace, an action potential is elicited at point A at the time indicated by the arrow. The size and shape of the action potential is exactly the same at points B and C further along the axon. The latency of activation at points B and C is longer, since there is a finite travel time for the action potential to reach these points from its site of initiation at point A. No matter how long the axon, the amplitude and shape of the action potential are maintained along the entire length.

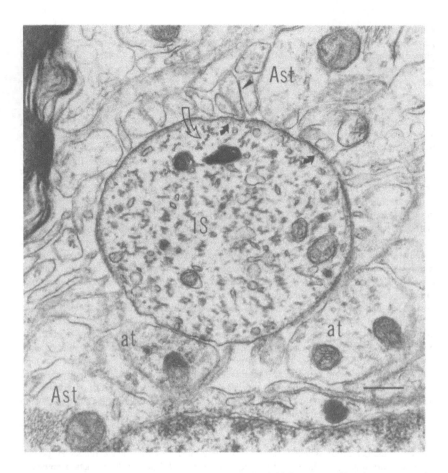

Figure 7. Electron micrograph of a section through the initial segment (IS) of an axon cut in cross section. The initial segment is characterized by a fuzzy undercoat associated with the cell membrane (solid arrows) and by fascicles of microtubules (large curved arrow). For the most part the initial segment is surrounded by astrocytic processes (Ast), some of which form gap junctions with each other (arrowhead). In addition a few axonal terminals containing synaptic vesicles (at) may contact the initial segment. The scale bar represents approximately 0.5 μm.

to near or distant parts of the nervous system. Axons may be extremely long and because of this, even though they are very thin, they often constitute most of the volume of the neuron. At the origin of the axon, a morphologically distinct region known as the initial segment (Figure 7), an action potential is generated when the excitatory inputs to the receptive surface of the neuron are sufficient to change the electrical potential difference across the membrane to threshold. A membrane that contains ion channels that are sensitive to changes in the membrane potential and that generates action potentials is termed electrically excitable. Thus, the axon is specialized for generating the all-or-none potentials that are propagated for long distances

without decrement. The larger the axon, the faster the velocity with which
the impulse is conducted. Larger axons are also encased in a myelin sheath
which further increases the velocity of conduction of the nervous impulse.
The myelin is not actually a part of the nerve cell but is formed by a
specialization of the membranes of glial cells, which are discussed below.

Synaptic Terminals

The terminals of axons are specialized for transmitting information to
other nerve cells, to muscles or to glands. In some cases, the dendrites also
transmit rather than receive inputs, but this does not appear to be very
common. At their terminals, the axons form connections with the receptive
surface of the next cell in the system at specializations known as synapses.
Synapses may be specialized for transmission by chemical or electrical means,
although in the mammalian nervous system, almost all synapses are chemical.
At chemical synapses, the arrival of the action potential at the axon terminal
stimulates the opening of electrically excitable calcium ion channels. The
resultant net flux of calcium ions into the terminal causes the release of a
chemical substance, the neurotransmitter, which is stored in synaptic vesicles
inside the terminal (Figure 8). A large number of putative neurotransmitters
have been identified, some of them excitatory and others inhibitory. By
diffusion, the transmitter travels across the narrow synaptic cleft to the
postsynaptic cell where it causes a graded electrical change in the receptive
surface of that cell. The effect is either excitatory or inhibitory depending
on the nature both of the transmitter and also of the postsynaptic receptor.
Thus the presynaptic terminal is not the exclusive determinant of synaptic
effect; there may be several types of receptors responsive to a given transmitter and the exact effects of the transmitter will depend on the particular
receptor involved.

Chemical Synapses

Chemical synapses are often classified as one of two morphological
types, although this should be regarded as an oversimplification. The first,
the type I or asymmetrical synapse, is illustrated in Figure 8A. The presynaptic axon terminals contain round synaptic vesicles and are separated from
the postsynaptic surface by a relatively widened extracellular space or synaptic cleft. The postsynaptic surface contains an accumulation of a dense
substance just beneath the membrane. The presence of this material in the
postsynaptic but not the presynaptic element gives rise to the characterization
of these synapses as asymmetrical. The second type, the type II or symmetrical synapse (Figure 8B), contains flattened vesicles and faces the postsynaptic cell over a relatively narrow cleft. The postsynaptic element lacks a
prominent postsynaptic density. In the few cases in which the effects of
synapses of these two types are known, the type I synapses have proven to

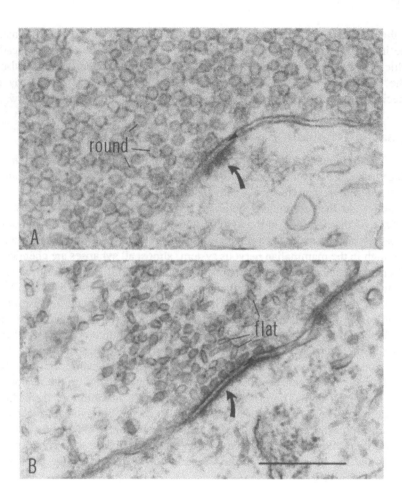

Figure 8. Electron micrographs of two types of synaptic complexes in the lateral superior olivary nucleus of the cat. A. The type I synapse is characterized by round synaptic vesicles in the presynaptic element, by a relatively widened synaptic cleft (arrowhead), and by a cytoplasmic density in the postsynaptic element (curved arrow). The fact that the density is more prominent in the postsynaptic element than in the presynaptic element gives rise to the name asymmetrical for this synaptic type. In general, synaptic complexes with type I morphology are believed to be excitatory. B. The type II synapse is characterized by flattened synaptic vesicles in the presynaptic element, by a relatively narrow synaptic cleft (arrowhead), and by a density in the postsynaptic element only slightly more prominent than a similar presynaptic density. The lack of a clear postsynaptic density gives rise to the name symmetrical for these synapses. In general synapses with type II morphology are believed to be inhibitory. The scale bar represents approximately 0.5 μm.

be excitatory and the type II synapses inhibitory. The types of synaptic terminals and their physiological effects cannot be categorized so simply, however, since, especially in the brain stem and spinal cord, synaptic terminals come in many varieties, perhaps indicating that they contain different transmitters. Moreover, many terminals contain more than one neuroactive compound, including recently discovered neuropeptides that appear to be released from synaptic terminals along with the more traditionally recognized neurotransmitters. Whether different substances are released from a given terminal under different circumstances is not known.

Electrical Synapses

A few synapses in the central nervous system are electrical rather than chemical. The synaptic cleft in these contacts is obliterated so that the two membranes meet with a small gap between them that is visible only in the electron microscope. Thus these structures are called gap junctions. It is thought that the gap junction provides a low resistance pathway for the spread of electrical activity from one cell to the next. Gap junctions are also quite commonly observed between adjacent glial cells (Figure 7). Their function there is not known.

Neurosecretory Neurons

Some axonal terminals do not form synaptic contacts with other cells. These terminals, which contain vesicles and several types of secretory products, come into close apposition with blood vessels. Upon activation they release their contents into the circulating blood and thus form a link between the nervous system and the endocrine system. The most common examples of these neurosecretory neurons are found in the hypothalamus of the brain.

Cell Body

All of the processes of the neuron, the dendrites, the axons and the axon terminals, depend on the cell body for their maintenance. If they are severed from the cell body, the processes die and degenerate. The cytoplasm of the cell body of a neuron is packed with ribosomes, whereas there are few or none in the axon and in the smaller dendrites. Therefore, proteins, such as the many enzymes concerned with the synthesis and degradation of transmitters, must be synthesized in the cell body and then transported to the axon and its terminals and to the dendrites. In fact, the cytoplasm in neurons is in constant motion and there is a net transport of materials toward the axon terminals via the phenomenon of anterograde axonal transport, which includes fast and slow components and apppears to depend on the integrity of microtubules in the axons. There is slow axonal transport in the other

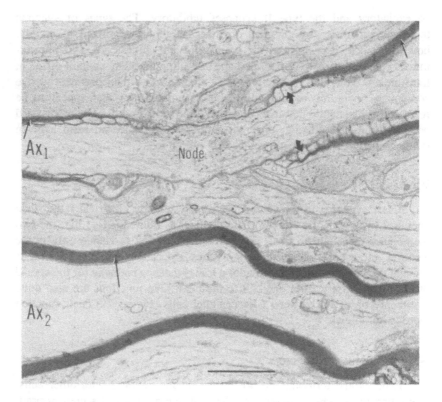

Figure 9. Electron micrograph of a section through two myelinated axons (Ax₁ and Ax₂) in the optic nerve of the frog. In the middle of Ax₁, the myelin layers disappear at a node of Ranvier (Node). It is this unmyelinated portion of the axon that contains a very high density of voltage-dependent sodium channels involved in the propagation of the action potential. The layers of glial cell membrane that form the myelin can be seen particularly clearly next to Ax₂ at the arrow. Arrows also indicate the internodal myelin around Ax₁. The small curved arrows indicate the edges of the sheet formed by the oligodendroglial cell. Small bits of glial cell cytoplasm are still present at the edges of the sheets, although the cytoplasm is completely extruded in the compact myelin of the internodal regions. Scale bar represents approximately 10 μm. Micrograph courtesy of Thomas E. Hughes.

direction, from the terminals to the soma, as well. The function of this retrograde axonal transport may be to keep the cell body informed, via some chemical messenger(s), of conditions at its terminals.

Supporting Cells

Neurons, as numerous as they are, are outnumbered considerably by the second cell type that populates nervous tissue, the supporting cells. Unlike neurons, supporting cells do not conduct axon potentials or form synapses. Also unlike neurons, they do continue to divide throughout life. They share with neurons a complex structrue which may vary from one part of the nervous system to another. Multiple functions have been suggested for these cells, some more well-documented than others. In the central nervous system, the main category of supporting cell is the glial cell, of which there are at least three types.

Glial Cells

Astrocytes

The astroglia or astrocytes (Figures 1A, 2 and 3) have small cell bodies from which extend a very large number of sinuous processes. The processes appear to surround and isolate neurons and groups of synapses and to separate blood vessels from nerve cells by forming end feet along the vessel walls. It has been suggested that the astrocytes provide passive structural support to the nervous system or that they have a nutritive function. A more intriguing possibility is that they function to regulate the ionic environment of neurons. They may shunt excess ions from sites of activity or take up transmitters after their release by nerve terminals. Both processes would serve to maintain the extracellular environment of active neurons more nearly constant. Astrocytes also appear to be phagocytes of debris associated with degeneration after injury. After trauma, they may proliferate in the area of injury and form a glial "scar."

Oligodendroglia

A second group of glial cells, the oligodendroglia or oligodendrocytes, are small cells with long processes that form the myelin sheaths of axons in the central nervous system (Figure 9). One oligodendrocyte may form up to forty or more myelin sheaths on a number of different axons. Like other cell membranes, the myelin-forming membrane of the oligodendroglial cell is made up of layers of lipid and protein. In the myelin sheath, the cytoplasm of the cell is squeezed out so that the two inner protein layers of the membrane come together. As these membranes spiral around the axon, the outer membrane layers also come together. The result is a sheath around the axon formed of many layers of membranes of the glial cell (Figure 9).

Microglia

A third class of glial cells, the microglia, are small, phagocytic cells that appear to be active only after injury or in disease. They resemble scavengers, moving in large numbers into damaged areas and actively taking up debris.

Ependymal Cells

Another group of supporting cells, the ependymal cells, are ciliated epithelial-like cells that line the ventricles of the brain. They also participate in the formation of the choroid plexus which secretes cerebrospinal fluid. A nonnervous cell type performing an important function in the nervous system is the endothelial cell that forms the walls of capillaries. Tight junctions between the membranes of adjacent endothelial cells are thought to be the anatomical basis of the blood-brain barrier, a barrier that prevents certain substances in the blood from entering the nervous system.

Peripheral Nervous System Supporting Cells

Schwann Cells

In the peripheral nervous system, the most prominent group of supporting cells are the Schwann cells. They form myelin sheaths around peripheral nerve fibers which resemble those formed by the oligodendroglial cells in the central nervous system. Each myelinated axon is encapsulated by a succession of single Schwann cells forming a chain. The Schwann cells also enfold non-myelinated axons in the peripheral nerves but do not form wrappings of myelin around them.

Cytoskeletal Proteins

Daniel M. Lapadula
Department of Toxicology
Schering-Plough Research
Lafayette, New Jersey

Mohamed B. Abou-Donia
Department of Pharmacology
Duke University Medical Center
Durham, North Carolina

CYTOSKELETAL ELEMENTS

Cytoskeletal elements of the nervous system have been identified as targets for many neurotoxic compounds. There are three main components to the neuronal cytoskeleton: 1) microtubules (24 nm in diameter), 2) neurofilaments (10 nm in diameter), and 3) microfilaments (7 nm in diameter) (Table 1; Ellisman and Porter, 1983). All three components share certain common characteristics (Table 2). Generally, they all consist of a core of acidic backbone proteins. They form helical filaments with chemically distinct head and tail regions whose functions are regulated by Ca^{++}. In addition, their interactions with themselves or other associated proteins may be regulated by phosphorylation with specific protein kinases. Cytoskeletal proteins provide a cell with its shape and freedom of movement. They allow the movement of materials within and from the cell.

Microtubules comprise approximately 10% of the total brain proteins. The microtubule subunit is an $\alpha\beta$ dimer, which forms upon linear polymerization. The α or β tubulin subunit protein has a molecular weight of approximately 55 kDa and binds GTP or GDP. The polymerization of tubulin dimers into microtubules is optimal in the presence of GTP, ATP, and Mg^{++}, in the absence of Ca^{++}, and at a temperature of 37°C (Weisenberg, 1972). Depolymerization takes place upon cooling to 4°C or with the addition of added Ca^{++}. The depolymerization in the presence of Ca^{++} is enhanced in the presence of calmodulin, a calcium binding protein that modulates a number

TABLE 1.
Properties of Cytoskeletal Filaments

	Diameter (nm)	Subunit MW (kDa)	Subunit Name	Cofactors For Assembly
Microfilaments	7–10	42	Actin	Mg^{++}, Ca^{++}, ATP
Neurofilaments	10–11	65, 140, 160	NFL, NFM, NFH	—
Microtubules	24–25	55	Tubulin	Mg^{++}, GTP

TABLE 2.
Characteristics of Cytoskeletal Proteins

1. All three proteins form filaments
2. The filaments are polar; their heads and tails are chemically distinct
3. The filaments are helical
4. They possess multiple equivalent binding sites
5. They exist in multiple forms
6. Calcium ion plays a role in the regulation and function of the three filament systems
7. All three proteins can associate with other proteins
8. Phosphorylation plays many roles in their interactions
9. All are acidic

TABLE 3.
Microtubule Associate Proteins (MAPS)

There are two major groups of MAPS:
1. High molecular weight (HMW) MAPs which are common in brain and are further subdivided to:
    a. MAP-1 with molecular weight of 350 kDa and consists of at least three proteins: MAP-1A, 1B, and 1C
    b. MAP-2 with molecular weight of 270 kDa and consists of two proteins: MAP-2A and 2B
2. Tau:
    This group of MAPs has a molecular weight ranging from 55 to 70 kDa

of cellular functions. There are also a number of microtubule-associated proteins (MAPs), which are classified (Table 3) into two major subgroups based on molecular weights (Vallee et al., 1984). High molecular weight (HMW) MAPs are common in neuronal cells. At least five different HMW MAPs have been identified (classified as MAPs 1–5). In addition there are apparently several different isomeric forms of several of these proteins. MAP-1 has a molecular weight of 350 kDa and has at least three isomers: MAP-1A, 1B, and 1C. MAP-2 has a molecular weight of approximately 270 kDa and has at least two isomers (MAP-2A, 2B). In addition a lower molecular weight (70 kDa) form of MAP-2 (MAP-2C) has been identified. The second major subgroup of MAPs are called Tau proteins. They have molecular weights in the range of 55 to 70 kDa. HMW MAPs and Tau proteins appear

TABLE 4.
Classification of Intermediate Filament Proteins (IFPs)

IFPs are grouped into five groups based on their presence:
1. *Cytokeratins:* MW 45 to 60 kDa; present in epithelial cells
2. *Neurofilaments:* Consist of three major polypeptides, MW 68 kDa, 145 kDa, and 220 kDa; present in neuronal cells
3. *Desmin or Skeleton:* MW 53 kDa; present in muscle cells
4. *Glial-Fibrillary Acidic Protein:* MW 55 kDa; present in glial cells
5. *Vimentin or Decamin:* MW 58 kDa; present in mesenchymal cells

to regulate the polymerization of microtubule proteins. The distribution of these proteins in the nervous system appears to be specific not only for particular neurons but also for different parts of the same neuron. For example, in the brain MAP-2 is found only in the cell body and dendrites; however, it appears to be present in axons in the peripheral nervous system. Unmyelinated axons contain more microtubules than myelinated ones. Microtubules appear to cluster around other organelles, i.e., mitochondria and endoplasmic reticulum, and are more uniformly distributed in smaller axons.

Neurofilaments are the intermediate filaments (Table 4) of the neuron. Neurofilaments are primarily made up of three subunit proteins designated NFL, NFM, and NFH (low, middle and high molecular weight neurofilament proteins). These designations are due to their anomalous behavior on sodium dodecyl sulfate-polyacryamide gel electrophoresis (SDS-PAGE; Scott et al., 1985). On SDS-PAGE they appear to have molecular weights of 70 kDa, 160 kDa, and 210 kDa; however, based on other procedures (including amino acid sequencing), their molecular weights are substantially lower (65 kDa, 140 kDa, and 160 kDa). This behavior is apparently due to their extreme acidity which apparently interferes with the binding of SDS to the subunits. Structurally, each subunit possesses a similar 40 kDa backbone middle region but are different in their head and tail regions (Weber et al., 1983). The head and tail pieces appear to be particularly rich in hydroxyl-containing amino acid residues (i.e., serine and threonine), as well as lysine and arginine residues. All of the subunit proteins are phosphorylated to some extent. The greatest number of phosphorylated residues are found on NFH (21 mol phosphate/mol polypeptide), while the least are on NFL (3 mol phosphate/mol polypeptide; Julien and Mushynski, 1982). Although all three subunits are individually able to self assemble under certain conditions, the prevailing concensus is that NFL serves as a backbone of the filament. Unlike the other two components of the cytoskeleton, neurofilaments do not appear to require specific types of accessory proteins, specific ions, or high energy phosphates or cofactors to assemble into filaments. Monoclonal antibodies to the phosphorylated and dephosphorylated epitopes of individual neurofilament proteins have demonstrated that neurofilaments are in a nonphosphorylated state in the cell body (Sternberger and Sternberger, 1983). Phosphorylation appears to be critical to stabilizing the neurofilament supramolecular structure, and neurofilaments are phosphorylated once they are beyond the proximal portion of the axon. The specific sites for neurofilament protein phosphorylation (and also MAP-2 phosphorylation) involve a very specific amino acid sequence. Lee et al. (1988) have found that the sequence lys-ser-pro-val occurs in several places in NFM and is found to be phosphorylated *in vivo*. In several other places in the neurofilament protein a lys-ser sequence, which is also a likely target for protein phosphorylation, is found in these proteins (Napolitano et al., 1987). The methods used to isolate neurofilament proteins usually depend on their insolubility in nonionic detergents. Further purification of the subunits entails column chromatography on either hydroxyapatite, DE-52 cellulose or gel filtration. The distribution of neurofilaments appears to be the inverse of that of microtubules, with myelinated axons containing more than unmyelinated ones.

The third class of cytoskeletal proteins are microfilaments. These filaments consist of actin. The monomeric form of actin is G-actin or globular actin, while polymerized actin is known as F-actin or filamentous actin. Actin is abundant in all cells, typically comprising between 5 and 10% of the total cellular protein. The subunit molecular weight of actin is approximately 42 kDa. Each actin monomer contains one ATP binding site, one high affinity and about 5 low affinity sites for divalent cations (Ca^{++}, or Mg^{++}) (Korn, 1982). Microfilaments form a network in the periphery of many animal cells. Actin filaments also produce stress fibers or bundles that are most common in cultured cells. In neurons, they are usually found in the axons and dendritic spines.

The microtubules appear to be linked to neurofilaments via crossbridge structures (Ellisman and Porter, 1983). These mixed complexes have been noted in cell processes where microtubules and neurofilaments run in parallel arrays. In most regions, particularly within the cell body, aggregates of neurofilaments are found. The "neurofibrillary bundles" encapsulate the nucleus and extend into axonal processes (Lasek, 1981). Microfilaments are confined to the peripheral zone of the cell and are distinct in the neuron from microtubule-neurofilament rich zones. The "core" neuronal cytoskeleton (neurofilaments and microtubules) is implicated in routing of synthesized components from the cell body into and down neuronal processes. Crossbridges often connect microtubules with microtubules, neurofilaments with neurofilaments, as well as microtubules with neurofilaments (Ellisman and Porter, 1980; Ishikawa and Tsukita, 1982). These crossbridges also link fibrous elements with membranous organelles such as multivesicular bodies or smooth membranous cisternae in the axon or other regions of the neuron. Protein phosphorylation has been proposed to play a critical role in the interaction of neurofilaments with microtubules (Julien and Mushynski, 1981, 1982; Julien et al., 1983; Leterrier et al., 1982, 1984; Vallano et al., 1985). Although microtubule-associated proteins were first identified by their co-purification with microtubules, it is now clear that these proteins also have a strong association with neurofilaments. The crossbridges between microtubules and neurofilaments are at least in part made up of these associated proteins. In addition, there are a number of protein kinases that may regulate the interaction of neurofilaments with microtubules, MAP-2 is known to be phosphorylated by cAMP-dependent, Ca^{++}/calmodulin-dependent (calmodulin kinase II) and Ca^{++}/phospholipid-dependent (protein kinase C) protein kinases.

AGENTS THAT INTERFERE WITH MICROTUBULES

Vinca Alkaloids

Many compounds are known to possess toxicity directly related to their ability to interact with different components of the cytoskeleton. A number of these compounds have clinical use in cancer chemotherapy. Compounds that disrupt microtubules and prevent cell division, i.e., vinca alkaloids, are useful chemotherapeutic agents. However, the side effects of these drugs include disruption of the cytoskeleton of the nervous system, often producing a neuropathy. The production of the neuropathy is the limiting factor during treatment with vincristine. There are two binding sites for vinca alkaloids on tubulin. One is a high-affinity site and the other is a moderate-affinity binding site (Bhattacharya and Wolff, 1976). At low concentrations there is a substoichiometric end "poisoning" of microtubule polymerization. At higher concentrations (m*M*) these drugs induce the formation of a paracrystalline lattice composed of microtubule proteins.

Colchicine

Colchicine binds with high affinity to the dimeric form of tubulin in a one-to-one complex (Dustin, 1984). It binds to a site that is different from that of vinca alkaloids. This characteristic of its binding has led to its usefulness in the measurement of the amount of tubulin present in a preparation. Tubulin, which is bound to colchicine, adds to the end of a microtubule and then prevents further addition of tubulin dimers to the microtubule. This gives it the property of substoichiometric end "poisoning" of microtubule polymerization. Colchicine also destabilizes already formed microtubules. In addition, tubulin bound with colchicine appears to form nonmicrotubule polymers. Nerves that are locally exposed to either colchicine or vinca alkaloids undergo a Wallerian-type degeneration starting at the site of application.

Nocodazole

This compound binds rapidly to tubulin to inhibit microtubule assembly and induce disassembly of existing microtubules. It is more readily reversible than either colchicine or vinca alkaloids.

Taxol

In contrast to the above compounds, taxol appears to stabilize the microtubule. Instead of binding to the microtubule subunit, it binds to the formed microtubule. It stabilizes the microtubule to depolymerization by cold, calcium, colchicine, vinca alkaloids, or nocodazole. It appears to decrease the critical concentration for assembly of microtubules, thereby shifting the equilibrium in favor of microtubule formation. It also stabilizes formed microtubules from destabilizing conditions.

Methyl Mercury

Recently it has been demonstrated that methyl mercury interferes with polymerization of microtubules by binding to sulfhydryl groups on the tubulin dimer (Sager et al., 1986). Cell division was found to be inhibited by methyl mercury exposure. The granule cells of the cerebellum in the developing nervous system appeared to be extremely sensitive to methyl mercury.

Heavy Metals

Heavy metals such as cadmium, lead, etc., have been demonstrated to depolymerize microtubules. The mechanism of this effect appears to be due to their ability to bind to the calcium-binding protein, calmodulin. Calmodulin lowers the concentration required for depolymerization of microtubules by calcium. Heavy metals have a high affinity for calmodulin and mimic the effect of calcium on the microtubules.

Organophosphorus Compounds

Exposure to certain organophosphorus esters produces a dying back neuropathy known as organophosphorus compound-induced neurotoxicity (OP-IDN) (Smith et al., 1930; Abou-Donia, 1981) in which there is an aggregation, accumulation, and partial condensation of neurofilaments and microtubules. Recent studies have indicated that calmodulin kinase II may be (in part) responsible for the development of this neuropathy (Abou-Donia et al., 1984, 1988). This kinase phosphorylates a number of cytoskeletal proteins including tubulin, MAP-2 and neurofilaments (Suwita et al., 1986). In animals treated with organophosphorus compounds, there is an increase in the phosphorylation in a calcium/calmodulin-dependent manner. This increase was greatest in cold stable microtubules (microtubules that are stable to cold depolymerization), with which calmodulin kinase II copurifies.

AGENTS THAT INTERFERE WITH NEUROFILAMENTS

β,β'-Iminodiproprionitrile (IDPN)

IDPN appears to have a specific effect on the organization of neurofilament proteins. Early studies on IDPN demonstrate a specific blockage of the transport of neurofilaments, with a lesser effect on tubulin and actin transport (Griffin et al., 1978). The pathology of this toxicant indicates there is a ballooning of the proximal regions of anterior horn cells of the spinal cord at the initial segment of the axon. These regions were filled with neurofilaments. There also appears to be a segregation of neurofilaments to the periphery of the axon with microtubules and other organelles more centrally located. The biochemical mechanism of this neuropathy is currently unknown.

Aliphatic Hexacarbons

Hexacarbons or other compounds which form γ-diketones produce a neuropathy in which there is an accumulation of neurofilaments in the more distal regions of the axon (Schaumberg and Spencer, 1976; Abou-Donia et al., 1982). This neuropathy was first recognized in fabric factory workers and glue sniffers who were exposed to high concentrations of methyl-*n*-butyl ketone, a metabolite of *n*-hexane. 2,5-Hexandione has been identified as the metabolite of *n*-hexane that is responsible for the neurotoxicity. In addition, other γ-diketones have been identified as more potent neurotoxicants than *n*-hexane. 3,4-Dimethyl-2,5-hexanedione is between 10 to 20 times more potent a neurotoxicant than 2,5-hexanedione (Anthony et al., 1983). This compound also produces an accumulation of neurofilaments; however, the accumulation is in the middle regions of the axon. The mechanism of action of these compounds may likely include the formation of pyrroles on lysine residues and the possible subsequent crosslinking of cytoskeletal proteins. The high proportion of lysine residues makes the neurofilaments likely targets of these toxicants.

Acrylamide

Acrylamide produces a neuropathy similar to that of hexacarbons. There is an accumulation of neurofilaments in the distal regions of the axons of peripheral nerves and spinal cord. However, there is also an increase in the

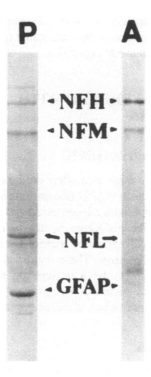

Figure 1. *In vitro* binding of [¹⁴C]acrylamide to spinal cord cytoskeletal protein, P, protein staining and A, autoradiographs. The three neurofilament triplet proteins are denoted by NFL (70 kDa), NFM (160 kDa), and NFH (210 kDa). GFAP is glial-fibrillary acidic protein.

smooth endoplasmic reticulum in the terminal regions of the nerves. Acrylamide binds to the cytoskeletal proteins of the nervous system, similar to that of hexacarbons without the concomitant crosslinking of these proteins (Figure 1). Acrylamide may interfere with the organization of the cytoskeleton by interference with the normal interactions of cytoskeletal proteins.

Dimethylaminoproprionitrile

This compound is structurally related to both IDPN and acrylamide, and is used as a polymerizing agent in acrylamide. It produces a neuropathy similar to the other two related compounds, with a notable increase in neurofilaments in the distal axon.

Aluminum

Aluminum is not normally thought of as a neurotoxin since it is debatable that dietary aluminum can cause a neuropathy. Recent evidence from patients with Alzheimer's disease suggest that the neurofibrillary tangles associated with this neurodegenerative disorder contain a high content of aluminum. Aluminum can be applied to the brain topically or by intrathecal injection. Histological examination of this tissue reveals neurons with neurofibrillary bundles in the cytoplasm that are quite similar to those of Alzheimer's disease patients. However, examination by electron microscopy reveals these to be rather normal 10 nm neurofilaments and not the paired helical filaments usually seen in Alzheimer's disease. Aluminum has been shown to alter the distribution of phosphorylated neurofilaments in axons and to increase the phosphorylation of MAP-2 and NFH (Troncoso et al., 1986; Johnson and Jope, 1988).

Carbon Disulfide (CS₂)

CS_2 produces a neuropathy that is virtually identical to that of 2,5-hexanedione. A direct binding of CS_2 to neurofilaments has been demonstrated in rats. The mechanism of action of this compound is currently unknown. The similarity of pathologies of these neurotoxicants may suggest a common mechanism of action.

To date no single mechanism accounts for the toxicity of these compounds. Several universal hypotheses have been proposed to account for the accumulation of neurofilaments. It has also recently been suggested that many naturally occurring disease states (e.g., Alzheimer's disease, giant axonal neuropathy, amytrophic lateral sclerosis, Parkinson's disease) may have an environmental component that contributes to the initiation and/or progression of the disease. Several common features are shared by many of the aforementioned compounds.

1. Many of them appear to be able to bind to ε-amino groups of lysine.
2. They all produce a neuropathy in which the common pathological feature is the accumulation of neurofilaments.
3. In most cases there is a segregation of microtubules and neurofilaments in the axon.

The mechanism of action of γ-diketones has been the most extensively studied of all of the above neurotoxicants. Several hypotheses have been proposed to account for this particular neuropathy. These hypotheses are subdivided into two major groups:

1. Indirect action on the neurofilaments. One of the first proposed mechanisms of action of 2,5-hexanedione was an inhibition of glycolysis by its reactivity to sulfhydryl groups. Specifically, phosphofructokinase and glyceraldehyde-3-phosphate dehydrogenase were proposed as enzymes that were inhibited by 2,5-hexanedione (Spencer et al., 1979). The inhibition of these enzymes led to an inhibition of axonal transport of neurofilament proteins. A universal hypothesis

for all neurofilamentous neuropathies was subsequently proposed based on this mechanism. The inhibition of glycolytic enzymes appears unlikely, since extremely high concentrations of 2,5-hexanedione were required to produce an inhibition of these enzymes *in vitro*. In addition, *in vivo* evidence indicates that glucose-dependent lactate production in the nerves of rats treated with 2,5-hexanedione was not altered.

2.   Direct action on the neurofilaments. The other proposed mechanisms of action all have a direct action on the neurofilaments. The reaction of 2,5-hexanedione with ε-amino groups of lysine is central to all these hypotheses. These hypotheses fall into several major categories:

a.   Pyrrole formation interferes with the hydrophobic interactions of neurofilaments with each other or other cytoskeletal proteins. The formation of pyrrole adducts on neurofilament proteins has been found to correlate with the neurotoxicity (DeCaprio et al., 1982). In addition, compounds that form a pyrrole more easily than 2,5-hexanedione (i.e., 3,4-dimethyl-2,5-hexanedione) are 10 to 20 times more neurotoxic than 2,5-hexanedione (Anthony et al., 1983).

b.   Crosslinking of cytoskeletal proteins. Crosslinking of neurofilament proteins was first proposed in 1980 (Graham, 1980; Graham and Abou-Donia, 1980). With the originally proposed mechanism, neurofilaments were crosslinked to each other through Schiff base formation. Subsequently, it was suggested that the pyrrole formation found by DeCaprio et al. (1982) could also crosslink neurofilaments through autooxidation of the pyrrole (Graham et al., 1982; Anthony et al., 1983). Protein crosslinking *in vitro* was demonstrated by Graham et al. (1984) with both 2,5-hexanedione and its 3,4-dimethyl derivative. *In vivo* crosslinking of neurofilaments was subsequently demonstrated in rats after exposure of 2,5-hexanedione (Lapadula et al., 1986). It is currently unknown if crosslinking is a prerequisite for the development of the neuropathy.

c.   Alteration of protein phosphorylation. Recently it has been demonstrated that protein phosphorylation is decreased in the spinal cords of animals treated with 2,5-hexanedione (Abou-Donia et al., 1988; Lapadula et al., 1988). The decreased phosphorylation occurred in a dose-dependent manner. The *in vivo* phosphorylation state of neurofilaments proteins was quantified by the use of monoclonal antibodies to phosphorylated and dephosphorylated epitopes on the neurofilament proteins. In rats treated with 2,5-hexanedione there was a decrease in both of these epitopes (Table 5). However, the decrease of the phosphorylated neurofilament protein epitopes was significantly greater than the dephosphorylated epitopes. Neurofilament pro-

TABLE 5. Comparison of Amounts of Phosphorylated and Nonphosphorylated Epitopes of Neurofilament Proteins From Rats Treated with 2,5-Hexanedione

Epitope	Dose				
	Control	0.1	0.25	0.5	1.0
Nonphosphorylated	100 ± 8^a	66 ± 6^a	63 ± 8^a	75 ± 6^a	53 ± 6^a
Phosphorylated	$100 \pm 13^{a,b}$	$42 \pm 8^{a,b}$	$32 \pm 5^{a,b}$	$40 \pm 4^{a,b}$	$26 \pm 4^{a,b}$

Note: All statistics were done with an ANOVA.

[a]Significantly different from control $p < 0.01$.
[b]Significantly different from corresponding nonphosphorylated epitope $p < 0.01$.

teins are known to have amino acid sequences in which there is a close proximity of lysine to serine residues (see above). Serine residues are known to be phosphorylated in neurofilament proteins and may be physically blocked by their close proximity to a lysine residue to which 2,5-hexanedione is bound.

d. Inhibition of proteolytic breakdown of neurofilaments. Proteolysis of neurofilaments occurs in the terminal regions of the axon and during their transport down the axon. Neurofilaments are degraded by calcium activated proteases which are thought to be primarily active in the nerve terminals. A diminished amount of neurofilament proteolytic breakdown products have been found in spinal cords of rats treated with 2,5-hexanedione. It is currently unknown if there is an actual inhibition of proteolysis or if the diminished proteolysis is merely an artifact of the crosslinked proteins not being a good substrate for the proteolytic enzymes.

AGENTS THAT INTERFERE WITH MICROFILAMENTS

There are two well-characterized agents that specifically interfere with actin filament formation: cytochalasins and phalloidin.

Cytochalasins

These naturally occurring compounds are fungal metabolites. They appear to bind to high-affinity sites on F-actin at a ratio of 1:1 per actin filament (Tannenbaum, 1978). This results in an inhibition in the rate of growth of the actin filament at low concentrations. The application of cytochalasins is usually dramatic, with a complete breakdown of the actin skeleton.

Phalloidin

These are naturally occurring products derived from toadstools. Their action is completely opposite to that of the cytochalasins, with a stabilization of the actin filaments (Lengsfeld et al., 1974). Their effects are similar to the effects of taxol on microtubules in that they lower the critical concentration for assembly, accelerate the rate of polymerization and prevent the breakdown of the filaments in the presence of destabilizing conditions.

SUMMARY

The axonal cytoskeleton is a dynamic structure with which numerous neurotoxicants interact. The role of the cytoskeleton in axonal transport will be dealt with in another chapter. In this chapter compounds were divided into classes dependent upon which cytoskeletal component they specifically interact; however, this separation is artificial. Since none of the individual filaments exist by themselves and they interact with each other, it is reasonable to assume that if a specific compound interacts with only one of these components, the other cytoskeletal components will also be affected. This was shown by studies on IDPN. Although there is a specific interaction with the neurofilaments, the transport rates of both tubulin and actin were also altered by IDPN. There are various accessory proteins that interact with more than one cytoskeletal component, and it would not be surprising if one of these proteins was a target for a specific toxicant. In addition, once the mechanism of action of a specific neurotoxicant is known, it could be used as a tool to study the neuronal cytoskeleton. Indeed, many of these compounds are being used for that purpose. Future studies on these neurotoxicants will not only lead to a greater understanding of their mechanism of action but also the role of the cytoskeleton in neurons.

REFERENCES

Abou-Donia, M.B. (1981). Organophosphorous ester-induced delayed neurotoxicity. *Annu. Rev. Pharmacol. Toxicol.* 21: 511–548.

Abou-Donia, M.B., H.-A.M. Makkawy, and D.G. Graham (1982). The relative neurotoxicities of *n*-hexane, methyl *n*-butyl ketone, 2,5-hexanediol, and 2,5-hexanedione following oral or intraperitoneal administration in hens. *Toxicol. Appl. Pharmacol.* 62: 369–389.

Abou-Donia, M.B., S.E. Patton, and D.M. Lapadula (1984). Possible role of endogenous protein phosphorylation in organophosphorus compound-induced delayed neurotoxicity. In *Cellular and Molecular Neurotoxicology*. T. Narahashi, Ed., Raven Press, New York, pp. 265–283.

Abou-Donia, M.B., D.M. Lapadula, and E. Suwita (1988). Cytoskeletal proteins as targets for organophosphorus compounds and aliphatic hexacarbon-induced neurotoxicity. *Toxicology* 49: 469–477.

Anthony, D.C., K. Boekelheide, C.W. Anderson, and D.G. Graham (1983). The effect of 3,4-dimethyl substitution on the neurotoxicity of 2,5-hexanedione. II. Dimethyl substitution accelerates pyrrole formation and protein crosslinking. *Toxicol. Appl. Pharmacol.* 71: 372–382.

Bhattacharya, B. and J. Wolff (1976). Tubulin aggregation and disaggregation-mediation by two distinct vinblastine-binding sites. *Proc. Natl. Acad. Sci. U.S.A.* 73: 2375–2378.

DeCaprio, A.P., E.J. Olajos, and P. Weber (1982). Covalent binding of a neurotoxic *n*-hexane metabolism: conversion of primary amines to substituted pyrrole adducts by 2,5-hexanedione. *Toxicol. Appl. Pharmacol.* 65: 440–450.

Dustin, P. (1984). *Microtubules*, 2nd ed. Springer-Verlag, Berlin.

Ellisman, M.H. and K.R. Porter (1980). The microtrabecular structure of the axoplasmic matrix: visualization of cross-linking structures and their distribution. *J. Cell Biol.* 87: 464–479.

Ellisman, M.H. and K.R. Porter (1983). Introduction to the cytoskeleton. In *Neurofilaments*. C.A. Marotta, Ed., University of Minnesota Press, Minneapolis, pp. 3–26.

Graham, D.G. (1980). Hexane neuropathy: a proposal for pathogenesis of a hazard of occupational exposure and inhalant abuse. *Chem. Biol. Interact.* 32: 339–345.

Graham, D.G. and M.B. Abou-Donia (1980). Studies on the molecular pathogenesis of hexane neuropathy. I. Evaluation of the inhibition of glyceraldehyde 3-phosphate dehydrogenase by 2,5-hexanedione. *J. Toxicol. Environ. Health* 4: 629–634.

Graham, D.G., D.C. Anthony, K. Boekelheide, N.A. Maschmann, R.G. Richards, J.W. Wolfram, and B.R. Shaw (1982). Studies on the molecular pathogenesis of hexane neuropathy. II. Evidence that pyrrole derivatization of lysyl residues leads to protein cross linking. *Toxicol. Appl. Pharmacol.* 64: 415–422.

Graham, D.G., G. Szakai-Quin, J.W. Priest and D.C. Anthony (1984). *In vitro* evidence that covalent crosslinking of neurofilaments occurs in γ-diketone neuropathy. *Proc. Natl. Acad. Sci. U.S.A.* 81: 4979–4982.

Griffin, J.W., P.N. Hoffman, A.W. Clark, P.T. Carroll, and D.L. Price (1978). Slow axonal transport of neurofilament proteins: impairment by β,β'-iminodiproprionitrile administration. *Science* 202: 633–635.

Ishikawa, H. and S. Tsukita (1982). Morphological and functional correlates of axoplasmic transport. In *Axoplasmic Transport*. G.G. Weiss, Ed. Springer-Verlag, Berlin, pp. 251–259.

Johnson, G.V.W. and R.S. Jope (1988). Phosphorylation of rat brain cytoskeletal proteins is increased after orally administered aluminum. *Brain Res.* 456: 95–103.

Julien, J.P. and W.E. Mushynski (1981). A comparison of *in vitro* and *in vitro*-phosphorylated neurofilament polypeptides. *J. Neurochem.* 37: 1579–1585.

Julien, J.P. and W.E. Mushynski (1982). Multiple phosphorylation sites in mammalian neurofilament polypeptide. *J. Biol. Chem.* 257: 10467–10470.

Julien, J.P., G.D. Smoluk, and W.E. Mushynski (1983). Characteristics of the protein kinase activity associated with rat neurofilament preparations. *Biochem. Biophys. Acta* 755: 25–31.

Korn, R.D. (1982). Actin polymerization and its regulation by proteins from nonmuscle cells. *Physiol. Rev.* 62: 672–737.

Lapadula, D.M., R.D. Irwin, E. Suwita, and M.B. Abou-Donia (1986). Cross-linking of neurofilament proteins of rat spinal cord in vivo after administration of 2,5-hexanedione. *J. Neurochem.* 46: 1843–1850.

Lapadula, D.M., E. Suwita, and M.B. Abou-Donia (1988). Evidence for multiple mechanisms responsible for 2,5-hexanedione-induced neuropathy. *Brain Res.* 458: 123–131.

Lasek, R.J. (1981). The dynamic ordering of neuronal cytoskeletons. *Neurosci. Res. Program Bull.* 19: 7–32.

Lee, V.M.-Y., L. Otvos, M.J. Carden, M. Hollosi, B. Dietzschold, and R.A. Larrarini (1988). Identification of the major multiphosphorylation site in mammalian neurofilaments. *Proc. Natl. Acad. Sci. U.S.A.* 85: 1988–2002.

Lengsfeld, A.M., I. Low, T. Wieland, P. Dancker, and W. Hasselbach (1974). Interaction of phalloidin with actin. *Proc. Natl. Acad. Sci. U.S.A.* 71: 2803–2807.

Leterrier, J.R., R.K.H. Liem, and M.L. Shelanski (1982). Interactions between neurofilaments and microtubule-associated proteins: a possible mechanism for intraorganellar binding. *J. Cell Biol.* 95: 982–986.

Leterrier, J.F., J. Wong, R.K.H. Liem, and M.L. Shelanski (1984). Promotion of microtubule assembly by neurofilament-associated microtubule-associated proteins. *J. Neurochem.* 43:1385–1391.

Napotitano, E.W., S.S. Chin, D.R. Colman, and R.M. Liem (1987). Complete amine acid sequence and *in vitro* expression of rat NF-M, the middle molecular weight neurofilament protein. *J. Neurosci.* 7: 2590–2599.

Sager, P.R. and T.L.M. Syversen (1986). Disruption of microtubules by methyl mercury. In *The Cytoskeleton: A Target for Toxic Agents.* T.W. Clarkson, P.R. Sager, and T.L.M. Syversen, Eds., Plenum, New York, pp. 97–116.

Schaumburg, H.H. and P.A. Spencer (1976). Central and peripheral nervous system degeneration produced by pure *n*-hexane: an experimental study. *Brain* 99: 183–192.

Scott, D., K.E. Smith, B.J. O'Brien, and K.J. Angelides (1985). Characterization of mammalian neurofilament triplet proteins. *J. Biol. Chem.* 260: 10736–10747.

Smith, M.I., E. Elvove, and W.H. Frazier (1930). The pharmacological action of certain phenol esters, with special reference to the etiology of so-called ginger paralysis. *Public Health Rep.* 45: 2509–2524.

Spencer, P.S., M.I. Sabri, H.H. Schaumburg, and C.L. Moore (1979). Does a defect of energy metabolism in the nerve fiber underlie axonal degenerations in poly-neuropathies. *Ann. Neurol.* 5: 501–507.

Sternberger, L.A. and N.H. Sternberger (1983). Monoclonal antibodies distinguish phosphorylated and nonphosphorylated forms of neurofilaments *in situ. Proc. Natl. Acad. Sci. U.S.A.* 80: 6126–6130.

Suwita, E., D.M. Lapadula, and M.B. Abou-Donia (1986). Calcium and calmodulin-enhanced *in vitro* phosphorylation of hen brain cold-stable microtubules and spinal cord neurofilament triplet proteins after a single oral dose of tri-*o*-cresyl phosphate. *Proc. Natl. Acad. Sci. U.S.A.* 83: 6174–6178.

Tannenbaum, S.W. (1978). *Cytochalasins: Biochemical and Cell Biological Aspects.* Elsevier, Amsterdam.

Troncoso, J.C., N.H. Sternberger, L.A. Sternberger, P.N. Hoffman, and D.L. Price (1986). Immunocytochemical studies of neurofilament antigens in the neurofibrillary pathology induced by aluminum. *Brain Res.* 364: 295–300.

Vallano, M.L., J.R. Goldenring, R.S. Lasher, and R.J. DeLorenzo (1985). Association of calcium/calmodulin dependent kinase with cytoskeletal preparations: phosphorylation of tubulin, neurofilament and microtubule-associated proteins. *Ann. N.Y. Acad. Sci.* 382: 3202–3206.

Vallee, R.B., G.S. Bloom, and W.E. Theurkauf (1984). Microtubule-associated proteins; subunits of the cytomatrix. *J. Cell Biol.* 99: 38s–44s.

Weber, K., G. Shaw, H. Osborn, E. Debos, and N. Geisler (1983). Neurofilaments, a subclass of intermediate filaments: structure and expression. *Cold Spring Harbor Symp. Quant. Biol.* 48: 717–729.

Weisenberg, R.C. (1972). Microtubule formation *in vitro* in solutions containing low calcium. *Science* 177: 1104–1105.

Neurotypic and Gliotypic Proteins as Biochemical Indicators of Neurotoxicity

James P. O'Callaghan
Neurotoxicology Division
United States Environmental Protection Agency
Research Triangle Park, North Carolina

INTRODUCTION

Exposure of the developing or mature nervous system to neurotoxic xenobiotics results in complex behavioral, physiological, and morphological alterations. With few exceptions, a biochemical basis for these toxicant-induced changes has yet to be described (Damstra and Bondy, 1982). This chapter will briefly address the issues to be considered when developing a biochemical approach to neurotoxicity assessment and will review the progress of work concerning one strategy: the use of neurotypic and gliotypic proteins as biochemical indicators of neurotoxicity.

What is Neurotoxicity?

Perhaps one of the most confusing issues confronting the student of neurotoxicology is the difference of opinion with respect to what constitutes a neurotoxic response. For the purposes of this review, it will be assumed that the common thread linking neurotoxic effects is nervous system damge initiated by an exogenous factor (Dayan, 1984). In this context, pharmacological effects, which by nature are short acting, reversible, and presumably nonharmful (unwanted side effects not withstanding) will not be considered as neurotoxic. Instead, neurotoxicity will be defined as the slowly reversible or irreversible effects of toxic agents manifested by alterations in the cellular and subcellular components of the developing or mature nervous system.

What Types of Agents are Neurotoxic?

On the basis of this operational definition of neurotoxicity one may next consider the types or classes of agents that are neurotoxic. Here, one is confronted with a dilemma of major proportions, for it is becoming increasingly apparent that an enormous chemical variety of environmental pollutants can affect nervous system structure and function (Spencer and Schaumburg, 1980). These can range from chemicals of known but widely dissimilar structure, such as pesticides and industrial chemicals, to chemical mixtures of unknown composition, such as those found in toxic waste dumps. Thus, the investigator who wishes to establish a biochemical basis for neurotoxicity must face the formidable challenge of choosing neurochemical indices which are sensitive to the effects of structurally diverse, and in some cases, unknown chemical toxicants.

What are Cellular Targets of Neurotoxicity?

Unfortunately, assessing neurotoxicity is complicated not only by the variety of potential neurotoxicants (exogenous variables), but also by the innate molecular and cellular heterogeneity of the nervous system itself (endogenous variables). This inherent complexity of nervous system tissue is of concern to the neurotoxicologist because neurotoxicant-induced injury to the developing or mature organism often is manifested by alterations in the cytoarchitecture of specific neuroanatomical regions (Spencer and Schaumburg, 1980). Moreover, within an affected area the response to chemical insult may encompass one or several types of neurons and glia, the major cellular elements of the nervous system. This regional and cell type-specific action of toxic agents is widely documented in the neuropathology literature and often is referred to as selective vulnerability (e.g., see Spencer and Schaumberg, 1980; Spencer et al., 1985; Price et al., 1986).

EXAMPLES OF SELECTIVE VULNERABILITY TO NEUROTOXICITY

To illustrate the concept of selective vulnerability, the extreme effects of three known neurotoxicants are shown in Figure 1. The first toxicant, trimethyltin (TMT), preferentially affects a region of the brain known as the hippocampus, where a specific type of neuron is selectively damaged; loss

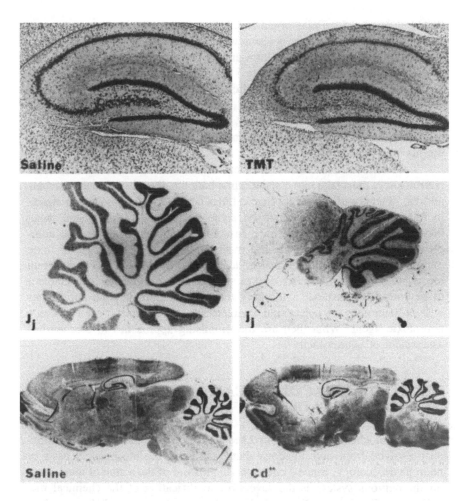

Figure 1. Examples of selective vulnerability to neurotoxicity. Micrographs were prepared from cresyl violet stained sagittal sections of rat hippocampus (upper panels), cerebellum (middle panels) and whole brain (lower panels). The neurotoxic agents were trimethyltin (TMT), bilirubin (the hyperbilirubinemic, jj, Gunn rat was used as a model) and cadmium (Cd^{++}). For a description of subjects and dosing regimens, see O'Callaghan and Miller, 1984 (TMT); O'Callaghan and Miller, 1985 (bilirubin); and O'Callaghan and Miller, 1986 (Cd^{++}).

of these cells in man is associated with deficits in learning and memory, a frequent symptom of toxicant exposure. In contrast to TMT, the second toxicant, bilirubin, has a pronounced effect on the development of the cerebellum; such disruption of cerebellar development in man is often an etiological factor in cerebral palsy. The third toxicant, cadmium, while sparing both the hippocampus and the cerebellum, destroys an area of the rat brain known as the neostriatum. Destruction of neurons in an equivalent brain area in man is associated with Parkinson's disease and Huntington's chorea. As these examples suggest, selective vulnerability to toxic insult may be manifested at levels of cellular organization as complex as the nervous system itself. Indeed, toxicants are known which preferentially affect not only a given brain region (e.g., cerebellum and hippocampus) and a given cell type (e.g., neurons, oligodendroglia, etc.) but also a given subclass of cell type (e.g., specific types of neurons) often at the level of specific subcellular components (e.g., neuronal perikarya, axons or synapses). Thus, to effect a biochemical detection of neurotoxicity will require the use of neurochemical indicators that reflects damage to diverse and unpredictable targets throughout the nervous system.

To summarize, it should be clear that a biochemical assessment of neurotoxicity must take into consideration both the diversity of neurotoxicants and the diversity of molecular and cellular targets. With these requirements in mind, the central question to be addressed can be simply stated: what biochemical indice(s) should be measured? As indicated above, the complexity of the nervous system and the complexity of its response to toxic exposures rule out a simple answer to this question. To put it in different terms, no single measure is likely to stand alone as a biochemical indicator of neurotoxicity; instead, it is likely that several multiendpoint approaches will be necessary to detect and characterize the effects of even a single neurotoxic agent. What then are the promising approaches to biochemical assessment of neurotoxicity?

NEUROPHARMACOLOGICAL APPROACH TO NEUROTOXICOLOGY

Most biochemical techniques used to detect and characterize neurotoxicity evolved from investigations that established the biochemical basis of neuropharmacology. Since a characteristic feature of the action of most drugs on the nervous system is the reversible modification of the chemical machinery mediating synaptic transmission, endpoints examined in neuropharmacology frequently reflect effects which are short acting and restricted to specific neurotransmitter systems. Thus, when applied to neurotoxicology, these measures may prove useful for assessing the effects of toxicants on neuronal systems sensitive to pharmacological intervention (i.e., drug-like effects); however, because this approach limits evaluations to acute effects on neurons, neurotoxicants that produce delayed effects or that affect primarily non-neuronal cell types (e.g., oligodendroglia, astrocytes, microglia) are likely to escape detection.

The successful application of a neuropharmacological approach to neurotoxicology is also hampered by the gap between our knowledge of how drugs act on the nervous system and our understanding of how toxicants cause their effects on the nervous system. For example, in neuropharmacology, the process of choosing a biochemical index is often made easier by 1) prior knowledge of the effects of agents similar to the drug in question and

2) the availability of pharmacological probes (e.g., structurally related agonists, inactive stereoisomers, specific antagonists, transmitter depletors, inhibitors of key enzymes, etc.) that affect specific neurotransmitter systems in a predictable manner (e.g., see Cooper et al., 1982). In neurotoxicology, frequently there is not an *a priori* basis for predicting which, if any, neurotransmitter system will be affected by a suspect neurotoxicant. Given the extensive number of putative neurotransmitter/neuromodulator systems (Barchas et al., 1978; Cooper et al., 1982), it then becomes exceedingly difficult to assess the integrity of more than a few following toxicant exposure. As a consequence, the decision to select a particular measure often is influenced more by the availability of appropriate expertise and equipment than by the need for a systematic and comprehensive approach to the determination of the locus of toxicant action.

NEUROTYPIC AND GLIOTYPIC PROTEINS AS BIOCHEMICAL INDICATORS OF NEUROTOXICITY

The limitations of a neuropharmacological approach to neurotoxicity assessment underscore the need for alternate strategies. The approach that we have adopted (O'Callaghan and Miller, 1983; O'Callaghan, 1988) takes advantage of the recent discovery of cell type-specific proteins in the nervous system. These proteins distinguish the major cellular and subcellular elements composing the nervous system of all mammalian species, including man (Raff et al., 1979; Schachner, 1982; Nestler et al., 1984; O'Callaghan (1988). Examples of these neuron-specific (neurotypic) and glia-specific (gliotypic) proteins are shown in Table 1. Some proteins, like the cell types they represent, are found in all areas of the nervous system while others are restricted to a single cell type found only in a discrete brain region. For example, the neurotypic protein, synapsin I, is present in virtually all nerve terminals of the CNS and PNS. In contrast, DARPP-32 (dopamine and cyclic AMP-regulated phosphoprotein of 32,000 molecular weight) is highly enriched in dopaminoceptive neurons that possess D1-type receptors. Similarly, cGMP-dependent protein kinase, its substrate (G-substrate) and PCPP-260 (Purkinje cell specific phosphoprotein of 260,000 molecular weight) are restricted to only a single type of neuron, the cerebellar Purkinje cell. Neurotypic and gliotypic proteins are associated with subcellular structures as specialized as the cell types that contain them. For example, the myelin sheath elaborated by oligodendroglia is distinguished by the presence of myelin basic protein, whereas the neuronal cytoskeleton is characterized by three of its component proteins, the neurofilament triplet proteins: p68, p160 and p200.

The availability of techniques for identifying (by immunohistochemistry) and, in most cases, measuring (by radioimmunoassay) these proteins makes it feasible to detect and characterize the cellular and subcellular responses to

TABLE 1. Examples of Neurotypic and Gliotypic Proteins Suitable For Use as Biochemical Indicators of Neurotoxicity

Protein	Regional Distribution	Cellular Distribution	Reference
Synapsin I	All neurons in CNS and PNS	Synaptic vesicles (cytoplasmic surface)	Nestler and Greengard, 1984
Protein III	All neurons in CNS and PNS and adrenal chromaffin cells	Synaptic vesicles chromaffin granules	Nestler and Greengard, 1984
P38	All neurons in CNS and PNS; tissue with secretory vesicles	Synaptic vesicles (integral membrane protein); secretory vesicles	Jahn et al., 1985
cGMP-dependent protein kinase	Cerebellar Purkinje cells	Cytosol	De Camilli et al., 1984a
G-substrate	Cerebellar Purkinje cells	Cytosol	Detre et al., 1984
PCPP-260	Cerebellar Purkinje cells	Membranes	Walaas et al., 1986
CaM kinase II	CNS neurons	Soma, dendrites, postsynaptic density	Ouimet et al., 1984
Neurofilament p68, 160, 200	All neurons	Axon intermediate filaments	Shaw et al., 1981
DARPP-32	Dopaminoceptive neurons that possess D-1 receptors	Cytosol	Nestler and Greengard, 1984
PCM	CNS neurons	Cytosol	Billingsley and Balaban, 1985
MAP-2	CNS neurons	Microtubules	De Camilli et al., 1984b
NSE	CNS neurons, peripheral neuroendocrine cells	Cytosol	Kirino et al., 1983

B-50	CNS neurons	Presynaptic membranes	Gispen et al., 1985
GFAP	Astrocytes	Intermediate filaments	Eng, 1985
Myelin basic protein	Oligodendrocytes	Myelin	Schwob et al., 1985
Proteolipid protein	Oligodendrocytes	Myelin	Schwob et al., 1985
Myelin associated glycoprotein	Oligodendrocytes	Myelin	Sternberger et al., 1979

Abbreviations: PCPP-260. Purkinje cell-specific phosphoprotein, M_r 260,000; CaM, calcium/calmodulin; DARPP-32, dopamine and cAMP-regulated phosphoprotein, M_r 32,000; PCM, protein-O-carboxymethyltransferase; MAP, microtubule associated protein, GFAP, glial fibrillary acidic protein.

chemical-induced injury. Thus, in the adult animal, changes in the amount of a particular protein would be indicative of damage to a particular cell type or its subcellular constituents. For example, exposure to neurotoxicants often results in distal (dying-back) neuropathy (Spencer et al., 1985). Because this process is associated first with loss of the presynaptic terminal and then the axon, this damage can be characterized by measuring the decrease in proteins associated with the synapse and the neuronal cytoskeleton. On the other hand, damage to the myelin sheath, another common target of neurotoxicants, can be monitored by measuring myelin basic protein. Perhaps the most recognized cellular response to toxicant- or trauma-induced injury of the CNS is proliferation and hypertrophy of astrocytes (Nathaniel and Nathaniel, 1981). Glial fibrillary acidic protein is a major protein in CNS astrocytes (Eng, 1985) and as such it represents a specific indicator for assessing astrocytic response to injury.

The developing nervous system has proven to be especially vulnerable to chemical-induced injury (Rodier, 1980; Suzuki, 1980). During both pre-natal and postnatal development, exposure to neurotoxicants can affect pro-cesses critical to normal nervous-system maturation, such as neurogenesis, gliogenesis, myelinogenesis, and synaptogenesis. These cell type-specific events can also be characterized by measuring neurotypic and gliotypic proteins. For example, the ontogeny of the nerve terminal protein, synapsin I (Lohmann et al., 1978) and the myelin-specific protein, myelin basic protein (Banik and Smith, 1977), coincide with two critical processes in postnatal maturation: synapse formation and myelination, respectively. Thus, the on-togeny of cellular and subcellular elements associated with specific devel-opmental processes, as well as neurotoxicant-induced changes in these events, can be detected and quantified by measuring the appropriately selected proteins.

Since samples of nervous system tissue are required for most biochemical analyses, neurotoxicity normally is assessed in experimental animals. It is conceivable, however, that by measuring specific proteins in accessible biological fluids, such as cerebrospinal fluid or blood, it may be possible to detect and quantify neurotoxicity in the exposed human population. One example of a protein that has been used to diagnose damage to a specific brain component is myelin basic protein; it has been found in the CSF of patients suffering from the demyelinating disease, multiple sclerosis (Cohen et al., 1980). Due to proteolytic degradation, neurotypic and gliotypic pro-teins released in CSF are unlikely to be found in blood in their intact form. Antigenic determinants unique to specific proteins, however, may be detected by antibodies recognizing these specific sites on proteolytic fragments of the native protein. If this turns out to be the case, it may be possible to develop a blood test for neurotoxicity based on the presence of unique protein deter-minants in the serum of individuals exposed to neurotoxicants.

In summary, it appears likely that neurotypic and gliotypic proteins represent versatile biochemical endpoints for detecting and characterizing neurotoxicity. Some of the advantages inherent to this approach are as follows:

1. Prior knowledge of the nature of the toxicant(s) being evaluated is not required. By measuring neurotypic and gliotypic proteins, one is not restricted to analysis of specific brain regions, or limited to specific neurotransmitter systems or tied to specific functional endpoints. In theory, this permits the effects of agents as diverse as unknown mixtures, such as hazardous wastes, or physical hazards, such as radiation, to be detected and characterized throughout the central and peripheral nervous systems.
2. Neurotypic and gliotypic proteins appear to be as numerous and diverse as the cell types they represent. Thus, when applied to neurotoxicology, one is not restricted to measuring a select number or type of protein. Instead, one can take advantage of current advances in neuroscience that continue to reveal the extensive cellular heterogeneity of the nervous system based on the discovery of novel neurotypic and gliotypic proteins (e.g., see DeBlas et al., 1984; Nestler and Greengard, 1984; Hockfield and McKay, 1985).
3. Because neurotypic and gliotypic proteins are intimately linked to nervous system cytoarchitecture, measurement of these proteins make it possible to establish an integrated molecular and cellular basis of neurotoxicity. For example, radioimmunoassays (RIAs) of specific proteins can be combined with immunohistochemical localization of these proteins using the same antibodies. In effect, this allows for a qualitative localization of the affected region to be combined with a quantitative analysis of the proteins associated with the affected cell type. By documenting the effects of toxicant exposure in this manner, a cellular and molecular basis for interpreting changes in specific aspects of whole animal function (behavioral or physiological) may ultimately be established.

EFFECTS OF KNOWN NEUROTOXICANTS ON NEUROTYPIC AND GLIOTYPIC PROTEINS

In order to validate the use of neurotypic and gliotypic proteins as biochemical indicators of neurotoxicity, we have administered known neurotoxicants to experimental animals to assess the effects of these agents on several previously characterized proteins. One neurotoxic chemical employed in our studies is TMT (Figure 1). As noted above, TMT preferentially damages the hippocampus (Brown et al., 1979), where it causes consistent dose-related decrements in pyramidal neurons. By employing TMT as a denervation tool to alter the morphology of the hippocampus, we have been able to demonstrate accompanying changes in neuron-specific proteins (O'Callaghan and Miller, 1984; Miller and O'Callaghan, 1984; Brock and O'Callaghan, 1987). The effects of TMT on one of these proteins, synapsin I, are shown in Figure 2. Three weeks following acute administration of

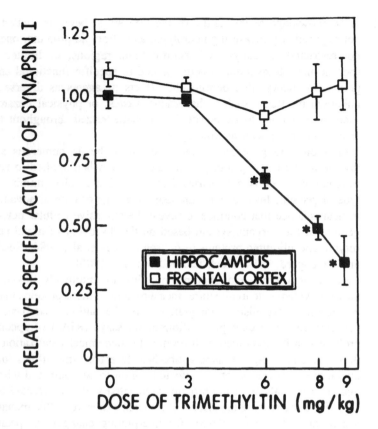

Figure 2. Effects of TMT on synapsin I in rat hippocampus and frontal cortex. Subjects were killed 21 days after treatment. Each value (N = 4–5) represents the mean relative specific activity ± S.E.M. expressed on a concentration (per mg total protein) basis. Values obtained for hippocampus-saline (0 mg/kg) were used to calculate a mean relative specific activity = 1.0 (100%). *Significantly different from corresponding saline control, p < 0.01. (Adapted from Brock and O'Callaghan, 1987).

TMT to the rat, large dose-related decreases in synapsin I were observed in hippocampus. Consistent with the region-selective effects of TMT, decrements in synapsin I were not observed in frontal cortex, a brain region that does not show overt cytopathological damage as a result of exposure to TMT. These findings suggest that TMT causes a loss of synaptic terminals in hippocampus as a consequence of pyramidal cell destruction without affecting synapses in frontal cortex.

Exposure to TMT, like other physical or chemical insults to the nervous system, has been reported to elicit as astrocytic response at the site of damage (Brown et al., 1979). To verify this response biochemically, we assayed the amount of GFAP, the astrocyte-specific protein, in the same tissue homogenates used for assays of synapsin I. These data are shown in Figure 3. TMT caused large dose-related increases in GFAP. These effects, unlike those on

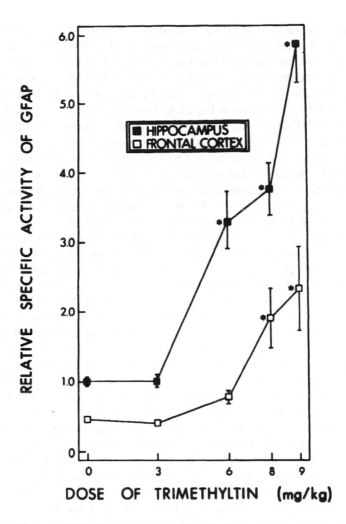

Figure 3. Effects of TMT on glial fibrillary acidic protein in rat hippocampus and frontal cortex. Subjects were killed 21 days after treatment. Each value (N = 4–5) represents the mean relative specific activity ± S.E.M. expressed on a concentration (per mg total protein) basis. Values obtained for hippocampus-saline (0 mg/kg) were used to calculate a mean relative specific activity = 1.0 (100%). *Significantly different from corresponding saline control, p < 0.01. (Adapted from Brock and O'Callaghan, 1987).

synapsin I, were present in frontal cortex as well as hippocampus, although to a lesser degree. These findings suggest that RIAs of GFAP may be used to monitor astrocyte response to injury and that TMT causes damage to a nonlimbic brain region (frontal cortex). Because histological data and assays of neuronal proteins were negative, the nature of the damage to frontal cortex remains to be characterized.

In order to confirm that TMT-induced increases in GFAP were localized to astrocytes, we subjected tissue sections from TMT-treated subjects to immunohistochemistry of GFAP. Astrocytic reactivity in these sections paralleled the increase in GFAP as determined by RIA. Thus, in comparison to GFAP positive cells in frontal cortex, immunostained cells in hippocampus were more intensely immunoreactive. Representative micrographs obtained from the region of hippocampus with the greatest neuronal degeneration (CA3 and CA4) are shown in Figure 4. These micrographs demonstrate that in saline- and TMT-treated subjects GFAP immunoreactivity delineates stellate-shaped astrocytes. In comparison to immunoreactive astrocytes present in saline-treated rats, those present in TMT-treated rats were characterized by markedly hypertrophied perikarya and processes. These data raise the possibility that, by combining assays for GFAP with immunohistochemistry of this protein, it may be possible to characterize the degree and extent of damage caused by agents with unknown toxicity profiles.

The cerebellum often has been shown to be susceptible to toxic insult and we have also characterized damage to this brain region using assays of neurotypic and gliotypic proteins. For these studies we employed the Gunn rat, an extensively studied model for toxicant-induced injury to the cerebellum.The Gunn rat is an autosomal recessive mutant that exhibits lifelong hyperbilirubinemia due to a lack of UDP-glucuronyl transferase (Blanc and Johnson, 1959). As shown in Figure 1, the elevated serum bilirubin in homozygous recessive Gunn rats results in marked cerebellar hypoplasia. Histological examinations indicate that cerebellar Purkinje cells are affected preferentially (Mikoshiba et al., 1980, 1982). A reduction in granule cell number and hypertrophy of astrocytes also have been reported (Mikoshiba et al., 1980, 1982). To characterize cerebellar neurotoxicity in the Gunn rat, we measured five neurotypic and gliotypic proteins associated with specific cell types as follows: the Purkinje cell specific proteins, G-substrate and PCPP-260; the synapse specific protein, synapsin I (which is associated predominantly with granule cells in cerebellum); the astrocyte protein, GFAP; and the myelin protein, MBP. In comparison to heterozygote (Jj) controls, homozygous (jj) rats showed alterations in the amounts of neurotypic and gliotypic proteins that were consistent with the neuropathological effects associated with the development of hyperbilirubinemia in the Gunn rat. Representative data are shown in Table 2. In jj rats with confirmed hyperbilirubinemia, cerebellar weight was reduced by 50% in comparison to corresponding Jj controls. Marked reductions in the Purkinje cell proteins, G-substrate and PCPP-260 and a lesser decrease in synapsin I accompanied the cerebellar hypoplasia. In contrast, a marked increase was observed in GFAP whereas values for MBP were only slightly increased. Since these data are expressed on a concentration basis, the alterations in specific proteins were not simply a reflection of reduced cerebellar weight in jj rats. Thus, these changes in neurotypic and gliotypic proteins serve as a quantitative index of a marked loss of cerebellar Purkinje cells, a reduction in granule cells or other cerebellar interneurons, and an astrocytic response to this neuronal damage with sparing of myelin.

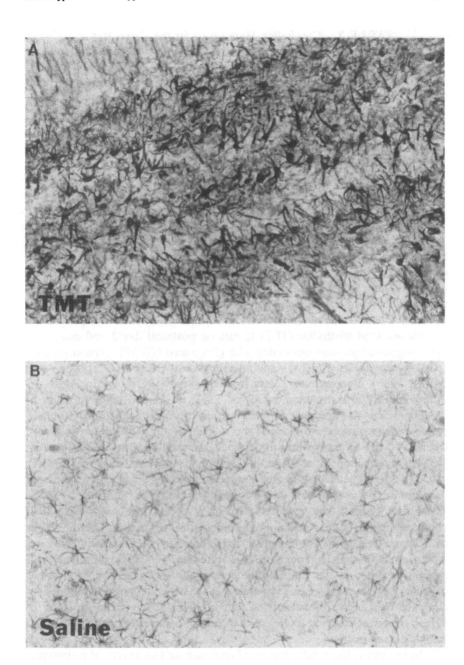

Figure 4. Astrocytic response to TMT-induced damage of the rat hippocampus (areas CA3c-CA4) as revealed by immunohistochemistry of GFAP. Subjects were killed 21 days after administration of 8.0 mg/kg of TMT. (Adapted from Brock and O'Callaghan, 1987).

TABLE 2.   Cerebellar Hypoplasia in the Gunn Rat is Accompanied by Changes in Neurotypic and Gliotypic Proteins

	Cerebellum Weight (g)	Relative Specific Activity				
		G-Substrate	PCPP-260	Synapsin I	GFAP	MBP
Jj	0.30 ± .01	1.00 ± .09	1.00 ± .03	1.00 ± .11	1.00 ± .06	1.00 ± .05
jj	0.15 ± .02[a]	0.36 ± .09[a]	0.40 ± .10[a]	0.63 ± 0.6[b]	3.81 ± .61[b]	1.23 ± 0.3[a]

Note: Each value represents the mean relative specific activity ± S.E.M. expressed on concentration (per milligram of protein) basis.

[a]Significantly different from Jj controls, p < 0.001.
[b]Significantly different from Jj controls, p < 0.01.

Adapted from O'Callaghan and Miller (1985).

In the above examples, measurements of neurotypic and gliotypic proteins were obtained from adult animals exposed to toxicants known to cause overt destruction of neurons. Given the susceptibility of the developing nervous system to the effects of toxic chemicals, we reasoned that measurements of neurotypic and gliotypic proteins might reveal alterations in developmental events in the absence of overt cell loss. To examine this possibility we administered triethyltin (TET) to rats on postnatal day 5 and measured the ontogeny of proteins associated with gliogenesis (GFAP), synaptogenesis (p38), neurogenesis (NF-200), and myelinogenesis (MBP) (O'Callaghan and Miller, 1988). Representative data from this study are shown in Figure 5. Although TET is a structural analog of TMT, its administration to adult or developing rats has not been reported to cause neuronal damage, but instead, is associated with perturbations of the myelin sheath. The reduced concentration of MBP in cerebellum throughout ontogeny confirms the effects of TET on myelination. In addition, however, early postnatal exposure to TET caused a dose-dependent increase in hippocampal GFAP and a corresponding decrease in p38 as well as a dose-dependent decrease in cerebellar NF-200. These data suggest that, in addition to an effect on myelinogenesis, TET causes an astrocytic response in hippocampus (GFAP) that may be the result of neuronal damage in this structure, as indicated by the permanent decrease in the synaptic vesicle protein, p38. Furthermore, in cerebellum, the permanent decrease in the concentration of NF-200 is suggestive of arrested neuronal development in this structure. Taken together, these data provide evidence that assays of neurotypic and gliotypic proteins can be used to reveal effects of toxicant exposure that are not apparent from classical histological examinations.

In summary, the results presented demonstrate that assays of neurotypic and gliotypic proteins can be used to detect and characterize cell type-specific responses to toxicant-induced injury of the nervous system. The sensitivity, specificity, and quantitative nature of this approach, when combined with immunohistochemical evaluations, should prove useful for characterizing the neurotoxicological profiles of broad classes of environmental pollutants.

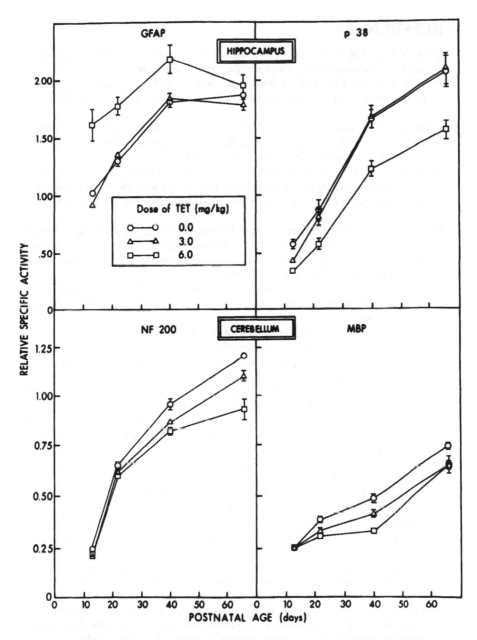

Figure 5. Effects of acute administration of triethyltin (TET) on glial fibrillary acidic protein (GFAP), p38, neurofilament 200 (NF 200) and myelin basic protein (MBP) in the developing rat. TET (0.0, 3.0 and 6.0 mg/kg) was administered on postnatal day 5 and neurotypic and gliotypic proteins were assayed on postnatal days 13, 22, 40 and 66. Each value (N = 8) represents the mean relative specific activity ± S.E.M. expressed on a concentration (per mg total protein) basis. For each protein, values obtained for forebrain saline controls (data not shown) were used to calculate a mean relative specific activity = 1.0 (100%). (Adapted from O'Callaghan and Miller, 1988).

REFERENCES

Banik, N.L. and M.E. Smith (1977). Protein determinants of myelination in different regions of developing rat central nervous system. *Biochem. J.* 162: 247–255.

Barchas, J.D., H. Akil, G.R. Elliott, R.B. Holman, and S.J. Watson (1978). Behavioral neurochemistry: neuroregulators and behavioral states. *Science* 200: 964–973.

Billingsley, M.L. and C.D. Balaban (1985). Protein-O-carboxylmethyltransferase in the rat brain: high regional levels in the substantia nigra, locus coeruleus and paraventricular nucleus. *Brain Res.* 358: 96–103.

Blanc, W.A. and L. Johnson (1959). Studies on kernicterus. Relationship with sulfonamide intoxication, report on kernicterus in rats with glucuronyl transferase deficiency and review of pathogenesis. *J. Neuropathol. Exp. Neurol.* 18: 165–189.

Brock, T.O. and J.P. O'Callaghan (1987). Quantitative changes in the synaptic vesicle proteins, synapsin I and p38, and the astrocyte specific protein, glial fibrillary acidic protein, are associated with chemical-induced injury to the rat central nervous system. *J. Neurosci.* 7: 931–942.

Brown, A.W., W.M. Aldridge, B.W. Street, and R.D. Verschoyle (1979). The behavioral and neuropathologic sequelae of intoxication by trimethyltin compounds in the rat. *Am. J. Pathol.* 97: 59–82.

Cohen, S.R., B.R. Brooks, B. Jubelt, R.M. Herndon, and G.M. McKhann (1980). Myelin basic protein in cerebrospinal fluid: index of active demyelination. In *Neurobiology of Cerebrospinal Fluid* (1). J.H. Wood, Ed., Plenum Press, New York, pp. 487–494.

Cooper, J.R., F.E. Bloom, and R.H. Roth, Eds. (1982). *The Biochemical Basis of Neuropharmacology*, 4th ed. Oxford University Press, New York.

Damstra, T. and S.C. Bondy (1982). Neurochemical approaches to the detection of neurotoxicity. In *Nervous System Toxicology*. C.L. Mitchell, Ed., Raven Press, New York, pp. 349–373.

Dayan, A.D. (1984). Training toxicologists: the move from learning to understanding. *Trends Pharmacol. Sci.* 5: 359–362.

De Blas, A.L., R.O. Kuljis, and H.M. Cherwinski (1984). Mammalian brain antigens defined by monoclonal antibodies. *Brain Res.* 322: 277–287.

De Camilli, P., P.E. Miller, P. Levitt, U. Walter, and P. Greengard (1984a). Anatomy of cerebellar Purkinje cells in the rat determined by a specific immunohistochemical marker. *Neuroscience* 11: 761–817.

De Camilli, P., P.E. Miller, F. Navone, W.E. Theurkauf, and R.B. Vallee (1984b). Distribution of microtubule-associated protein 2 in the nervous system of the rat studied by immunofluorescence. *Neuroscience* 11: 819–846.

Detre, J.A., A.C. Nairn, D.W. Aswad, and P. Greengard (1984). Localization in mammalian brain of G-substrate, a specific substrate for guanosine 3', 5'-cyclic monophosphate-dependent protein kinase. *J. Neurosci.* 4: 2843–2849.

Eng, L.F. (1985). Glial fibrillary acidic protein (GFAP): the major protein of glial intermediate filaments in differentiated astrocytes. *J. Neuroimmunol.* 8: 203–214.

Gispen, W.H., J.L.M. Leunissen, A.B. Oestreicher, A.J. Verkleij, and H. Zwiers (1985). Presynaptic localization of B-50 phosphoprotein: the (ACTH)-sensitive protein kinase substrate involved in rat brain polyphosphoinositide metabolism. *Brain Res.* 328: 381–385.

Hockfield, S. and R.D.G. McKay (1985). Identification of major cell classes in the developing mammalian nervous system. *J. Neurosci.* 5: 3310–3328.

Jahn, R., W. Schiebler, C. Ouimet, and P. Greengard (1985). A 38,000-dalton membrane protein (p.38) present in synaptic vesicles. *Proc. Natl. Acad. Sci. U.S.A.* 82: 4137–4141.

Kirino, T., M.W. Brightman, W.H. Oertel, D.E. Schmechel, and P.J. Marangos (1983). Neuron-specific enolase as an index of neuronal regeneration and reinnervation. *J. Neurosci.* 3: 915–923.

Lohmann, S.M., T. Ueda, and P. Greengard (1978). Ontogeny of synaptic phosphoproteins in brain. *Proc. Natl. Acad. Sci. U.S.A.* 75: 4037–4041.

Mikoshiba, K., K. Adagawa, K. Takamatsu, and Y. Tsukada (1982). Neurochemical studies on the cerebellar hypoplasia of Gunn rat (hereditary hyperbilirubinemic rat). *J. Neurochem.* 39: 1028–1032.

Mikoshiba, K., S. Kohsaka, K. Takamatsu, and Y. Tsukada (1980). Cerebellar hypoplasia in the Gunn rat with hereditary hyperbilirubinemia: immunohistochemical and neurochemical studies. *J. Neurochem.* 35: 1309–1318.

Miller, D.B. and J.P. O'Callaghan (1984). Biochemical, functional and morphological indicators of neurotoxicity: effects of acute administration of trimethyltin to the developing rat. *J. Pharmacol. Exp. Ther.* 231: 744–751.

Nathaniel, E.J.H. and D.R. Nathaniel (1981). The reactive astrocyte. In *Advances in Cellular Neurobiology*, Vol. 2. S. Federoff and L. Hertz, Eds., Academic Press, New York, pp. 249–301.

Nestler, E.J. and P. Greengard (1984). *Protein Phosphorylation in the Nervous System*. John Wiley & Sons, New York.

Nestler, D.J., S.I. Walaas, and P. Greengard (1984). Neuronal phosphoproteins: physiological and clinical implications. *Science* 225: 1357–1364.

O'Callaghan, J.P. and D.B. Miller (1983). Nervous-system specific proteins as biochemical indicators of neurotoxicity. *Trends Pharmacol. Sci.* 4: 388–390.

O'Callaghan, J.P. and D.B. Miller (1984). Neuron-specific phosphoproteins as biochemical indicators of neurotoxicity: effects of acute administration of trimethyltin to the adult rat. *J. Pharmacol. Exp. Ther.* 231: 736–743.

O'Callaghan, J.P. and D.B. Miller (1985). Cerebellar hypoplasia in the Gunn rat is associated with quantitative changes in neurotypic and gliotypic proteins. *J. Pharmacol. Exp. Ther.* 234: 522–533.

O'Callaghan, J.P. and D.B. Miller (1986). Diethyldithiocarbamate increases distribution of cadmium to brain but prevents cadmium-induced neurotoxicity. *Brain Res.* 370: 354–358.

O'Callaghan, J.P. (1988). Neurotypic and gliotypic proteins as biochemical markers for neurotoxicity. *Neurotoxicol. Teratol.* 10: 445–452.

O'Callaghan, J.P. and D.B. Miller (1988). Acute exposure of the neonatal rat to triethyltin results in persistent changes in neurotypic and gliotypic proteins. *J. Pharmacol. Exp. Ther.* 244: 368–378.

Ouimet, C.C., T.L. McGuinness and P. Greengard (1984). Immunocytochemical localization of calcium/calmodulin-dependent protein kinase II in rat brain. *Proc. Natl. Acad. Sci. U.S.A.* 81: 5604–5608.

Price, D.L., P.J. Whitehouse, and R.G. Struble (1986). Cellular pathology in Alzheimer's and Parkinson's diseases. *Trends Neurosci.* 9: 29–33.

Raff, M.C., K.L. Fields, S. Hakomori, R.M. Pruss, and J. Winter (1979). Cell type-specific markers for distinguishing and studying neurons and the major classes of glial cells in culture. *Brain Res.* 174: 283–308.

Rodier, P. (1980). Chronology of neuron development: animal studies and their clinical implications. *Dev. Med. Child Neurol.* 22: 525–545.

Schachner, M. (1982). Cell-type specific surface antigens in the mammalian nervous system. *J. Neurochem.* 39: 1–8.

Schwob, V.S., H.B. Clar, D. Agrawal, and H.C. Agrawal (1985). Electron microscopic immunocytochemical localization of myelin proteolipid protein and myelin basic protein to oligodendrocytes in rat brain during myelination. *J. Neurochem.* 45: 559–571.

Shaw, G., M. Osborn, and K. Weber (1981). An immunofluorescence microscopical study of the neurofilament triplet proteins, vimentin and glial fibrillary protein within the adult rat brain. *Euro. J. Cell. Biol.* 26: 68–82.

Spencer, P.S. and H.H. Schaumburg, Eds. (1980). *Experimental and Clinical Neurotoxicology.* Williams & Wilkins, Baltimore.

Spencer, P.S., J. Arezzo, and H. Schaumburg (1985). Chemicals causing disease of neurons and their processes. In *Neurotoxicity of Industrial and Commercial Chemicals,* Vol. 11. J.L. O'Donoghue, Ed., CRC Press, Boca Raton, FL, pp. 1–14.

Sternberger, N.H., R.H. Quarles, Y. Itoyama, and H. deF. Webster (1979). Myelin-associated glycoprotein demonstrated immunocytochemically in myelin and myelin-forming cells of developing rat. *Proc. Natl. Acad. Sci. U.S.A.* 76: 1510–1514.

Suzuki, K. (1980). Special vulnerabilities of the developing nervous system to toxic substances. In *Experimental and Clinical Neurotoxicology.* P.S. Spencer and H.H. Schaumburg, Eds., Williams & Wilkins, Baltimore, pp. 48–61.

Walaas, S.I., A.C. Nairn, and P. Greengard (1986). PCPP-260, a Purkinje cell-specific cyclic AMP-regulated membrane phosphoprotein of Mr 260,000. *J. Neurosci.* 6: 954–961.

Axonal Transport

Jeffry F. Goodrum[1] and Pierre Morell[2]
Brain and Development Research Center[1,2]
Departments of Pathology[1] and Biochemistry[2]
University of North Carolina
Chapel Hill, North Carolina

INTRODUCTION

Neurons, with their unique morphology, have a highly asymmetric distribution of their cytoplasm and plasma membrane. In neurons of long central nervous system (CNS) tracts and peripheral nervous system (PNS) nerves the axonal volume and surface area may exceed that of the cell body by several orders of magnitude. While the neuronal cell body may contain only a small fraction of the total cell volume, it contains most of the metabolic machinery for the synthesis and degradation of macromolecules and organelles. Thus, the normal functioning of the axon and the nerve terminal are critically dependent on the process of axonal transport for the continued delivery of materials to and from the cell body.

Axonal transport presumably involves mechanisms similar to those which all eucaryotic cells utilize to transport materials from their sites of synthesis to their sites of utilization, and subsequently to their sites of degradation. The unique morphology of neurons increases the prominence and highlights the importance of these processes. There have been many reviews of the various aspects of axonal transport (Lasek, 1970; Jeffrey and Austin, 1973; Lubinska, 1975; Schwartz, 1979; Wilson and Stone, 1979; Bisby, 1980; Grafstein and Forman, 1980; Ochs, 1982; Weiss, 1982; Schwab and Thoenen, 1983); the reader is referred to these for a more detailed summary of the literature. The goals of this chapter are to present the rationales behind the methodologies currently used for the study of axonal transport and to introduce the reader to the current research concerning the influence of toxicants on axonal transport. The reader is warned at the outset that a satisfactory understanding of the mechanisms of axonal transport remains to be achieved and therefore, although it is clear that toxicants may perturb axonal transport, knowledge of the mechanisms involved is very limited.

TECHNIQUES FOR QUANTIFYING AXONAL TRANSPORT

There are a variety of methods used for examining axonal transport, each providing a different perspective on the properties of the process involved. In Figure 1 a schematic of a neuron is provided to illustrate the following brief outline of axonal transport. Proteins and other molecules which are to be committed to the process of axonal transport are synthesized in the cell body (with the exception of the limited independent capacity of mitochondria to synthesize certain mitochondrial proteins). The various size arrows pointing in the anterograde (cell body to nerve ending) direction are intended to indicate that a variety of particles are moving at different rates within the axon. A retrograde (nerve ending to cell body) component is also indicated. Figure 1 also illustrates that axonal transport not only supplies material destined for the nerve ending, but that some transported material must also be deposited along the length of the axon to account for metabolic turnover and maintenance of the vast expanse of membrane which forms the axon. It is important to keep in mind in what follows that nerves (associated axons in the PNS) and tracts (parallel assemblies of axons in the CNS) contain not only neuronal processes but also the cell bodies and processes of auxiliary cells, predominantly the glial cells which synthesize the myelin that envelops most larger axons.

Morphological Approaches

Some of the earliest quantitative studies of axonal transport utilized light microscopic autoradiography to visualize the location of labeled macromolecules following application of radioactive precursors in the vicinity of neuronal cell bodies. These studies, utilizing amino acid precursors, localized protein synthesis to the neuronal cell bodies. Over a period of days the autoradiographic silver grains appeared sequentially over the axon hillock, proximal axon regions and distal axon regions, both in the CNS and PNS (Droz and Leblond, 1963). The higher resolution available with electron microscopic autoradiography shows that there is, in addition, a rapid appearance of labeled material in nerve endings. The movement of this radioactivity can be quantitated by counting of silver grains at various levels of the axon to give a quantitative measure of transport. However, this meth-

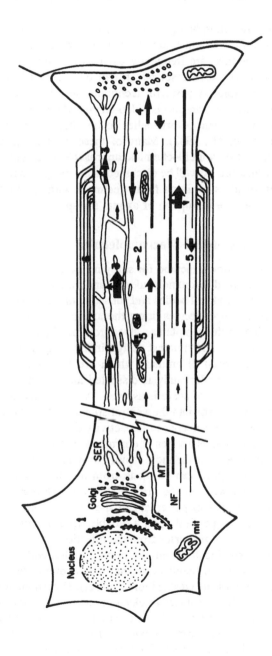

Figure 1. Schematic of a neuron indicating several general features of axonal transport. In this figure the upper half depicts the membranous compartments and the lower half the major cytoskeletal compartments of the axon. The numbers point out several major features of axonal transport. 1. All protein (and other macromolecular) synthesis is restricted to the cell body. 2. Various size arrows pointing in the anterograde (cell body to terminal) direction represent a variety of cellular components moving at a variety of rates. (The arrows are not meant to be associated with any particular cytological structures in the figure.) 3. Branching arrows indicate that much transported material replenishes axonal constituents. 4. Other transported macromolecules reach the nerve terminals to supply their needs. 5. Some material carried in the anterograde direction is returned to the cell body by a retrograde transport. 6. Many axons of the CNS and PNS are myelinated by glial cells.

odology is much too tedious to be used routinely for such studies and is more frequently used to define morphologically the subcellular location of the transported material.

There are a variety of other morphological techniques used to examine axonal transport. Exogenous tracers, which are taken up by nerve terminals and carried toward the cell bodies, are frequently used as an indicator of retrograde transport. Horseradish peroxidase, which can be detected histochemically, is a common example. A powerful technique now being used very effectively in examining axonal transport is video-enhanced microscopy. This technique enables the movement of organelles to be observed within dissected living axons (Brady et al., 1982). Many of these morphological techniques are used in combination with biochemical approaches to elucidate the mechanisms of axonal transport. As will be discussed later, morphological studies of the axonal abnormalities resulting from treatment with toxicants can give clues as to the reasons for disruption of axonal transport.

Downflow of Radioactively Labeled Macromolecules

In recent years, the most common method for studying axonal transport has involved labeling biosynthesized macromolecules by applying radioactive precursors in the vicinity of neuronal cell bodies and subsequently monitoring the movement of these radioactive molecules down the axons of these neurons. Macromolecules studied include proteins, glycoproteins, lipids, glycolipids, and glycosaminoglycans. The simplest design involves application of radioactive precursors to a group of animals which are then killed at various times. The nerve or tract under study is dissected and precipitable radioactivity in consecutive segments of nerve is assayed. A commonly used protocol involves injection of precursor into the ventral spinal cord or dorsal root ganglion, and monitoring the movement of radioactivity down the sciatic nerve. The data generated are plotted as a series of downflow patterns (Figure 2). Calculation of the transport rate for the most rapidly moving component (called rapid, or fast axonal transport) can be made from the displacement of either the leading foot of the wave or the crest of the wave (Figure 2). The calculation of rate based on the displacement of the leading foot with time is the most accurate, but is dependent upon a reliable determination of the baseline. The variable leakage of precursor from the injection site (allowing for incorporation of precursor locally along the nerve by glial cells) can lead to considerable variability in the baseline, so that it is common to see rates calculated from the position of the crest (Figure 2, inset). Determination of the position of the crest also has a subjective element since deposition of the labeled transported material in the axon leads to attenuation and broadening of the crest. It is essential that several downflow patterns at different time points be utilized to establish rapid rates of transport; calculations based on only a single determination will fail to account for an initial time lag

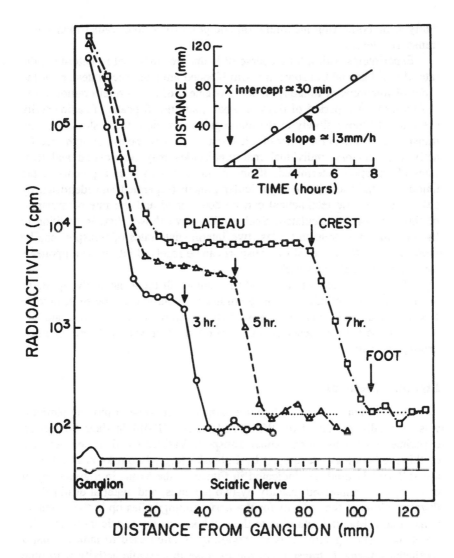

Figure 2. Downflow patterns in the sciatic nerve following injection of radiolabeled precursor into the dorsal root ganglion. Patterns are shown for 3, 5 and 7 hours following injection (adapted from data obtained in our laboratory). Arrows indicate the position of the crest of the advancing wave of radioactivity. The inset shows that the crest is moving at 13 mm/h (312 mm/ day). The regression line intersects the x-axis at about 30 min, indicating the uptake, incorporation and processing lag prior to the initiation of transport.

during which incorporation of precursor and processing of the macromolecules occurs in the cell body prior to transport (arrow, Figure 2, inset). Determinations based on single time points will therefore give a considerable underestimate of the rapid transport rate (and comparisons of transport rates between toxicant-treated and control animals based on single time points are likely to be comparing incorporation and processing times rather than actual transport rates).

Experiments with a time course of hours are sufficient to determine the rate of rapid axonal transport in warm blooded animals; such rates are of the order of hundreds of mm/day. If the movement of total protein radioactivity is followed for a period of days or weeks, additional peaks of radioactivity are found to move slowly in the anterograde direction; these slower protein transport rates are of the order of a few mm/day. With the use of high specific activity precursors, individual macromolecules may be isolated and their rates of transport determined. If an amino acid is used as a precursor, the amount of radioactivity in a particular protein (separated and identified by gel electrophoretic techniques) can be determined and the rate of transport of this component calculated. Such studies have shown that, in addition to the fast and slow transport rates, two intermediate rates of transport can be identified. Furthermore, slow transport can be resolved into two components (see Slow Axonal Transport).

Many investigators have noted that almost all fucosylated glycoproteins are rapidly transported; thus many studies of rapid axonal transport involve labeling with radioactive fucose so that the above noted intermediate and slow transport components do not become a complicating factor in interpreting the data.

Ligature Methods

The technique of using a ligature to dam up the flow of moving particles was originally presented by Weiss and Hiscoe (1948) in their pioneering experiments demonstrating axonal transport. Various sophisticated versions of this methodology, developed by Lubinska (e.g., Lubinska et al., 1964), are still widely used. As indicated in Figure 3, the axons in a given region of nerve are compressed, usually by two ligatures, and material destined for distal portions of the axon or for the nerve endings piles up at the proximal constriction. Material returning to the cell body via retrograde transport piles up at the distal ligature. This methodology is often used to indicate that a particular enzyme is transported; the increase in enzymic activity with time in a segment of nerve immediately proximal or distal to a ligature is presumptive evidence of anterograde or retrograde axonal transport, respectively, of this enzyme. In order to make a quantitative estimate of the transport velocity, two further pieces of information are necessary: the enzyme concentration in a control segment of the same nerve and the percentage of that enzyme activity which is actually "mobile." A numerical example may make this issue easier to grasp. Figure 3 indicates that, in the control situation (segment 1, many millimeters from the ligatures or in the contralateral nerve), each 5-mm segment of nerve contains 25 units of enzyme activity "X" (i.e., 5 units/mm). If a ligature is applied for 3 hours, 115 units of enzyme activity

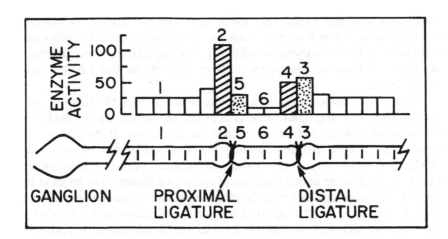

Figure 3. The accumulation of transported enzyme activity at ligatures placed on a nerve. The lower portion of the figure illustrates the placement of the ligatures and the 5-mm segments of nerve taken to assay enzyme activity. The upper half illustrates the enzyme activity in each nerve segment 3 h after tying the ligatures. The shaded bars (2 and 4) represent the anterograde accumulation segments and the stippled bars (3 and 5) represent the retrograde accumulation segments. Segment 1 is used to measure the normal enzyme activity in the nerve and segment 6 to measure the amount of nonmobile enzyme. See the text for a discussion of transport rate calculations using these data.

are found in the 5-mm segment proximal to the proximal ligature (segment 2). Because there is a net accumulation of 90 units (115–25 units) of enzyme activity, 30 units/h must be moving into the defined segment. At 5 units/mm this represents a transport rate of 6 mm/h (30 units/h ÷ 5 units/mm). A similar calculation using segment 3 yields a retrograde rate of 2 mm/h. However, there is an additional factor to consider; what if only a fraction of the enzyme "X" is free to move over the time period of the experiment? In fact, it does appears that most transported components do not remain in the mobile phase of axonal transport and thus a correction is usually necessary in experiments of this type. An estimate of the fraction of enzyme in the nonmobile phase can be made from the segments between the ligatures, because the combination of accumulation of anterograde moving material at the distal ligature (segment 4) and retrograde moving material at the proximal ligature (segment 5) depletes the central region between ligatures (segment 6) of the mobile enzyme. Enzyme activity measured in that interior segment therefore corresponds to a nonmobile fraction. In Figure 3, 40% of the enzymic activity between the ligatures is not mobile during the three hours. Forty percent of the activity moves in the anterograde and 20% in the retrograde directions, respectively. When these values are used to correct the calculations above, the transport rates become 15 mm/h and 10 mm/h for the anterograde and retrograde components, respectively. Note that in the example given the higher transport rate and larger mobile fraction in the anterograde direction means that much more enzymic activity is moving down the axon than is returning. Thus one would conclude that in this case there must be loss of enzyme activity (e.g., by degradation or inactivation) at some distal location; this appears to be the general case (for a review, see Bisby, 1980).

The above methodology was developed for studies of transport and accumulation of endogenous enzyme activities. However, this method can be combined with the isotope technology detailed earlier. If the macromolecules being synthesized and committed to transport in the cell body are labeled with radioactive precursors for several hours prior to application of ligatures, the pile up of radioactivity adjacent to a ligature can be used to study transport of any particular labeled component. If retrograde transport is to be examined, prelabeling for a long period of time—on the order of a day—is often necessary (radioactive material must be committed to transport, complete its journey to the nerve ending, and already be returning to the cell body before it is possible to measure accumulation of retrograde moving material at the ligature). In these cases, data are frequently expressed as "relative accumulation," where the amount of radioactivity accumulated per unit of time adjacent to a ligature is divided by the "background" of preexisting radioactivity in the accumulaton segment, calculated from the average amount of radioactivity in the surrounding nerve segments (Bisby and Bulger, 1977).

"Stop-Flow" Method

A technique related to ligation which is increasingly being used takes advantage of the fact that cooling of the nerve will block both anterograde and retrograde transport. This cold block is reversible; rewarming the nerve allows transported material to continue its movement through the axon. If cooling is applied to a small local region of nerve, then transported material will accumulate adjacent to the cooled region; when the cooled region is rewarmed the accumulated material will move down the nerve as a large pulse. Both the rate of movement and the attenuation of the moving front (presumably because of deposition into stationary structures) can be studied (Figure 4). This local cooling and rewarming is known as the "stop-flow" technique (Brimijoin, 1975).

NEURONAL PATHWAYS USED FOR STUDYING AXONAL TRANSPORT

Investigations of two nervous system pathways, the sciatic nerve and the optic system, account for much of the research on axonal transport. The sciatic nerve is a favorite subject of study because of its great length and

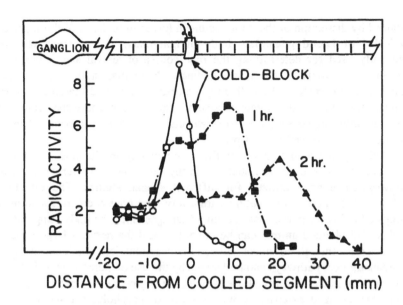

Figure 4. Axonal transport demonstrated by a reversible cold block. Transport is blocked by decreasing the temperature of a local region of nerve to <10°C for several hours. Transported radiolabeled material accumulates proximal to the cooled region. Upon rewarming the nerve, this accumulated material moves down the nerve, in this example at about 12 mm/h. Note that the crest is continuously attenuated as material is deposited along the axons. (Radioactivity is plotted on an arbitrary scale.)

ready accessibility. The sciatic motor axons which project from cell bodies in the ventral horn of the spinal cord can be labeled by injection of precursors into the spinal cord. Axons of sensory neurons, which convey information from the periphery to the central nervous system, can also be labeled by injection of the dorsal root ganglia which contain the relevant cell bodies. Both these injections require dorsal laminectomies to be performed under long acting anesthesia. *In vitro* preparations of sciatic nerve are also useful; a spinal cord—dorsal root ganglion—sciatic nerve preparation from frogs has been utilized for a number of different types of studies (Dravid and Hammerschlag, 1975), as well as sciatic nerve preparations from warm blooded species (Ochs and Smith, 1975). *In vitro* preparations allow for control of such variables as temperature and external chemical environment. Of special interest in the context of the present chapter is that *in vitro* systems allow for careful control of toxicant application.

For studies of the CNS, the optic pathway is of particular utility. Retinal ganglion cells can easily be labeled by intraocular injection—a simple and rapid procedure which can be carried out under light, short acting anesthesia. The length and anatomical accessibility of the optic pathway allows for several sections of the optic nerve and tract to be collected. For certain studies it is sufficient to obtain the nerve ending region of the pathway (easily obtained by dissection of the optic tectum in lower animals, or of the superior colliculus or lateral geniculate body of mammals). These structures are frequently used for determining the time course of arrival and subsequent loss of radioactivity at the nerve endings. (Note that in this case a rate calculation will include the cell body processing time and therefore underestimate the rate.) This system is also advantageous because the region of nerve endings can be subjected to subcellular fractionation to better identify the organelles involved in transport.

Although the sciatic nerve and the optic pathway are the most commonly used systems, it should be noted that many other defined neuroanatomical pathways have been utilized for different special studies. The olfactory pathway of the garfish (a creature with a very long nose and correspondingly elongated olfactory nerve) has certain advantages since the body temperature of this cold blooded animal can be controlled and the nerve is large enough and discrete enough to make subcellular fractionation possible. Specific neuroanatomical pathways can be labeled by the use of stereotaxic technique, making it possible to study, for example, the nigro-striatal tract, which has been characterized extensively with respect to morphological and pharmacological properties. Examples of other pathways that have been utilized include the auditory, vagus, hypoglossal, and other cranial nerves; geniculocortical tract; median forebrain bundle; hippocampal commissure; septohippocampal tract; and autonomic nerves.

Several invertebrate species have unusually large neurons or axons and this characteristic has been exploited very successfully in studying axonal transport. Giant ganglion cells in the sea slug *Aplysia* can be individually injected with radioactive precursors so that axonal transport can be examined in single identified neurons (Ambron et al., 1974). The squid and lobster have giant axons in which transport can be examined *in vitro*, using a variety of combined biochemical and morphological techniques (Brady et al., 1982; Vale et al., 1985).

PROPERTIES OF AXONAL TRANSPORT

Currently, five discrete rates of anterograde transport and one of retrograde transport have been recognized, each carrying a distinct set of macromolecules associated with specific cytological structures (Table 1). The characteristics of each of these transport components are briefly described below.

Fast Axonal Transport

The very rapid movement of material down axons has been quantitated in a large number of vertebrate and invertebrate nerves; in mammals, the

TABLE 1. Axonal Transport Components and Their Compositions

Transport Rate	Component	Rate (mm/d) in Mammals	Identified Components	Postulated Cytological Structures Involved
Anterograde				
Fast	I	200–400	Proteins, glycoproteins, phospholipids, glycolipids, neurotransmitters, and associated enzymes	Vesicles, agranular endoplasmic reticulum
Intermediate	II	50–200	F_1 ATPase, diphosphatidylglycerol	Mitochondria
	III	15	Myosin-like actin-binding protein	?
Slow	SCb/IV	2–5	Actin, clathrin, calmodulin, enolase	Microfilaments, axoplasmic matrix
	SCa/V	0.2–1	Tubulin, MAPS, neurofilament triplet	Microtubules, neurofilaments
Retrograde		100–200	Similar to fast component and including lysosomal hydrolases	Pre-lysosomal structures

Adapted from Lasek, R.J., 1981.

rate is around 400 mm/day. This rate appears quite constant over a wide range of axon lengths and types and is the same in poikilotherms when temperature is accounted for, assuming a Q_{10} of between 2 and 3.

When tissues labeled by radioactive, rapidly transported material are subjected to subcellular fractionation, most of the radioactivity sediments with the particulate material. Morphological studies have shown this transported material to be associated with the axonal smooth endoplasmic reticulum (SER) and associated vesiculotubular structures, but it appears that it is primarily the vesiculotubular structures that are transported (Ellisman and Lindsey, 1983). The axonal SER appears to be stationary and is presumably labeled by either macromolecular exchange with or reversible fusion with these vesiculotubular structures. Fast axonal transport thus provides the proteins and lipids needed to maintain the axonal and nerve terminal membranes. It is generally assumed that the function of much of rapid axonal transport is to move material to the nerve ending, the region at which a chemically based transfer of information takes place from the neuron to a postsynaptic structure. The subcellular localization of transported material at this location may give clues as to its functional significance (e.g., much of the transported material presumably serves to replenish synaptic vesicles or enzymes of neurotransmitter metabolism). Because of this general membrane maintenance function it is not surprising that the labeled protein and glycoprotein composition of rapidly transported material is found to be very similar in a variety of nerves and tracts (Padilla et al., 1979).

While the shapes of the downflow curves in Figure 2 are the archetype for rapid transport, the actual shape of a downflow curve depends greatly on the amount of labeled material deposited along the axon, the cell body pool size and the metabolic stability of the labeled molecules, all of which vary greatly between molecular species. If there is extensive deposition of the labeled molecules as they are transported, there is no identifiable crest—a proximal to distal gradient of radioactivity is established and gradually there is equilibration along the nerve. For proteins it has been estimated that about 1.5% of the rapidly transported radioactivity is deposited in each 5-mm segment of cat sciatic nerve (Munoz-Martinez et al., 1981). There can also be a differential deposition of protein; in the garfish olfactory nerve the composition of the plateau material is somewhat different from that in the crest (Cancalon and Beidler, 1977). In some systems, especially the optic system, it appears that most of the glycoproteins are destined exclusively for the nerve endings while a few are also deposited in the axons (Goodrum et al., 1989). In general, lipids appear to be deposited along axons to a greater extent than glycoproteins, but among lipids there is a large variation in the extent of deposition during transport (Toews et al., 1986a).

Slow Axonal Transport

In general, movement of material in axons at less than 5 mm/day in mammals is considered "slow" axonal transport. In contrast to the material carried in the fast component, at least half of the material moving slowly remains soluble upon homogenization and centrifugation of nerve tissue. Slow transport involves two groups of proteins, termed slow components a and b (SCa and SCb) (Hoffman and Lasek, 1975) or IV and V (Willard et al., 1974; Willard and Hulebak, 1977), which move at different rates and have different compositions. SCa contains primarily tubulin, microtubule-associated proteins, and the neurofilament triplet proteins, which move at a rate between 0.20 and 1.0 mm/day. SCb contains many proteins including actin and a number of soluble enzymes, which move down the axon with a rate of about 2 to 5 mm/day. This distinction between the two components is not absolute; there is considerable overlap in compositions in some axons (McQuarrie et al., 1986). It was originally postulated that slow transport was the movement of the assembled cytoskeletal framework of the cell, SCa containing the microtubules and neurofilaments, and SCb containing the "axoplasmic lattice" to which so called "soluble" macromolecules are bound (Lasek and Brady, 1982). However, a popular current interpretation of the available data is that the individual polymers or subunits are the moving forms of tubulin and neurofilament proteins, and that the cytoskeleton is stationary (for a review, see Hollenbeck, 1989). While it appears that the rate of rapid transport is constant in adult nerves, the rates of slow transport are probably a function of the length and/or the diameter of the axon. This phenomenon is clearly illustrated in the bipolar dorsal root ganglion cell where the rates of SCa and SCb are faster in the longer peripheral branch than in the central branch (Mori et al., 1979).

Note that there is an assumption of considerable metabolic stability for slowly transported material. In larger mammals it may take many months for slowly transported molecules to reach the terminals of long peripheral nerves. Thus, much of the tubulin in whole rat brain turns over with a half life of around seven days (Forgue and Dahl, 1978) (presumably tubulin present in cell cytoplasm and nuclear structures), while tubulin in the rat optic nerve and tract must be stable for at least a month to arrive intact at the nerve endings. Upon reaching the nerve terminals most of the slow component material is rapidly degraded, presumably by specific calcium-activated proteases (Lasek and Hoffman, 1976; Nixon, 1983).

Intermediate Rates of Axonal Transport

Gel electrophoretic analysis of the composition of transported proteins reveals two distinct sets of labeled proteins appearing in nerve terminals at times corresponding to rates around 100 mm/day and 10 mm/day (Willard et al., 1974). Subcellular fractionation studies suggest that the material moving at 100 mm/day is associated with mitochondrial fractions while the material moving at 10 mm/day is distributed throughout several subcellular fractions (Lorenz and Willard, 1978).

Most of the proteins in these two intermediate transport components are unidentified. However, one component in the faster moving group of proteins has characteristics similar to the F_1 ATPase of mitochondria (Lorenz and Willard, 1978). This correlates with the observation that this transport component also carries the mitochondria-specific lipid, diphosphatidylglycerol (Blaker et al., 1981). One component of the slower moving intermediate group of proteins is also partially characterized as a myosin-like actin binding protein (Willard, 1977).

Retrograde Axonal Transport

In most cases the rate of retrograde transport has been found to be around 50 to 70% of that of the anterograde fast component, although more rapid rates are reported (for a review, see Bisby, 1980). The composition of radiolabeled retrogradely transported material is very similar to that committed to rapid anterograde transport with respect to protein (Bisby, 1981) and lipid (Armstrong et al., 1985; Aquino et al., 1985). The retrograde component also carries a significant portion of the labeled material delivered by the fast anterograde component (as much as 50% of labeled protein in motoneurons; see Bisby, 1982). It is therefore assumed that much of this retrogradely transported material results from the reversal of the rapid anterograde stream at the nerve terminals, a process called "turnaround."

Electron microscopic examination of the material accumulating distal to nerve ligatures and coldblocks reveals primarily large multivesicular structures and other lysosomal-like structures. This composition is in contrast to the vesiculotubular structures of the anterograde fast component (Tsukita and Ishikawa, 1980). Other morphological studies using exogenous tracers have shown that transported material returned to the cell body enters the lysosomal system (Smith, 1980). Thus it appears that retrograde transport serves to return material from the axons and terminals for degradation in the cell body.

MECHANISMS OF AXONAL TRANSPORT

Initiation of Axonal Transport

The antecedents of axonal transport include the synthesis of macromolecules in the cell body and their subsequent commitment to transport. These processes are best understood in connection with the fast axonal transport component (Hammerschlag and Stone, 1982). The synthesis of proteins and lipids destined for rapid axonal transport takes place in the rough endoplasmic reticulum, and these molecules remain membrane associated throughout their lifetimes (this class includes proteins within the lumin of the transported membrane vesicles, some of which may be secreted at their destinations).

Following their synthesis, it appears that most, if not all, membrane components destined for axonal transport must pass through the Golgi apparatus. The next step involves association with the smooth endomembrane system of the axon, a process shown to be Ca^{2+} dependent, inhibited by Co^{2+}, and possibly involving coated vesicles (Stone et al., 1984). The biogenesis, sorting and routing of membranes are processes common to all eucaryotic cells, and detailed studies in nonneuronal cells are providing many insights into the process of rapid transport initiation (Hammerschlag, 1984).

The initiation processes for the intermediate and slow components of transport are less well understood. *In vitro* experiments suggest that the cell body is the site of assembly of the cytoskeletal structures involved in the slow transport (Black et al., 1986). Axonal microtubules however, lack certain "microtubule associated proteins"; it is presumed that the segregation of axonal from perikaryal and dendritic microtubules must occur in the cell body (Brady and Black, 1986). In addition, tubulin and neurofilament proteins have several posttranslational modifications which distinguish axonal from perikaryal forms. Some of these posttranslational modifications occur at the site of assembly (Burgoyne and Norman, 1986) while others are acquired within the axon (Nixon et al., 1982; Brown et al., 1982).

Energetic and Ionic Requirements of Axonal Transport

The metabolic demands of transport are best understood in the case of the rapid transport component (for a review, see Grafstein and Forman, 1980). Once material enters the rapid transport vector within axons, its movement is not immediately dependent on the cell body. Following either inhibition of protein synthesis in the cell body or axotomy, normal fast transport continues for several hours within the axon. Fast transport is also independent of the electrical activity of the axon. Rapid axonal transport is, however, critically dependent on local oxidative metabolism within the axon for a continued supply of ATP, and on extracellular Ca^{+2}. Rapid axonal transport is not inhibited by Co^{2+}, indicating that the Co^{2+} sensitive mechanism involved in initiation does not play a continued role following commitment to transport (Hammerschlag and Stone, 1982). Less is known about the requirements for retrograde transport of material, but in general they are the same as for anterograde transport. Retrograde transport is dependent upon a local supply of ATP and Ca^{+2}, independent of the cell body, and its rate is altered as a function of temperature (for a review, see Bisby, 1980).

The energy and ionic requirements for the intermediate and slow components are less well known. In the goldfish optic system the movement of the slow components is almost immediately arrested distal to the site of axotomy or following addition of protein synthesis blockers (Grafstein and Alpert, 1982). In contrast, it has been reported that slow transport continues for weeks in the garfish olfactory nerve following disconnection from the cell bodies (Cancalon, 1982). The energy and ion requirements of slow transport have not been measured directly, primarily because nerves cannot be kept *in vitro* long enough to perform the appropriate tests. In goldfish optic nerve (Grafstein et al., 1972) and hibernating ground squirrels (Kemper and Matzke, 1981) the rate of slow transport appears unaffected by temperature, while in the garfish the Q_{10} for slow transport is reported to be greater than 3 (Cancalon, 1979). The failure to find any temperature effects in the goldfish is likely to be due to the short axons examined and the indirect means of measuring slow transport (arrival at the nerve terminals). The lack of any difference in rate between hibernating and nonhibernating ground squirrels may be a unique adaptation of hibernating animals to allow continuation of slow transport at very low temperatures. (In contrast, it is reported that, at least *in vitro*, rapid transport retains its temperature sensitivity in hibernating rodents; Bisby and Jones, 1978).

Postulated Mechanisms of Axonal Transport

While there is much information concerning the characteristics of axonal transport, the molecular mechanisms responsible are not well understood. However, there is growing evidence that all phases of transport are dependent on some or all of the cytoskeletal structures connected to driving forces utilizing ATP. It is now thought that rapid anterograde transport is driven by a microtubule-dependent ATPase named kinesin which links the transported membrane vesicles to microtubules and propels them along. For retrograde transport both kinesin and cytoplasmic dynein have been postulated to be the motor (for reviews see McIntosh and Porter, 1989; Brady, 1991). Likewise, it is postulated that a cytoplasmic dynein may propel slow transport (Cleveland and Hoffman, 1991).

TOXICANT-INDUCED DISORDERS OF AXONAL TRANSPORT

Any major metabolic or traumatic insult to neurons will eventually disrupt axonal transport. Thus it is almost a trivial conclusion that a progressive neuropathy induced by a toxicant will eventually result in a perturbation of axonal transport. Many toxicants cause primarily axonal pathology; such disorders have been termed "axonopathies" to indicate the primary target (Spencer and Schaumburg, 1984). Since in these cases the axons are preferentially involved, disturbances of axonal transport have been a major focus of investigation. A key question concerning toxicants causing axonopathies is whether any of these agents have a direct effect on axonal transport which occurs before morphological alterations or clinical symptoms (i.e., can disturbances of axonal transport be responsible for the axonopathy?). If a toxicant can be shown to produce such early effects, then a case for causality can be made. A toxicant might produce such an effect in several ways. One possibility is an interaction with some step in the mechanism of axonal transport. In the absence of a good understanding of transport mechanisms (and consequent inability to assay the effect of toxicants on steps in this process) much of the work in this area is phenomenological. A second possibility is that a toxicant might directly affect movement of a specific transported component (e.g., by chemically modifying this component) without altering axonal transport per se. A third possibility is that a toxicant might cause some more general biochemical lesion, such as the disruption of energy metabolism, which in turn alters axonal transport. It should be stressed that even if axonal transport, or some specific component of transport, is shown to be altered, either directly or secondarily to some other lesion, it still remains to be demonstrated that this disruption of transport is a causal factor in the subsequent neuropathy.

In no case is there conclusive evidence that a toxicant causes an axonopathy through a direct effect on the mechanism of axonal transport, as opposed to secondary effects on transport such as general metabolic insufficiency or structural damage to the axons. In a few cases there is evidence that modification of a particular transported component may be involved. There have been many reviews of the involvement of axonal transport in toxicant-induced neuropathies (Mendell and Sahenk, 1980; Droz and Chretien, 1980; Brimijoin, 1982, 1984; Jakobsen et al., 1983, 1986; Kristensson and Gustafsson, 1984; Samson and Donoso, 1984; Spencer et al., 1985; McLean et al., 1985; Ochs,1987; Griffin and Watson, 1988).

A major problem encountered in attempting to elucidate the mechanisms of action of a neurotoxicant is that the severity of the pathology, its time course, and even the sequence of cellular manifestations can depend greatly on the dose, dosing schedule, route of administration, species and neural pathway examined. This variability probably accounts for at least some of the apparent conflicts in the literature concerning both the morphological and biochemical alterations caused by toxicants. For the sake of brevity, we have not reiterated the importance of these factors in the sections which follow, except for one or two examples. Even within a defined model system, there is frequently a sufficient large animal-to-animal variability in response to a toxicant that ideally one should correlate the biochemical alterations with the extent of morphologically defined alterations within the same animal. This correlation is rarely done because of technical problems and the sheer magnitude of morphological work involved. More frequently, the biochemical alterations have been correlated with the clinical symptoms—a much easier task.

For convenience, we have grouped the major axonopathies into two broad categories based on morphologic criteria with relevance to axonal transport. The first category includes those neuropathies which affect the cytoskeletal structures of the axon, particularly the neurofilaments, and thus may involve alterations in slow transport. The second category includes those neuropathies which appear to involve primarily disruption of the membranous elements of the axon, and thus may involve alterations in rapid axonal transport. This division is somewhat arbitrary and artificial as the effects of most toxicants are not as specific as these categories imply. In addition to the above, several other neuropathies which fall outside these categories but which may cause alterations in axonal transport are also discussed.

Toxicants Affecting Neurofilaments

Morphology

Exposure to certain toxicants induces focal accumulations of neurofilaments within neurons and their processes. (A review by Kosik and Selkoe, 1983, discusses these in detail.) Accumulation may be within the soma, or at proximal, middle, or distal regions of axons. Within axons these accumulations of neurofilaments generally occur at multifocal, paranodal sites in myelinated axons. Examples of toxicants causing somal accumulation are the vinca alkaloids and aluminum salts. Proximal, giant axonal swellings are induced by intoxication with β,β'-iminodipropionitrile (IDPN). Hexacarbons (of presumably 2,5-hexanedione [2,5-HD], their common toxic metabolite) can cause proximal, middle, or distal accumulations depending on the specific derivative used. Carbon disulfide and acrylamide cause distal axonal accumulations of neurofilaments. This common feature of multifocal accumulations of neurofilaments suggests that these toxicants share a common mechanism of action. The specific effects of these compounds on axonal transport, and proposed mechanisms of action are discussed below.

Slow Transport

The accumulations of neurofilaments described above suggest that some aspect of slow transport may be altered. Therefore many of the compounds causing such neuropathies have been studied with respect to whether they cause alterations in the slow components of transport. It has been shown that following either acute (intraperitoneal) or chronic (present in drinking water) administration of IDPN, transport of neurofilament proteins is blocked at the proximal axonal swellings. In contrast, transport of tubulin and other slow component proteins is affected to a much smaller extent, and significantly slowed only after prolonged chronic dosing (Griffin et al., 1978; Yokayama et al., 1980). Furthermore, if IDPN is administered several weeks after radiolabeling of anterior horn cells, transport of labeled neurofilament proteins already in transit down the sciatic nerve is halted (Griffin et al., 1978). These experiments involved injection of radioactive methionine into the lumbar spinal cord and gel electrophoretic analysis of radioactive proteins at various times in sequential segments of sciatic nerve.

Because tubulin transport is normal, IDPN probably does not directly affect the slow transport mechanism but rather has a direct chemical interaction with neurofilaments which renders them nontransportable (see Toxicant-Induced Disorders of Axonal Transport, Postulated Mechanisms). Whatever this alteration may be, it apparently does not disrupt processing in the cell body or the initiation of slow neurofilament transport since the neurofilaments move normally into the proximal axon where they accumulate to form the swellings. The posttranslational modifications (phosphorylation) of neurofilament proteins which take place within the proximal axon have been suggested as a possible site of action for IDPN (Sayre et al., 1985; Eyer et al., 1989). Although this inhibition of neurofilament transport accounts for the pathogenesis of the axonal swellings, its relationship to the neurological abnormalities is less clear. The inhibition of slow transport recovers in a few weeks whereas the neurological abnormalities are long lasting (Komiya et al., 1987).

The interpretation of studies of aluminum salt lesions are much less clear than those involving IDPN. Local injection of aluminum chloride into spinal cord induced swellings in greater than 85% of anterior horn cells of rabbits but did not prevent labeled neurofilament proteins from entering the proximal axons. In addition, distal downflow patterns in sciatic nerve indicated a normal amount and rate of transported neurofilaments (Kosik et al., 1985). In contrast, when a similar study in hypoglossal neurons of rabbits was performed (Bizzi et al., 1984), local injection of aluminum chloride was shown to block export of labeled neurofilament proteins into the proximal axons; there was no effect on the movement of other slow component proteins. Morphologically, the region just distal to the swellings showed a nearly complete loss of neurofilaments, while more distal regions of the axons were normal. This suggests that filament transport is maintained distally, resulting in a clearing of filaments from the segments just distal to the swellings. No adequate explanation has been offered for the discrepancy between these two studies. It is not likely that the molecular mechanisms of aluminum chloride neurotoxicity varies from nerve to nerve. Rather, the effective dose at the site of action was probably different. In a third study, some retardation of neurofilament transport within the motor axons of the rabbit sciatic nerve was observed following aluminum chloride intoxication (Troncoso et al., 1985).

In the cases of the hexacarbons causing midway and distal axonal accumulations of neurofilaments, it appears that the slow transport of neurofilaments proximal to the region of swellings is actually accelerated in chronically treated animals showing clinical signs of neuropathy (Monaco et al., 1985; Braendgaard and Sidenius, 1986b; Monaco et al., 1989a,b). Transport of other slow component proteins is unaffected. This accelerated transport is suggested to be the cause of the morphologically observed loss of neurofilaments and decrease in diameter of the proximal regions of axons.

The 3,4-dimethylated derivative, which causes proximal axonal swellings, blocks slow transport of neurofilaments much like IDPN (Griffin et al., 1984).

Carbon disulfide produces effects very similar to the hexacarbons (Papolla et al., 1987). The neuropathies caused by most of the other compounds listed above are less clearly defined as neurofilamentous axonopathies and have little or no effect on neurofilament transport. The vinca alkaloids cause a peripheral neuropathy characterized by increased numbers of neurofilaments but also by a disruption of microtubules. At high acute doses these compounds are able to block both fast and slow transport, presumably due to the extensive disruption of microtubules. At the chronic doses that cause the distal neuropathy, however, there is little evidence that a blockage of transport is the cause (Mendell and Sahenk, 1980). Finally, acrylamide causes a constellation of effects, only one of which is an accumulation of neurofilaments, and evidence suggests this compound belongs in a different category (see Toxicants Affecting Smooth Membranes, Morphology).

Rapid Transport

Despite the extensive reorganization of the cytoskeletal framework and the large prenodal accumulations of neurofilaments, neither IDPN nor aluminum-salt treatment perturb rapid anterograde transport or retrograde transport (Griffin et al., 1978; Liwnicz et al., 1974). In the case of IDPN-treated animals it has been shown that rapidly transported membrane components continue to move normally through the central core of the axon which contains the microtubules, a fact used in support of the hypothesized role of microtubules in fast transport (Griffin et al., 1983; Papasozomenos et al., 1982).

There are conflicting reports on the effects of hexacarbons on rapid transport. Mendell et al. (1977) noted a progressive slowing of the rapid transport rate with distance down the nerve that correlated with the severity of the neuropathy caused by methyl *n*-butylketone. In more recent studies of chronic 2,5-HD toxicity, however, no effect on rapid transport rate or amount was found in proximal regions of sciatic nerve in rats exhibiting severe clinical neuropathy (Braendgaard and Sidenius, 1986a,b). In this latter study the authors argue that an uncontrolled temperature drop in the limbs during anesthesia was responsible for the transport deficits reported in the earlier study. In the earlier study measures were taken to avoid temperature fluctuations but no body temperatures are reported to verify their effectiveness. In a related study, however, these authors report that these measures were adequate to maintain body temperature (Sahenk and Mendell, 1981). Temperature fluctuation is a factor frequently overlooked in studies of axonal transport. Anesthesia can lower body temperature, and in the presence of a toxicant this effect may be greatly exacerbated (Padilla and Lyerly, 1986). Another possibility is that the differences in these sets of studies may be accounted for by differences in the magnitude of the pathology. While similar clinical neurological signs are present in each case, no morphological examinations were performed to assess the actual condition of the axons. Not only were different hexacarbon compounds used, but also different strains of rat. Both of these differences can result in the axons being more severely damaged in the earlier study. In contrast to the above studies, Sickles (1989b) reported that single doses of 2,5-HD cause a modest decrease in rapid transport rate and a large reduction in the quantity of protein transported. The estimates of rate and amount were based on a single time point, however, and thus the reported alterations are likely to be due to changes in the cell body uptake, synthesis, and processing, rather than transport itself.

Consistent with the observations above concerning possible alterations in rapid anterograde transport by chronic exposure to hexacarbons, there is a disruption of the axonal SER and accumulation of vesiculotubular membranous elements occurring simultaneously with the accumulation of neurofilaments (Sahenk and Mendell, 1983). This temporal relationship and the late onset of reported alterations in rapid anterograde transport suggest that these deficits are a result of, rather than a cause of, the disorder. The large accumulations of neurofilaments caused by the hexacarbons may cause focal, partial blockages of anterograde transport which could account for the observed retardation of rate and decreased amount of material in this transport component. Support for this explanation comes from evidence that partial ligation of nerves results in the same anterograde transport abnormalities as seen with hexacarbons (Mendell and Sahenk, 1980). (Note that no effects on rapid transport have been found with the IDPN-induced neurofilamentous swellings. This difference is presumably due to the sparing of a central core of the axon in the IDPN-induced neurofilament accumulations, as mentioned above.)

Alterations in retrograde transport have also been observed following hexacarbon administration. In animals having hindlimb paralysis from chronic 2,5-HD administration, Sahenk and Mendell (1981) examined the amount of labeled protein accumulating distal to sciatic nerve ligatures following injection of precursors into the ventral spinal cord. They concluded that there was a decreased rate of retrograde transport while the amount of material transported was normal. This interpretation is questionable, however, because the onset of accumulation at the ligature does not appear to be significantly delayed in the treated animals, whereas the amount accumulating appears to be decreased. More recently, Braendgaard and Sidenius (1986a) reported a large decrease in the amount of labeled material accumulating distal to sciatic nerve ligatures at times well before clinical symptoms or the earliest reported morphological changes. This effect was also dose related. There is no obvious explanation for this early deficit in the amount of material returning via retrograde transport. Because there were no significant changes in the fraction of synthesized material committed to anterograde transport, the height of the advancing anterograde crest, or the span of the anterograde wave front in the treated animals compared to controls, it seems unlikely that an increased deposition of anterogradely transported material in the proximal axons could account for the diminished retrograde component.

The early onset of these effects suggests that altered retrograde transport may play a role in the development of the axonopathy. In conflict with this conclusion, a histochemical study in the visual system using horseradish peroxidase found retrograde transport affected only in axons of the very large neurons, although all axons showed giant neurofilamentous swellings (Pasternak et al., 1985). This observation suggests there is little correlation between the axonopathy and effects on retrograde transport.

Postulated Mechanisms

Many of the toxicants causing accumulations of neurofilaments react covalently with proteins, including neurofilament proteins. Based on this observation and the generally similar pathology caused by these agents, a single "unified" hypothesis has been advanced to account for this class of neuropathies (Sayre et al., 1985). Briefly, it is suggested that each of these toxicants causes a specific covalent modification of neurofilament proteins (with the exception of aluminum which is suggested to be chelated by proteins). These modifications are postulated to destablize the cytoskeletal framework, uncoupling microtubule and neurofilament transport. Freed of this supramolecular cytoskeletal structure, neurofilaments could be transported faster. As the neurofilaments become increasingly modified they would eventually become nontransportable and accumulate. The exact position along the nerve at which transport fails would depend on the reactivity, concentration, and time course of administration of the agent. The somewhat different ultrastructural pathology seen with each of these agents is presumed to be a result of the specific covalent modification of neurofilaments each toxicant makes.

This hypothesis is best supported in the case of the hexacarbons. Hexacarbons can react both *in vitro* and *in vivo* with protein lysine ϵ-amine moieties to yield pyrrole adducts, and secondary auto-oxidative crosslinking is possible. Therefore, it has been postulated that pyrrolyation and secondary crosslinking of neurofilaments are the initiating events in hexacarbon neuropathy (De Caprio et al., 1982, 1983; Graham et al., 1982). In support of this possibility, it has been reported that neurofilaments can become crosslinked *in vitro* and *in vivo* following 2,5-HD treatment (De Caprio and O'Neill, 1985; Lapadula et al., 1986, 1988; Genter St. Clair et al., 1989), that methylated derivatives of 2,5-HD with different reactivities produce swellings at different locations along nerves (Anthony et al., 1983; Monaco et al., 1984; Rosenburg et al., 1987), and that a nonreactive methylated derivative of 2,5-HD causes no neuropathy (Sayre et al., 1986). While the pyrrolation of neurofilaments by 2,5-HD is generally accepted, the issue of secondary crosslinking of filaments is still debated (Monaco et al., 1990). While this model fits very elegantly the behavior of the hexacarbons, at least in relation to their effects on neurofilaments, it offers no explanation for the early effects on retrograde transport. A similar model may also account for the effects of IDPN on neurofilaments.

The morphologic alterations and transport abnormalities of the other agents the authors include in this "unified" hypothesis are not as easily reconciled to the model. The effects of aluminum salts on slow transport are unresolved. In addition, the postulated chelation of aluminum by glutamic acid carboxylates in neurofilaments is undemonstrated. Acrylamide, as stated earlier and as suggested by Sayre et al. (1985), may belong in a different category.

In addition to the above model, it has been proposed that the hexacarbons act through inhibiting local energy production in the axon, an action which could account for their reported effects on rapid anterograde and retrograde transport (Schoental and Cavanagh, 1977; Spencer et al., 1979; Sabri et al., 1979a,b). A shortage of axonal ATP would lead to an impairment of rapid axonal transport and accumulation of transported material. There is evidence that the hexacarbons can, in some circumstances, inhibit several glycolytic enzymes (Sabri et al., 1979a,b), but there is also evidence that lactate production in nervous tissue of treated animals is not impaired, suggesting normal glucose utilization (LoPachin et al., 1984). This hypothesis of hexacarbon action is supported by the fact that some agents known to disrupt energy metabolism, as well as certain vitamin deficiencies, cause clinical signs of peripheral neuropathy. The morphologic alterations associated with these conditions have not been well studied, however. Thus it is not possible to say how similar these conditions are to each other or to that of the hexacarbons. This model (and others invoking some general metabolic insult) accounts for the distal axonal pathology by suggesting that when enzymes are inhibited along the length of the axon the cell body cannot keep up with the increased demand for these enzymes. Some glycolytic enzymes are carried by slow transport (SCb). For these enzymes it would take many weeks to resupply the distal axon with new, functional enzymes. In this situation slowly transported enzymes of energy metabolism would be utilized to meet the needs of the proximal portion of the axon, leaving the distal axon without sufficient enzyme to meet its energy needs. This model requires that the enzyme inhibition be of some irreversible nature. Note that mitochondrial enzymes can be replenished from the cell body too rapidly for there to be any long term distal axonal deficit.

Although they have flaws, models like the two above are useful in that they provide a framework for testable hypotheses regarding mechanisms of toxicant induction of neuropathies.

Toxicants Affecting Smooth Membranes

Morphology

Several neurotoxicants cause distal axonal swellings composed partly or entirely of accumulations of vesiculotubular membrane structures presumably derived from the SER. These neuropathies generally manifest a distal "dying-back" degeneration of nerve fibers. Zinc pyridinethione (ZPT) causes a disruption of the SER and accumulation of membranes in motor nerve terminals (Sahenk and Mendell, 1979). This disruption progresses proximally into the axons as the neuropathy develops. p-Bromophenylacetylurea (BPAU) causes accumulations of membranes in terminal axons of both the PNS and CNS, leading to distal nerve degeneration (Ohnishi and Ikeda, 1980). Many of the organophosphorous esters cause distal, nonterminal varicosities containing accumulations of vesiculotubular membranes (Bouldin and Cavanagh, 1979). In contrast to the fairly specific effects of the above compounds, acrylamide causes a constellation of morphologic alterations in axons, only one of which is accumulations of membranes. Nevertheless, Cavanagh (1985) concluded that acrylamide intoxication leads to a major disruption of the somal and axonal SER which could be responsible for the distal neuropathy. Despite the relatively similar pathology caused by these compounds, no unified mechanism of action has been proposed for these compounds. The specific effects of these compounds on axonal transport are discussed below.

Slow Transport

The compounds described above appear to have little or no effect on slow axonal transport, except perhaps when the neuropathy is severe and axons may be degenerating. An early autoradiographic study suggested that chronic dosing with acrylamide blocked slow axonal transport in cat sciatic nerve, while organophosphorous compounds had no effect (Pleasure et al., 1969). A more recent biochemical study using rats failed to reproduce this result for acrylamide (Sidenius and Jakobsen, 1983). Instead, moderate decreases in the amount of material carried in the slow component and a small increase in rate were reported at high cumulative doses where the neuropathy was advanced. However, it has been reported that a single high dose of acrylamide will cause large proximal neurofilamentous swellings and disruption of slow transport (Gold et al., 1985). An effect on the rate of slow transport has also been reported following exposure to BPAU, but again only in severely disabled rats (Jakobsen and Brimijoin, 1981; Nagata et al., 1987).

Rapid Transport

The accumulation of membranous structures associated with these toxicants suggests that some aspect of either fast anterograde or retrograde transport may be affected. Of the compounds listed above acrylamide is the most thoroughly studied in this regard and the studies of this compound have been reviewed (Miller and Spencer, 1985).

Some authors have reported that chronic acrylamide treatment induces a moderate decrease in rapid transport rate (Bradley and Williams, 1973; Weir et al., 1978). In these reports, however, the transport studies were initiated only after the animals were clearly impaired. Under these conditions the effects on axonal transport could be secondary to other axonal lesions. More recent studies suggest that chronic acrylamide treatment, even at levels sufficient to cause ataxia, does not decrease the rapid transport rate (Sidenius and Jakobsen, 1983; Harry et al., 1989). At very high cumulative doses (500 mg/kg), a slight decrease in the amount of material accumulating proximal to a ligature was found. A possible explanation suggested for these differing results is that in the earlier studies steps may not have been taken to prevent a possible decrease in body temperature during anesthesia in the acrylamide treated animals. This issue was also raised in reference to the effects of hexacarbons on rapid transport (see Toxicant-Induced Disorders of Axonal Transport, Toxicants Affecting Neurofilaments, Rapid Transport). Although the rapid transport rate is apparently not affected, there may be other abnormalities in rapid transport. Harry et al. (1989) have reported that there is an apparent increased deposition of transported glycoproteins (but not proteins in general) along sensory axons in acrylamide-treated rats which occurs before any morphological signs of neuropathy.

Recently, it was reported that a single dose of acrylamide reduces the rapid transport rate as well as the amount of rapidly transported material in rat sensory axons (Sickles, 1989a). However, several methodological and analytical problems with this study make these conclusions questionable. For example, transport rates were estimated from a single time point, making it likely that the observed rate differences were due to differences in somal processing rather than transport rate.

Autoradiographic studies of rapidly transported material in the ciliary ganglion of chickens treated with acrylamide demonstrate focal, intense labeling underneath the axolemma in 20 to 30% of the axons. Labeling in the nerve endings was decreased substantially (Souyri et al., 1981). Ultrastructurally, this label was associated with accumulations of disorganized SER (Chretien et al., 1981). The authors concluded that acrylamide intoxication causes a multifocal disruption of the axonal SER which is responsible for the focal retention of rapidly transported proteins in these axons. Note that these results are not inconsistent with the unaffected transport rates and amounts discussed above. First, electron microscopic autoradiography allows for detection of alterations in individual axons—an observation lost in the averaging inherent in biochemical studies. Second, the biochemical rate determinations described above were made in proximal and presumably morphologically normal regions of nerve. These studies point out the different kinds of data and perspectives provided by biochemical and morphological methods of examining axonal transport (see Techniques for Quantifying Axonal Transport).

When retrograde axonal transport of radiolabeled proteins was examined in acrylamide treated animals using double ligature techniques (see Techniques for Quantifying Axonal Transport, Ligature Methods for a description of this technique), a decreased amount of labeled material accumulating just distal to the distal ligature was found (Sahenk and Mendell, 1981; Jakobsen and Sidenius, 1983). At the highest cumulative doses used in these studies (500 to 800 mg/kg) there was some indication that the retrograde transport rate was affected. Accumulation in the anterograde direction was normal except at the very highest doses. In one case this deficit in retrograde transport was found to be dose dependent and to begin before neurological signs of neuropathy; morphological changes were not assessed, however (Jakobsen and Sidenius, 1983). With the exception of this latter study, the reports described above on possible deficits in both anterograde and retrograde transport are consistent with the notion that some distal structural alteration causes a partial blockage and retention of transported material rather than acrylamide having a direct effect on transport itself.

Retrograde transport has also been reported to be slowed in animals treated acutely with acrylamide (Miller and Spencer, 1984, 1985; Moretto and Sabri, 1988). In these studies ^{125}I-labeled nerve growth factor or tetanus toxin was injected into the gastrocnemius muscles of mice or hens and the retrograde transport up the sciatic nerve and accumulation of radioactivity in neurons of dorsal root ganglion (DRG) and spinal cord followed. Single doses of acrylamide injected after the tetanus toxin reduced the amount of label accumlating in spinal cord motor neurons in a dose related manner. In the sensory neurons of the DRG there was no clear dose dependence of the effect. In mice, a single dose of 100 mg/kg of acrylamide administered 3 hours before the tetanus toxin completely blocked the movement of label in the distal sciatic nerve, and lower doses caused an apparent slowing of the rate in a dose related fashion.

These acute-dosing studies suggest that there is some alteration of retrograde transport rate, but they are difficult to reconcile with the data from double-ligature studies in chronically treated animals, where even very large cumulative doses had little or no effect on retrograde transport rate. One possibility is that rate effects may be transient, lasting for only a few hours after each administration of the acrylamide. Another possibility may be that acute dosing acts in a different way than the chronic dosing which causes a neuropathy. For example, Gold et al. (1985) have reported that a large, single dose of acrylamide causes proximal nerve swellings containing accumulations of neurofilaments, in contrast to the distal swellings found with chronic dosing.

As is the case with studies of acrylamide, the effects of ZPT on fast transport are not well understood. Decreases in the rapid transport rate were reported, but the effect observed was variable and not correlated with the severity of the neuropathy (Sahenk and Mendell, 1980). In companion double-ligature studies of retrograde transport the appearance of label distal to the distal ligature was delayed compared to control animals and the amount of material accumulating was decreased. The authors concluded that retrograde rate was not slowed but that the defect was in the turnaround at the nerve terminals. This interpretation is consistent with the initial appearance of pathological alterations in the terminals. Failure of turnarond at the terminals would cause a backing up of anterograde material as it arrived at the terminals, and account for the proximal progression of the pathology.

Bradley and Williams (1973) reported that the organophosphorous compound triorthocresyl phosphate caused a moderate decrease in fast transport rate in cats, but the rate calculations were based on downflow curves from a single time point. Possible alterations in somal processing were therefore not eliminated as an explanation for the apparent decrease in rate. More recently, Reichert and Abou-Donia (1980) reported that when different organophosphorous esters were applied locally to the retina, those compounds which are known to produce a neuropathy substantially reduced the amount of labeled protein in the optic nerve and tract following intraocular injection of labeled proline. The magnitude of the effect was correlated with the potency of the toxicant, and parathione, which does not produce a neuropathy, had no effect. This decreased delivery of transported material was not due to decreased retinal synthesis of total protein; this was actually stimulated in many cases (note that this does not rule out decreased synthesis in one class of retinal cells, the ganglion cells which are the origin of the optic nerve axons). The authors speculate that these compounds give rise to a local disruption of axonal transport in distal axons which could lead to the focal axonal degeneration these agents produce.

Axonal transport in BPAU-intoxicated animals has been studied in some detail by Brimijoin and coworkers (Jakobsen and Brimijoin, 1981; Nagata and Brimijoin, 1986; Nagata et al., 1987). These studies involved injection of dorsal root ganglia or ventral horn with labeled fucose and methionine to examine rapid transport, and in conjunction with ligatures to examine retrograde transport. Accumulation of several enzyme activities at ligatures was also examined. Several anterograde and retrograde transport abnormalities were described. In sensory axons, rapid transport rate was normal, but the time for initiation of transport was doubled. Furthermore, while the initial accumulation of rapidly transported protein at a ligature was normal, the total long term accumulation was decreased. Taken together, these two observations suggest some abnormality in somal protein synthesis and processing. There was also a nearly 60% reduction in labeled protein initially accumulating distal to a set of double ligatures while the long term distal accumulation was normal. In contrast to these results, when motor axons were examined, the time for initiation of transport was shortened and the initial amount of transported material was increased. Furthermore, in motor axons the dose dependence of the effects was questionable while in sensory axons doses of BPAU which caused no clinical or morphological evidence of neuropathy affected retrograde accumulation just as severely as doses causing severe neuropathy. These differences between sensory and motor axons are unexplained.

This last observation that severe alterations in retrograde transport can exist without signs of neuropathy suggests that the deficits in retrograde transport are not directly responsible for the BPAU neuropathy, and that similar deficits in retrograde transport with compounds like zinc pyridine-thione and acrylamide should be interpreted with caution. Such results point out that demonstrating an early and apparently direct effect of a toxicant on axonal transport does not demonstrate causality in terms of the neuropathy, and call into question any proposed relationship between disturbances of axonal transport and toxicant-induced axonopathies.

Postulated Mechanisms

Acrylamide has been reported to inhibit glycolytic enzymes, and thus it has been proposed that acrylamide perturbs axonal transport and causes neuropathy by disrupting glycolysis (Spencer et al., 1979; Howland, 1981). In addition to effects on glycolysis, it has also been reported that acrylamide decreases the reducing potential in neurons but not in nonneural tissues, suggesting that oxidative energy production may be compromised (Sickles and Goldstein, 1986; Sickles, 1987). However, acrylamide's effects on gly-colytic and oxidative enzymes has been challenged (Ross et al., 1985; LoPachin et al., 1984; Brimijoin and Hammond, 1985). Thus the role of acrylamide's inhibition of energy metablism in producing axonal neuropathy is still unclear.

Miller and Spencer (1985) have drawn attention to the large number of similarities in the sensory neuron's response to acrylamide and its regener-ative response to axotomy. They have proposed that acrylamide mimics axon transsection, perhaps by blocking retrograde transport, and at the same time interferes with a normal cellular response, resulting in a faulty axonal repair. Support for this hypothesis of a faulty regenerative response in the presence of acrylamide is provided by the observation that if sciatic nerve is crushed in a morphologically normal proximal region of a treated animal, large amounts of membrane accumulate at the cut but normal sprouting fails to take place (Griffin et al., 1977). The axons proximal to the cut subsequently degenerate (Cavanagh and Gysbers, 1980; Sharer and Lowndes, 1985). A similar inhibition of sprouting has been demonstrated in the nerves that innervate the rat sternocostalis muscles (Kemplay and Cavanagh, 1984). In addition it has been demonstrated that changes in the composition of trans-ported proteins in acrylamide neuropathy resemble those seen in regenerating axons (Bisby and Redshaw, 1987). The alterations in retrograde transport that have been reported with acrylamide may indeed account for the apparent regenerative response of the cell body, but how acrylamide might render that response ineffective is not known.

Another hypothesis to account for the axonal transport abnormalities caused by acrylamide has been presented by Harry et al. (1989). They suggest that the apparent increase in deposition of transported glycoproteins into stationary axonal structures in acrylamide-treated animals may be explained as a compensatory response of the neurons to replace specific axonal components damaged by acrylamide. An increased deposition of rapidly transported material may account for the deficits in retrograde transport reported by others.

For the other compounds included here no specific mechanisms have been proposed other than that they, along with acrylamide, might interfere with the turnaround process at the axon terminals (Brimijoin, 1982).

Other Toxicants

Methyl Mercury

An early study of chronic methyl mercury intoxication reported modest increases in transport rate and amount of transported material in sensory axons of rat sciatic nerve (Wakabayshi et al., 1976). Only a single transport time point was examined, leaving open the possibility of a shortened processing time. More recently, Aschner and coworkers (Aschner, 1987; Aschner et al., 1986, 1987) have examined axonal transport in the visual system after either systemic or local exposure to methyl mercury. Local administration to the vitreous chamber greatly reduced the protein synthesis in the retina and export into the optic pathway. In contrast, systemic exposure produced an increased synthesis of proteins in the retina and an increased export. Although the rapid transport rate was also reported to be increased, this conclusion is questionable because of the unconventional way in which transport was calculated, leading to an abnormally low rate estimate for the control animals. These authors also report that there was an increased synthesis of growth-associated proteins in the retina, suggesting a regenerative response of retinal neurons.

The methyl mercury neuropathy has been classified as neuronopathy (as opposed to an axonopathy) based on a variety of criteria (Spencer and Schaumburg, 1980). The results reviewed above are consistent with its classification. The changes reported appear to reflect changes in somal synthesis and processing rather than changes in axonal transport. In this regard, methyl mercury is similar to trimethyltin, which causes similar effects in retina (Toews et al., 1986b).

Experimental Diabetic Neuropathy

Streptozotocin causes a destruction of the β cells of the pancreas, leading to a decrease in insulin production and elevated serum glucose (Junod et al., 1969). Rats treated with streptozotocin display a peripheral neuropathy similar to that in human diabetes and which is reversed by insulin treatment (Mayer and Tomlinson, 1983).

It has been reported that streptozotocin treatment slows fast axonal transport in the motor axons of rat sciatic nerve (Meri and McLean, 1982; Laduron and Janssen, 1986), but does not alter this parameter in the sensory axons of this nerve (Jakobsen and Sidenius, 1980). Abnormalities in the accumulation at ligatures of material carried by both anterograde (Mayer and Tomlinson, 1983) and retrograde transport (Jakobsen and Sidenius, 1979; Laduron and Janssen, 1986) have been described in both sensory and motor axons. In addition, the rate of slow axonal transport has been reported to be slowed in rats treated with streptozotocin (Jakobsen and Sidenius, 1980; Medori et al., 1985).

In most of the cases above reporting alterations in anterograde and retrograde transport, insulin treatment can prevent, as well as reverse, the effects of the streptozotocin on axonal transport. In addition, Sidenius and Jakobsen (1980) reported that galactose feeding, which also produces a peripheral neuropathy, resulted in similar transport abnormalities. These results indicate that the transport abnormalities are due to the hyperglycemia rather than a direct toxic effect of the streptozotocin on axonal transport.

Demyelinating Neuropathies

There are a wide range of toxicants that cause demyelination of peripheral nerves following either local or systemic administration. For some of these agents it has been reported that axonal transport may be affected secondary to the demyelination. These results have been interpreted to indicate either an influence of ensheathing cells on transport or a dependence of axonal transport on an intact myelin sheath.

Focal demyelination, induced by local application of Simliki Forest virus or the induction of allergic neuritis by immunization with white matter, has been reported to increase the amount of labeled protein transported in the fast component in cranial nerves (Pessoa and Ikeda, 1984; Rao et al., 1981). In contrast, local injection of diphtheria toxin, which also produces a focal region of demyelination surrounding the injection site, is reported to cause a partial blockage of rapid transport in the region of the demyelination (Kidman et al., 1978a,b). However, it is quite possible that features other than demyelination were causative of the axonal transport perturbations reported in these studies. A recent study utilized an immunological method to create a very specific, focal, primary segmental demyelination; no alterations in rapid axonal transport of glycoproteins through the demyelinated region of sciatic nerve were observed (Armstrong et al., 1987). This suggests that perturbations reported in the previous studies may involve treatments which have a direct effect on the axons.

SUMMARY AND CONCLUSIONS

Because axonal transport is necessary for the maintenance and functioning of axons, it is usually assumed that an alteration in axonal transport will necessarily lead to neuropathy (Sabri and Spencer, 1990). Decreases in either rate or amount of anterograde transport are expected to lead to insufficient material reaching the axon and its ending, and therefore to decreased function and eventually neuropathy. Alterations in the content of the material transported are expected to deprive the nerve terminal of essential components. Likewise, alteration in retrograde transport would initially disrupt feedback of information from the nerve ending to the perikaryon. The resultant loss of information as to what material the perikaryon should commit to transport could compromise metabolic support of the nerve ending. Furthermore, the accumulation of anterogradely transported material at the nerve terminal would presumably physically disrupt the terminal. While these are reasonable assumptions, few direct tests of these assumptions have been made. Thus, even if a neurotoxicant can be shown to have a direct effect on some aspect of axonal transport, it remains to be shown that this alteration leads to the neuropathy.

Axonal transport is indeed compromised by many neurotoxicants, but such alterations are implicated in the causation of axonopathy in only a few cases. The best example is the giant neurofilamentous axonopathy caused by 2,5-HD where a covalent modification of neurofilaments may interfere with their transport and be responsible for the pathology observed. For most other neuropathies, many more questions remain. Acrylamide, 2,5-HD, BPAU and zinc pyridinethione all are reported to affect retrograde transport relatively early in the progression of their neuropathies but much more correlative work must be done to establish whether those effects are causative, especially in light of the studies of BPAU and 2,5-HD (see Toxicant-Induced Disorders of Axonal Transport, Toxicants Affecting Neurofilaments, Rapid Transport and Toxicants Affecting Smooth Membranes, Rapid Transport).

That the rates and amounts of material carried in the various transport components are not obviously altered by neurotoxicants (or if altered, do not appear causally related to the neuropathy) does not rule out axonal transport as a factor in the progression of their neuropathies. For example, there may be alterations in the delivery and deposition of specific, essential macromolecules into stationary axonal and terminal structures (Harry et al., 1989) which may lead to structural and functional failure of the axon. In addition, the focus on the axon as the site of action of these toxicants may have tended to draw attention away from other aspects of these neuropathies. Although these disorders are considered axonopathies, cell body alterations are frequently seen as early, if not earlier, than the axonal changes (Sterman, 1982; Sterman and Sposito, 1985). Thus the complex processes of transport initiation in the cell body may be useful events to examine in more detail in these neuropathies. Finally, it should be stressed that whether alteration of axonal transport is a cause or an effect of neuropathy, disruption of this process is part of the cellular pathology determining the subsequent course of the neuropathy. Thus information about axonal transport is an important factor in the understanding of the pathogenesis and progression of the toxicant-induced neuropathies.

REFERENCES

Ambron, R.T., J.E., Goldman, and J.H. Schwartz (1974). Axonal transport of newly synthesized glycoproteins in a single identified neuron of *aplysia californica*. *J. Cell Biol.* 61: 665–675.

Anthony, D.C., K. Boekelheide, and D.G. Graham (1983). The effect of 3,4-dimethyl substitution on the neurotoxicity of 2,5-hexanedione. I. Accelerated clinical neuropathy is accompanied by more proximal axonal swellings. *Toxicol. Appl. Pharmacol.* 71: 362–371.

Aquino, D.A., M.A. Bisby, and R.W. Ledeen (1985). Retrograde axonal transport of gangliosides and glycoproteins in the motoneurons of rat sciatic nerve. *J. Neurochem.* 45: 1262–1267.

Armstrong, R., A.D. Toews, and P. Morell (1987). Rapid axonal transport in focally demyelinated sciatic nerve. *J. Neurosci.* 7: 4044–4053.

Armstrong, R., A.D. Toews, R.B. Ray, and P. Morell (1985). Retrograde axonal transport of endogenous phospholipids in rat sciatic nerve. *J. Neurosci.* 5: 965–969.

Aschner, M. (1987). Changes in axonally transported proteins in the mature and developing rat nervous system during early stages of methyl mercury exposure. *Pharmacol. Toxicol.* 60: 81–85.

Aschner, M., P.M. Rodier, and J.N. Finkelstein (1986). Reduction of axonal transport in the rat optic system after direct application of methyl mercury. *Brain Res.* 381: 244–250.

Aschner, M., P.M. Rodier, and J.N. Finkelstein (1987). Increased axonal transport in the rat optic system after systemic exposure to methyl mercury, differential effects in local vs. systemic exposure conditions. *Brain Res.* 401: 132–141.

Bisby, M.A. (1980). Retrograde axonal transport. *Adv. Cell Neurobiol.* 1: 69–117.

Bisby, M.A. (1981). Reversal of axonal transport, similarity of proteins transported in anterograde and retrograde directions. *J. Neurochem.* 36: 741–745.

Bisby, M.A. (1982). Functions of retrograde transport. *Fed. Proc. Fed. Am. Soc. Exp. Biol.* 41: 2307–2311.

Bisby, M.A. and V.T. Bulger (1977). Reversal of axonal transport at a nerve crush. *J. Neurochem.* 29: 313–320.

Bisby, M.A. and D.L. Jones (1978). Temperature sensitivity of axonal transport in hibernating and nonhibernating rodents. *Exp. Neurol.* 61: 74–83.

Bisby, M.A. and J.D. Redshaw (1987). Acrylamide neuropathy, changes in the composition of proteins of fast axonal transport resemble those observed in regenerating axons. *J. Neurochem.* 48: 924–928.

Bizzi, A., R.C. Crane, L. Autilio-Gambetti, and P. Gambetti (1984). Aluminum effect on slow axonal transport, a novel impairment of neurofilament transport. *J. Neurosci.* 4: 722–731.

Black, M.M., P. Keyser, and E. Sobel (1986). Interval between the synthesis and assembly of cytoskeletal proteins in cultured neurons. *J. Neurosci.* 6: 1004–1012.

Blaker, W.D., J.F. Goodrum, and P. Morell (1981). Axonal transport of the mitochondria-specific lipid, diphosphatidylglycerol, in the rat visual system. *J. Cell Biol.* 89: 579–584.

Bouldin, T.W. and J.B. Cavanagh (1979). Organophosphorous neuropathy. II. A fine-structural study of the early stages of axonal degeneration. *Am. J. Pathol.* 94: 253–262.

Bradley, W.G. and M.H. Williams (1973). Axoplasmic flow in axonal neuropathies. *Brain* 96: 235–246.

Brady, S.T. and M.M. Black (1986). Axonal transport of microtubule proteins, cytotypic variation of tubulin and MAPs in neurons. *Ann. N.Y. Acad. Sci.* 466: 199–217.

Brady, S.T., R.J. Lasek, and R.D. Allen (1982). Fast axonal transport in extruded axoplasm from squid giant axon. *Science* 218: 1129–1131.

Brady, S.T. (1991). Molecular motors in the nervous system. *Neuron* 7: 521–533.

Braendgaard, H. and P. Sidenius (1986a). The retrograde fast component of axonal transport in motor and sensory nerves of the rat during administration of 2,5-hexanedione. *Brain Res.* 378: 1–7.

Braendgaard, H. and P. Sidenius (1986b). Anterograde components of axonal transport in motor and sensory nerves in experimental 2,5-hexanedione neuropathy. *J. Neurochem.* 47: 31–37.

Brimijoin, S. (1975). Stop-flow, a new technique for measuring axonal transport, and its application to the transport of dopamine-β-hydroxylase. *J. Neurobiol.* 6: 379–394.

Brimijoin, S. (1984). The role of axonal transport in nerve disease. In *Peripheral Neuropathology*, 2nd ed. P.J. Dyck, P.K. Thomas, E.H. Lambert, and R. Bunge, Eds., W.B. Saunders, Philadelphia, pp. 477–493.

Brimijoin, W.S. (1982). Abnormalities of axonal transport. Are they a cause of peripheral nerve disease? *Mayo Clin. Proc.* 57: 707–714.

Brimijoin, W.S. and P.I. Hammond (1985). Acrylamide neuropathy in the rat, effects on energy metabolism in sciatic nerve. *Mayo Clin. Proc.* 60: 3–8.

Brown, B.A., R.A. Nixon, and C.A. Marotta (1982). Posttranslational processing of α-tubulin during axoplasmic transport in CSN axons. *J. Cell Biol.* 94: 159–164.

Burgoyne, R.D. and K.M. Norman (1986). Alpha-tubulin is not detyrosylated during axonal transport. *Brain Res.* 381: 113–120.

Cancalon, P. (1979). Influence of temperature on the velocity and on the isotope profile of slowly transported labeled proteins. *J. Neurochem.* 32: 997–1007

Cancalon, P. (1982). Slow flow in axons detached from their perikarya. *J. Cell Biol.* 95: 989–992.

Cancalon, P. and L.M. Beidler (1977). Differences in the composition of the polypeptides deposited in the axon and the nerve terminals by fast axonal transport in the garfish olfactory nerve. *Brain Res.* 121: 215–227.

Cavanagh, J.B. (1985). Peripheral nervous system toxicity, a morphological approach. In *Neurotoxicology*. K. Blum and L. Manzo, Eds., Marcel Dekker, New York, pp. 1–45.

Cavanagh, J.B. and M.F. Gysbers (1980). "Dying back" above a nerve ligature produced by acrylamide. *Acta Neuropathol.* 51: 169–177.

Chretien, M., G. Patey, F. Souyri, and B. Droz (1981). 'Acrylamide-induced' neuropathy and impairment of axonal transport of proteins. II. Abnormal accumulations of smooth endoplasmic reticulum as sites of focal retention of fast transported proteins. Electron microscope radioautographic study. *Brain Res.* 205: 15–28.

Cleveland, D.W. and P.N. Hoffman (1991). Slow axonal transport models come full circle: evidence that microtubule sliding mediates axon elongation and tubulin transport. *Cell* 67: 453–456.

De Caprio, A.P. and E.A. O'Neill (1985). Alterations in rat axonal cytoskeletal proteins induced by *in vitro* and *in vivo* 2,5-hexanedione exposure. *Toxicol. Appl. Pharmacol.* 78: 235–247.

De Caprio, A.P., E.J. Olajos, and P. Weber (1982). Covalent binding of a neurotoxic n-hexane metabolite, conversion of primary annines to substituted pyrrole adducts by 2,5-hexandione. *Toxicol. Appl. Pharamcol.* 65: 440–450.

De Caprio, A.P., N.L. Strominger, and P. Weber (1983). Neurotoxicity and protein binding of 2,5-hexandione in the hen. *Toxicol. Appl. Pharmacol.* 68: 297–307.

Dravid, A.R. and R. Hammerschlag (1975). Axoplasmic transport of proteins *in vitro* in primary afferent neurons of frog spinal cord, effect of Ca^{2+}-free incubation conditions. *J. Neurochem.* 24: 711–718.

Droz, B. and M. Chretien (1980). Axonal flow and toxic neuropathies. In *Mechanisms of Toxicity and Hazard Evaluation*. B. Holmstedt, R. Lauwerys, M. Mercier, and M. Roberfroid, Eds., Elsevier/North Holland, Amsterdam, pp. 13–25.

Droz, B. and C.P. Leblond (1963). Axonal migration of proteins in the central nervous system and peripheral nerves as shown by radioautography. *J. Comp. Neurol.* 12: 325–346.

Ellisman, M.H. and J.D. Lindsey (1983). The axoplasmic reticulum within myelinated axons is not transported rapidly. *J. Neurocytol.* 12: 393–411.

Eyer, J., W.G. Mclean, and J.F. Leterrier (1989). Effect of a single dose of β,β'-iminodipropionitrile in vivo on the properties of neurofilaments in vitro: comparison with the effect of iminodipropionitrile added directly to neurofilaments in vitro. *J. Neurochem.* 52: 1759–1765.

Forgue, S.T. and J.L. Dahl (1978). The turnover of tubulin in rat brain. *J. Neurochem.* 31: 1289–1297.

Genter St. Clair, M.B., D.C. Anthony, C.J. Wikstrand, D.G. Graham (1989). Neurofilament protein crosslinking in γ-diketone neuropathy: in vitro and in vivo studies using the seaworm *Myxicola infundibulum*. *Neurotoxicology* 10: 743–756.

Gold, B.G., J.W. Griffin, and D.L. Price (1985). Slow axonal transport in acrylamide neuropathy, different abnormalities produced by single-dose and continuous administration. *J. Neurosci.* 5: 1755–1768.

Goodrum, J.F. and P. Morell (1982). Axonal transport, deposition and metabolic turnover of glycoproteins in the rat optic pathway. *J. Neurochem.* 36: 696–704.

Goodrum, J.F., G.C. Stone, and P. Morell (1989). Axonal transport and intracellular sorting of glycoconjugates. In *Neurobiology of Glycoconjugates*. R.N. Margolis and R.K. Margolis, Eds., Plenum Press, New York, pp. 277–308.

Grafstein, B. and R.M. Alpert (1982). Properties of slow axonal transport, studies in goldfish optic axons. In *Axonal Transport*. D.G. Weiss, Ed., Springer-Verlag, New York, pp. 226–231.

Grafstein, B. and D.S. Forman (1980). Intracellular transport in neurons. *Physiol. Rev.* 60: 1167–1283.

Graftstein, B., D.S. Forman, and B.S. McEwen (1972). Effects of temperature on axonal transport and turnover of protein in goldfish optic system. *Exp. Neurol.* 34: 158–170.

Graham, D.G., D.C. Anthony, K. Boekelheide, N.A. Maschmann, R.G. Richards, J.W. Wolfram and B.R. Shaw (1982). Studies of the molecular pathogenesis of hexane neuropathy. II. Evidence that pyrrole derivatization of lysyl residues leads to protein crosslinking. *Toxicol. Appl. Pharmacol.* 64: 415–422.

Griffin, J.W. and D.F. Watson (1988). Axonal transport in neurological disease. *Ann. Neurol.* 23: 3–13.

Griffin, J. W., D.C. Anthony, K.E. Fahnestock, P.N. Hoffman, and D.G. Graham (1984). 3,4-dimethyl-2,5-hexanedione impairs the axonal transport of neurofilament proteins. *J. Neurosci.* 4: 1516–1526.

Griffin, J.W., K.E. Fahnestock, D.L. Price, and P.N. Hoffman (1983). Microtubule-neurofilament segregation produced by β,β'-iminodipropionitrile, evidence for the association of fast axonal transport with microbutules. *J. Neurosci.* 3: 557–566.

Griffin, J.W., P.N. Hoffman, A.W. Clark, P.T. Carroll, and D.L. Price (1978). Slow axonal transport of neurofilament proteins, impairment by β,β'-iminodipropionitrile administration. *Science* 202: 633–635.

Griffin, J.W., D.L. Price, and D.B. Drachman (1977). Impaired axonal regeneration in acrylamide intoxication. *J. Neurobiol.* 8: 355–370.

Hammerschlag, R. (1984). How do neuronal proteins know where they are going? Speculations on the role of molecular address markers. *Dev. Neurosci.* 6: 2–17.

Hammerschlag, R. and G.C. Stone (1982). Membrane delivery by fast axonal transport. *Trends Neurosci.* 5: 12–15.

Harry, G.J., J.F. Goodrum, T.W. Bouldin, A.D. Toews, and P. Morell (1989). Acrylamide-induced increase in deposition of axonally transported glycoproteins in rat sciatic nerve. *J. Neurochem.* 52: 1240–1247.

Hoffman, P.N. and R. J. Lasek (1975). The slow component of axonal transport, identification of major structural polypeptides of the axon and their generality among mammalian neurons. *J. Cell Biol.* 66: 351–366.

Hollenbeck, P. J. (1989). The transport and assembly of the axonal cytoskeleton. *J. Cell Biol.* 108: 223–227.

Howland, R.D. (1981). The etiology of acrylamide neuropathy, enolase, phosphofructokinase, and glyceraldehyde-3-phosphate dehydrogenase activities in peripheral nerve, spinal cord, brain, and skeletal muscle of acrylamide-intoxicated cats. *Toxicol. Appl. Pharmacol.* 60: 324–333.

Jakobsen, J. and S. Brimijoin (1981). Axonal transport of enzymes and labeled proteins in experimental axonopathy induced by p-bromophenylacetylurea. *Brain Res.* 229: 103–122.

Jakobsen, J., S. Brimijoin, and P. Sidenius (1983). Axonal transport in neuropathy. *Muscle Nerve* 6: 164–166.

Jakobsen, J. and P. Sidenius (1979). Decreased axonal flux of retrogradely transported glycoprotein in early experimental diabetes. *J. Neurochem.* 33: 1055–1060.

Jakobsen, J. and P. Sidenius (1980). Decreased axonal transport of structural proteins in streptozotocin diabetic rats. *J. Clin. Invest.* 66: 292–297.

Jakobsen, J. and P. Sidenius (1983). Early and dose-dependent decrease of retrograde axonal transport in acrylamide-intoxicated rats. *J. Neurochem.* 40: 447–454.

Jakobsen, J., P. Sidenius, and H. Braendgaard (1986). A proposal for a classification of neuropathies according to their axonal transport abnormalities. *J. Neurol. Neurosurg. Psychiatry* 49: 986–990.

Jeffrey, P.L. and L. Austin (1973). Axoplasmic transport. *Prog. Neurobiol.* 2: 205–225.

Junod, A., A.E. Lambert, W. Stauffacher, and A.E. Renold (1969). Diabetogenic action of streptozotocin: relationship of dose to metabolic response. *J. Clin. Invest.* 48: 2129–2139.

Kimper, G.B. and H.A. Matzke (1981). Slow axoplasmic transport in the hibernating and nonhibernating ground squirrel. *Exp. Neurol.* 71: 649–660.

Kimplay, S. and J.B. Cavanaugh (1984). Effects of acrylamide and some other reagents on spontaneous and pathologically induced terminal sprouting of motor end-plates. *Muscle Nerve* 7: 101–109.

Kidman, A.D., W.C. Baker, and H.J. Sippe (1978a). Effect of diphtheritic demyelination on axonal transport in the sciatic nerve and subsequent muscle changes in the chicken. *Adv. Exp. Med. Biol.* 100: 439–452.

Kidman, A.D., L. Dolan, and H.J. Sippe (1978b). Blockade of fast axonal transport by diphtheritic demyelination in the chicken sciatic nerve. *J. Neurochem.* 30: 57–61.

Komiya, Y., N.A. Cooper, and A.D. Kidman (1987). The recovery of slow axonal transport after a single intraperitoneal injection of β,β'-iminodiproprionitrite in the rat. *J. Biochem.* 102: 869–873.

Kosik, K.S., A.H. McCluskey, F.X. Walsh, and D.J. Selkoe (1985). Axonal transport of cytoskeletal proteins in aluminum toxicity. *Neurochem. Pathol.* 3: 99–108.

Kosik, K.S. and D.J. Selkoe (1983). Experimental models of neurofilamentous pathology. In *Neurofilaments*. C.A. Marotta, Ed., University of Minnesota Press, Minneapolis, pp. 155–195.

Kristensson, K. and M. Gustafsson (1984). Toxic actions on axonal transport mechanisms. *Acta Neurol. Scand.* 70: 27–32.

Laduron, P.M. and P.F.M. Janssen (1986). Impaired axonal transport of opiate and muscarinic receptors in streptozotocin rats. *Brain Res.* 380: 359–362.

Lapadula, D.M., R.D. Irwin, E. Suwita, and M.B. Abou-Donia (1986). Cross-linking of neurofilament proteins of rat spinal cord *in vivo* after administration of 2,5-hexanedione. *J. Neurochem.* 46: 1843–1850.

Lapadula, D.M., E. Suwita and M.B. Abou-Donia (1988). Evidence for multiple mechanisms responsible for 2,5-hexandione-induced neuropathy. *Brain Res.* 458: 123–131.

Lasek, R.J. (1970). Protein transport in neurons. *Int. Rev. Neurobiol.* 13: 289–321.

Lasek, R.J. and P.N. Hoffman (1976). The neuronal cytoskeleton, axonal transport and axonal growth. In *Cell Motility,* Book C. R. Goldman, T. Pollard, and J. Rosenbaum, Eds., Cold Spring Harbor Laboratory, Cold Spring Harbor, NY, pp. 1021–1049.

Lasek, R.J. and S.T. Brady (1982). The structural hypothesis of axonal transport, two classes of moving elements. In *Axoplasmic Transport*. D. Weiss, Ed., Springer-Verlag, Berlin, pp. 397–405.

Liwnicz, B.H., K. Kristensson, H.M. Wisniewski, M.L. Shelanski, and R.D. Terry (1974). Observations on axoplasmic transport in rabbits with aluminum-induced neurofibrillary tangles. *Brain Res.* 80: 413–420.

LoPachin, R.M., R.W. Moore, L.A. Menahan, and R.E. Peterson (1984). Glucose-dependent lactate production by homogenates of neuronal tissues prepared from rats treated with 2,4-dithiobiuret, acrylamide, *p*-bromophenylacetylurea and 2,5-hexanedione. *Neurotoxicology* 5: 25–36.

Lorenz, T. and M. Willard (1978). Subcellular fractionation of intra-axonally transported polypeptides in the rabbit visual system. *Proc. Natl. Acad. Sci. U.S.A.* 75: 505–509.

Lubinska, L. (1975). On axoplasmic flow. *Int. Rev. Neurobiol.* 17: 241–256.

Lubinska, L., B. Niemierko, B. Oderfield-Nowak, and L. Szwarc (1964). Behavior of acetylcholinesterase in isolated nerve segments. *J. Neurochem.* 11: 493–503.

Mayer, J.H. and D.R. Tomlinson (1983). Axonal transport of cholinergic transmitter enzymes in vagus and sciatic nerves of rats with acute experimental diabetes mellitus; correlation with motor nerve conduction velocity and effects of insulin. *Neuroscience* 9: 951–957.

McIntosh, J.R. and M.E. Porter (1989). Enzymes for microtubule-dependent motility. *J. Biol. Chem.* 264: 6001–6004.

McLean, W.G., M. Frizell, and J. Sjostrand (1985). Pathology of axonal transport. In *Handbook of Neurochemistry*, 2nd ed., Vol. 9. A. Lajtha, Ed., pp. 67–86.

McQuarrie, I.G., S.T. Brady, and R.J. Lasek (1986). Diversity in the axonal transport of structural proteins, major differences between optic and spinal axons in the rat. *J. Neurosci.* 6: 1593–1605.

Medori, R., L. Autilio-Gambetti, S. Monaco, and P. Gambetti (1985). Experimental diabetic neuropathy, impairment of slow transport with changes in axon cross-sectional area. *Proc. Natl. Acad. Sci. U.S.A.* 82: 7716–7720.

Mendell, J.R. and Z. Sahenk (1980). Interference of neuronal processing and axoplasmic transport by toxic chemicals. In *Experimental and Clinical Neurotoxicology*. P.S. Spencer and H.H. Shaumburg, Eds., Williams & Wilkins, Baltimore, pp. 139–160.

Mendell, J.R., Z. Sahenk, K. Saida, H.S. Weiss, R. Savage, and D. Kouri (1977). Alterations of fast axoplasmic transport in experimental methyl n-butyl ketone neuropathy. *Brain Res.* 133: 107–118.

Merri, K.F. and W.G. McLean (1982). Axonal transport of protein in motor fibers of experimentally diabetic rats—fast anterograde transport. *Brain Res.* 238: 77–88.

Miller, M.S., M.J. Miller, T.F. Burks, and I.G. Sipes (1983). Altered retrograde axonal transport of nerve growth factor after single and repeated doses of acrylamide in the rat. *Toxicol. Appl. Pharmacol.* 69: 96–101.

Miller, M.S. and P.S. Spencer (1984). Single doses of acrylamide reduce retrograde transport velocity. *J. Neurochem.* 43: 1401–1408.

Miller, M.S. and P.S. Spencer (1985). The mechanisms of acrylamide axonopathy. *Annu. Rev. Pharmacol. Toxicol.* 25: 643–666.

Monaco, S., L. Autilio-Gambetti, D. Zabel, and P. Gambetti (1985). Giant axonal neuropathy, acceleration of neurofilament transport in optic axons. *Proc. Natl. Acad. Sci. U.S.A.* 82: 920–924.

Monaco, S., L. Autillo-Gambetti, R.J. Lasek, M.J. Katz, and P. Gambetti (1989a). Experimental increase of neurofilament transport rate, decreases in neurofilament number and in axon diameter. *J. Neuropathol. Exp. Neurol.* 48: 23–32.

Monaco, S., J. Jacob, H. Jenrich, A. Patton, L. Autilio-Gambetti, and P. Gambetti (1989b). Axon transport of neurofilament is accelerated in peripheral nerve during 2,5-hexanedione intoxication. *Brain Res.* 491: 328–334.

Monaco, S., T. Wongmongkolrit, L.M. Sayre, L. Autilio-Gambetti, and P. Gambetti (1984). Giant axonopathy in 3-methyl-2,5-hexanedione (3-M-2,5-HD) effect on morphology and slow axonal transport. *J. Neuropathol. Exp. Neurol.* 43: 304.

Monaco, S.O., T. Wongmongkolrit, C.M. Shearson, A. Patton, B. Schaetzle, L. Autilio-Gambetti, P. Gambetti, and L.M. Sayre (1990). Giant axonopathy characterized by intermediate location of axonal enlargements and acceleration of neurofilament transport. *Brain Res.* 519: 73–81.

Moretto, A. and M.I. Sabri (1988). Progressive deficits in retograde axon transport precede degeneration of motor axons in acrylamide neuropathy. *Brain Res.* 440: 18–24.

Mori, H., Y. Komiya, and M. Kurokawa (1979). Slowly migrating axonal polypeptides. *J. Cell Biol.* 82: 174–184.

Munoz-Martinez, E.J., R. Nunez, and A. Sanderson (1981). Axonal transport, a quantitative study of retained and transported protein fraction in the cat. *J. Neurobiol.* 12: 15–26.

Nagata, H. and S. Brimijoin (1986). Axonal transport in the motor neurons of rats with neuropathy induced by *p*-bromophenylacetylurea. *Ann. Neurol.* 19: 458–464.

Nagata, H., S. Brimijoin, P. Low, and J.D. Schmeizer (1987.). Slow axonal transport in experimental hypoxia and in neuropathy induced by *p*-bromophenylacetylurea. *Brain Res.* 422: 319–326.

Nixon, R.A. (1983). Proteolysis of neurofilaments. In *Neurofilaments*. C.A. Marotta, Ed., University of Minnesota Press, Minneapolis, pp. 117–154.

Nixon, R.A., B.A. Brown, and C.A. Marotta (1982). Posttranslational modification of a neurofilament protein during axoplasmic transport, implications for regional specialization of CNS axons. *J. Cell Biol.* 94: 150–158.

Ochs, S. (1982). *Axoplasmic Transport and Its Relation to Other Nerve Functions.* John Wiley & Sons, New York.

Ochs, S. (1987). The action of neurotoxins in relation to axoplasmic transport. *Neurotoxicology* 8: 155–166.

Ochs, S. and C. Smith (1975). Low temperature slowing and cold-block of fast axoplasmic transport in mammalian nerves *in vitro*. *J. Neurobiol.* 6: 85–102.

Ohnishi, A. and M. Ikeda (1980). Morphometric evaluation of primary sensory neurons in experimental *p*-bromophenylacetylurea intoxication. *Acta Neuropathol.* 52: 111–118.

Padilla, S. and D. Lyerly (1986). Effects of hypothermia on the *in vivo* measurement of rapid axonal transport in the rat, a precautionary note. *J. Neurochem.* 46: 1227–1230.

Padilla, S.S., L.J. Roger, A.D. Toews, J.F. Goodrum, and P. Morell (1979). Comparison of proteins transported in different tracts of the central nervous system. *Brain Res.* 176: 407–411.

Papasozomenos, S.C., M. Yoon, R. Crane, L. Autilio-Gambetti, and P. Gambetti (1982). Redistribution of proteins of fast axonal transport following administration of β,β'-iminodipropionitrile, a quantitative autoradiographic study. *J. Cell Biol.* 95: 672–675.

Papolla, M., R. Penton, H.S. Weiss, C.H. Miller Jr., Z. Sahenk, L. Autilio-Gambetti and P. Gambetti (1987). Carbon disulfide axonopathy. Another experimental model characterized by acceleration of neurofilament transport and distinct changes of axonal size. *Brain Res.* 424: 272–280.

Pasternak, T., D.G. Flood, T.A. Eskin, and W.H. Merigan (1985). Selective damage to large cells in the cat retinogeniculate pathway by 2,5-hexanedione. *J. Neurosci.* 5: 1641–1652.

Pessoa, V.F. and H. Ikeda (1984). Increase in axonal transport in demyelinating optic nerve fibers in the mouse infected with Semliki Forest virus. *Brain* 107: 433–446.

Pleasure, D.E., K.C. Mishler, and W.K. Engel (1969). Axonal transport of proteins in experimental neuropathies. *Science* 166: 524–525.

Rao, N.A., J. Guy, and P.S. Sheffield (1981). Effects of chronic demyelination on axonal transport in experimental allergic optic neuritis. *Inv. Ophthalmol. Vis. Sci.* 21: 606–611.

Reichert, B.L. and M.B. Abou-Donia (1980). Inhibition of fast axoplasmic transport by delayed neurotoxic organophosphorus esters, a possible mode of action. *Mol. Pharmacol.* 17: 56–60.

Rosenburg, C.K., M.B. Genter, G. Szakal-Quin, D.C. Anthony, and D.G. Graham (1987). dl- Versus meso-3,4-dimethyl-2,5-hexanedione, a morphometric study of the proxima-distal distribution of axonal swellings in the anterior root of the rat. *Toxicol. Appl. Pharmacol.* 87: 367–373.

Ross, S.M., M.I. Sabri, and P.S. Spencer (1985). Action of acrylamide on selected enzymes of energy metabolism in denervated cat peripheral nerves. *Brain Res.* 340: 189–191.

Sabri, M.I. and P.S. Spencer (1990). Acrylamide impairs fast and slow axonal transport in rat optic sytem. *Neurochem. Res.* 15: 603–608.

Sabri, M.I., K. Ederle, C.E. Holdsworth, and P.S. Spencer (1979a). Studies on the biochemical basis of distal axonopathies. II. Specific inhibition of fructose-6-phosphate kinase by 2,5-hexanedione and methyl-butyl ketone. *Neurotoxicol.* 1: 285–297.

Sabri, M.I., C.L. Moore, and P.S. Spencer (1979b). Studies on the biochemical basis of distal axonopathies. I. Inhibition of glycolysis by neurotoxic hexacarbon compounds. *J. Neurochem.* 32: 683–689.

Sabri, M.I., W. Dairman, M. Fenton, L. Juhansz, T. Ng, and P.S. Spencer (1989). Effect of exogenous pyruvate on acrylamide neuropathy in rats. *Brain Res.* 483: 1–11.

Sahenk, Z. and J.R. Mendell (1979). Ultrastructural study of zinc pyridinethione-induced peripheral neuropathy. *J. Neuropathol. Exp. Neurol.* 38: 532–550.

Sahenk, Z. and J.R. Mendell (1980). Axoplasmic transport in zinc pyridinethione neuropathy, evidence for an abnormality in distal turn-around. *Brain Res.* 186: 343–353.

Sahenk, Z. and J.R. Mendell (1981). Acrylamide and 2,5-hexanedione neuropathies, abnormal bidirectional transport rate in distal axons. *Brain Res.* 219: 397–405.

Sahenk, Z. and J.R. Mendell (1983). Studies on the morphologic alterations of axonal membranous organelles in neurofilamentous neuropathies. *Brain Res.* 268: 239–247.

Samson, F.E. and J.A. Donoso (1984). Pharmacology and toxicology of axoplasmic transport. *Prog. Drug Res.* 28: 53–81.

Sayre, L.M., L. Autilio-Gambetti, and P. Gambetti (1985). Pathogenesis of experimental giant neurofilamentous axonopathies, a unified hypothesis based on chemical modification of neurofilaments. *Brain Res. Rev.* 10: 69–83.

Sayre, L.M., C.M. Shearson, T. Wongmongkolrit, R. Medori, and P. Gambetti (1986). Structural basis of γ-diketone neurotoxicity, non-neurotoxicity of 3,3-dimethyl-2,5-hexanedione, a γ-diketone incapable of pyrrole formation. *Toxicol. Appl. Pharm.* 84: 36–44.

Schoental, R. and J.B. Cavanagh (1977). Mechanisms involved in the 'dying-back' process—an hypothesis implicating coenzymes. *Neuropathol. Appl. Neurobiol.* 3: 145–157.

Schwab, M.E. and H. Thoenen (1983). Retrograde axonal transport. In *Handbook of Neurochemistry*, 2nd ed. A. Lajtha, Ed., Plenum Press, New York, pp. 381–404.

Schwartz, J.H. (1979). Axonal transport, components, mechanisms, and specificity. *Annu. Rev. Neurosci.* 2: 467–504.

Sharer, L.R. and H.E. Lowndes (1985). Acrylamide-induced ascending degeneration of ligated peripheral nerve, effect of ligature location. *Neuropathol. Appl. Neurobiol.* 11: 191–200.

Sickles, D.W. (1987). Neural specificity of acrylamide upon enzymes associated with oxidative energy-producing pathways. I. Histochemical analysis of HADH-tetrazolium reductase activity. *Neurotoxicology* 8: 623–630.

Sickles, D.W. (1989a). Toxic neurofilamentous axonopathies and fast anterograde axonal transport. I. The effects of single doses of acrylamide on the rate and capacity of transport. *Neurotoxicology* 10: 91–102.

Sickles, D.W. (1989b). Toxic neurofilamentous axonopathies and fast anterograde axonal transport. II. The effects of single doses of neurotoxic and non-neurotoxic diketones and β,β'-iminodiproprionitrile (IDPN) on the rate and capacity of transport. *Neurotoxicology* 10: 103–112.

Sickles, D.W. and B.D. Golstein (1986). Acrylamide produces a direct, dose-dependent and specific inhibition of oxidative metabolism in motoneurons. *Neurotoxicology* 7: 187–196.

Sidenius, P. and J. Jakobsen (1980). Axonal transport in rats after galactose feeding. *Diabetologia* 19: 229–233.

Sidenius, P. and J. Jakobsen (1983). Anterograde axonal transport in rats during intoxication with acrylamide. *J. Neurochem.* 40: 697–704.

Smith, R.S. (1980). The short term accumulation of axonally transported organelles in the region of localized lesions of single myelinated axons. *J. Neurocytol.* 9: 39–65.

Souyri, F., M. Chretien, B. Droz (1981). 'Acrylamide-induced' neuropathy and impairment of axonal transport of proteins. I. Multifocal retention of fast transported proteins at the periphery of axons as revealed by light microscope radioautography. *Brain Res.* 205: 1–13.

Spencer, P.S. and H.H. Schaumburg (1980). Classification of neurotoxic disease, a morphologic approach. In *Experimental and Clinical Neurotoxicology*. P.S. Spencer and H.H. Schaumburg, Eds., Williams & Wilkins, Baltimore, pp. 92–99.

Spencer, P.S. and H.H. Schaumburg (1984). Experimental models of primary axonal disease induced by toxic chemicals. In *Peripheral Neuropathy*, 2nd ed. P.J. Dyck, P.K. Thomas, E.H. Lambert, and R. Bunge, Eds., W.B. Saunders, Philadelphia, pp. 636–649.

Spencer, P.S., M.S. Miller, S.M. Ross, B.W. Schuab and M.I. Sabri (1985). Biochemical mechanisms underlying primary degeneration of axons. In *Handbook of Neurochemistry*, 2nd ed., Vol. 9. A. Lajtha, Ed., Plenum Press, New York, pp. 31–65.

Spencer, P.S., M.I. Sabri, H.H. Schaumburg, and C.L. Moore (1979). Does a defect of energy metabolism in the nerve fiber underlie axonal degeneration in polyneuropathies? *Ann. Neurol.* 5: 501–507.

Sterman, A.B. (1982). The role of the neuronal cell body in neurotoxic injury. *Neurobehav. Toxicol. Teratol.* 4: 493–494.

Sterman, A.B. and N. Sposito (1985). 2,5-hexanedione and acrylamide produce reorganization of motor neuron perikarya. *Neuropathol. Appl. Neurobiol.* 11: 201–212.

Stone, G.C., R. Hammerschlag, and J.A. Bobinski (1984). Involvement of coated vesicles in the initiation of fast axonal transport. *Brain Res.* 291: 219–228.

Toews, A.D., R. Armstrong, J. Holshek, R.M. Gould, and P. Morell (1986a). Unloading and transfer of axonally transported lipids. In *Neurology and Neurobiology,* Vol. 25. R.S. Smith and M.A. Bisby, Eds., Alan R. Liss, New York, pp. 327–346.

Toews, A.D., R.B. Ray, N.D. Goines, and T.W. Bouldin (1986b). Increased synthesis of membrane macromolecules is an early response of retinal neurons to trimethyltin intoxication. *Brain Res.* 398: 298–304.

Troncoso, J.C., P.N. Hoffman, J.W. Griffin, K.M. Hess-Kozlow, and D.L. Price (1985). Aluminum intoxication, a disorder of neurofilament transport in motor neurons. *Brain Res.* 342: 172–175.

Tsukita, S. and H. Ishikawa (1980). The movement of membranous organelles in axons, electron microscopic identification of anterogradely and retrogradely transported organelles. *J. Cell Biol.* 84: 513–530.

Tytell, M., M.M. Black, J.A. Garner, and R.J. Lasek (1981). Axonal transport, each major rate component reflects the movement of distinct macromolecular complexes. *Science* 214: 179–181.

Vale, R.D., B.J. Schnapp, T.S. Reese, and M.P. Sheetz (1985). Movement of organelles along filaments dissociated from the axoplasm of the squid giant axon. *Cell* 40: 449–454.

Wakabayashim, M., K. Araki, and Y. Takahashi (1976). Increased rate of fast transport in methylmercury-induced neuropathy. *Brain Res.* 117: 524–528.

Weir, R.L., G. Glaubiger, and T.N. Chase (1978). Inhibition of fast axoplasmic transport by acrylamide. *Environ. Res.* 17: 251–255.

Weiss, P. and H.B. Hiscoe (1948). Experiments on the mechanism of nerve growth. *J. Exp. Zool.* 107: 315–395.

Willard, M. (1977). The identification of two intra-axonally transported polypeptides resembling myosin in some respects in the rabbit visual system. *J. Cell Biol.* 75: 1–11.

Willard, M., W.M. Cowan, and P.R. Vagelos (1974). The polypeptide composition of intra-axonally transported proteins, evidence for four transport velocities. *Proc. Natl. Acad. Sci. U.S.A.* 71: 2183–2187.

Willard, M.B. and K.L. Hulebak (1977). The intra-axonal transport of polypeptide H, evidence for a fifth (very slow) group of transported proteins in the retinal ganglion cells of the rabbit. *Brain Res.* 136: 289–306.

Wilson, D.L. and G.C. Stone (1979). Axoplasmic transport of proteins. *Annu. Rev. Biophys. Bioeng.* 8: 27–45.

Yokoyama, K., S. Tsukita, H. Ishikawa, and M. Kurokowa (1980). Early changes in the neural cytoskeleton caused by β,β'-iminodipropionitrile, selective impairment of neurofilament peptides. *Biomed. Res.* 1: 537–547.

Neurotransmitter Receptors

Stephen C. Bondy
Department of Community and Environmental Medicine
University of California
Irvine, California

Syed F. Ali
Division of Neurotoxicology
National Center for Toxicological Research
Jefferson, Arkansas

INTRODUCTION

The selective attachment of small molecules such as neurotransmitters, neuromodulators, or hormones to distinctive sites on the cell membrane is the basic mechanism by which cells communicate with each other. Methods have been developed that allow the convenient measurement of such binding interactions (Yamamura et al., 1984). The addition of this procedure to modern enzymatic and analytical methodology of neurochemistry has led to new insights concerning brain function.

The action of many of neuropharmacological agents is now known to be mediated by their attachment to receptor sites within the brain. Such binding may mimic or inhibit a preexisting interaction between an endogenous ligand and the receptor. Pharmacological agents which are thought to act by virtue of their agonistic or antagonistic effect at specific receptors include α- and β-adrenergic blockers, neuroleptic drugs, tricyclic antidepressants, benzodiazepines, opiates, strychnine, LSD, atropine, and the snake venom α-bungarotoxin. In many cases, a dose-response correlation between binding affinity and pharmacological potency has been demonstrated (Snyder, 1985).

Many diseases are related to inappropriate communication between cells (Rubenstein, 1980), and the receptors on the cell surface play a crucial role in such intercellular exchanges. Within the nervous system, several neurological disorders are thought to be correlated with derangements of synaptic events and thus with altered levels of transmitter binding. Such diseases include myasthenia gravis, schizophrenia, posttraumatic seizure onset, and Huntington's disease. Changes in receptor responsiveness may also be responsible for gradual loss of effectiveness of bronchodilators when used in the chronic treatment of asthma, and in failure of epinephrine to restore blood pressure in terminal shock. Receptor responses often adjust in a homeostasis-maintaining direction, and thus act as biological buffers (Schwarcz et al., 1978a).

In view of the involvement of receptors in many pharmacological and pathological phenomena, it seems reasonable to suppose that some effects of neurotoxic agents are mediated by receptor-related events. Receptor density frequently bears an inverse relation to the extent of transmitter release into the synaptic cleft. These adjustments may achieve homeostasis by assuring compensatory alteration in response to changes in neuronal firing rates. Thus, if a toxic agent selectively changes the activity of a certain neuronal circuit, such changes may be detected by altered binding capacity within a particular nerve circuit.

Other approaches to the chemistry of neurotoxicity have included measurements of levels of enzymes and of content of transmitter chemicals and proteins (Dewar and Moffet, 1979). Materials that are largely confined to nerve tissue such as neurotransmitters and myelin- and transmitter-related enzymes are most likely to be the targets of agents that are selectively neurotoxic. Other parameters of brain metabolism that may be targets are quantitatively unique, although not confined to nerve tissue. These include the unusually high glucose and oxygen requirements of central nerve tissue. Since even a temporary lack of nutrients usually has a more devastating and irreversible effect upon the brain, neurotoxic symptoms may follow an interruption of cerebral anabolic processes caused by a nonspecific toxic process such as interruption of normal cell respiration by carbon monoxide or cyanide.

Relatively nonspecific chemicals, such as alcohol, can cause alterations in activity of some, but not all receptors (Tabakoff et al., 1979; Karobath et al., 1980; Ticku and Burch, 1980). Toxic effects on receptors have to be mediated by endocrine or other events, since receptors have been shown to be responsive to a wide range of parameters. These include seizures (Gillespie et al., 1980), environmental and chemical stress (Uchida et al., 1978; Braestrup et al., 1979a; Frey et al., 1987), endocrine influences (Wilkinson et al., 1979), and learning experiences (Rose et al., 1980). The responsiveness and behavior of receptors in aged animals and humans has frequently been found to be abnormal (Maggi et al., 1979). These receptor changes may be the cause rather than the consequence of many observed behavioral changes that follow chemical or physiological modulation of the nervous system (Gee et al., 1979). Receptor alterations may have subtle as well as gross behavioral sequelae. The benzodiazepine receptor, for example, seems to play a role in emotional behavior (Robertson, 1979).

UTILITY OF RECEPTOR STUDY

The addition of receptor binding studies to the classical biochemical tools used in neurotoxicological investigation has several attractive features:

1. Receptor binding levels can respond rapidly to physiological change. The benzodiazepine receptor has been shown to be altered within 15 min of seizure induction or within 20 min after commencement of sleep (Paul and Skolnick, 1978; Poddar et al., 1980). This receptor is also rapidly responsive to diazepam administration (Speth et al., 1979). The striatal muscarinic receptor is altered within 1 hour of morphine, atropine, or amphetamine administration (Pelham and Munsat, 1979).
2. Levels of binding activity can change in the absence of detectable change in the concentration of the corresponding neurotransmitter. Thus, chronic phenothiazine administration causes an increased level of dopamine receptors, whereas dopamine tissue concentrations are not altered (Creese et al., 1978).
3. Binding activity can be modified in response to a variety of altered neuronal metabolic events. Thus, chemicals that alter firing rates or rates of transmitter release or reuptake from the synaptic cleft are likely to cause change in transmitter binding. Other compounds that block enzymes for neurotransmitter synthesis or breakdown can also cause such changes. For example, the inhibition of acetylcholinesterase will specifically reduce the binding of a muscarinic agonist (quinuclidinyl benzilate) to brain membranes (Ali et al., 1984). This effect is presumably mediated by an elevation of cerebral acetylcholine levels. Similarly, the depression of β-adrenergic receptor capacity by tricyclic antidepressants is the result of impaired catecholamine reuptake mechanisms, leading to elevated norepinephrine levels within the synaptic cleft (Banerjee et al., 1977). Subsensitivity to norepi-

nephrine has been postulated as a cause of depression (Stone, 1979).
The above instances of subsensitivity can be paralleled by examples
of supersensitivity induction, such as the elevation of catecholamine
binding found in reserpinized animals with depleted levels of catecholamines (U'Pritchard and Snyder, 1978; Buelke-Sam et al., 1989).
Neuronal death can also result in supersensitivity (i.e., enhanced
transmitter receptor activity) of the denervated areas.

4. Although the exact locus and mechanism of many binding interactions
is unknown, this does not preclude use of this rapid and relatively
simple technique. By assay of a series of labeled ligand receptor
interactions, an objective screen for deranged neuronal circuits can
be easily performed, and defects in a single brain region or specific
transmitter class may be detected. Such a preliminary survey can
then be followed by more detailed evaluation of those transmitters
whose binding sites are disturbed by a toxic agent. This kind of broad
screen has been carried out in some neurological disorders, such as
Alzheimer's disease and Huntington's disease (Enna et al., 1976;
Reisine et al., 1978), and also in psychiatric disorders (Bennett et
al., 1979).

5. Neurotransmitter binding sites are relatively stable, and this offers
opportunity for postmortem analysis of tissue (Camus et al., 1986).
However, some receptors have been reported to change their properties soon after death (Tunnicliff and Matheson, 1980).

6. Very specific chemical and regional changes can be detected by
receptor studies. Thus, the neonatal administration of 6-hydroxydo-
pamine depresses β-adrenergic binding in the cerebellum, while the
same sites are elevated in the cerebral cortex (Jonsson and Hallman,
1978; Harden et al., 1979). Electroconvulsive shock treatment decreases β-adrenergic but not α-adrenergic sensitivity in rat brain
(Pandey et al., 1979). Even the α_1 and α_2 and β_1 and β_2 subclasses
of adrenergic receptor are independently regulated and can be sepa-
rately assayed (Hoffmann et al., 1979; Minneman et al., 1979).

7. Behavioral changes can sometimes be precisley correlated with alterations in the levels of receptor binding (Lippa et al., 1979; Ali et al.,
1986a).

Potential Limitations of Receptor-Ligand Interactions

Since neurotransmitter binding is a relatively new area of research, there
are several fundamental issues that are currently unresolved. It is important
to be aware of these, and to realize that fresh findings from the areas of
basic investigation may necessitate reinterpretation or redesign of neurotox-
icologically related receptor research.

1. The issue of whether receptors exist in interchangeable agonist and antagonist forms (Burt et al., 1976) is not clear, since in the case of dopamine, these two receptor types have been partially separated by centrifugal methods (Titeler et al., 1978) and exhibit different thermolability (Lew and Goldstein, 1979).

   Receptors for most transmitters are very heterogeneous. Thus, each transmitter has several types of receptor protein which can be divided into classes and then subclasses. For example, several types of muscarinic acetylcholine receptors exist (Bridsall et al., 1978). The location of some of these classes may be presynaptic (Szerb et al., 1979) or within non-synapse-containing areas of the neuron (Marquis et al., 1977). At least two dopamine agonist receptors exist which have a heterogenous regional distribution and differing sensitivity to pharmacological agents (Thal et al., 1978). This may be significant in distinguishing the role of dopamine in movement control from its psychic-related role in modulation of effect. Pre- and postsynaptic dopamine receptors have been reported (Schwarcz et al., 1978). Similarly, presynaptic GABA receptors exist (Mitchell and Martin, 1978). Presynaptic adaptations to change in neuronal activity may be in an opposite direction to corresponding responses to postsynaptic receptors (Lee et al., 1983).

2. The diversity of receptor populations often reflects differences in receptor location. Receptors can be found at different neuronal loci, as discussed above, but there is also evidence of nonneuronal localization of receptors. Serotonin receptor levels may decline during maturation at a time when neuronal differentiation is increasing (Uzbekov et al., 1979). Acetylcholine receptors are present in the retina before synapses appear (Vogel and Nirenberg, 1976), and the benzodiazepine receptor is also found in nerve tissue before synaptogenesis (Candy and Martin, 1979). This latter receptor is reduced in the cerebellum of a mutant mouse strain lacking Purkinje cells (Lippa et al., 1979), suggesting a neuronal location. However, the receptors are also present in glial cell lines (Henn and Henke, 1978). Multiple benzodiazepine receptors with various morphological loci exist (Klepner et al., 1979). The nonneuronal location of receptors may be an especially significant feature during ontogenesis when the classical synaptic receptors have not yet differentiated. This may bear some relation to the finding that the behavioral and transmitter binding responses of adult and immature animals differ after treatment with pharmacological (Rosengarten and Friedhoff, 1979) or toxic agents (Agrawal et al., 1981a). The anxiolytic effect of benzodiazepines is probably exerted cortically, whereas their anticonvulsive effects may be exerted in other brain regions (Lippa et al., 1979).

3. This diversity of receptor populations mentioned above can make the interpretation of binding data complicated. The specificity of the radioactive materials used in such studies toward the appropriate receptors within the cell membrane is not absolute. Frequently, more than one receptor species will complex with a given pharmacological ligand. Thus, spiroperidol, which is often assumed to bind solely to dopamine receptors, may attach itself to serotonin receptors, especially in regions where there are relatively few dopaminergic terminals, such as the cerebral cortex (Leysen et al., 1978).

4. There exist within the brain both inhibitory and stimulatory factors that can influence binding. GABA will enhance benzodiazepine binding, whereas an inhibitor protein also exists in nerve tissue (Toffano et al., 1978). The regulation of benzodiazepine binding by GABA suggests that the two receptor species may be coupled (Braestrup et al., 1979b). Uncharacterized inhibitory materials that alter muscarinic (Acton et al., 1979) and dopaminergic (Lahti et al., 1978) binding have also been reported. Guanine nucleotides are known to regulate dopamine binding (Creese et al., 1979). Any alteration in the levels of such modulatory compounds by toxic agents may modify the receptor profile and behavioral responses of an animal.

5. Neurotransmitter species interact in a complex manner so that disturbance of one circuit will ultimately affect a wider range of transmitters and regions. A particular transmitter may have a regulatory effect upon a synapse of another transmitter. Both glutamate and GABA have been reported to stimulate the synaptic release of dopamine (Roberts and Sharif, 1978), whereas endorphins reduce catecholamine and acetylcholine release (Pollard et al., 1977). Cholinergic receptors exist presynaptically on dopaminergic terminals (DeBelleroche et al., 1979). All of these presynaptic modulations presumably involve transmitter-binding phenomena. Presynaptic receptors often have opposite effects on postsynaptic sites, since they play an autoregulatory role in the modulation of transmitter release (Skirboll et al., 1979).

6. The correlation of binding and behavioral data has several hazards. Animals that are chronically treated with agents that alter transmitter binding levels may possess behavioral compensatory mechanisms. Thus, after repeated anticholinesterase administration, rats will in time regain largely normal responses although muscarinic receptors remain depressed (Schallert et al., 1980). Furthermore, drug tolerance can be achieved without changes in receptor properties (Braestrup et al., 1979c). Behavioral and binding results may not show a clear linear relation (Scalzo et al., 1989a,b). The maximal anxiolytic effect of benzodiazepines is achieved when only approximately 15% of the corresponding receptors are occupied (Lippa et al., 1979).

7. Many biological responses to toxic or pharmacological agents may oscillate with time, and this must be taken into account in biochemical-behavioral studies. Thus, β-adrenergic receptors are initially reduced following ethanol withdrawal. However, in time there is a rebound effect and these receptors become elevated in number (Banerjee et al., 1978). It is this latter process that appears to account for many of the behavioral sequelae of withdrawal, such as tremor, agitation, and increased blood pressure. Similarly, acutely administered amphetamine will increase the number of striatal dopamine receptors (as measured using ^{3}H-spiroperidol for assay), whereas chronically administered amphetamine decreased striatal spiroperidol binding (Howlett and Nahorski, 1979). Attempts to correlate an altered receptor profile with a behavioral change are difficult. The screening of a range of transmitter binding-site ligands may enhance the objectivity of such an approach. Receptor levels also show a circadian flux and this should be considered when designing experiments (Por and Bondy, 1981).

THE ASSAY OF RECEPTOR-LIGAND BINDING

The development of a series of binding assays which can be used for both *in vivo* and *in vitro* evaluation of neurotoxicity requires, as a first step, the concise delineation of the binding phenomenon for each ligand to be used (Yamamura et al., 1984). These basic criteria include demonstration of:

1. The specificity of the binding reaction, as judged by competition with varous pharmacological agents.
2. An appropriate regional distribution of binding sites within the brain.
3. The attainment of a reversible equilibrium of the receptor-ligand interaction.
4. The extent of binding should be proportional to the amount of membrane in the incubation tube.
5. The isotope used should remain stable throughout the incubation.
6. The binding reaction should have a relatively low dissociation constant, and the receptor sites should be saturable.
7. If possible, receptor binding should be coupled to, and correlated with, another relevant biological event. This could include stimuation of adenyl cyclase (in the case of catecholamine agonist binding) or modulation of the rate of transmitter release from an isolated synaptosomal preparation.

We have developed a receptor-ligand screening system by which a variety of binding reactions can be studied at close to optimal conditions in the same tissue preparation, using essentially parallel procedures. This allows the simultaneous assay of several receptors in an objective and nonprejudiced way. This approach is essential if one is to find a transmitter or modulator species that is selectively altered by a toxic agent. The underlying hypothesis is that some neural circuits are more sensitive than others to metabolic disturbances. This is in contrast to a specifically predictive approach by which the interaction of a toxicant with only a single transmitter species is studied.

Briefly, rat brain regions are dissected out by the method of Iversen and Glowinski (1966) homogenized in nineteen volumes 0.32 M sucrose and centrifuged (40,000 × g for 10 min). The precipitate is resuspended in the same volume of deionized water and recentrifuged (40,000 × g, 10 min). This final pellet can be stored frozen ($-40°C$) or suspended in 50 mM tris pH 7.4 for immediate use. In the case of GABA and diazepam assays, the final pellet is washed twice more in tris buffer to assure removal of endogenous inhibitory materials. This preparation consists of crude membrane. We have chosen to use these rather than purified synaptosomal membranes so that binding in tissues that may be in differing states can be compared. The degree of myelination of a tissue is likely to alter the proportion of synaptosomes recovered. Treated animals may have brains that are abnormal with respect to their water or myeline content, or may be developmentally delayed. In such cases, subcellualr fractionation may make comparison with tissues from control animals unreliable.

Binding measurement is carried out in a 1-ml incubation mixture containing 40 mM tris pH 7.4 and appropriate labeled and unlabeled pharmacological agents. The amount of tissue per tube corresponds to 5 mg original wet tissue and contains around 400 μg protein. Incubation is for 15 min at 37°C. In all cases, equilibrium is reached in this period. At the end of incubation, samples are filtered on glass fiber filters (25 mm diameter, 0.3 μm pore size) and washed rapidly three times with 5 ml of tris buffer. In the case of the strychnine assay, filters are only washed twice, since the receptor-ligand complex formed has a very rapid rate of dissocation. The stability of the isotope solution is important, and fresh isotope dilutions should be made every few days. Filters are then dried and counted in 5 ml of a scintillation mixture.

Table 1 shows the concentrations of tritiated ligand that we use. The incubations are carried out in the presence and absence of a large excess of a competing chemical whose concentrations are also presented in Table 1. The differences between these two values is taken as an index of specific binding. Specific binding constitutes over 65% of total binding in all our assays. Preliminary criteria of binding have been established for many of the above listed assays. Some of these are illustrated using the binding of ^{3}H-spiroperidol to brain membranes as an example. This interaction is proportional to the amount of membrane in the incubation tube, is stereospecific

TABLE 1.
Pharmacological Agents Used in Neurotransmitter Receptor Binding Studies

Labeled Ligand	Specific Activity (Ci/mmole)	Conc. (nM)	Unlabeled Competitor	Conc. (mM)	Incubation Condition	Neurotransmitter Receptor Species Assayed
DL-[Benzilic-4,4'-³H]-Quinuclidinyl Benzilate	30.1	1.0	Atropine	1.0	60 min/37°C	Muscarinic Cholinergic
[1-Phenyl-4,-³H]-Spiroperidol	24.5	1.0	(+)Butaclamol	1.0	20 min/37°C	Dopamine
[Methylene-³H(N)]-Muscimol	20.6	2.0	GABA	2.0	30 min/4°C	GABA
[Methyl-³H]-Diazepam	83.0	0.75	Diazepam/Flunitrazepam	1.0	45 min/4°C	Benzodiazepine
[1,2-³H(N)]-Serotonin	26.3	3.0	Serotonin	1.0	15 min/37°C	Serotonin
[G-³H]-Strychnine Sulfate	15.0	4.0	Strychnine	1.0	20 min/4°C	Glycine
9-10[9,10-³H(N)]-Dihydro-α-Ergocryptine	23.0	1.3	Ergocryptine	1.0	30 min/25°C	α-adrenergic
Levo-[proply-1,2,3-³H]-Dihydroalprenolol	47.5	0.7	Alprenolol	1.0	30 min/25°C	β-adrenergic
[N-allyl-2,3-³H]-Naloxone	50.2	1.0	Levalorphan	1.0	45 min/25°C	Opiate
Opiate Subtype						
[Tyrosyl-3,5-³H(N)]-DAGO	40.1	1.0	Etorphine	1.0	45 min/25°C	Mu
[9-³H(N)]-Bremazocine	33.3	1.0	Etorphine	1.0	45 min/25°C	Kappa
[Tyrosyl-3,5-³H(N)]-Enkephalin	30.5	1.0	Etorphine	1.0	45 min/25°C	Delta
[Piperidyl-3,4-³H(N)]-TCP	42.2	5.0	Phencylidine	5.0	45 min/25°C	Sigma (PCP)

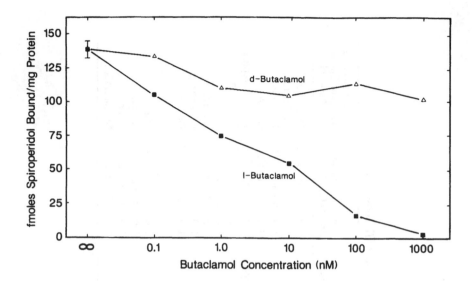

Figure 1. Binding of ³H-spiroperidol to rat brain striatal membranes; inhibition by varying amounts of d- and l-butaclamol.

with respect to competition with the isomers of butaclamol (Figure 1), and reaches a relatively rapid equilibrium which is reversible (Figure 2). Binding is heterogeneous within brain regions, being highest in the striatum, a region rich in dopaminergic terminals (Figure 3). As newer ligands with narrower specificity become available, the series of compounds listed requires continual modification.

The use of pharmacological agents rather than endogenous biological molecular species has several advantages.

1. These synthetic chemicals often have a higher affinity for the receptor site than the endogenous substance. This frequently accounts for their pharmacological efficacy.
2. The synthetic chemicals are generally less likely to be degraded by catabolic enzymes, such as monoamine oxidase, and are also less prone to oxidation than are catecholamines and indolamines.
3. The specificity of pharmacological agents can be much greater than that of neurotransmitters. Thus, norepinephrine will bind to α_1, α_2, β_1, and β_2 adrenergic sites, although these can be distinguished by use of drugs. Homogenization of a tissue destroys the normal anatomical separation of sites, but careful design of incubation conditions will allow partial analysis of subclasses of receptors.

The specificity of binding reactions can be improved by selection of brain regions. Thus, ³H-spiroperidol binding in striatal membranes is largely

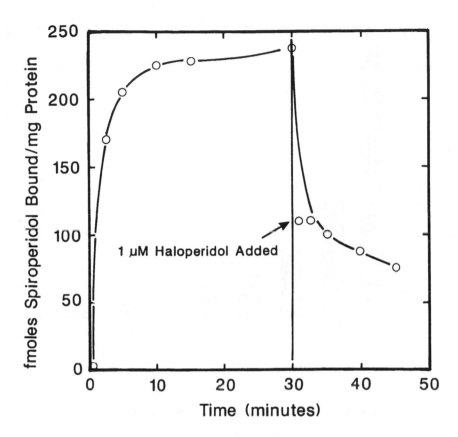

Figure 2. Time course of ³H-spiroperidol binding to striatal membranes and effect of addition of 10⁻⁶M haloperidol.

to dopamine D2 receptors, but in frontal cortex serotinin receptors are largely involved (Leysen et al., 1978). Another means of enhancing specificity is by utilization of appropriate unlabeled agents. For example, ³H-spiroperidol in the presence of 1 μ*M* ketanserin, will largely bind only to dopamine receptors and conversely the presence of 10 μ*M*(⁻) sulpiride will ensure a predominance of interaction of the labeled ligand with serotonin receptors (Altar et al., 1985).

Scatchard Analysis

The Scatchard plot (Scatchard, 1949) is one of the most popular methods to represent the graphical analysis of radioligand receptor binding studies. With the increasing number of neurotransmitter as well as nonneurotransmitter receptor binding studies in neurosciences, it is very important to use this method correctly or to analyze and interpret the data appropriately.

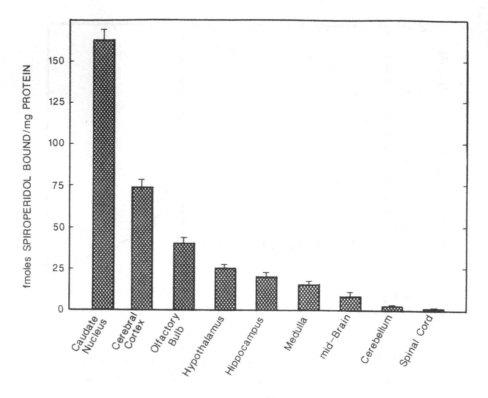

Figure 3. ³H-spiroperidol binding to various brain regions of the rat.

Usually, when the set of binding data are plotted on a Scatchard graph there is an enormous temptation to fit all the points to a straight line, either by eye or by least squares methods, so that the number of receptor sites (Bmax) and binding affinity (K_D) can be calculated. But, if we take the same data from the Scatchard plot and replot the same values on a semilogarithmic graph, in most of the cases, one finds that the conclusion derived from the Scatchard analysis is not really valid due to the fact that the saturation curve is not complete.

1. In a receptor binding experiment, before one goes to the Scatchard analysis, it may be convenient to use a single point assay at a saturation concentration (about 10 times K_D) of ligand. The results from this experiment will reveal whether there are any changes in receptor binding consequent to treatment with a chemical agent. This is the case whether Bmax or K_D are altered. Practically speaking, this type of experiment does not require a major commitment of resources as six independent samples can readily be performed in triplicate in a day.

2. Some investigators may prefer to go directly to a complete Scatchard analysis. In this case, one should always use a wide range of concentrations of ligand. When the same data are used to plot a saturation curve, the shape of the curve should be S-shaped. Although from a practical point of view it is very hard to cover a complete range of concentrations, it is important to at least achieve a saturation plateau.

3. Investigators should successively repeat the entire binding curve so that independent Bmax and K_D values between replicates can be compared. However, these values are derivative and not as reliable as the simple saturation curve which makes no claim concerning receptor number or binding affinity.

4. In a receptor binding experiment, one should always use approximately the same amount of protein in all groups of membrane preparation. Using different concentrations of protein may change the shape of the curve and the values of Bmax and K_D (Figure 4) (Ali and Peck, 1985).

5. There is a lot of controversy for using the Scatchard analysis and calculation of Bmax and K_D by Scatchard plot (Munson and Rodbard, 1983; Klotz, 1983; Laduron, 1983; Burgisser, 1984). In the field of neurobiology, the Munson and Rodbard (1980) method for computer extraction of Bmax and K_D is well accepted. In our laboratory we routinely used the "ligand" program developed by Munson and Rodbard (1980) to analyze receptor binding data. However, the attempt to fit such plots into one or more straight lines may be thwarted by the multiplicity and heterogeneity of binding sites.

Autoradiography and Image Analysis

Autoradiography is a technique by which one can study the localization of a specific radioligand in the central nervous system. In this way, one can visualize defined binding sites in various brain regions. Autoradiography has the advantage over techniques involving homogenization in that the precise anatomy of the CNS is preserved allowing a much clearer delineation of binding sites. On the other hand, this technique has some disadvantages. It is slower and technically more difficult to use compared to other techniques and computerized quantitation of grain density is rather expensive.

The autoradiographic visualization of receptors requires selective labeling and subsequent localization of receptor molecules with a radioactive ligand. The method is an indirect one, since it is the ligand which is being detected, and not the receptor. It is therefore important to develop methods whereby a high degree of selective labeling of receptors is obtained, which is then maintained throughout autoradiography. Selection of an appropriate ligand is probably the most important step in the entire procedure of autoradiography. For example, quinuclidinyl benzilate is one of the best known

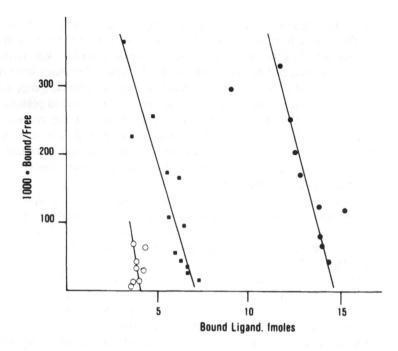

Figure 4. Scatchard analysis of ^{3}H-spiroperidol binding to various concentrations of membranes from anterior pituitary. Saturation analyses were performed on three dilutions of these purified membranes, approximately 25 (open circles), 50 (closed squares), and 100 (closed circles) µg of membrane protein per assay tube.

ligands for muscarinic cholinergic receptors and this possesses high affinity binding properties and a high specific to nonspecific binding ratio. This ligand has been shown to display saturable binding to receptors in the CNS with a K_D in the subnanomolar range.

Image processing is a computer based video image processing system which can quantify the binding of the different neurotransmitter receptors. Previously, the computer assisted analysis of ^{3}H-ligand autoradiography (Herkenham and Pert, 1982; Palacios et al., 1981) used a scanning densitometer based image processor originally developed for the analysis of [^{14}C]-2-deoxyglucose autoradiography (Goochee et al., 1980). But today, the autoradiograph can be recorded by a video camera and image digitizer that can simultaneously display several images from a multiple memory plane. The image processing system can store or continuously compare multiple images (Walter and Berns, 1981). Recently, Altar and coworkers (1984, 1985) and Geary et al. (1985) reported a new method to quantify the image analysis system. Using this method, the digitized autoradiograph image is adjusted for geometric distortion and unevenness of background illumination resulting from the light source or video camera. By using a calibration curve generated from ^{3}H-containing standards, the digitized image intensity (gray value) of each picture element (pixel) is converted to a new gray value that is a linear function of the quantity of bound ^{3}H-ligand. After this linearization step, images of specific and percent specific binding are generated mathmatically from superimposed autoradiography images of brain sections incubated with or without a drug that displaces the radioligand from the binding site(s) of interest. The perception of variation in ^{3}H-ligand binding in discrete regions is increased by converting the gray values to color using a pseudocolor transformation routine. The image processing system consists of a video digitizer, image processor and microprocessor.

To produce quantitative images of total, nonspecific, specific, and percent specific ^{3}H-ligand binding, uncorrected digital autoradiographic images have to be modified by the image processor, by a procedure described by Altar et al. (1984). In brief, one has to take the following steps:

1. Correction for shading distortion in the video camera (called the "ratio" program)
2. Correction for geometric distortion (called the "Y-magnification" program)
3. Removing the area surrounding the section (called the "mask" program)
4. Conversion of gray values to amount of ^{3}H-ligand binding (called the "linearization" program)
5. Superimposition of adjacent sections and image subtraction
6. Image division
7. Quantification of ^{3}H-ligand bound in brain areas
8. Histogram analysis
9. Contrast enhancement
10. Photography

By this means, the computer assisted image processing method facilitates the quantification of ^{3}H-ligand concentration in autoradiographs in which total or nonspecific binding is represented, and permits the derivation of images of specific and percent specific binding.

The results derived from image processing system can be misleading for the following reasons.

1. The image entered in the digitizer can be manipulated before subtraction from the nonspecific binding to get the specific or percent of specific binding.
2. The image can be adjusted for darkness or lightness before the calculation utilizing the standards. Results from such an adjustment can be inappropriate.
3. Statistical problem: from one rat one can produce at least 40 to 60 slides of coronal sections. How many separate animals are necessary to draw a conclusion from a study.

THE EFFECTS OF NEUROTOXICANTS UPON HIGH AFFINITY BINDING INTERACTIONS

Binding studies can be performed by screening of receptor sites within dosed animals or by measurement of the effects of toxicants added directly to the incubation tube. The former approach permits the discovery of secondary, indirect effects of environmental situation. In the case of cadmium, *in vitro* binding measurements have been correlated with receptor changes following *in vivo* dosing studies (Hedlund et al., 1979). However in most situations, receptor changes reflect altered neuronal activity rather than a direct effect of a toxic agent on the neurotransmitter binding site.

Few reports exist on the effect of neurotoxic agents on transmitter binding. The nicotinic acetylcholine receptor has been suggested as the site of action of the insecticide 2-isothiocyanatoethyl-trimethyl ammonium iodide, since this agent blocks α-bungarotoxin binding (Gepner et al., 1978). The effect of various heavy metals on the muscarinic acetylcholine receptor has also been described (Aronstam and Eldefrawi, 1979).

Pyridinium oximes are dramatically effective antidotes to organophosphate poisons such as soman, which are powerful anticholinesterases. The mechanism of this therapy appears to involve the interaction of the oxime with the muscarinic acetylcholine receptor. This direct receptor block reduces the cholinergic hyperactivity caused by failure of acetylcholine catabolism (Amitai et al., 1980).

Other receptor studies have been performed following treatment of animals with toluene (Celani et al., 1983), p-chloromercuribenzoate (Birdsall et al., 1978), methyl mercury (Corda et al., 1981), DDT (Eriksson et al., 1984), various insecticides (Lawrence and Casida, 1984), and lead (Lucchi et al., 1981). These investigations have generally been confined to a single receptor species in order to confirm a hypothesis by deduction, rather than develop one by induction.

Triorthocresyl Phosphate

An initial illustration of the value of receptor analysis will be made using two neurotoxic agents where the vulnerable neuronal species are already well understood.

Tri-ortho-cresyl phosphate is known to have an inhibitory effect on acetylcholinesterase and thus causes cholinergic hyperactivity (Abou-Donia, 1981). Avian species are especially vulnerable to this industrial product and we have assayed receptor binding intensity in the forebrains of TOCP-treated hens (Ali et al., 1984). The muscarinic cholinergic receptor binding capacity was depressed in treated hens while no change was detected in six other receptor species (Table 2).

This implied that the firing rate of the muscarinic cholinergic neurons was elevated or that the persistance of acetylcholine in the synaptic cleft was

TABLE 2.
High Affinity Binding in Forebrains of Hens
Exposed to TOCP

Receptor	Control	TOCP
Muscarinic, cholinergic	147.0 ± 8.0	$116.0^a \pm 4.0$
GABA	21.2 ± 3.9	23.0 ± 1.5
Dopamine	65.0 ± 5.6	71.6 ± 8.4
Benzodiazepine	14.2 ± 0.8	11.4 ± 1.0
Serotonin	35.1 ± 3.9	31.5 ± 1.0
β-Adrenergic	14.7 ± 3.4	12.6 ± 1.6
Glycine	55.3 ± 7.0	56.9 ± 4.4

Note: Receptor binding expressed as pmole ligand bound/g
protein $\pm$ S.E.

[a]$p < 0.01$ that value differs from control value. TOCP dose
was 750 mg/kg body weight administered once orally 21 d
prior to assay.

Adapted from Ali et al., 1984.

enhanced due to a reduced rate of catabolism. If less were known about
TOCP, such data could lead to a closer examination as to what the cause of
such cholinergic hyperactivity might be.

Manganese

Another neurotoxic agent whose locus of damage is somewhat understood is manganese. This metal appears to selectively damage dopaminergic
circuitry of treated animals (Donaldson et al., 1981). Exposure of humans
to excessive amounts of this element results initially in a reversible, schiz-
ophrenic-like state and later in a permanent Parkinsonian-like condition (Cook
et al., 1974). This suggests an initially high level of dopaminergic activity
followed by death of dopamine neurons (Cotzias et al., 1974). Various
receptor species were examined in brain regions of rats repeatedly dosed with
manganese (Seth et al., 1981a). Striatal spiroperidol binding was elevated in
exposed rats and this was attributed to damaged dopamine neurons. Cerebellar GABA receptors showed increased binding capacity, perhaps reflecting
activation of inhibitory neurons in response to reduced dopaminergic activity.
High pressure liquid chromatography of serotonin, dopamine, and their metabolites suggested increased turnover of these monoamines in exposed rats
(Hong et al., 1984). The apparent contradiction between dopamine and
serotonin metabolite concentrations (suggesting increased neuronal activity)
and receptor binding data (suggesting decreased activity) has a parallel in
haloperidol treated rats. Under certain conditions, the dopamine receptors
can be elevated in rats while behavioral tests indicate dopaminergic hyperactivity (Fuxe et al., 1980).

There are several possible reasons that may account for apparent discrepancies of this nature:

(1) If a proportion of a neuronal group is damaged or destroyed, the surviving neurons may compensate by becoming excessively active. Thus, levels of receptors and metabolites of neurotransmitters may not always be altered in a consonant manner.

(2) The direction in which presynaptic receptor levels respond to altered rates of neuronal firing, in the case of adrenergic neurons, is in a direction opposite to that of postsynaptic responses (Lee et al., 1983). However, it is premature to state that this can be considered a general phenomenon. Presynaptic receptors are involved in the regulation of stimulus induced secretion of neurotransmitters by feedback inhibition. Therefore, their downregulation in a circuit with usually low activity may increase the presynaptic release rate of transmitters within the synaptic cleft and thus potentiates postsynaptic effects. Thus, the regulation of both pre- and postsynaptic receptors is in a direction tending to maintain homeostasis.

(3) Many biological responses are multiphasic and overlapping. Thus patients chronically treated with neuroleptics may simultaneously present tarditive dyskinesia and stereotypy (implying dopaminergic hyperactivity) and Parkinsonian symptoms (implying abnormally reduced activity of dopamine neurons) (Barnes et al., 1980). These complications make the transmitter species affected by a toxic agent more readily detected than the directionality of changes in firing rate.

Acrylamide

A series of studies conducted with acrylamide will be used as an example of the application of high affinity binding studies to neurotoxicology. This widely used industrial chemical has applications in grouting, soil stabilization and plastics production. Acrylamide is known to cause peripheral neuropathy (Schaumburg et al., 1974) and more recently has been found to damage the central nervous system (Schotman et al., 1978). Acrylamide poisoning leading to distinct CNS behavioral deficits and encephalopathy has also been reported (Igisu et al., 1975).

Many of the signs of acrylamide poisoning appear to involve excess drive on the nervous system (Auld and Bedwell, 1967). These include pupil dilation, excess salivation, distension of the urinary bladder, and tremor (Thomann et al., 1974). For this reason, we examined the effect of acrylamide on a series of CNS receptor sites including a central catecholamine system involved in motor control, the striatal dopaminergic system (Agrawal et al., 1981a–c). There was a significant increase in ^{3}H-spiroperidol bringing to

TABLE 3.
Effect of 24-Hour Exposure to Acrylamide on Cerebral Receptors

Receptor	Region	Dose (mg/kg)			
		0	25	50	100
Dopamine	Striatum	334 ± 13	413 ± 11[a]	417 ± 11[a]	481 ± 32[a]
Muscarinic-cholinergic	Striatum	527 ± 26	490 ± 25	479 ± 21	549 ± 33
Benzodiazepine	Frontal cortex	76 ± 6	79 ± 6	63 ± 5	80 ± 4
GABA	Cerebellum	640 ± 64	768 ± 48	576 ± 48	544 ± 32[a]
Glycine	Medulla	558 ± 96	636 ± 24	654 ± 42	690 ± 35[a]
5HT	Frontal cortex	66 ± 5	72 ± 5	70 ± 4	84 ± 7[a]

Note: Binding expressed as pmole/100 mg protein ± S.E.M.

[a]Differs from zero dose (p < 0.05)

Adapted from Agrawal et al., 1981b.

striatal membranes (Table 3) 24 hours after a single administration of acrylamide to 6-week old rats (Agrawal et al., 1981a). This effect showed a proportional dose-response characteristic. Analysis of binding in the presence of a range of spiroperidol concentrations by the method of Scatchard showed that the affinity of receptors from treated animals was $0.40 \times 10^{-9}M$, significantly greater than the control value of $0.56 \times 10^{-9}M$. Treated rats also exhibited a lesser increase in receptor site density (48 pmole bound/100 mg protein, as compared to control values of 42 pmole/100 mg protein).

The specificity of this increased spiroperidol binding was examined by measurement of the high affinity binding of several labeled ligands to membranes from various brain regions of control and treated aniamls (Agrawal et al., 1981b) (Table 3).

No other receptor species measured was as sensitive to disturbance by a 24-hour exposure to acrylamide as was the dopamine site. An increase in strychnine and serotonin binding was found only at the highest level of acrylamide studied, whereas the GABA, benzodiazepine, and muscarinic cholinergic receptors were not significantly altered by an acrylamide dose. The lack of effect upon the striatal muscarinic receptor demonstrated that acrylamide was not acting by an indiscriminant interference with this area.

Acrylamide at low doses effects a selective alteration in the properties of receptors specific for a single transmitter. At higher doses or with more prolonged dosing (Bondy et al., 1981), an increasing number of nerve systems appear to become involved. The value of receptor binding methodology in diagnosis of aberrant neural circuitry is thus greatest at low doses of neurotoxic agent, when the primary deficit may be isolated in the absence of secondary changes. This relatively low dose approach also has the advantage that it may more closely reflect human environmental exposures. Once the dopamine system was identified as potentially sensitive to acrylamide-induced damage, this formed the basis for more detailed studies. The acrylamide-induced elevation of dopamine binding sites was shown reversible,

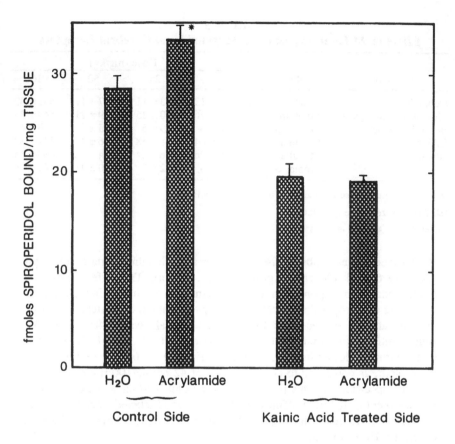

Figure 5. Effect of acrylamide administration of ³H-spiroperidol binding to rat striatum. Rats received a unilateral injection of 1 μg kainic acid. Two weeks later, acrylamide (20 mg/kg/day) was given orally for 10 days to one group of rats, the controls receiving water. Rats were killed 24 hours after the last acrylamide dose. *P < 0.05 that acrylamide value differs from corresponding control value.

implying that cell death was not involved at the doses used (Agrawal et al., 1981a). By intrastriatal injection of kainic acid, destroying intrinsic neurons but leaving synaptic termini of dopaminergic input from the substantia nigra intact, it was shown that receptor changes were attributable to postsynaptic rather than presynaptic binding sites (Hong et al., 1982). Since kainate was injected unilaterally, the contralateral striatum in these rats served as a useful control (Figure 5).

Another relevant question is whether the active agent affecting these changes is acrylamide directly or a metabolite thereof. When acrylamide catabolism was retarded by pretreatment of rats with a mixed function oxidase inhibitor, SKF 525A (2-diethylaminoethyl 2,2-diphenyl valerate), the upregulation of dopamine receptors in the striatum was blocked in acrylamide-

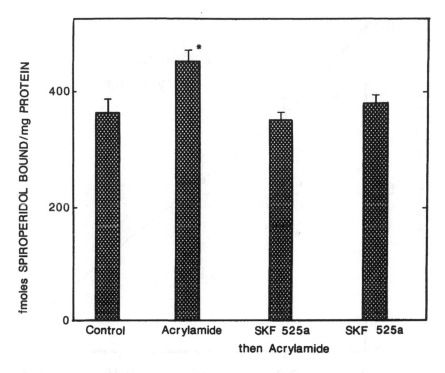

Figure 6. Effect of SKF525a upon acrylamide-induced elevation of spiroperidol binding. A single dose (100 mg/kg body weight) of acrylamide was given and binding assays were performed 24 hours later. 50 mg/kg SKF525a was injected intraperitoneally 18 and 24 hours before acrylamide dosing in some cases. *P < 0.05 that value differs from control value.

treated rats (Agrawal et al., 1981b; Figure 6). This implied that acrylamide-induced effects were mediated by a metabolite of this agent.

Delineation of a biochemical change raises the question whether this has any behavioral relevance. This has been shown to be the case with acrylamide in a study by Tilson and Squibb (1982). A dose of 12.5 mg/kg acrylamide had no effect on a lever-pressing food reinforcement schedule. However, animals treated in this way showed an unusual susceptibility to apomorphine-induced suppression of this response (Figure 7).

Apomorphine is known to act upon dopaminergic receptors and this synergism implied that acrylamide also affects dopamine circuits. A similar effect was found using d-amphetamine, but not with agents acting at non-dopamine loci, such as an α-adrenergic agonist, clonidine or a benzodiaze-pine, chlordiazepoxide. Physiologic changes induced by neurotoxic agents may be masked by homeostatic metabolic adaptations, including alterations of receptor density. However, the use of pharmacological agents with known specificity may cause a double stress to the organism and thereby elicit behavioral abnormality. This pharmacological challenge is very useful in conjunction with biochemical receptor studies (Walsh and Tilson, 1986).

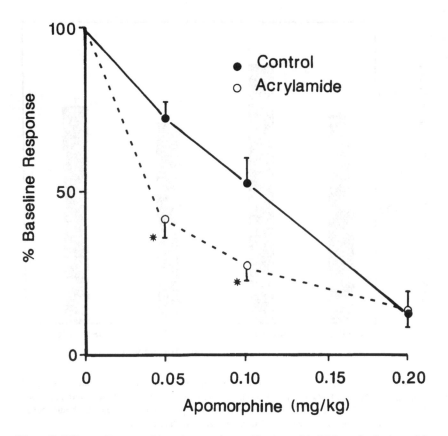

Figure 7. Effects of apomorphine alone or in combination with 12.5 mg/kg body weight acrylamide upon response to variable interval schedule reinforcement. Acrylamide dosing preceded testing by 24 hours while apomorphine was administered 16 min before testing. *P < 0.05 that value differs from corresponding group not receiving acrylamide.

    Further confirmation of the susceptibility of dopamine neurons to acrylamide came from assay of levels of catecholamine and serotonin metabolites in brain regions of acrylamide-treated rats (Ali et al., 1983). Cortical dopamine and dihydroxyphenylacetic acid levels were not altered in the striatum of treated rats but were depressed in the frontal cortex. It may be that receptor assay can in some cases be a more sensitive index of deranged circuitry than levels of neurotransmitters.

    Comparison of receptor responses in adult animals treated with toxic agents, with responses of gestationally or neonatally exposed animals provides a surprising contrast. In the case of acrylamide, gestationally treated rats show a downregulation of the striatal dopamine receptor, the opposite of the adult case (Agrawal and Squibb, 1981). Similarly, opposite responses of dopamine receptor binding have been reported for at least three other agents—chlordecone (Seth et al., 1981b), manganese (Seth and Chandra, 1984), and neuroleptics (Miller and Friedhoff, 1986a,b). It may be that a high level of presynaptic neuronal activity during ontogenesis plays an inductive role in postsynaptic receptor formation, rather than effecting receptor downregulation (Miller and Friedhoff, 1988). Developmental studies are complicated by this possibility and circuit impairment during maturation may be qualitatively different from effects on adult animals. Deficits incurred during ontogenesis may be very persistent. For example, prenatal exposure of mice to a polychlorinated biphenyl, 3,4,3',4' tetrachlorobiphenyl and reserpine to rats has been shown to decrease dopamine receptor binding capacity in the striatum of the mature animal (Agrawal et al., 1981c; Ali et al., 1986; Buelkc-Sam et al., 1989).

CONCLUSION

A promising approach in the evaluation of neurotoxicity is the screening of a range of transmitter and neuromodulator binding systems. This area of neuroscience is relatively new, and there is little consensus as to practical and interpretative issues. A bewildering diversity of radiolabeled ligands is available, and there is rapid obsolescence as more specific pharmacological agents are developed. The mechanisms underlying the physiological consequences of agonist and antagonist binding are partially understood in only a few cases. However, this method of scanning for disturbances of neural circuitry offers an opportunity to pinpoint sites of action of neurotoxicants. Disturbances of binding phenomena are likely to be detected in many toxic situations. For example, exposure to lead during development severely reduces cortical synaptic density, delayed synaptogenesis and deficits in synapses formation, while leaving cell number intact (Petit and LeBoutillier, 1979; McCauley et al., 1982; Cookman et al., 1987; Lasley and Lane, 1988). Changes in the extent of dendritic branching are also related to other, more subtle, environmental factors (Globus et al., 1973). Such anatomical changes might reflect the differential plasticity and susceptibility of certain synapses specific for a single transmitter.

The utility of this system in examination of nontransmitter binding sites should be noted. Thus, neurotoxicological studies involving analysis of calcium channels by labeled nitrendipine or the ouabain binding site of N,K ATPase can be carried out with analogous methodology (Bondy and Hall, 1984).

The purpose of this chapter is to put forward the concept of utilizing alterations in cerebral high affinity binding sites in order to detect changes in the brain caused by toxic substances. Such a preliminary survey could suggest biochemical and anatomical areas that might be candidates for more detailed metabolic studies. The advantages and limitations of this approach have been described. The modulation of receptors may be an inescapable concomitant of neurotoxicity. It is hoped that receptor studies could provide the basis for relating disturbances of cerebral function to specific metabolic derangement.

REFERENCES

Abou-Donia, M.B. (1981). Organophosphorus ester-induced delayed neurotoxicity. *Annu. Rev. Pharmacol. Toxicol.* 21: 511–548.

Acton, G., J.W. Dailey, S.W. Morris, and L. McNatt (1979). Evidence for an endogenous factor interfering with antagonist binding at the muscarinic cholinergic receptor. *Pharmacology* 58: 343–344.

Agrawal, A.K. and R.S. Squibb (1981). Effects of acrylamide during gestation on dopamine receptor binding in rat pups. *Toxicology Lett.* 7: 233–238.

Agrawal, A.K., R.E. Squibb, and S.C. Bondy (1981a). The effects of acrylamide treatment upon the dopamine receptor. *Toxicol. Appl. Pharmacol.* 58: 88–99.

Agrawal, A.K., P.K. Seth, R.E. Squibb, H.A. Tilson, L.L. Uphouse, and S.C. Bondy (1981b). Neurotransmitter receptors in brain regions of acrylamide-treated rats. I. Effects of a single exposure to acrylamide. *Pharmacol. Biochem. Behav.* 14: 527–531.

Agrawal, A.K., H.A. Tilson, and S.C. Bondy (1981c). 3,4,3'4,4' tetrachlorobiphenyl given to mice prenatally produces long-term decreases in striatal dopamine and receptor binding site in the caudate nucleus. *Toxicology Lett.* 7: 417–424.

Ali, S.F., M.B. Abou-Donia, and S.C. Bondy (1984). Modulation of avian muscarinic cholinergic high affinity binding sites by a neurotoxic organophosphate. *Neurochem. Pathol.* 2: 267–275.

Ali, S.F., J.S. Hong, W.E. Wilson, L.L. Uphouse, and S.C. Bondy (1983). Effect of acrylamide on neurotransmitter metabolism and neuropeptide levels in several brain regions and upon circulating hormones. *Arch. Toxicol.* 52: 5–43.

Ali, S.F. and E.J. Peck (1985). Modulation of anterior pituitary dopamine receptors by estradiol 17-: dose response relationship. *J. Neurosci. Res.* 13: 497–507.

Ali, S.F., J. Buelke-Sam, G.D. Newport, and W. Slikker, Jr. (1986a). Early neurobehavioral and neurochemical alterations in rats prenatally exposed to imipramine. *Neurotoxicology* 7: 365–380.

Ali, S.F., J. Buelke-Sam, and W. Slikker, Jr. (1986b). Prenatal reserpine exposure in rats decreases caudate nucleus dopamine receptor binding in female offspring. *Toxicology Lett.* 31: 195–201.

Altar, C.A., R.J. Walter, Jr., K.A. Neve, and J.F. Marshall (1984). Computer-assisted video analysis of (^{3}H)-spiroperidol binding autoradiographs. *J. Neurosci. Methods* 10: 173–188.

Altar, C.A., S. O'Neil, R.J. Walter, Jr., and J.F. Marshall (1985). Brain dopamine and serotonin receptor sites revealed by digital substraction autoradiography. *Science* 228: 597–600.

Amitai, G., Y. Kloog, P. Balderman, and M. Sokolonsky (1980). The interaction of bis-pyridinium oximes with mouse brain muscarinic receptor. *Biochem. Pharmacol.* 29: 483–488.

Aronstam, R.S. and M.E. Eldefrawi (1979). Transition and heavy metal inhibition of ligand binding to muscarinic acetylcholine receptors from rat brain. *Toxicol. Appl. Pharmacol.* 48: 489–496.

Auld, R.B. and S.F. Bedwell (1967). Peripheral neuropathy and sympathetic overactivity from industrial contact with acrylamide. *Can. Med. Assoc. J.* 96: 652.

Banerjee, S.P., L.S. Kung, S.J. Riggi, and S.K. Chanda (1977). Development of adrenergic subsensitivity by antidepressants. *Nature* 268: 455–456.

Banerjee, S.P., V.K. Sharma, and J.M. Khanna (1978). Alterations in adrenergic receptor binding during ethanol withdrawal. *Nature* 276: 407–408.

Barnes, T.R.E., T. Kidger, T. Trauer, and P. Taylor (1980). Reclassification of the tarditive dyskinesia syndrome. *Adv. Biochem. Psychopharm.* 24: 565–566.

Bennett, J.P., S.J. Enna, D.B. Bylund, J.C. Gillian, R.J. Wyatt, and S.H. Snyder (1979). Neurotransmitter receptor in frontal cortex of schizophrenics. *Arch. Gen. Psychiatry* 36: 927–934.

Birdsall, N.J.M., A.S.V. Burgen, and E.C. Hulme (1978). The binding of agonists to brain muscarinic receptors. *Mol. Pharmacol.* 14: 723–736.

Bondy, S.C., H.A. Tilson, and A.K. Agrawal (1981). Neurotransmitter receptors in brain regions of acrylamide treated rats. II. Effect of extended exposure to acrylamide. *Pharmacol. Biochem. Behav.* 4: 533–537.

Bondy, S.C. and D.L. Hall (1984). Effect of acute triethyl lead treatment on metalloenzymes and binding characteristics of rat brain hippocampus. *Neurochem. Pathol.* 2: 251–266.

Braestrup, C., M. Nielsen, E.B. Nielson, and M. Lyon (1979a). Benzodiazepine receptors in the brain as affected by different environmental stresses: the changes are small and not unidirectional. *Psychopharmacology* 65: 273–277.

Braestrup, C., M. Nielsen, P. Krogsgard-Larsen, and E. Falch (1979b). Partial agonists for brain GABA/benzodiazepine receptor complex. *Nature* 280: 331–333.

Braestrup, C., M. Nielsen, and R.F. Squires (1979c). No changes in rat benzodiazepine receptors after withdrawal from continuous treatment with Lorazepam and Diazepam. *Life Sci.* 24: 347–350.

Buelke-Sam, J., S.F. Ali, G.L. Kimmel, W. Slikker, Jr., G.D. Newport, and J.R. Harmon (1989). Postnatal function following prenatal reserpine exposure in rats: neurobehavioral toxicity. *Neurotox. Teratol.* 11: 515–522.

Burgisser, E. (1984). Radioligand receptor binding studies: what's wrong with Scatchard analysis. *Trends Pharmacol. Sci.* 5: 142–144.

Burt, D.R., I. Creese, and S.H. Snyder (1976). Properties of ^{3}H-haloperidol and ^{3}H-dopamine binding associated with dopamine receptors in the calf brain membranes. *Mol. Pharmacol.* 12: 800–812.

Camus, A., F. Javoy-Agid, A. Dubois and B. Seatton (1986). Autoradiographic localization and quantification of dopamine D_2 receptors in normal human brain with [^{3}H]-N-n-propylnoromorphine. *Brain Res.* 375: 135–149.

Candy, J.M. and I.L. Martin (1979). The postnatal development of the benzodiazepine receptor in the cerebral cortex and cerebellum of the rat. *J. Neurochem.* 32: 655–658.

Celani, M.F., K. Fuxe, L.F. Agnati, K. Andersson, T. Hansson, J.A. Gustavsson, N. Battistini, and P. Eneroth (1983). Effects of subacute treatment with toluene on central monoamine receptor in the rat. *Toxicology Lett.* 17: 275–281.

Cook, D.G., S. Fahn, and K.A. Brait (1974). Chronic manganese intoxication. *Arch. Neurol.* 30: 59–64.

Cookman, G.R., W. King, and C.M. Regan (1987). Chronic low-level lead exposure impairs embryonic to adult conversion of the neural cell adhesion molecule. *J. Neurochem.* 49: 399–403.

Corda, M.G., A. Concas, Z. Rossetti, P. Guarneri, F.P. Corongiu, and G. Biggio (1981). Methyl mercury enhances ³H-diazepam binding in different areas of the rat brain. *Brain Res.* 229: 264–269.

Cotzias, G.C., P.S. Papavasiliou, I. Mena, L.C. Tang, and S.T. Miller (1974). Manganese and catecholamine. *Adv. Neurol.* 5: 235–243.

Creese, I., D.R. Burt, and S.H. Snyder (1978). Biochemical actions of neuroleptic drugs: focus on the dopamine receptor. In *Handbook of Psychopharmacology 10.* L. Iversen, S.D. Iversen, and S.H. Snyder, Eds., Plenum Press, New York, pp. 37–51.

Creese, I., T.B. Usdin, and S.H. Snyder (1979). Dopamine receptor binding regulated by guanine nucleotides. *Mol. Pharmacol.* 16: 69–75.

Damstra, T.E. and S.C. Bondy (1980). Neurochemical assay systems for assessing toxicity. In *Experimental and Clinical Neurotoxicology.* P.S. Spencer and H.H. Schaumberg, Eds., Williams & Wilkins, Baltimore, pp. 820–833.

DeBelleroche, J., Y. Luqmani, and H.F. Bradford (1979). Evidence for presynaptic cholinergic receptors on dopaminergic terminals: degeneration studies with 6-hydroxydopamine. *Neuroscience Lett.* 11: 209–213.

Dewar, A.J. and B.J. Moffett (1979). Biochemical methods for evaluating neurotoxicity—a short review. *Pharmacol. Ther.* 5: 545–562.

Donaldson, J., F.S. Labella, and D. Gesser (1981). Enhanced autooxidation of dopamine as a possible basis of manganese neurotoxicity. *Neurotoxicology* 2: 53–64.

Enna, S.J., J.P. Bennett, D.B. Bylund, I. Creese, E.D. Bird, and L.I. Iversen (1976). Alterations of brain neurotransmitter receptor binding in Huntington's chorea. *Brain Res.* 116: 531–537.

Eriksson, P., Y. Falkeborn, A. Nordberg, and P. Slanina (1984). Effects of DDT on muscarine- and nicotine-like binding sites in CNS of immature and adult mice. *Toxicology Lett.* 22: 329–334.

Frey, J.M., W.S. Morgan, M.K. Ticku, and R.D. Huffman (1987). A dietary haloperidol regimen for inducing dopamine receptor supersensitivity in rats. *Pharmacol. Biochem. Behav.* 26: 661–669.

Fuxe, K., H. Hall, and C. Kohler (1979). Evidence for an exclusive localization of ³H-ADTN binding sites to postsynaptic nerve cells in the striatum of the rat. *Eur. J. Pharmacol.* 58: 515–551.

Geary, W.A. II, A.W. Toga, and G.F. Wooten (1985). Quantitative film autoradiography for tritium: methodological considerations. *Brain Res.* 337: 99–108.

Gee, K.W., M.A. Hallinger, J.F. Bowyer, and E.K. Killam (1979). Modification of the dopaminergic receptor sensitivity in rat brain after amygdaloid kindling. *Exp. Neurol.* 66: 771–777.

Gepner, J.I., L.M. Hall, and D.B. Sattelle (1978). Insect acetylcholine receptors as a site of insecticide action. *Nature* 276: 188–190.

Gillespie, D.D., D.H. Manier, and F. Sulser (1980). Electroconvulsive treatment: rapid subsensitivity of the norepinephrine receptor coupled adenylate cyclase system in brain linked to down regulation of β-adrenergic receptors. *Commun. Psychopharmacol.* 3: 191–195.

Globus, A., M.R. Rosenzweig, E.L. Bennett, and M.C. Diamond (1973). Effects of differential experience on dendritic spine counts. *J. Comp. Physiol. Psychol.* 82: 175–181.

Goochee, C., W. Rasband, and M.L. Sokoloff (1980). Computerized densitometry and color coding of (^{14}C)-deoxyglucose autoradiographs. *Ann. Neurol.* 7: 359–370.

Harden, T.K., R.B. Mailman, R.A. Mueller, and G.R. Breese (1979). Noradrenergic hyperinnervation reduces the density of β-adrenergic receptors in rat cerebellum. *Brain Res.* 166: 194–198.

Hedlund, B., M. Gamarra, and T. Bartfai (1979). Inhibition of muscarinic receptors *in vivo* by cadmium. *Brain Res.* 168: 216–218.

Henn, F.A. and D.J. Henke (1978). Cellular localization of ^{3}H-diazepam receptors. *Neuropharmacology* 17: 985–988.

Herkenham, M. and C.B. Pert (1982). Light microscopic localization of brain opiate receptors: a general autoradiographic method which preserves tissue quality. *J. Neurosci.* 2: 1129–1149.

Hoffman, B.B., A. Delean, C.L. Wood, P.D. Schocken, and R.J. Lefkowitz (1979). Alpha-adrenergic receptor subtypes: quantitative assessment by ligand binding. *Life Sci.* 24: 1739–1746.

Hong, J.S., H.A. Tilson, A.K. Agrawal, F. Karoum, and S.C. Bondy (1982). Postsynaptic location of acrylamide-induced modulation of striatal ^{3}H-spiroperidol binding. *Neurotoxicology* 3: 108–112.

Hong, J.S., C.R. Hung, P.K. Seth, G. Mason, and S.C. Bondy (1984). The result of manganese treatment of the levels of neurotransmitters, hormones, and neuropeptides: interaction of stress with such effects. *Environ. Res.* 34: 242–249.

Howlett, D.R. and S.R. Nahorski (1979). Acute and chronic amphetamine treatments modulate striatal dopamine receptor sites. *Brain Res.* 161: 173–178.

Igisu, H., I. Goto, Y. Kawamura, M. Kato, and K. Izumi (1975). Acrylamide encephalopathy due to well water polution. *J. Neurol. Neurosurg. Psychiatry* 38: 581–584.

Iversen, L. L. and J. Glowinski (1966). Regional studies of catecholamines in the rat brain. I. The disposition of ^{3}H-norepinephrine, ^{3}H-dopamine and ^{3}H-dopa in various regions of the brain. *J. Neurochem.* 13: 655–669.

Jonsson, G. and H. Hallman (1978). Changes in β-receptor binding sites in rat brain after neonatal 6-hydroxydopamine treatment. *Neuroscience Lett.* 9: 27–32.

Karobath, M., J. Rogers, and F.E. Bloom (1980). Benzodiazepine receptors remain unchanged after chronic ethanol administration. *Neuropharmacology* 19: 125–128.

Klepner, C.A., A.S. Lippa, D.I. Benson, M.C. Sano, and B. Beer (1979). Resolution of two biochemically and pharmacologically distinct benzodiazepine receptors. *Pharmacol. Biochem. Behav.* 11: 457–462.

Klotz, I.M. (1983). Ligand receptor interaction: what we can and cannot learn from binding measurements. *Trends Pharmacol. Sci.* 4: 253–255.

Laduron, P. (1983). More binding, more fancy. *Trends Pharmacol. Sci.* 4: 333–335.

Lahti, R.A., D.D. Gay, and C. Barshum (1978). Inhibition of ^{3}H-spiroperidol by an endogenous material from brain. *Eur. J. Pharmacol.* 53: 117–118.

Lasley, S.M. and J.D. Lane (1988). Diminished regulation of mesolimbic dopaminergic activity in rat after chronic inorganic lead exposure. *Toxicol. Appl. Pharmacol.* 95: 474–483.

Lawrence, L.J. and J.E. Casida (1984). Interactions of lindane, toxaphene and cyclodienes with brain-specific t-butybicyclophosphorothionate receptor. *Life Sci.* 35: 171–178.

Lee, C.M., J.A. Javitch, and S.H. Snyder (1983). Recognition sites for norepinephrine uptake: regulation by neurotransmitter. *Science* 220: 626–629.

Lew, J.Y. and M. Goldstein (1979). Dopamine receptor binding to agonists and antagonists in thermal exposed membranes. *Eur. J. Pharmacol.* 55: 429–433.

Leysen, J.E., C. Niemegeers, J.P. Tollenaere, and P.M. Laduron (1978). Serotonergic component of neuroleptic receptors. *Nature* 272: 168–171.

Lippa, A.S., D. Critchett, M.C. Sano, C.A. Klepner, E.N. Greenblatt, J. Couplet, and B. Beer (1979). Benzodiazepine receptors: cellular and behavioral characteristics. *Pharmacol. Biochem. Behav.* 10: 831–843.

Lucchi, L., M. Memo, M.L. Airaghi, P.F. Spano, and M. Trabucchi (1981). Chronic lead treatment induces in rat, a specific and differential effect on dopamine receptors in different brain areas. *Brain Res.* 213: 397–404.

Maggi, A., M.J. Schmidt, B. Ghetti, and S.J. Enna (1979). Effect of aging on neurotransmitter receptor binding in rat and human brain. *Life Sci.* 24: 367–374.

Marquis, K.K., D.C. Hilt, and H.G. Mautner (1977). Direct binding studies of ^{125}I-α-bungarotoxin and ^{3}H-quinuclidinyl benzilate interaction with axon plasma membrane fragments. Evidence for nicotinic and muscarinic binding sites. *Biochem. Biophys. Res. Commun.* 78: 476–480.

McCauley, P.T., R.J. Bull, A.P. Tonti, S.D. Lutkenhoff, M.V. Meister, J.V. Doerger, and J.A. Stober (1982). The effect of prenatal and postnatal lead exposure on neonatal synaptogenesis in rat cerebral cortex. *J. Toxicol. Environ. Health* 10: 639–651.

Miller, J.C. and A.J. Friedhoff (1986a). Development of specificity and stereoselectivity of rat brain dopamine receptors. *Int. J. Dev. Neurosci.* 4: 21–26.

Miller, J.C. and A.J. Friedhoff (1986b). Prenatal neuroleptic exposure alters postnatal striatal cholinergic activity in the rat. *Dev. Neurosci.* 8: 11–116.

Miller, J.C. and A.J. Friedhoff (1988). Prenatal neurotransmitter programming of postnatal receptor function. *Prog. Brain Res.* 73: 509–522.

Minneman, K.P., M.D. Dibner, B.B. Wolfe, and P.B. Molinoff (1979). β_1 and β_2 adrenergic receptors in rat cerebral cortex are independently regulated. *Science* 204: 866–868.

Mitchell, P.R. and I.L. Martin (1978). Is GABA release modulated by presynaptic receptors. *Nature* 274: 904–905.

Munson, P.J. and D. Rodbard (1980). Ligand: a versatile computerized approach for characterization of ligand binding system. *Anal. Biochem.* 107: 220–233.

Munson, P.J. and D. Rodbard (1983). Number of receptor sites from Scatchard and Klotz graphs: a constructive critique. *Science* 220: 979–981.

Palacios, J.M., D.C. Niehoff, and M.J. Kuhar (1981). (^{3}H)-spiperone binding sites in brain: autoradiographic localization of multiple receptors. *Brain Res.* 213: 277–289.

Pandey, G.N., W.J. Heinze, B.D. Brown, and J.M. Davis (1979). Electroconvulsive shock treatment decreases β-adrenergic receptor sensitivity in rat brain. *Nature* 280: 234–235.

Paul, S.M. and P. Skolnick (1978). Rapid changes in brain benzodiazepene receptors after experimental seizures. *Science* 202: 892–894.

Pelham, R.W. and T.L. Munsat (1979). Identification of direct competition for, and indirect influences on, striatal muscarinic cholinergic receptors: *in vivo* [³H]quinuclidinyl benzilate binding in rats. *Brain Res.* 171: 473–480.

Petit, T.L. and J.C. LeBoutillier (1979). Effect of lead exposure during development on neocortical dendritic and synaptic structure. *Exp. Neurol.* 64: 482–492.

Poddar, M.K., D.A. Urquhart, and A.K. Sinha (1980). Diazepam binding in brain after sleep and wakefulness. *Brain Res.* 193: 519–528.

Por, S.B. and S.C. Bondy (1981). Regional circadian variations of acetylcholine muscarinic receptors in the rat brain. *J. Neurosci. Res.* 6: 315–318.

Pollard, H., C. Lorens-Cortes, and J.C. Schwartz (1977). Enkephalin receptors on dopaminergic neurones in rat striatum. *Nature* 268: 745–747.

Reisine, T.D., H.I. Yamamura, E.D. Bird, E. Spokes, and S.J. Enna (1978). Pre- and postsynaptic neurochemical alterations in Alzheimer's disease. *Brain Res.* 159: 477–481.

Roberts, P.J. and N.A. Sharif (1978). Effects of L-glutamate and related amino acids upon the release of [³H] dopamine from rat striatal slices. *Brain Res.* 157: 391–195.

Robertson, H.A. (1979). Benzodiazepine receptors in "emotional" and "nonemotional" mice: comparison of four strains. *Eur. J. Pharmacol.* 56: 163–166.

Rose, S.P.R., M.E. Gibbs, and J. Hambley (1980). Transient increase in forebrain muscarinic cholinergic receptor binding following passive avoidance learning in the young chick. *Neuroscience* 5: 169–172.

Rosengarten, H. and A.J. Friedhoff (1979). Enduring changes in dopamine receptor cells of pups from drug administration to pregnant and nursing rats. *Science* 203: 1133–1135.

Rubenstein, E. (1980). Disease caused by impaired communication among cells. *Sci. Am.* 242: 102–123.

Scalzo, F.M., S.F. Ali, and R.R. Holson (1989a). Behavioral effects of prenatal haloperidol exposure. *Pharmacol. Biochem. Behav.* 34: 721–725.

Scalzo, F.M., R.R. Holson, B.J. Gough, and S.F. Ali (1989b). Neurochemical effects of prenatal haloperidol exposure. *Pharmacol. Biochem. Behav.* 34: in press.

Scatchard, G. (1949). The attraction of proteins for small molecules and ions. *Ann. N.Y. Acad. Sci.* 51: 660–672.

Schallert, T., D.H. Overstreet, and H.I. Yamamura (1980). Muscarinic receptor binding and behavioral effects of atropine following chronic catecholamine depletion of acetylcholinesterase inhibition in rats. *Pharmacol. Biochem. Behav.* 13: 187–192.

Schaumburg, H.H., H.M. Wisniewski, and P.S. Spencer (1974). Ultrastructural studies of the dying back process. I. Peripheral nerve terminal and axon degeneration in systemic acrylamide intoxication. *J. Neuropath. Exp. Neurol.* 33: 260–284.

Schotman, P., L. Gipon, F.G.I. Jennekens, and W.J. Gispen (1978). Polyneuropathies and CNS protein metabolism. III. Changes in protein synthesis induced by acrylamide intoxication. *J. Neuropath. Exp. Neurol.* 37: 820–837.

Schwarcz, R., I. Creese, J.T. Coyle, and S.H. Snyder (1978). Dopamine receptors localized on cortical afferents to rat corpus striatum. *Nature* 271: 766–768.

Schwartz, J.C., J. Costentin, M.P. Martres, P. Protais, and M. Baudry (1978). Modulation of receptor mechanisms in the CNS: hyper- and hyposensitivity to catecholamines. *Neuropharmacology* 17: 665–685.

Seth, P.K., J.S. Hong, C. Kilts, and S.C. Bondy (1981a). Alteration of cerebral neurotransmitter receptor function by exposure of rats to manganese. *Toxicology Lett.* 9: 247–254.

Seth, P.K., A.K. Agrawal, and S.C. Bondy (1981b). Biochemical changes in the brain subsequent to dietary exposure of developing and mature rats to chlordecone. *Toxicol. Appl. Pharmacol.* 59: 262–267.

Seth, P.K. and S.V. Chandra (1984). Neurotransmitters and neurotransmitter receptors in developing and adult rats during manganese poisoning. *Neurotoxicology* 5: 67–76.

Skirboll, L.R., A.A. Grace, and B.S. Bunney (1979). Dopamine auto and post-synaptic receptors: electrophysiological evidence for differential sensitivity to dopamine agonists. *Science* 206: 80–82.

Snyder, S.H. (1985). Drugs and neurotransmitter receptor in the brain. In *Neurosciences*, P.H. Abelson, E. Butz, and S.H. Snyder, Eds., The American Association for the Advancement of Sciences Publication, Washington, D.C.

Speth, R.C., N. Bresolin, and H.I. Yamamura (1979). Acute diazepam administration produces rapid increases in brain benzodiazepene receptor density. *Eur. J. Pharmacol.* 59: 159–160.

Stone, E.A. (1979). Subsensitivity to norepinephrine as a link between adaptation to stress and antidepressant therapy—an hypothesis. *Res. Commun. Psychol. Psychiatr. Behav.* 4: 241–255.

Szerb, J.C., P. Hadhazy, and J.D. Dudar (1979). Release of [³H]acetylcholine from rat hippocampal slices: effect of septal lesion and of graded concentrations of muscarinic agonists and antagonists. *Brain Res.* 128: 285–291.

Tabakoff, B., M. Munoz-Marcus, and J.Z. Fields (1979). Chronic ethanol feeding produces an increase in muscarinic cholinergic mouse brain. *Life Sci.* 25: 2173–2180.

Thal, L.J., M.H. Makman, H.S. Ahn, R.K. Mishra, S.G. Horowitz, B. Dvorkin, and R. Katzman (1978). ³H-spiroperidol binding and dopamine-stimulated adenylate cyclase: evidence for multiple classes of receptors in primate brain regions. *Life Sci.* 23: 629–634.

Thomann, P., W.P. Koella, G. Krinke, H. Peterman, F. Zak, and R. Hess (1974). The assessment of peripheral neurotoxicity in dogs—comparative studies with acrylamide and clioquinol. *Agents Actions* 4: 47–53.

Ticku, M.K. and T. Burch (1980). Alterations in γ-aminobutyric acid receptor sensitivity following acute and chronic ethanol treatments. *J. Neurochem.* 34: 417–423.

Tilson, H.A. and R.E. Squibb (1982). The effects of acrylamide on the behavioral suppression produced by psychoactive agents. *Neurotoxicology* 3: 113–120.

Titeler, M., P. Seeman, and F. Hess (1978). Differential centrifugation of ³H-apomorphine and ³H-spiroperidol binding sites. *Eur. J. Pharmacol.* 51: 459–460.

Toffano, G., A. Guidotti, and E. Costa (1978). Purification of an endogenous protein inhibitor of the high affinity binding of γ-aminobutyric acid to synaptic membranes of rat brain. *Proc. Natl. Acad. Sci. U.S.A.* 75: 4024–4028.

Tunnicliff, G. and G.K. Matheson (1980). Postmortem increases in GABA receptor binding to membranes of cat central nervous system. *Can. J. Neurol. Sci.* 7: 19–21.

Uchida, S., Y. Takeyasu, Y. Naguchi, H. Yoshida, T. Hata, and T. Kita (1978). Decrease in muscarinic acetylcholine receptors in the small intestine of mice subjected to repeated cold stress. *Life Sci.* 22: 2197–2206.

U'Pritchard, D.C. and S.H. Snyder (1978). ^{3}H-catecholamine binding to α-receptors in rat brain: enhancement by reserpine. *Eur. J. Pharmacol.* 51: 145–155.

Uzbekov, M.G., S. Murphy, and S.P.R. Rose (1979). Ontogenesis of serotonin "receptors" in different regions of rat brain. *Brain Res.* 168: 195–199.

Vogel, Z. and M. Nirenberg (1976). Localization of acetylcholine receptors during synaptogenesis in retina. *Proc. Natl. Acad. Sci. U.S.A.* 73: 1806–1810.

Walsh, T.J. and H.A. Tilson (1986). The use of pharmacological challenges in neurotoxicology. In *Behavioral Toxicology,* Z. Annau, Ed., Johns Hopkins University Press, Baltimore, pp. 244–267.

Walter, R.J. and M.W. Berns (1981). Computer-assisted video microscopy: digitally processed microscope image can be produced in real time. *Proc. Natl. Acad. Sci. U.S.A.* 78: 6927–6931.

Wilkinson, M., H. Herdon, M. Pearce, and C. Wilson (1979). Radioligand binding studies on hypothalamic noradrenergic receptors during the estrous cycle or after steroid injection in ovariectomized rats. *Brain Res.* 168: 652–655.

Yamamura, H.I., S.J. Enna, and M.J. Kuhar (1984). *Neurotransmitter Receptor Binding,* Raven Press, New York.

Part
III

Mechanisms of Neurotoxicity

Section B. Electrophysiological Studies

Cellular Electrophysiology
T. Narahashi

Electromyographic Methods
R.J. Anderson

Brain Slices
P.G. Aitken

Cellular Electrophysiology

Toshio Narahashi
Department of Pharmacology
Northwestern University Medical School
Chicago, Illinois

INTRODUCTION

Since electrical signals generated by nerve and muscle cells are the primary manifestation of excitation, it is imperative to utilize electrophysiological techniques to study the mechanisms of action of toxicants on the nerve and muscle system. Excitation is associated with membrane potential changes and ionic fluxes across the membrane and occurs in a matter of milliseconds, so that conventional biochemical and isotope measurements are too slow to follow the event. Substantial developments in the field of cellular electrophysiology have made it possible to apply advanced technology to the study of neurotoxicology. This chapter gives a brief account of the basic concepts of nerve excitation and electrophysiological techniques, and a few examples of the current studies.

NERVE EXCITATION AND BASIC APPROACHES

Mechanism of Nerve Excitation

It has been well established that nerve excitation takes place as a result of changes in membrane permeabilities to various cations such as sodium, potassium and calcium. The basic theory was advanced using squid giant axons as a model (Hodgkin and Huxley, 1952d), and is applicable to many

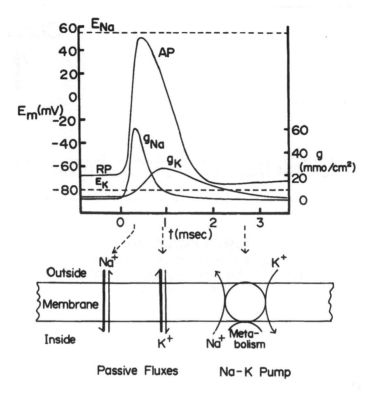

Figure 1. Mechanism of action potential generation. Upper half illustrates changes in membrane Na conductance (g_{Na}) and K conductance (g_K) during an action potential (AP). Resting potential (RP) is close to the K equilibrium potential (E_K), and the peak of the action potential approaches the Na equilibrium potential (E_{Na}). Lower half illustrates ionic fluxes during and after the action potential. See text for further explanation. E_m, membrane potential; g, conductance. (From Narahashi, 1984b.)

other excitable cells in slightly modified forms. More recent studies have clearly demonstrated that the observed changes in ionic permeabilities of the membrane are the results of opening and closing of ion channels. In the following account, squid giant axons are used as a prototype of excitable cells.

In resting conditions, the cell interior is electrically negative with respect to the cell exterior by approximately 70 mV. This is resting membrane potential (Figure 1). At rest the membrane is permeable to potassium, but not so permeable to other ions such as sodium and chloride. Potassium concentration is higher inside than outside of the cell. Thus a potential difference is created across the membrane with the inside more negative than the outside, and the value approaches the equilibrium potential for potassium (E_K) as calculated by the Nernst equation:

$$E_K = \frac{RT}{F} \ln \frac{[K]_o}{[K]_i} \tag{1}$$

where $[K]_o$ and $[K]_i$ represent the potassium concentrations (strictly speaking activities) outside and inside of the cell, respectively, and R, T and F represent the gas constant, absolute temperature, and Faraday constant, respectively. Because of slight contributions of permeabilities to sodium and other ions, the membrane potential is usually slightly less negative than the calculated potassium equilibrium potential.

When the cell is stimulated by a depolarizing pulse, the membrane sodium permeability increases rapidly, making the membrane almost exclusively permeable to sodium. Unlike potassium, sodium is usually more concentrated in the external phase than in the internal phase. Thus the membrane potential changes rapidly from a value close to the potassium equilibrium potential to a value close to the sodium equilibrium potential (E_{Na}) which takes a positive value according to the Nernst equation for sodium:

$$E_{Na} = \frac{RT}{F} \ln \frac{[Na]_o}{[Na]_i} \tag{2}$$

where $[Na]_o$ and $[Na]_i$ are sodium concentrations (strictly speaking activities) outside and inside the cell, respectively. This potential change corresponds to the rising phase of an action potential (Figure 1). The increased sodium permeability starts decreasing quickly, and the potassium permeability starts increasing beyond its resting value. Therefore, the membrane again becomes almost exclusively permeable to potassium, changing the membrane potential close to the sodium equilibrium potential to that close to the potassium equilibrium potential. This results in the falling phase of the action potential (Figure 1). The increased potassium permeability subsides as the membrane potential approaches its resting level.

Since the action potential thus generated is a result of membrane permeability changes to sodium and potassium, ionic fluxes occur. During the rising phase of the action potential while the sodium permeability is high, sodium ions enter the cell according to their electrochemical gradient. During the falling phase while the potassium permeability is predominant, potassium ions leave the cell according to their electrochemical gradient. However, the amount of ionic fluxes is very small. For instance, for an axon with a diameter of 1 μm, the changes in internal sodium and potassium concentrations as a result of one action potential are approximately 1/1000 of the respective internal concentration. Despite such small changes in internal ionic concentrations, it is imperative that some mechanism is operative to restore the imbalance. The Na-K pump is stimulated by an increase in internal sodium concentration, and excludes the extra sodium ions and absorbs potassium ions from the cell exterior to restore the ionic imbalance (Figure 1). The Na-K pump is operated by metabolic energy derived from ATP. However, it is very important to recognize that the ionic permeability changes caused by stimulation (membrane depolarization) are totally independent of metabolic energy. This is shown by the absence of the effect of metabolic inhibitors such as cyanide and azide on the resting and action potentials in giant axons despite their potent inhibitory effect on ion pump.

It should be emphasized that a variety of toxicants and drugs affect the metabolism-independent permeability mechanisms thereby causing toxic or therapeutic effects. Thus, we will have to study the mechanisms whereby these agents interact with membrane ionic permeabilities or ion channels. Since the action potential is associated with complicated ionic permeability changes as described above, we will have to measure permeability changes directly. This can be accomplished by a method called voltage clamp.

Voltage Clamp: Principle and Technique

The voltage clamp technique was originally developed by Cole (1949) and then extensively used in the study of ionic permeability changes in squid giant axons (Hodgkin and Huxley, 1952a–d; Hodgkin et al., 1952). The technique is based on two principles, i.e., the Ohm's law and space clamp. The Ohm's law calls for electric conductance being given by electric current as divided by electric potential. Thus, if we can measure the currents carried by sodium (I_{Na}) and potassium (I_K), the membrane potential (E_M), and the respective equilibrium potentials, the membrane conductances to sodium (g_{Na}) and potassium (g_K) are given by

$$g_{Na} = I_{Na}/(E_M - E_{Na}) \qquad (3)$$

$$g_K = I_K/(E_M - E_K) \qquad (4)$$

However, the complex structure of an axon causes a problem. To understand the situation, the electrical properties of an axon must be analyzed. A portion of an axon can be represented by an electrical equivalent circuit as is shown in Figure 2. The external and internal phases of the axon are represented by external resistance (r_o) and internal resistance (r_i). The membrane also has resistance (r_m). In addition, the membrane has capacity (c_m). Therefore, when a current is injected to the axon, the current and the resultant membrane potential change are distributed along the axon in a nonuniform manner (Figure 2, top). The distribution of current and membrane potential change can be made uniform if the external and internal resistances are eliminated. This can be accomplished easily by placing a large, longitudinal metal electrode outside the axon and by inserting a similar electrode inside (Figure 2, middle). This is called space clamp condition. However, this manipulation leaves the membrane capacitance intact, still causing a complex flow of current across the membrane. The capacity current can be eliminated if the membrane current is measured while keeping the membrane potential

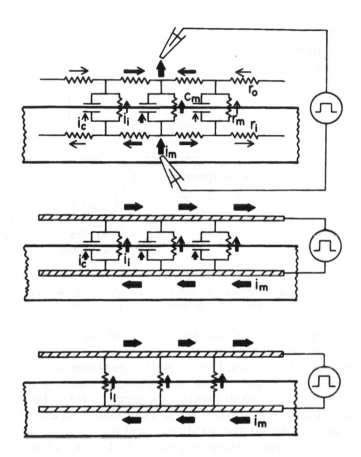

Figure 2. Current flow in an axon preparation. Top: Current is applied to the axon through internal and external microelectrodes. The membrane current and longitudinal current are not uniform along the axon. Middle: Current is applied through internal and external wire electrodes. The membrane current and longitudinal current are uniform along the axon (space clamp). Bottom: When the axon in space clamp condition is voltage clamped, no capacitive current (i_c) flows while the membrane potential is maintained at a constant level, making it possible to measure the ionic current (i_i). i_m, total membrane current; r_o, external resistance; r_i, internal resitance; r_m, membrane resistance; c_m, membrane capacity. (From Narahashi, 1971.)

constant (Figure 2, bottom). This is called voltage clamp. Thus, the space clamp is a prerequisite for the voltage clamp measurement of membrane ionic currents.

The basic principle of voltage clamp arrangement and circuitory is illustrated in Figure 3 for an internally perfused squid axon. The membrane potential is recorded by a glass capillary electrode inserted longitudinally to the axon and another glass capillary reference electrode placed outside the axon. The recorded membrane potential is fed into a control amplifier to which a command pulse generated from a pulse generator is applied. The

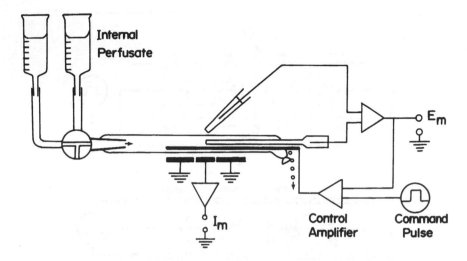

Figure 3. Voltage clamp of an internally perfused squid giant axon. See text for further explanation. E_m, membrane potential; I_m, membrane current. (From Narahashi, 1984b.)

difference between the membrane potential and the command pulse is then amplified by the control amplifier, and a current is generated from the output of the amplifier. This current flows across the membrane via a longitudial internal wire electrode and large external metal electrodes in such a way as to make the membrane potential equal to the command pulse. Thus with the aid of the voltage clamp feedback circuit, the membrane potential is maintained at the level desired (command pulse), and the membrane current (I_M) necessary for the voltage clamp can be recorded by a current-recording amplifier.

Internal Perfusion

Internal perfusion is not required but very important for voltage clamp experiments. It allows us to control the internal as well as external environment, including ionic composition, pH and test compound. For example, if sodium currents are to be measured, potassium currents can be easily eliminated by perfusing the axon externally and internally with K-free media. Certain drugs are ionized depending on the pH of the medium, and this can be controlled easily using internal perfusion techniques. Any test compound can be applied either externally or internally. This is also important, because certain chemicals act only from one side of the membrane. Therefore, internal perfusion techniques can greatly broaden the spectrum of measurements and analyses in voltage clamp experiments.

Internal perfusion techniques were originally developed for squid giant axons by two groups, Baker et al. (1961) in England and Oikawa et al.

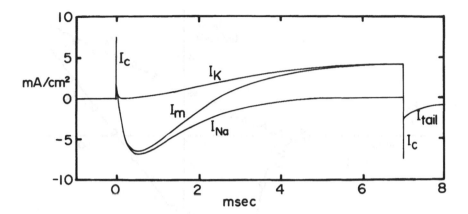

Figure 4. Membrane current (I_m) and its sodium current (I_{Na}), potassium current (I_K), and capacitive current (I_c) components associated with a step depolarization of the nerve membrane under voltage clamp condition. (From Narahashi, 1981.)

(1961) in the United States. The British group's method is based on squeezing out axoplasm using a small roller and inflating and perfusing the empty preparation with an internal perfusate. The U.S. group sucked out axoplasm by a glass capillary inserted longitudinally to the axons and perfused it with an internal perfusate. Both techniques have since been used widely. In our laboratory, the roller method was slightly modified (Narahashi, 1963; Narahashi and Anderson, 1967), and has been used routinely since then.

Membrane Ionic Currents and Conductances

Membrane currents associated with a step depolarizing pulse under voltage clamp condition are illustrated in Figure 4. A step depolarization generates a capacitive current (I_c) which is large in the initial amplitude but decays quickly. The capacitive current is followed by an inward current (downward deflection). This current attains a peak and is followed by a steady-state outward current (upward deflection). Upon termination of the depolarizing step, there appears another capacitive current which is the same as but opposite to the capacitive current that appears at the beginning of the pulse. Then a tail current (I_{tail}) follows and declines toward the original zero current level. The transient inward current is carried largely by sodium ions, while the steady state outward current is carried largely by potassium ions. Thus, the peak and steady-state currents represent, as a first approximation, sodium and potassium currents, respectively.

The peak sodium current and the steady-state potassium current can be plotted as a function of the membrane potential to construct a current-voltage (I-V) relationship (Figure 5). When the membrane is depolarized from the holding potential (E_h) to various levels, the peak inward sodium current

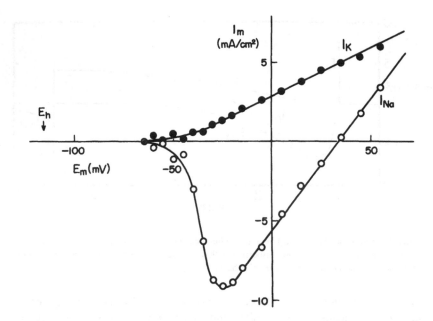

Figure 5. Current-voltage (I_m-E_m) relationships for the peak sodium current (I_{Na}) and for the steady-state potassium current (I_K) in a lobster giant axon. E_h, holding membrane potential. (From Narahashi, 1964.)

increases in amplitude with increasing depolarization up to around -25 mV. However, the amplitude of current decreases with further increasing depolarization, reverses its polarity around $+35$ mV, and becomes outward with larger depolarizations. The membrane potential where the sodium current reverses its polarity is close to the sodium equilibrium potential. The sodium conductance calculated by Equation (3) increases with increasing depolarization and attains a maximum at around -25 mV. The outward current increases with depolarization.

Modifications of Voltage Clamp

The original axial wire voltage clamp technique as applied to squid giant axons has been modified in various ways to apply to various excitable cells. Three such modifications are worth describing for the purpose of applying to neurotoxicological studies. One is the sucrose-gap voltage clamp as applied to smaller "giant" axons such as those from crayfish and lobster (80–250 μm in diameter). Second is the intracellular microelectrode voltage clamp as applied to either muscle endplates or neurons. Third is the patch clamp as applied to a variety of cell types.

The sucrose-gap voltage clamp technique was originally developed by Julian et al. (1962a,b) as applied to the giant axons of the lobster. We have

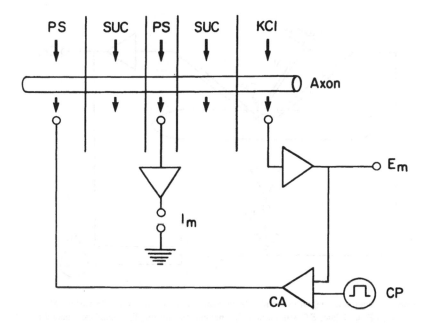

Figure 6. Schematic diagram of sucrose-gap voltage clamp of a giant axon. PS, physiological saline solution; SIC, isotonic sucrose solution; KCl, isotonic KCl solution; I_m, membrane current; E_m, membrane potential; CA, control amplifier; CP, command pulse. See text for further explanation.

been using this technique since then, and have also applied it to the internally perfused crayfish giant axon (Lund and Narahashi, 1981a). In short, a central portion of an isolated, cleaned giant axon is insulated from the rest of the axon by means of two streams of isotonic sucrose solutions (Figure 6). The central portion called "artificial node" is exposed to the physiological saline solution. One of the side pools contains the physiological saline solution, while the other contains the isotonic KCl solution to depolarize the membrane completely. All solutions are continuously flowing. The artificial node is small in width and a space clamp condition is established there. Thus, the membrane potential is measured between the artificial node and the side pool containing KCl solution, while current is injected through the other side pool containing the physiological saline solution. The sucrose-gap voltage clamp is an excellent technique, as the possible polarization of the electrodes such as that encountered in the axial wire voltage clamp experiment in which large and prolonged currents have to be dealt with. Year-round availability of crayfish and applicability of internal perfusion technique make it possible to apply to the study of various neurotoxicants.

Intracellular microelectrodes can be used for voltage clamp of endplates (Figure 7). This technique was first developed by Takeuchi and Takeuchi (1959). Two glass capillary microelectrodes are inserted in the endplate

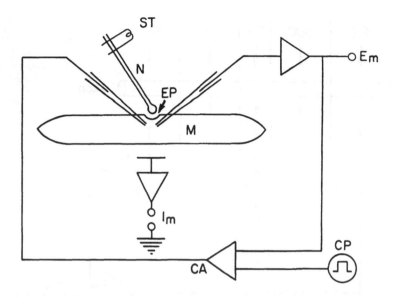

Figure 7. Schematic diagram of endplate voltage clamp by two microelectrodes. M, muscle; N, nerve; EP, endplate; ST, stimulation; I_m, membrane current; E_m, membrane potential; CA, control amplifier; CP, command pulse. See text for further explanation.

region, one for potential recording and the other for current delivery. Since the endplate is restricted, space clamp conditions are established by microelectrodes. This is an excellent technique to study the effect of toxicants on ion channels associated with receptors in the endplate region. Basically, the same technique has also been used for voltage clamp of a neuron. As long as the neuron is neither too large nor too small, this is a convenient voltage clamp technique. However, if the neuron is too large, space clamp conditions are difficult to establish. If the neuron is too small, inserting two microelectrodes would cause rupture of the cell soon. As a variation of this method, a single microelectrode voltage clamp technique was invented (Wilson and Goldner, 1975). However, this method does not give accurate current records when the time course is fast.

Patch Clamp

In 1976, Neher and Sakmann in Germany placed a glass capillary electrode onto the membrane surface to successfully record opening and closing of individual ion channels associated with acetylcholine receptors of denervated skeletal muscle. The technique was later improved to record single channel activity more clearly. The improvement was due largely to a very high seal resistance (10–100 GΩ) at the orfice of the capillary electrode which was placed onto the surface of the membrane (Hamill et al., 1981). The patch clamp technique has since been used for two purposes, one being recording of currents from the whole cell membrane and the other being recording of currents from single ion channels (Figure 8). Several references are available for detailed techniques (Hamill et al., 1981; Sakmann and Neher, 1983).

The whole cell patch clamp can accomplish essentially the same purpose as that by the axial wire voltage clamp as applied to the squid giant axon. However, the remarkable advantage of the whole cell patch clamp is that it can be applied practically to any type of cell, small cells and neurons in particular. Thus, the applicability of voltage clamp has been greatly broadened, and the patch clamp is now being used for the study of central neurons, secretory cells, lymphocytes, and even red blood cells.

The advantage of the patch clamp single channel recording technique is quite obvious. It permits recording opening and closing of individual ion channels (Figure 11). Thus, not only the gating kinetics of single channels, but also the interaction of a test compound with single channels can be analyzed in a straightforward manner. The method may be divided into two types. One is single channel recording from a portion (patch) of the intact cell, and the other is single channel recording from a membrane patch isolated from the cell (Figure 8). The single channel recording from the intact cell presents a unique opportunity to be able to examine the effect of permeant second messenger as acting on the channels from the interior of the cell, because the second messenger has access to the channels under examination only from inside.

The single channel recording from an isolation membrane patch can be performed in two different ways. Depending on the conditions, one can obtain either a patch with the original internal membrane surface facing outside (inside-out patch) or a patch with the original external surface facing outside (outside-out patch) (Figure 8). Since only the bathing medium can be changed freely and easily, one can use either of the patch depending on to which membrane side a test compound is to be applied.

MECHANISM OF ACTION OF NEUROTOXICANTS

History of Nerve Membrane Pharmacology

Whereas voltage clamp techniques were developed and applied to the basic physiological study of squid giant axons in the early 1950s (Hodgkin et al., 1952; Hodgkin and Huxley, 1952a–d), it was not until 1957 that the technique was applied to the study of drug action on nerve membranes. Tasaki and Hagiwara (1957) found that tetraethylammonium (TEA), when injected to the squid axon, blocked the steady-state potassium current leading to a prolongation of action potential. Shortly after that time, two groups studied the effects of procaine and cocaine on squid giant axons and found

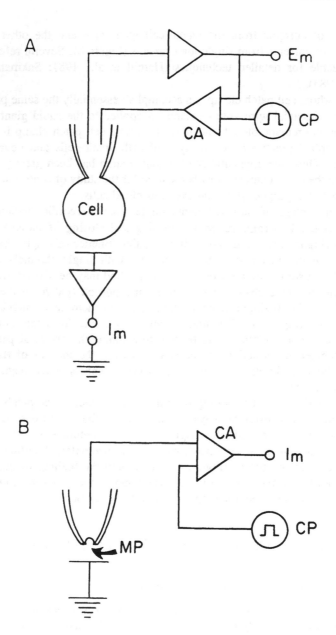

Figure 8. Schematic diagram of patch clamp. A, whole cell patch clamp; B, single channel recording from an isolated membrane patch; I_m, membrane current; E_m, membrane potential; CA, control amplifier; CP, command pulse; MP, membrane patch. See text for further explanation.

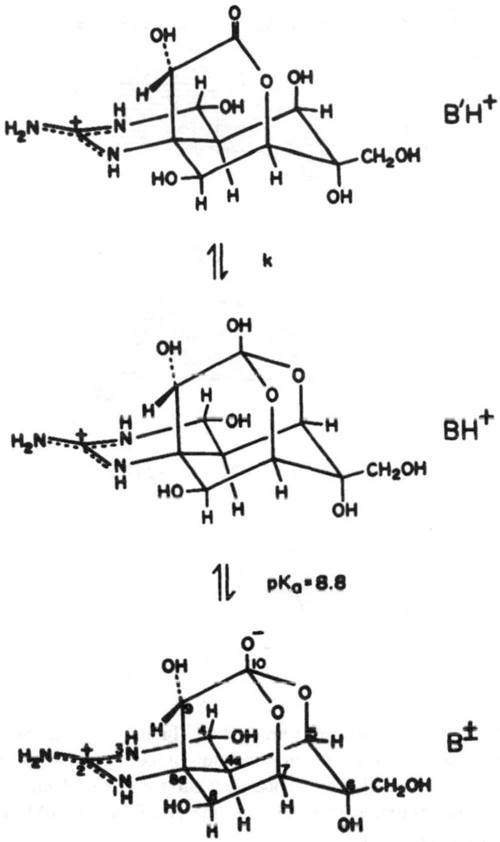

Figure 9. Structures of tetrodotoxin. (From Narahashi et al., 1969.)

that both sodium and potassium currents were inhibited (Taylor 1959; Shanes et al., 1959).

However, it was the discovery of the highly selective and potent blocking action of the puffer fish poison, tetrodotoxin (TTX) (Figure 9), on the sodium channel that triggered a widespread interest in using certain toxins as tools and in applying voltage clamp techniques to the excitable membrane pharmacology. In 1960, Narahashi et al. published a paper in which the TTX block of the sodium channel was strongly suggested based on the current clamp experiments using frog skeletal muscles. The hypothesis was later demonstrated by sucrose-gap voltage clamp experiments using the lobster giant axon (Narahashi et al., 1964). TTX blocked the sodium current without any effect on the potassium current (Figure 10).

A large number of studies have since been conducted for the purpose of elucidating the mechanisms of action of various toxins and drugs on ion channels (see reviews, Narahashi, 1974, 1984a, 1987a,b, 1988; Catterall,

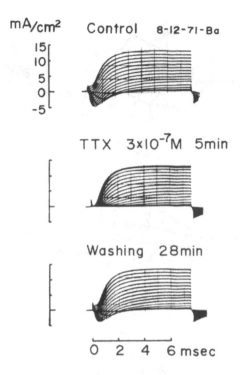

Figure 10. Block of sodium currents by tetrodotoxin. In the control, the membrane current associated with a step depolarization to various levels is composed of a transient inward or outward sodium current and a steady-state potassium current. TTX blocks the sodium currents without any effect on the potassium currents, and the effect is reversible after washing with TTX-free solution. (From Narahashi, 1975.)

1980; Ritchie, 1979). Voltage clamp experiments have now become one of the routine approaches to the study of drug action.

Classification of Drug Actions on Ion Channels

Chemicals and drugs that act on various ion channels may be classified into several groups according to the affinity for a particular channel and type of the effect. Table 1 gives some examples including voltage-dependent or voltage-activated channels such as sodium, potassium and calcium channels, and chemically-activated channels such as acetylcholine (ACh)-activated channels, glutamate-activated channels and GABA-activated channels. Drugs and chemicals are divided into blockers and modulators. The latter includes agents that modify the gating kinetics of the activation and/or inactivation mechanism.

TABLE 1.
Examples of Blockers and Modulators Acting on Voltage-Activated
and Chemically-Activated Ion Channels

	Blockers	Modulators
Voltage-Activated Channels		
Sodium	Tetrodotoxin	Batrachotoxin
	Saxitoxin	Grayanotoxin
	Local anesthetics	Veratridine
	Pancuronium	Pyrethroids
	N-alkylguanidines	DDT
		Goniopora toxin
		Sea anemone toxins
		Scorpion toxins
		Pronase
		N-bromoacetamide
Potassium	Tetraethylammonium	
	Aminopyridines	
	Cesium	
	Local anesthetics	
Calcium	Dihydropyridines	Bay K8644
	Diltiazem	
	Verapamil	
	Enkephalins	
	Phenytoin	
	Polyvalent cations	
Chemically-Activated Channels		
Acetylcholine receptor	Local anesthetics	
	Histrionicotoxin	
	Amantadine	
	N-alkylguanidines	
1-Glutamate receptor	Magnesium	
	2-amino-5-phosphonovaleric acid (2-APV)	
	γ-d-glutamylglycine	
	Barbiturates	
GABA receptor	Bicuculline	Baclofen
	Picrotoxin	Barbiturates
	Saclofen	Benzodiazepines
		Muscimol

Channel Blockers

The mechanisms of action of channel blockers have been reviewed extensively (e.g., Narahashi, 1974, 1984a,b; Ritchie, 1979; Yeh, 1982). Only a brief outline of the channel blocking mechanisms will be described here. Several questions may be asked regarding the kinetics of channel block: 1) dependence of block on membrane potential and/or membrane current, 2) block of channels in their open and/or closed configuration, and 3) dependence of block on channel activity. These three questions are related to each other, and the detailed kinetic schemes are discussed in a previous review (Narahashi, 1984a,b). In short, gating kinetics and interactions of a drug are summarized in the following scheme for the sodium channel:

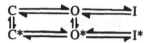

where C, O and I refer to the closed, open and inactivated states of the normal sodium channel, respectively, and C*, O* and I* refer to the closed, open but nonconducting, and inactivated states of the drug-bound channel, respectively.

Closed channel block with the activation and inactivation gating kinetics intact (C*⇌O*⇌I*) is observed in the presence of TTX or saxitoxin (STX). The gating currents associated with a step depolarizing pulse are kept unchanged after application of TTX (Armstrong, 1981; Almers, 1978), while the sodium current is completely blocked (Narahashi et al., 1964). This indicates that the sodium channel bound by TTX undergoes opening and inactivation although it is not conducting. Early in 1965, Kao and Nishiyama proposed a model in which the TTX molecule occludes the sodium channel at its exterior mouth. This model was later elaborated by Hille (1975). One-to-one stoichiometric block of the sodium channel by TTX was demonstrated by analyses of the dose-response curve (Cuervo and Adelman, 1970). Recent experiments with the aid of the single channel recording technique (Quandt et al., 1985) have clearly demonstrated that at a concentration of 3 nM, which is closed to the apparent dissociation constant, TTX does not alter the amplitude and open time of individual sodium channel currents while decreasing the probability of observing channel currents to approximately 50% (Figure 11). This indicates an all-or-none block of individual sodium channels.

Examples of open channel block can be found in many local anesthetics and other blocking agents acting on sodium and/or potassium channels. In the squid giant axon treated internally with either pancuronium or octylguanidine, the peak amplitude of sodium current is decreased and its falling phase is accelerated (Figure 12). The latter change is not due to an acceleration of the sodium inactivation kinetics, because in the axon internally perfused with pronase to destroy the sodium inactivation gate these agents still cause a rapid decline of sodium current (Figure 13). These changes in sodium current kinetics can be accounted for in terms of block of sodium channels as they start opening. Tetraethylammonium and its derivatives block the potassium channel in its open state (Armstrong, 1969, 1971). Thus, these compounds also cause an apparent inactivation of potassium current.

At the single channel level, open channel block is often manifested by an appearance of bursts while a channel is open. This has been shown in local anesthetic block of ACh-activated ion channels (Neher and Steinbach,

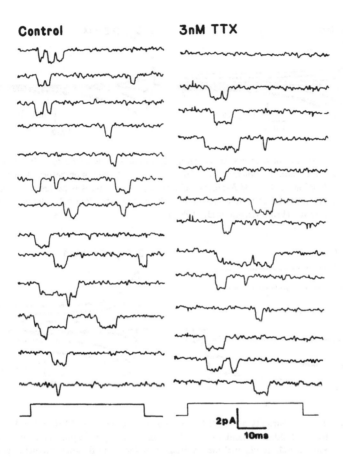

Figure 11. Effects of 3 n*M* tetrodotoxin on single sodium channel currents recorded from an outside-out membrane patch of neuroblastoma cell in response to a series of depolarizing pulses. The membrane was depolarized repetitively for 40 ms from a holding potential of −90 mV to −30 mV at 3 s intervals at 10°C. Downward steps in each trace represent inward currents due to the opening of individual sodium channels. After application of TTX, the appearance of individual currents remains normal, but many traces show no current as represented by the top trace. (From Quandt et al., 1985.)

1978). If a blocking chemical binds to and unbinds from a receptor site at a rate slow enough to be recorded, single channel current would show rapid on and off resulting in a burst. However, if the rates of binding and unbinding are too rapid to be resolved by the recording system, the net result would be a mere decrease in the single channel current amplitude as an average of open and closed states.

Channel Modulators

Most studies of ion channel modulators have been performed with the sodium channels. Perhaps the simplest form of sodium channel modulation

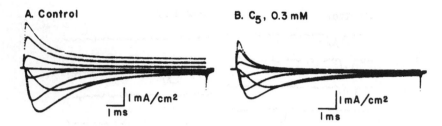

Figure 12. Sodium currents associated with depolarizing steps of 60 to 180 mV in 20 mV increments from a holding potential of −90 mV in a squid giant axon before (A) and after (B) internal application of 0.3 mM N-amylguanidine (C$_5$). The axon was perfused externally with K-free, $^1/_2$ Na artificial sea water and internally with K-free internal solution to eliminate potassium currents. (From Kirsch et al., 1980.)

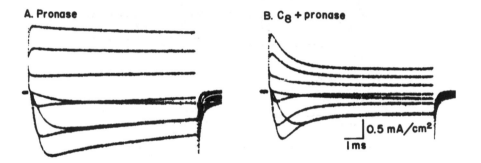

Figure 13. Sodium currents associated with depolarizing steps of 40 to 180 mV in 20 mV increments from a holding potential of −90 mV in a squid giant axon treated internally with pronase to remove sodium channel inactivation. A, before, and B, after internal application of 20 μM N-octylguanidine (C$_8$). External and internal perfusates are the same as those used for Figure 12. C$_8$ causes a time-dependent decrease of sodium currents even after destruction of the sodium inactivation mechanism by pronase. (From Kirsch et al., 1980.)

is a result of elimination of the sodium inactivation mechanism. This can be accomplished by internal perfusion with pronase which destroys the inactivation gate (Armstrong et al., 1973) or with N-bromoacetamide (NBA) which presumably alters tyrosine residues of the inactivation gate (Oxford et al., 1978). In either case, the sodium current associated with a step depolarizing pulse is maintained at a steady level (Figure 14).

The sodium channel inactivation has also been shown to be inhibited by a variety of natural toxins including scorpion toxins, sea anemone toxins, and *Goniopora* toxin. Figure 15 shows that the sodium currents associated with step depolarizations to various levels in a crayfish giant axon are greatly prolonged in time course and increased in amplitude after external application of a polypeptide toxin isolated from the coral *Goniopora* (Muramatsu et al., 1985).

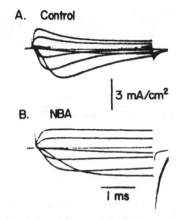

A. Control

3 mA/cm²

B. NBA

I ms

Figure 14. Elimination of sodium inactivation by *N*-bromoacetamide (NBA) in a squid giant axon. Sodium currents associated with step depolarization of 40 to 160 mV from a holding potential of − 80 mV in 20 mV increments. (From Oxford et al., 1978.)

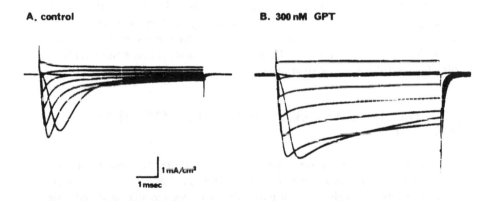

A. control   B. 300 nM GPT

1 mA/cm²
1 msec

Figure 15. Families of sodium currents associated with step depolarizations (50–190 mV, 20 mV steps) before (A) and 20 min after external application of 300 n*M* Goniopora toxin (B) in a crayfish giant axon. The holding potential was − 100 mV. (From Muramatsu et al., 1985.)

More complex modulation of the sodium channel gating has been found with batrachotoxin (BTX), grayanotoxin (GTX) and veratridine (see reviews, Narahashi, 1984a,b; Khodorov, 1978, 1985). These toxins modify not only the kinetics of the sodium channel inactivation but also those of the activation. The sodium channels can be kept open at large negative potentials, resulting in a large depolarization of the membrane.

An example of such an experiment is illustrated by Figure 16. The steady-state component of sodium current at − 10 mV is greatly increased by grayanotoxin I (GTX I). In the axon treated with GTX I, a steady-state sodium current is generated at − 70 mV where no sodium current is produced before drug treatment.

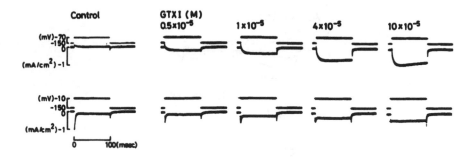

Figure 16. Sodium currents associated with step depolarizations from a holding potential of − 150 mV to − 70 mV (upper tracings) and to − 10 mV (lower tracings) in a squid giant axon before and during internal perfusion of various concentrations of grayanotoxin I. Internal perfusate contained 20 m*M* tetraethylammonium to block the potassium channels. (From Seyama and Narahashi, 1981.)

Striking changes have been observed at the single channel level as a result of BTX treatment (Quandt and Narahashi, 1982). Figure 17 shows that the open time of single sodium channels of a neuroblastoma cell is greatly prolonged after exposure to BTX. The probability of channel openings during a step depolarization follows the time course of the sodium current recorded from the whole cell, showing the presence of inactivation before BTX application and the absence of it after BTX.

MECHANISMS OF ACTION OF PYRETHROIDS AND DDT

Pyrethrins are the active ingredients contained in the flowers of *Chrysanthemum cinerariaefolium* which have been used as natural insecticides. A large number of derivatives of pyrethrins have been synthesized and tested for insecticidal activity, and some of them are widely being used as commercial insecticides. Pyrethroids and DDT are kwown to be potent neuropoisons, and their mechanisms of action on the nervous system have been studied extensively.

Pyrethrins have been found to stimulate nerve preparations to cause repetitive discharges (Lowenstein, 1942; Welsh and Gordon, 1947; Yamasaki and Ishii [Narahashi], 1952). DDT also exerts very similar effects (Welsh and Gordon, 1947; Roeder and Weiant, 1946, 1948; Shanes, 1949; Yamasaki and Ishii [Narahashi], 1952). However, it was not until 1967 that DDT and pyrethroids were subject to voltage clamp analyses for elucidating their mechanisms of action at the membrane and ion channel level (Narahashi and Haas, 1967; Narahashi and Anderson, 1967). Much of the early literature has been reviewed by Narahashi (1971, 1976). In the present chapter, only recent studies will be introduced to illustrate the most up-to-date version of the concept. The scope of the present chapter is limited to the aspects of pyrethroid and DDT actions on ion channels. Review articles have been published recently (Narahashi, 1981, 1984c, 1985, 1987b, 1988; Wouters and van der Bercken, 1978; Woolley, 1981; Ruigt, 1984).

Chemical Structures

The chemical structures of DDT, its derivatives and pyrethroids are shown in Figures 18 and 19. An oxetane derivative of p,p'-DDT, 2,2-bis-(p-ethoxyphenyl)-3,3-dimethyloxetane (EDO) (Holane, 1971), has been used experimentally as it exerts very similar effects to p,p'-DDT with a higher potency. Pyrethroids can be divided into two groups. Type I pyrethroids include a number of conventional compounds without a cyano group at the α position. Type II pyrethroids are represented by some of newer compounds and contain a cyano group at the α position.

Effects on Impulse Conduction and Synaptic Transmission

DDT causes repetitive discharges in various regions of the nervous system, including sensory neurons, nerve fibers, synapses, and neuromuscular junctions (Narahashi, 1971). In the crayfish neuromuscular preparation treated with EDO or DDT, a single stimulus applied to the presynaptic nerve evokes repetitive responses in the muscle (Farley et al., 1979). Careful inspection of the repetitive excitatory junctional potentials (EJPs) as recorded by an external microelectrode revealed that each EJP was preceded by an action potential of the presynaptic nerve terminal. Thus repetitive responses at synaptic junctions are initiated in the presynaptic element. Repetitive discharges were indeed evoked by a single stimulus in the nerve fiber treated with DDT (Narahashi, 1971).

Type I pyrethroids also cause repetitive discharges both in nerve fibers and at synapses (Narahashi, 1985; Lund and Narahashi, 1983; Takeno et al., 1977). As in the case of DDT, repetitive responses in the postsynaptic element originate in the presynaptic nerve (Evans, 1976; Wouters et al., 1977). In support of this notion, repetitive discharges are induced by a single stimulus in the nerve fiber preparation treated with type I pyrethroids (Narahashi, 1971, 1985; Lund and Narahashi, 1983).

Type II pyrethroids usually do not cause repetitive responses in nerve fibers. Instead, the membrane is gradually depolarized and impulse conduction is blocked (Lund and Narahashi, 1983).

Intracellular microelectrode experiments have revealed that repetitive after-discharges in the giant axon poisoned by DDT or type I pyrethroids are produced when the depolarizing after-potential is elevated to the threshold membrane potential for excitation (Figure 20) (Narahashi, 1971, 1976, 1985; Lund and Narahashi, 1981a, 1983). Therefore, the next question is how the

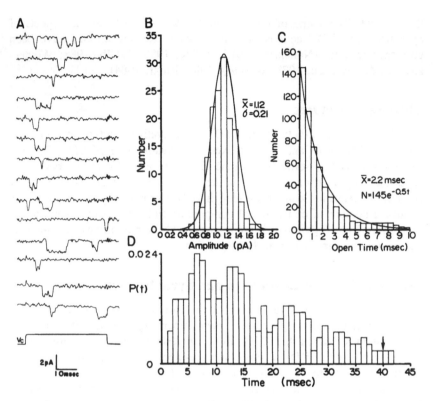

Figure 17. Single sodium channel activity of neuroblastoma cell membranes under normal conditions and in the presence of batrachotoxin. From Quandt and Narahashi, 1982. (*Continued on p. 177*)

depolarizing after-potential is elevated by these insecticides. This question can be examined by means of voltage clamp experiments.

Effects on Ion Channels

The major change in membrane ionic current of giant axons caused by DDT is a slowing of the falling phase of the sodium current (Narahashi and Haas, 1967, 1968; Wu et al., 1980; Lund and Narahashi, 1981b; Hille, 1968; Vijverberg et al., 1982). Type I pyrethroids also have a very similar effect on the sodium current (Figure 21) (de Weille et al., 1988; Lund and Narahashi, 1981a,c, 1983; Vijverberg et al., 1982). The prolonged inward sodium current thus produced would increase the depolarizing after-potential. There is another remarkable change in sodium current in the presence of DDT or type I pyrethroids, i.e., a large increase and prolongation of tail current associated with a step repolarization. When the normal nerve membrane is step depolarized and then step repolarized after the sodium current is largely inactivated, the tail current associated with the repolarization is small in

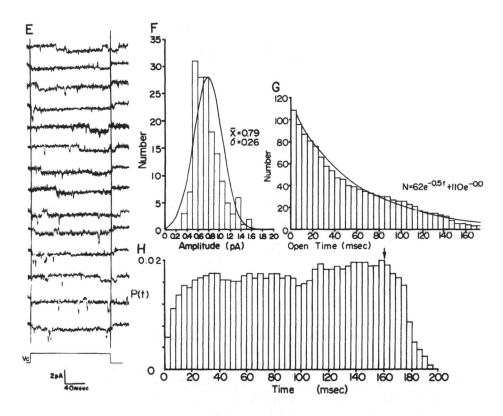

Figure 17. *(Continued)*

amplitude and decays quickly (Figure 21). After application of tetramethrin, however, the tail current becomes very large in amplitude and decays very slowly (Figure 21). The decay phase of tail current in pyrethroids is fitted by two or three exponential functions depending on the kind of pyrethroid and the kind of axon preparation. In any case, the decay of tail current represents the time course of channel closing at the repolarized membrane potential. The peak transient current and steady-state (slow) current during a depolarizing pulse and the tail current following a repolarization disappear after application of TTX, indicating that these currents flow through the sodium channels.

Type II pyrethroids such as fenvalerate and deltamethrin also prolong the sodium current (Brown and Narahashi, 1987; de Weille et al., 1988; Lund and Narahashi, 1983; Salgado et al., 1989). The amplitude of peak transient current is decreased somewhat, but the most remarkable change is observed in tail current. The time course of tail current decay is extremely prolonged, with a time constant as long as several minutes (Figure 22). Again, TTX blocks all of these currents indicating that the sodium channels are involved.

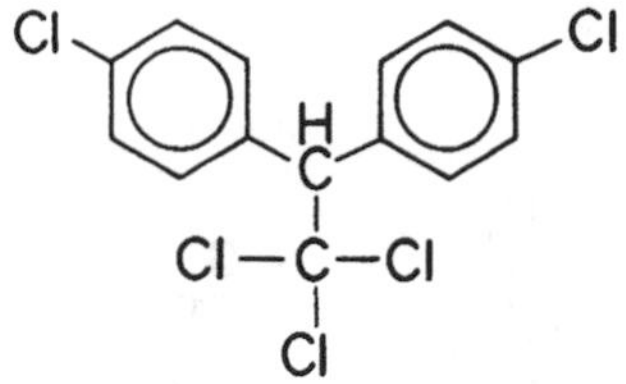

DDT

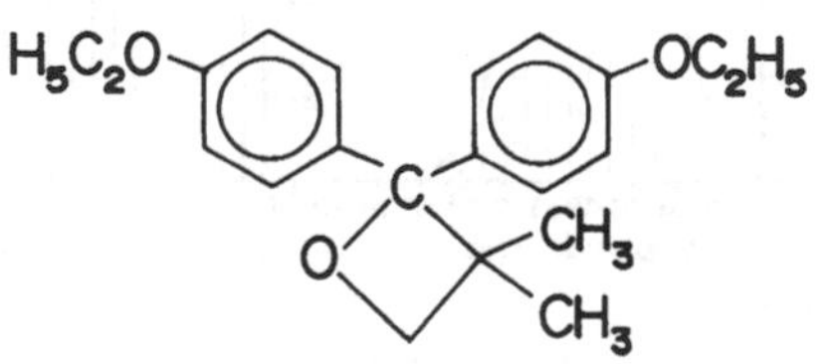

EDO

Figure 18. Structures of *p,p'*-DDT and EDO.

Since the time constant of tail current decay reflects the nature of modification of sodium channels by DDT and pyrethroids, a variety of compounds has been compared for their effects on tail current (Lund and Narahashi, 1983). As shown in Table 2, all of the DDT and pyrethroid type compounds tested can be arranged in a continuous spectrum based on the time constant of tail current decay. The time constant is shortest for DDT, intermediate for tetramethrin and longest for cyphenothrin. Thus the differences among these insecticides in their mechanisms of action on the sodium channel are quantitative rather than qualitative.

Modification of Individual Sodium Channels

The experiments briefly quoted in the preceding section are based on the observation of sodium current which is derived from a large number of sodium channels contained in the nerve membrane preparation. Modifications of individual sodium chanels caused by pyrethroids have been studied by patch clamp techniques (Yamamoto et al., 1983; 1984; Chinn and Narahashi, 1986, 1989; Holloway et al., 1984, 1989). Cultured neuroblastoma cells (N1E-115 line) were used as convenient material.

In normal cells, a step depolarization causes openings of individual sodium channels (Figure 23A). Openings are represented by downward deflections of short duration. After exposure to tetramethrin, each sodium channel is seen to open for a much longer period of time while the amplitude

PYRETHROIDS

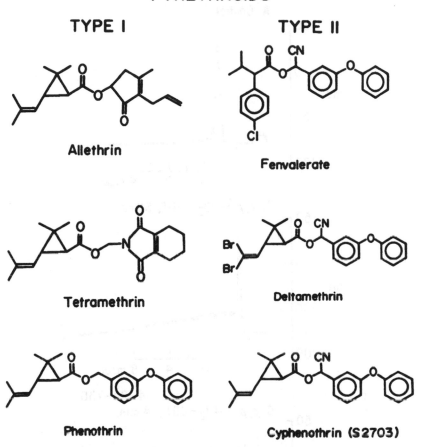

Figure 19. Structures of type I and type II pyrethroids. (From Narahashi, 1985.)

of current remained unchanged (Figure 23B). The histogram clearly shows the absence of change in the amplitude (Figures 23C and D). Open time distribution is expressed by a single exponential function in the control (Figure 23E), whereas it follows a dual exponential function after exposure to tetramethrin (Figure 23F). The short phase of time constant in tetramethrin coincides with the time constant of control preparations, indicating that there are two populations of sodium channels in the presence of tetramethrin, one being normal or unmodified and the other being modifieid channels.

Type II pyrethroids also prolong the open time of sodium channels, but to a much greater extent than that of type I pyrethroids (Holloway et al., 1984, 1989; Chinn and Narahashi, 1986, 1989). For example, in the neuroblastoma cell treated with fenvalerate, individual sodium channels are kept open for as long as several seconds.

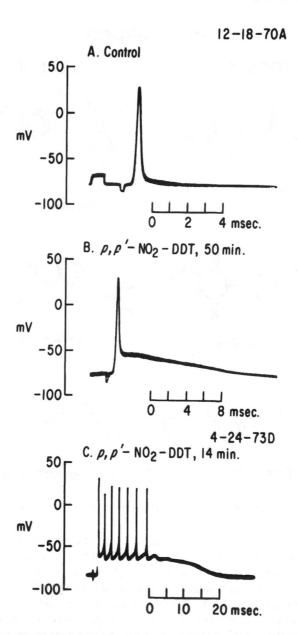

Figure 20. Effects of p,p'-NO$_2$-DDT, a DDT derivative, on the action potential of a crayfish giant axon. A, control; B, 50 min after exposure to 50 μM NO$_2$-DDT. The depolarizing after-potential is increased; C, 14 min after exposure to 100 μM NO$_2$-DDT. The increased depolarizing after-potential reaches the threshold membrane potential for repetitive after-discharges to be produced. (From Wu et al., 1975.)

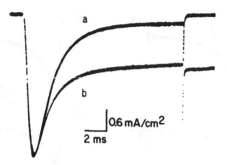

Figure 21. Effects of 1 μM (+)-trans allethrin, a type I pyrethroid, on the sodium current of a squid giant axon. The membrane was step depolarized to −20 mV from a holding potential of −100 mV in K-free external and internal perfusates. In the control (a), the peak transient sodium current is followed by a small slow current during a depolarizing step, and the tail sodium current upon step repolarization decays quickly. After application of allethrin (b), the peak transient sodium current remains unchanged, but the slow current and tail current are increased in amplitude and the latter decays very slowly. (From Narahashi, 1984b.)

The prolonged opening of individual sodium channels in the presence of pyrethroids accounts, at least partially, for the prolongation of sodium current in axon preparations. Another important feature of the change brought about by pyrethroids is the ability of individual sodium channels to open late during a step depolarizing pulse (Figures 23A and B). This clearly indicates that the sodium channel inactivation is largely removed or inhibited by pyrethroids.

From Channel to Animal

Modification of sodium channels caused by DDT and pyrethroid type insecticides can account for the symptoms of poisoning at the animal level. Type I pyrethroids such as tetramethrin and allethrin modify the sodium channel to open longer and late during a depolarizing pulse. This modification would prolong the sodium current recorded from a whole cell such as a giant axon. The prolonged sodium current would then elevate and prolong the depolarizing after-potential which would reach the threshold for generation of action potentials. Thus repetitive after-discharges would be evoked by a single stimulus, which in normal preparation would generate only a single action potential. Repetitive nerve activity thus induced would cause severe symptoms of poisoning in the animal as represented by hyperactivity, ataxia, convulsions and eventual paralysis.

One of the most important toxicological considerations is that the toxic effect is greatly amplified from channel to animal. Percentage of sodium channels that must be modified by tetramethrin to increase the depolarizing after-potential to the level of threshold for action potential generation has been calculated to be only one percent or less (Lund and Narahashi, 1982). The concentration of tetramethrin required for this modification is in the

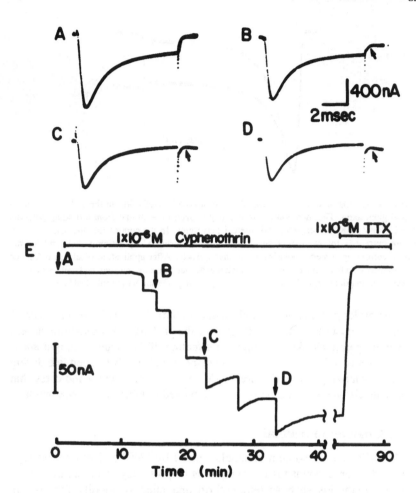

Figure 22. Sodium current recorded in response to an 8 ms depolarizing step to −20 mV from a holding potential of −100 mV before (A) and after (B-D) treatment with 1 µ*M* cyphenothrin, a type II pyrethroid. The arrows in B, C, and D indicate a small increase in inward tail current. The change in holding current with time is shown in a chart record (E), in which each increase in negative current was elicited by an 8 ms depolarization to −20 mV. The arrows mark the times at which records A-D were taken. (From Lund and Narahashi, 1983.)

order of nanomolar. Thus modifications of a small fraction of sodium channel population caused by a very low concentration of tetramethrin is greatly amplified at the animal level to produce severe symptoms of poisoning. This is why the pyrethroids are so potent.

The mechanisms by which symptoms of poisoning are produced by type II pyrethroids are less clear. Extreme prolongation of sodium channel opening would prolong the whole cell sodium current, which in turn would cause a depolarization of nerve membrane as has actually been observed. A small depolarization would be enough to modulate the activity of sensory neurons

TABLE 2.
Time Constants of Tail Currents Associated With
Step Repolarizations of the Membrane to
the Levels Indicated in Crayfish Giant Axons
Treated With Various Compounds

Compound	Conc. (M)	Tail Current Time Constant (msec)		
		-160 mV	-120 mV	-100 mV
DDT	1×10^{-4}	3.0	6.1	9.5
Plifenate	3×10^{-6}	9.4	14.8	17.0
EDO	1×10^{-4}	16	44	86
Tetramethrin	2×10^{-5}	30	225	620
Phenothrin	3×10^{-5}	200	750	1340
GH401	1×10^{-4}	700	1450	2220
Cyphenothrin	1×10^{-6}	(min)	(min)	(∞)
Fenvalerate	1×10^{-6}	(min)	(min)	(∞)
Deltamethrin	1×10^{-6}	(min)	(min)	(∞)

From Lund and Narahashi, 1983.

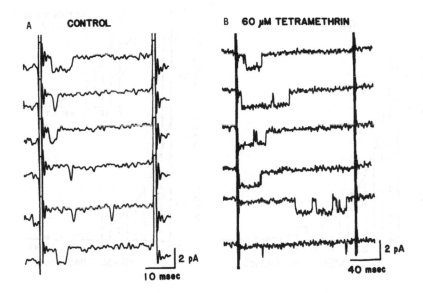

Figure 23. Effects of 60 μM (+)-trans tetramethrin, a type I pyrethroid, on single sodium channels in an inside-out membrane patch excised from a neuroblastoma cell. A, sample records of sodium channel currents (inward deflections) associated with step depolarizations from −90 mV to −50 mV. B, as in A, but after application of tetramethrin to the internal surface of the membrane. C, current amplitude histogram in the control. D, as in C, but after application of tetramethrin. E, channel open time distribution in the control. F, as in E, but after application of tetramethrin. Inset shows the distribution of short open times. From Yamamoto et al., 1983. (*Continued on p. 184*)

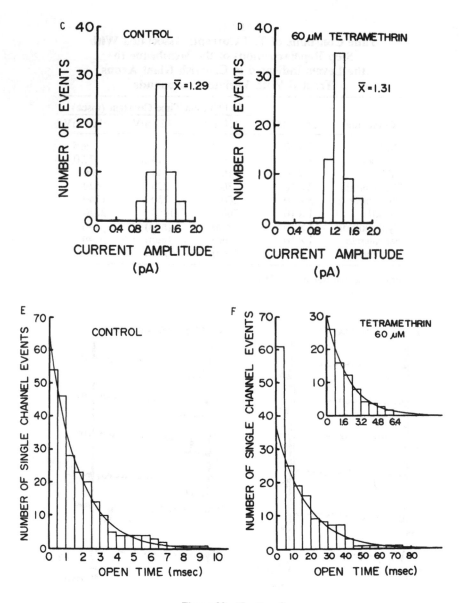

Figure 23. *(Continued)*

to discharge repetitively and of presynaptic nerve terminals to increase transmitter release. These increases in sensory and synaptic activities would alter
the function of the entire nervous system. Thus, the symptoms of poisoning
as characterized by hypersensitivity, choreoathetosis, tremors and paralysis
would be produced. Again, very low concentrations of type II pyrethroids
would be enough to modify a fraction of sodium channel population to
produce symptoms of poisoning in animals.

ACKNOWLEDGMENT

The author's studies quoted in this chapter were supported by NIH Grants NS14143 and NS14144. Thanks are due to Janet Henderson for secretarial assistance.

REFERENCES

Almers, W. (1978). Gating currents and charge movements in excitable membranes. *Rev. Physiol. Biochem. Pharmacol.* 82: 96–190.

Armstrong, C.M. (1969). Inactivation of the potassium conductance and related phenomena caused by quaternary ammonium ion injection in squid axons. *J. Gen. Physiol.* 54: 553–575.

Armstrong, C.M. (1971). Interaction of tetraethylammonium ion derivatives with the potassium channels of giant axons. *J. Gen. Physiol.* 58: 413–437.

Armstrong, C.M. (1981). Sodium channels and gating currents. *Physiol. Rev.* 61: 644–683.

Armstrong, C.M., F. Benzanilla, and E. Rojas (1973). Destruction of sodium conductance inactivation in squid axons perfused with pronase. *J. Gen. Physiol.* 62: 375–391.

Baker, P.F., A.L. Hodgkin, and T.I. Shaw (1961). Replacement of the protoplasm of a giant nerve fibre with artificial solutions. *Nature* 190: 885–887.

Brown, L.D. and T. Narahashi (1987). Activity of tralomethrin to modify the nerve membrane sodium channel. *Toxicol. Appl. Pharmacol.* 89: 305–313.

Catterall, W A (1980). Neurotoxins that act on voltage-sensitive sodium channels in excitable membranes. *Annu. Rev. Pharmacol. Toxicol.* 20: 15–43.

Chinn, K. and T. Narahashi (1986). Stabilization of sodium channel states by deltamethrin in mouse neuroblastoma cells. *J. Physiol.* 380: 191–207.

Chinn, K. and T. Narahashi (1989). Temperature-dependent subconducting states and kinetics of deltamethrin-modified sodium channels of neuroblastoma cells. *Pflügers Arch.* 413: 571–579.

Cole, K.S. (1949). Dynamic electrical characteristics of the squid axon membrane. *Arch. Sci. Physiol.* 3: 253–258.

Cuervo, L.A. and W.J. Adelman, Jr. (1970). Equilibrium and kinetic properties of the interaction between tetrodotoxin and the excitable membrane of the squid giant axon. *J. Gen. Physiol.* 55: 309–335.

de Weille, J.R., H.P.M. Vijverberg, and T. Narahashi (1988). Interactions of pyrethroids and octylguanidine with sodium channels of squid giant axons. *Brain Res.* 445: 1–11.

Evans, M.H. (1976). End-plate potentials in frog muscle exposed to a synthetic pyrethroid. *Pestic. Biochem. Physiol.* 6: 547–550.

Farley, J.M., T. Narahashi, and G. Holan (1979). The mechanism of action of a DDT analog on the crayfish neuromuscular junction. *Neurotoxicology* 1: 191–207.

Hamill, O.P., A. Marty, E. Neher, B. Sakmann, and F.J. Sigworth (1981). Improved patch-clamp techniques for high-resolution current recording from cells and cell-free membrane patches. *Pflügers Arch.* 391: 85–100.

Hille, B. (1968). Pharmacological modifications of the sodium channels of frog nerve. *J. Gen. Physiol.* 51: 199–219.

Hille, B. (1975). The receptor for tetrodotoxin and saxitoxin. A structural hypothesis. *Biophys. J.* 15: 615–619.

Hodgkin, A.L. and A.F. Huxley (1952a). Currents carried by sodium and potassium ions through the membrane of giant axon of *Loligo. J. Physiol.* 116: 449–472.

Hodgkin, A.L. and A.F. Huxley (1952b). The components of membrane conductance in the giant axon of *Loligo. J. Physiol.* 116: 473–496.

Hodgkin, A.L. and A.F. Huxley (1952c). The dual effect of membrane potential on sodium conductance in the giant axon of *Loligo. J. Physiol.* 116: 497–506.

Hodgkin, A.L. and A.F. Huxley (1952d). A quantitative description of membrane current and its application to conduction and excitation in nerve. *J. Physiol.* 117: 500–544.

Hodgkin, A.L., A.F. Huxley, and B. Katz (1952). Measurement of current-voltage relations in the membrane of the giant axon of *Loligo. J. Physiol.* 116: 424–448.

Holan, G. (1971). Rational design of degradable insecticides. *Nature* 232: 644–647.

Holloway, S.F., V.L. Salgado, C.H. Wu, and T. Narahashi (1984). Maintained opening of single Na channels by fenvalerate. *Soc. Neurosci. Abstr.* 10: 864.

Holloway, S.F., T. Narahashi, V.L. Salgado, and C.H. Wu (1989). Kinetic properties of single sodium channels modified by fenvalerate in mouse neuroblastoma cells. *Pflügers Arch.* 414: 613–621.

Julian, F.J., J.W. Moore and D.E. Goldman (1962a). Membrane potentials of the lobster giant axon obtained by use of the sucrose-gap technique. *J. Gen. Physiol.* 45: 1195–1216.

Julian, F.J., J.W. Moore, and D.E. Goldman (1962b). Current-voltage relations in the lobster giant axon membrane under voltage clamp conditions. *J. Gen. Physiol.* 45: 1217–1238.

Kao, C.Y. and A. Nishimaya (1965). Actions of saxitoxin on peripheral neuromuscular systems. *J. Physiol.* 180: 50–66.

Khodorov, B.I. (1978). Chemicals as tools to study nerve fiber sodium channels: effects of batrachotoxin and some local anesthetics. In *Membrane Transport Processes,* Vol. 2. D.C. Tosteson, A.O. Yu, and R. Latorre, Eds., Raven Press, New York, pp. 153–174.

Khodorov, B.I. (1985). Batrachotoxin as a tool to study voltage-sensitive sodium channels of excitable membrane. *Prog. Biophys. Mol. Biol.* 45: 57–148.

Kirsch, G.E., J.Z. Yeh, J.M. Farley, and T. Narahashi (1980). Interaction of n-alkylguanidines with the sodium channels of squid axon membranes. *J. Gen. Physiol.* 76: 315–335.

Lowenstein, O. (1942). A method of physiological assay of pyrethrum extract. *Nature* 150: 760–762.

Lund, A.E. and T. Narahashi (1981a). Modification of sodium channel kinetics by the insecticide tetramethrin in crayfish giant axons. *Neurotoxicology* 2: 213–229.

Lund, A.E. and T. Narahashi (1981b). Interaction of DDT with sodium channels in squid giant axon membrane. *Neuroscience* 6: 2253–2258.

Lund, A.E. and T. Narahashi (1981c). Kinetics of sodium channel modification by the insecticide tetramethrin in squid axon membranes. *J. Pharmacol. Exp. Ther.* 219: 464–473.

Lund, A.E. and T. Narahashi (1982). Dose-dependent interaction of the pyrethroid isomers with sodium channels of squid axon membranes. *Neurotoxicology* 3: 11–24.

Lund, A.E. and T. Narahashi (1983). Kinetics of sodium channel modification as the basis for the variation in the nerve membrane effects of pyrethroids and DDT analogs. *Pestic. Biochem. Physiol.* 20: 203–216.

Muramatsu, I., M. Fujiwara, A. Miura, and T. Narahashi (1985). Effects of Goniopora toxin on crayfish giant axons. *J. Pharmacol. Exp. Ther.* 234: 307–315.

Narahashi, T. (1963). Dependence of resting and action potentials on internal potassium in perfused squid giant axons. *J. Physiol.* 169: 91–115.

Narahashi, T. (1964). Excitable membrane and calcium. *Seibutsu Butsuri* (Biophysics) 4: 101–114.

Narahashi, T. (1971). Effects of insecticides on excitable tissues. In *Advances in Insect Physiology*, Vol. 8. J.W.L. Beament, J.E. Treherne, and V.B. Wiggleworth, Eds., Academic Press, London, pp. 1–93.

Narahashi, T. (1974). Chemicals as tools in the study of excitable membranes. *Physiol. Rev.* 54: 813–889.

Narahashi, T. (1975). Mode of action of dinoflagellate toxins on nerve membranes. In *Proceedings of the First International Conference on Toxic Dinoflagellate Blooms.* V.R. LoCicero, Ed., Massachusetts Science and Technology Foundation, Wakefield, pp. 395–402.

Narahashi, T. (1976). Effects of insecticides on nervous conduction and synaptic transmission. In *Insecticide Biochemistry and Physiology.* C.F. Wilkinson, Ed., Plenum Press, New York, pp. 327–352.

Narahashi, T. (1981). Mode of action of chlorinated hydrocarbon pesticides on the nervous system. In *Halogenated Hydrocarbons: Health and Ecological Effects.* M.A.Q. Khan, Ed., Pergamon Press, Elmsford, NY, pp. 222–242.

Narahashi, T. (1984a). Pharmacology of nerve membrane sodium channels. In *Current Topics in Membranes and Transport, Vol. 22, The Squid Axon.* P.F. Baker, Ed., Academic Press, New York, pp. 483–516.

Narahashi, T. (1984b). Drug-ionic channel interactions: single channel measurements. In *Basic Mechanisms of the Epilepsies. Annals of Neurology*, Vol. 16, pp. S39–S51.

Narahashi, T. (1984c). Nerve membrane sodium channels as the target of pyrethroids. In *Cellular and Molecular Neurotoxicology.* T. Narahashi, Ed., Raven Press, New York, pp. 85–108.

Narahashi, T. (1985). Nerve membrane ionic channels as the primary target of pyrethorids. *Neurotoxicology* 6 (2): 3–22.

Narahashi, T. (1987a). Effects of toxic agents on neural membranes. In *Electrophysiology in Neurotoxicology*, Vol. I. H.E. Lowndes, Ed., CRC Press, Boca Raton, FL, pp. 23–44.

Narahashi, T. (1987b). Neuronal target sites of insecticides. In *ACS Symposium Series 356, Sites of Action for Neurotoxic Pesticides.* R.M. Hollingworth and M.B. Green, Eds., American Chemical Society, Washington, D.C., pp. 226–250.

Narahashi, T. (1988). Molecular and cellular approaches to neurotoxicology: past, present and future. In *Neurotox '88: Molecular Basis of Drug and Pesticide Action.* G.G. Lunt, Ed., Elsevier, Amsterdam, pp. 269–288.

Narahashi, T. and N.C. Anderson (1967). Mechanism of excitation block by the insecticide allethrin applied externally and internally to squid giant axons. *Toxicol. Appl. Pharmacol* 10: 529–547.

Narahashi, T. and H.G. Haas (1967). DDT: Interaction with nerve membrane conductance changes. *Science* 157: 1438–1440.

Narahashi, T. and H.G. Haas (1968). Interaction of DDT with the components of lobster nerve membrane conductance. *J. Gen. Physiol.* 51: 177–198.

Narahashi, T., T. Deguchi, N. Urakawa, and Y. Ohkubo (1960). Stabilization and rectification of muscle fiber membrane by tetrodotoxin. *Am. J. Physiol.* 198: 934–938.

Narahashi, T., J.W. Moore, and W.R. Scott (1964). Tetrodotoxin blockage of sodium conductance increase in lobster giant axons. *J. Gen. Physiol.* 47: 965–974.

Narahashi, T., J.W. Moore, and D.T. Frazier (1969). Dependence of tetrodotoxin blockage of nerve membrane conductance on external pH. *J. Pharmacol. Exp. Ther.* 169: 224–228.

Neher, E. and B. Sakmann (1976). Single-channel currents recorded from membrane of denervated frog muscle fibres. *Nature* 260: 779–802.

Neher, E. and J.H. Steinbach (1978). Local anaesthetics transiently block currents through single acetylcholine-receptor channels. *J. Physiol.* 277: 153–176.

Oikawa, T., C.S. Spyropoulos, I. Tasaki, and T. Teorell (1961). Methods for perfusing the giant axon of *Loligo pealii*. *Acta Physiol. Scand.* 52: 195–196.

Oxford, G.S., C.H. Wu, and T. Narahashi (1978). Removal of sodium channel inactivation in squid giant axons by *N*-bromoacetamide. *J. Gen. Physiol.* 17: 227–247.

Quandt, F.N. and T. Narahashi (1982). Modification of single Na^+ channels by batrachotoxin. *Proc. Natl. Acad. Sci. U.S.A.* 79: 6732–6736.

Quandt, F.N., J.Z. Yeh, and T. Narahashi (1985). All or none block of single Na^+ channels by tetrodotoxin. *Neuroscience Lett.* 54: 77–83.

Ritchie, J.M. (1979). A pharmacological approach to the structure of sodium channels in myelinated axons. *Annu. Rev. Neurosci.* 2: 341–362.

Roeder, K.D. and E.A. Weiant (1946). The site of action of DDT in the cockroach. *Science* 103: 304–306.

Roeder, K.D. and E.A. Weiant (1948). The effect of DDT on sensory and motor structures in the cockroach leg. *J. Cell. Comp. Physiol.* 32: 175–186.

Ruigt, G.S.F. (1984). An Electrophysiological Investigation into the Mode of Action of Pyrethroid Insecticides, Ph.D. dissertation, University of Utrecht, The Netherlands.

Sakmann, B. and E. Neher (1983). *Single-Channel Recording*, Plenum Press, New York, 503 pp.

Salgado, V.L., M.D. Herman, and T. Narahashi (1989). Interactions of the pyrethroid fenvalerate with nerve membrane sodium channels: temperature dependence and mechanism of depolarization. *Neurotoxicology* 10: 1–14.

Seyama, I. and T. Narahashi (1981). Modulation of sodium channels of squid nerve membranes of grayanotoxin I. *J. Pharmacol. Exp. Ther.* 219: 614–624.

Shanes, A.M. (1949). Electrical phenomena in nerve. II. Crab nerve. *J. Gen. Physiol.* 33: 75–102.

Shanes, A.M., W.H. Freygang, H. Grundfest, and E. Amatniek (1959). Anesthetic and calcium action in the voltage clamped squid giant axon. *J. Gen. Physiol.* 42: 793–802.

Takeno, K., K. Nishimura, J. Parmentier, and T. Narahashi (1977). Insecticide screening with isolated nerve preparations for structure-activity relationships. *Pestic. Biochem. Physiol.* 7: 484–499.

Takeuchi, A. and N. Takeuchi (1959). Active phase of frog's end-plate potential. *J. Neurophysiol.* 22: 395–411.

Tasaki, I. and S. Hagiwara (1957). Demonstration of two stable potential states in the squid giant axon under tetraethylammonium chloride. *J. Gen. Physiol.* 40: 859–885.

Taylor, R.E. (1959). Effect of procaine on electrical properties of squid axon membrane. *Am. J. Physiol.* 196: 1071–1078.

Vijverberg, H.P.M., J.M. van der Zalm, and J. van den Bercken (1982). Similar mode of action of pyrethroids and DDT on sodium channel gating in myelinated nerves. *Nature* 295: 601–603.

Welsh, J.H. and H.T. Gordon (1947). The mode of action of certain insecticides on the arthropod nerve axon. *J. Cell. Comp. Physiol.* 30: 147–172.

Wilson, W.A. and M.M. Goldner (1975). Voltage clamping with single microelectrode. *J. Neurobiol.* 6: 411–422.

Woolley, D.E. (1981). The neurotoxicity of DDT and possible mechanisms of action. In *Mechanisms of Neurotoxic Substances.* K.N. Prasal and A. Vernadakis, Eds., Raven Press, New York, pp. 95–141.

Wouters, W. and J. van den Bercken (1978). Action of pyrethroids. *Gen. Pharmacology* 9: 387–398.

Wouters, W., J. van den Bercken, and A. van Ginneken (1977). Presynaptic action of the pyrethroid insecticide allethrin in the frog motor end plate. *Eur. J. Pharmacol.* 43: 163–171.

Wu, C.H., J. van den Bercken, and T. Narahashi (1975). The structure-activity relationship of DDT analogs in crayfish giant axons. *Pestic. Biochem. Physiol.* 5: 142–149.

Wu, C.H., G.S. Oxford, T. Narahashi, and G. Holan (1980). Interaction of a DDT analog with the sodium channel of lobster axon. *J. Pharmacol. Exp. Ther.* 212: 287–293.

Yamamoto, D., F.N. Quandt, and T. Narahashi (1983). Modification of single sodium channels by the insecticide tetramethrin. *Brain Res.* 274: 344–349.

Yamamoto, D., J.Z. Yeh, and T. Narahashi (1984). Voltage-dependent calcium block of normal and tetramethrin-modified single sodium channels. *Biophys. J.* 45: 337–344.

Yamasaki, T. and T. Ishii (Narahashi) (1952). Studies on the mechanism of action of insecticides. IV. The effects of insecticides on the nerve conduction of insects. *Oyo-Kontyu (J. Nippon Soc. Appl. Entomol.)* 7: 157–164.

Yeh, J.Z. (1982). A pharmacological approach to the structure of the Na channel in squid axon. In *Proteins in the Nervous System: Structure and Function.* B. Haber, J. Preg-Polo, and J. Coulter, Eds., Alan R. Liss, New York, pp. 17–49.

Electromyographic Methods

Rebecca J. Anderson
Boehringer Ingelheim Pharmaceuticals, Inc.
Ridgefield, Connecticut

ANATOMY AND PHYSIOLOGY OF THE NEUROMUSCULAR JUNCTION

The final common neuronal pathway controlling movement arises in the spinal cord from large alpha motor neurons (25 μm in diameter) which send out large myelinated axons (17 μm in diameter) whose conduction velocities are among the fastest in the body (50–200 m/sec) and activate muscle contraction.

The physiological chain of events leading to the contraction and relaxation of a muscle fiber is propagation of an action potential down the motor nerve axon; depolarization of the presynaptic nerve terminal; increased intraneuronal calcium; release of acetylcholine from vesicles into the synapse; binding of acetylcholine to postsynaptic receptor sites; conductance change (sodium and potassium ion flux) of the postsynaptic muscle membrane; depolarization of the muscle membrane; generation of an action potential in twitch muscle fibers; spread of the depolarization along the muscle fiber membrane and down into the transverse tubules; release of calcium ions from the longitudinal tubules, mediated in some way via the triads connecting the two types of tubules; binding of calcium ions to troponin molecules on the thin filaments which initiates the sliding of the thin and thick filaments past each other; reuptake of calcium ions into the longitudinal tubules; and relaxation of the muscular contraction.

The motor nerve is myelinated up to its terminal where it may branch into 5 to 100 separate endings. Each branch synapses onto a separate skeletal muscle fiber. The motor axon and the muscle fibers it innervates constitute one motor unit. Within the nerve terminal are vesicles containing the neurotransmitter, acetylcholine.

Acetylcholine is synthesized in the nerve terminal from choline and acetyl CoA. The rate limiting step in the reaction is the high affinity uptake of choline into the nerve terminal by an active process which can be blocked by the drug, hemicholinium. Acetyl CoA is readily available from its location in the mitochondrial membrane. Packaging of acetylcholine into vesicles may be an active process since ATP is also found in association with them. The machinery for manufacturing acetylcholine is all found in the nerve terminal, except for the mitochondria which transport acetyl CoA from the cell soma by axonal transport. Therefore, large and essentially continuous stocks of acetylcholine are available in the nerve terminal. Each vesicle contains approximately 10,000 molecules of acetylcholine.

Acetylcholine is released in quantal packets (probably the dumping of the contents of individual vesicles), both spontaneously and in response to nerve stimulation. When an action potential invades the motor nerve terminal, a small gating current permits the influx of calcium ions. The increased calcium concentration causes the vesicles to fuse with the presynaptic nerve membrane and release their contents of acetylcholine into the synaptic cleft. Regular rows of attachment sites on the presynaptic membrane are thought to be the binding sites for the vesicles, forming an "active zone" in close proximity to the synapse. The amount of transmitter released is dependent upon the inward gradient of calcium flux, but typically 100–200 quanta (vesicles) are released as a result of a nerve action potential.

On the corresponding postsynaptic side, the muscle membrane is specialized into a series of irregular folds which contain a high density of acetylcholine receptors (about 7500 per μm^2) on the crests of the junctional folds. The junctional folds are held in place by a basal lamina which also contains the enzyme acetylcholinesterase which destroys the transmitter in the synaptic cleft. The endplate region of the muscle membrane contains about 10^7 acetylcholine receptors and an equal number of acetylcholinesterase molecules.

The acetylcholine receptors on the muscle side of the neuromuscular junction are responsible for recognizing the presence of released transmitter in the synaptic cleft and initiating the muscle's contractile response. A great deal of recent research has been devoted to characterizing the acetylcholine receptor. It is a glycoprotein with a molecular weight of about 250,000 and consisting of five subunits: two identical alpha subunits (40,000 Da each) and one each of the beta, gamma, and delta subunits (50,000, 60,000 and 65,000 Da, respectively). The amino acid sequence is highly conserved between the subunits. It is generally thought that one acetylcholine molecule binds to each of the two alpha subunits resulting in optimal opening of a channel which permits the flow of potassium and sodium ions across the membrane. The resultant release of about 10^6 acetylcholine molecules during nerve stimulation trigger the opening of 250,000 ion channels. Each channel is open for about 1 msec, permitting about 10,000 sodium and potassium ions to cross the membrane. In the aggregate they thus generate a current of about -400 nAmp and account for the often recorded endplate potential (EPP).

Closing of these ion channels presumably is a result of dissociation of acetylcholine from the receptor sites. Excess acetylcholine in the synapse as well as that which dissociates from receptor sites is rapidly destroyed by acetylcholesterase in the cleft, making the transmission process from start to finish in less than 100 μsec.

The spontaneous release of acetylcholine into the synaptic cleft takes place continuously and is apparently not dependent upon external calcium. The release of single vesicles of transmitter results in a much smaller endplate current (about -4 nAmp). This is recorded experimentally as the miniature endplate potential (mepp).

The acetylcholine triggered opening of sodium and potassium channels at the motor endplate leads to a graded depolarization of the postsynaptic membrane directly associated with the nerve terminal, in proportion to the amount of transmitter binding to acetylcholine receptors. When the postsynaptic (muscle) membrane reaches threshold, it initiates a muscle action potential which propagates in much the same way along the muscle as a nerve action potential along the nerve axon. Propagation of the muscle action potential is facilitated by a network of transverse and longitudinal tubules which provide a low resistance conduction path to each of the individual muscle fibers. The coupling of this excitatory stimulus to mechanical contraction requires the presence of calcium ions. The action potential induced release of calcium ions into the sarcoplasmic reticulum rapidly bind to troponin sites on the thin filaments of the myofibril. This attachment of calcium to troponin exposes the active binding site, leading to the association of actin with myosin forming a crossbridge. The rotation of the crossbridge is an active process requiring the release of an ATP molecule (which was bound to the myosin) and the release of energy resulting in the formation of inorganic phosphate and ADP. The binding of another molecule of ATP dissociates the actin and myosin to prepare it for another cycle of attachment. In a series of ratchet movements dictated by the cycling of ATP binding and release, the sliding filaments result in muscular contraction. ATP is also required for the pumping of calcium ions back into the longitudinal tubules. The dissociation of calcium from its binding sites on the myofilaments and clearing of the calcium from the sarcoplasmic reticulum thus leads to a gradual relaxation of the muscle fiber.

The timecourse of the twitch is far slower than the action potentials generated by the nerve and muscle, thus permitting a subsequent nerve impulse to release more acetylcholine (and therefore also the release of more muscle calcium) and resulting in summation of the contractile events. The rate of nerve stimulation thus determines the rate and amount of muscle tension generated until so much calcium is released that all of the troponin molecules are saturated, at which point maximum muscle tension (or tetanus) is reached.

DRUGS, TOXINS, AND POISONS

Because the peripheral nerve/neuromuscular junction/skeletal muscle complex is readily accessible, many pharmacological and toxicological studies of neuromuscular function have been conducted. Among the most common isolated perparations are the rat phrenic nerve hemidiaphragm, the frog sciatic nerve/sartorius muscle, the crayfish neuromuscular preparation, and electroplax from the electric eel. Among the most common *in situ* preparations are the cat and rat sciatic nerve/triceps surae (medial gastrocnemius, lateral gastrocnemius, and soleus) muscles. A large number of drugs, chemicals, and toxins have been reported to have specific effects on the nerve, neuromuscular synapse, or skeletal muscle.

Uptake of choline into the presynaptic nerve terminal (the synthetic rate limiting step for transmitter production) is competitively blocked by hemicholinium-3 and triethylcholine. Excessive release of acetylcholine from the motor nerve terminals occurs with snake venom (β-bungarotoxin) and black widow spider venom (α-latrotoxin). Acetylcholine release is blocked by botulinum toxin and cytochalasin B.

Postsynaptically, disruption of the acetylcholine receptors can occur in several ways. Reversible pharmacologic blockade of the skeletal muscle acetylcholine receptors can be produced by curare, decamethonium, and succinylcholine. Irreversible blockade of the receptors occurs with the toxins α-bungarotoxin and rabies virus. Specific disruption of various parts of the ion channel of the acetylcholine receptor complex can occur following treatment with a number of substances including histrionicotoxin, amantadine, and quinacrine.

In addition, both alteration of neuromuscular transmission and (following chronic administration) myopathies can occur as a result of inhibition of acetylcholinesterase. Short acting acetylcholinesterase inhibitors such as metrifonate (trichlorfon) produce little more than an increase in repetitive activity of the evoked muscle action potential, probably a reflection of the increased concentration of synaptic acetylcholine. However, long term administration of reversible cholinesterase inhibitors such as neostigmine and pyridostigmine results in myopathy particularly of the neuromuscular junction area. Although the accumulation of transmitter probably contributes to this pathology, it has been shown that at least part of the pathology is due to a direct, postsynaptic action of the drug. Organophosphorus agents are irreversible cholinesterase inhibitors. These substances produce a lethal accumulation of acetylcholine, depolarization blockade of neuromuscular transmission, and muscle paralysis.

Disruption of excitation-contraction coupling is the most common type of toxicity directly on skeletal muscle. Dantrolene, which limits calcium availability in skeletal muscle, is a direct muscle relaxant and has been used therapeutically for that purpose. Aside from the myopathies which result indirectly from denervation and those attributed to cholinesterase inhibitors such as pyridostigmine, few toxic chemicals have been reported to produce myopathy. One example is triethyltin which has been demonstrated experimentally to produce muscle weakness by a direct action following chronic exposure.

ELECTROMYOGRAPHIC RECORDING TECHNIQUES

Electromyography (EMG) measures the electrical events in the muscle which lead to contraction. Most commonly bipolar surface electrodes are placed over the belly of the muscle to record its electrical activity. For deep muscles or small muscles which cannot be adequately recorded in this manner, fine needle electrodes are imbedded in the muscle of interest.

The amplitude of an individual muscle fiber action potential ranges from 0.2–4 mV, depending on the muscle fiber diameter, the distance between the muscle fiber and the recording site, and the filtering properties of the recording electrode. The amplitude of this action potential is directly proportional to the muscle fiber diameter and inversely proportional to the distance of the fiber from the electrode. The filtering properties of the bipolar recording electrode are a function of the size of the contacts, the distance between the contacts, and the chemical properties of the metal-electrolyte surface. The duration of the action potentials vary inversely with the conduction velocity of the muscle fiber and range from 2–15 msec. This poses a significant and variable time delay, starting from the point of initiation of the depolarization at the neuromuscular junction, propagating along muscle fibers and ending with depolarization of the muscle fiber. In addition, the shape and frequency characteristics of the muscle fiber action potentials will be affected by the tissue between the muscle fiber and the recording electrode. This is a significant factor when surface electrodes are used. The depolarization of one muscle fiber overlaps with others in the same motor unit, forming a motor unit action potential which is the spatial-temporal superposition of the contributing individual action potentials. The unit action potentials from many motor units contribute to form the electromyograph, which is the spatial-temporal superposition of the contributing motor units in the bulk muscle.

The recorded EMG signal is an ongoing record of the occurrence of muscle action potentials from all muscle fibers in the vicinity of the recording electrodes and represents their electrical excitability. At rest the motor units are not active and no EMG response is recorded. Weak voluntary muscle contraction may result in the recording of a single motor unit which will discharge in a rhythmic fashion. With increased voluntary contraction, additional motor units will be recruited, each discharging rhythmically and independently, making the EMG record progressively more complex.

Electrical stimulation of the motor nerve evokes a massive simultaneous activation of many motor units and results in a large, discrete muscle action potential which is time locked to the electrical stimulus. By using various experimental manipulations, electromyographic techniques have been used to examine characteristics of nerve, muscle, and neuromuscular junction function.

Stimulus evoked activation of a muscle action potential has commonly been used to measure the conduction velocity of peripheral motor nerves. Motor nerve conduction velocity is most accurately measured by stimulating a peripheral motor nerve at two sites and recording the onset of the evoked muscle action potentials. Maximal conduction velocity is calculated from the distance between the two stimulating electrodes and the difference in the two muscle action potential latency times. The muscle action potential recorded in this way is approximatley 1000 times larger than the nerve action potential because the multiple muscle fibers of the motor unit amplify the effect of stimulation of each motor nerve. The response therefore can be recorded with relatively insensitive surface electrodes. However, subcutaneous fat causes reduction of the response; muscle fibers closest to the surface contribute disproportionately to the signal; and the recording electrodes may detect distant muscles which differ in latency from the muscle under study. In atrophic muscle greater temporal dispersion contributes to the inaccuracy of determining the onset of the muscle action potential. This can be alleviated somewhat by the use of needle electrodes which give a more discrete measure of conduction onset but sample fewer muscle fibers only in the vicinity of the electrode.

The accuracy of making motor nerve conduction velocity calculations is affected by inaccuracy in measurement of conduction distance, temperature variability and age, which can lead to errors of up to 30%. Different nerves propagate action potentials at different rates. The motor conduction velocity measured from the nerves in the arm range from 50–65 m/sec, whereas those in the leg range from 40–55 m/sec. Motor conduction velocity decreases by approximately 1 m/sec for every 10 years of age between 20 and 60 years. Conduction velocity proportionately slows with decreases in temperature. This poses a troublesome problem when recording from deep nerves which may have a different and variable temperature from that recorded on the surface.

Maximal motor conduction velocity is greatly slowed in segmental demyelination. Slower conduction during reinnervation is directly related to the smaller diameter of the regenerating nerve fibers. Moderate slowing is observed in some cases of progressive muscular atrophy, due to selective destruction of large diameter, fast conduction nerve fibers, and retention of only slower conducting motor fibers.

Since measurement of maximal conduction velocity only provides information on the fastest fibers in the nerve bundle, investigators have also used amplitude as a measure of the total number of nerve fibers conducting. The size of the muscle action potential is directly proportional to the number of motor units activated and thus to the number of motor nerves propagating impulses. However, amplitude of the muscle action potential is much more subject to variation than measurement of conduction velocity. Among the most common sources of variation are improper electrode placement, movement of the electrodes during stimulation, change in skin resistance, intramuscular temperature, and characteristics of the amplifiers and electronic filters. These can lead to a three- to fourfold difference in amplitude between recording sessions in the same subject. Some clinical investigators have increased the sensitivity and decreased the variability in conduction velocity studies by measuring the amplitude ratios of the muscle action potentials generated by stimulation at the two nerve stimulation sites. This gives a measure of the extent of damage to nerve fibers situated between the two stimulating electrodes. Comparison of the duration of the evoked muscle action potentials generated from the two stimulation sites has also been used as a measure of nerve function.

The changes in the muscle compound action potential resulting from single and repeated stimulation provide information about the viability of the neuromuscular junction and the muscle contractile process. There is normally an increase in the size of the muscle action potential following short maximal voluntary contraction (postactivity facilitation). This is the result of enhanced release of neurotransmitter and the recruitment of more muscle fibers contributing to contraction. The decrements in the muscle action potential following extended muscle activity (postactivity exhaustion) are the result of muscle fatigue. Repeated electrical stimulation can also result in fatigue, in which there is a progressive decline in the amplitude of the potentials as fewer muscle fibers respond to the stimulus. Unusual fatigability is one of the hallmarks of myasthenia gravis but also occurs in cases where muscles are weakened by poliomyelitis, amyotrophic lateral sclerosis, and syringomyelia. Increased rate of fatigue may also occur during reinnervation.

Strength-duration curves measure the minimal stimulus intensity required to elicit a muscle action potential at each of various stimulus durations. This measurement is used to assess the threshold for activation of the motor nerve but is also affected by changes in the neuromuscular synapse and the muscle. In normal muscle the nerve is the most excitable component to threshold activation. Therefore, the normal strength-duration curve reflects nerve excitability. The method provides an earlier assessment of denervation in progressive peripheral neuropathy than the appearance of spontaneous muscle fibrillation. However, the technique is not as sensitive as the EMG in detecting early evidence of reinnervation.

From recordings of single muscle fiber EMG, measurements of jitter can be made by measuring the variability in the interval between two action potentials belonging to the same motor unit. Jitter is largely due to variability in the neuromuscular transmission time and is increased in cases with disturbed neuromuscular transmission even before impulse blockade. Fiber density is a measure of the average number of muscle fibers from one motor unit within the recording range of the single fiber EMG recording electrode, obtained from 20 recording sites. During reinnervation, collateral sprouting causes changes in the organization of muscle fibers in the motor unit and results in an increased fiber density.

In experimental settings, the use of more elaborate, invasive electro-physiologic recording techniques has provided more specific measurement of the responsiveness of the nerve terminal, neuromuscular junction, and muscle. Single axons can be isolated from both sensory and motor peripheral nerves for direct measurement of conduction. Microelectrode technology is now used routinely to study the characteristics of the neuromuscular junction. Intracellular recordings of muscle fibers in the endplate region of muscles provides a direct measure of the endplate responsiveness to neurotransmitter release. Alteration in the miniature endplate potential frequency and amplitude suggest changes in the spontaneous release and quantal content of transmitter in the motor nerve terminal, respectively. Changes in the endplate potential indicate disruption of the evoked release of neurotransmitter during nerve stimulation. However, these parameters could also change as a result of postsynaptic effects on the acetylcholine receptors on the muscle. Additional information, such as the resting membrane potential and membrane resistance of the recorded muscle fiber serve to localize the site of action. Changes in these later features are suggestive of denervation.

Direct intracellular recording from muscle fibers is hindered by the physiologic response of these cells. Contraction, especially in large muscles, generates enough tension and movement of the muscle that electrodes cannot be maintained in place without breaking or tearing the muscle fiber. In smaller muscles and in preparations which minimize movement of the muscle, these intracellular techniques have provided a great deal of information about the contractile properties of the muscle.

DIAGNOSIS

Electrophysiologic assessment of nerve and muscle function is indicated in patients with symptoms which suggest disturbance of the peripheral nervous system. These symptoms include flaccid paresis, muscular atrophy, muscular weakness, muscle spasm, parethesis and pain in the limbs, and abnormal fatigability. There is no routine procedure for an EMG examination. The scope of testing largely depends upon the clinical signs presented by individual patients. Clinicians look for changes in the EMG upon insertion of the electrodes, the spontaneous activity associated with voluntary contraction, and the characteristics of the unit potentials.

Normally insertion of needle electrodes produces a brief burst of electrical activity during the movement of the needles. Prolonged, repetitive activity can occur for long periods after insertion in denervated muscle fibers, muscle fibers in myotonia, and muscle fibers in the early stages of denervation or reinnervation. A reduction or absence of insertion activity occurs in severely atrophied muscle.

Normally there is no spontaneous EMG activity when the muscle is at rest. In neuromuscular disorders, however, there may be muscle fibrillation or fasciculation potentials. Fibrillation potentials are associated with the discharging of individual muscle fibers. They have an amplitude of 25–200 μV and a duration of 0.5 to 1.5 msec. They discharge at a slow, regular rate of 2 to 10 per second. Typically they are associated with degeneration of lower motor neurons. Fasciculations, on the other hand, are associated with the spontaneous discharges of whole motor units. Although some normal patients exhibit fasciculations, they are characteristic of anterior horn cell diseases such as amyotrophic lateral sclerosis and progressive muscular atrophy.

Amplitude and duration of motor unit action potentials are pathologically diminished by processes that reduce the number or density of activated muscle fibers in the motor unit or reduce the circumference of active muscle fibers in the motor unit. Examples of these situations include: (1) changes in muscle fibers such as muscle fiber degeneration (Duchenne dystrophy), muscle fiber atrophy without degeneration and not suggesting conventional denervation (myotonic dystrophy), muscle fiber smallness without degeneration due to presumed faulty formation or maturation (congenital myopathies); (2) impaired muscle membrane excitability (periodic paralysis); (3) faulty neuromuscular junction transmission (myasthenia gravis, botulism); and (4) neuropathic processes restricted to a fraction of the terminal arborizations within a motor unit. Each of these processes reduces the tension generated by single contractions of an affected motor unit, which must discharge more rapidly to produce the desired summated tension. If a sufficient number of motor units are involved, but no loss of operational motor unit occurs, the EMG exhibits excessively recruited, pathologically small motor unit potentials.

The duration and amplitude of motor unit action potentials is increased in diseases affecting anterior horn cells, such as amyotrophic lateral sclerosis, poliomyelitis, and syringomyelia. This is the result of a synchronous discharge of a large number of muscle fibers. Enlarged motor unit action potentials may be the result of nerve terminal sprouting to create enlarged motor units or reorganization of the anterior horn cells which synchronously activate multiple motor units.

Diagnosis of peripheral neuropathies and myopathies is made possible by an understanding of the pathological processes which take place during nerve and/or muscle damage. In the extreme, for instance when the peripheral nerve is severed, a number of homeostatic and compensitory processes are initiated. There is a reduction in the number of synaptic connections onto the motor neuron cell soma, which also undergoes chromatolysis. Severe chromatolysis can result in cell death. There is a thinning of the axon and a reduction in conduction velocity along the intact proximal segment. The muscle fiber deprived of its motor innervation becomes responsive to acetylcholine along its entire extent. This is the result of a vast increase in the density of acetylcholine receptors which cover the muscle fiber in addition to just the endplate region. Initially, there is no EMG in response to voluntary contraction, insertion activity is normal, and no fibrillation potentials are present. After 3–4 days electrical stimulation of the nerve below the lesion will not evoke a response indicating denervation. Abnormal fibrillation potentials first appear in the EMG after 8–14 days. Chronically denervated muscle is characterized by fibrillation potentials, which occur spontaneously 2–4 weeks after injury and are sustained until either reinnervation or severe muscle atrophy occurs. Reinnervation of the muscle reverses these processes (nerve conduction returns close to normal, the acetylcholine receptors again become restricted to the endplate region, and the fibrillation potentials cease). Often the earliest positive evidence of reinnervation is the appearance of motor unit action potentials in the EMG during voluntary contraction effort.

EMG activity during voluntary contraction provides the best discriminator of primary muscle disease versus that which is secondary to denervation. In myopathy there is a decreased size of motor unit action potentials. The number of action potentials, relative to the degree of contraction, may even be greater than normal. On the other hand, in peripheral neuropathy, the size of the motor unit action potentials is normal. Contractile weakness in this case is due to a decreased number of units being activated.

In diseases affecting the neuromuscular junction, such as myasthenia gravis, there is increased fatigability of the muscle. Upon repetitive stimulation of the motor nerve, the amplitude of the muscle action potentials progressively decrease. The EMG in muscle diseases is characterized by decreases in the amplitude and duriation of the motor unit potentials. The decreased functional capacity is a direct reflection of fewer viable muscle fibers contributing to the contractile process.

Neurogenic diseases involving the lower motor neuron result in an increase in duration and perhaps also the amplitude of motor unit potentials. There is increased spontaneous EMG activity which results from fibrillation and fasciculations of the denervated muscle. There is also a decrease in the number of recruitable motor units during maximal voluntary contraction.

Denervation (especially incomplete denervation) can also lead to muscle hypertrophy. This is thought to be the result of an extra load imposed on the surviving fibers, a physiological mechanism similar to that seen during extended exercise in normal muscle.

When there is destruction or disruption of the peripheral nerve, conduction velocity is slowed. However, depending on the extent of damage, the length of illness and the accuracy of the measurements, there may be only subtle changes in nerve conduction. Simultaneous collateral sprouting occurs concurrently with neuropathy in mild cases. The net result is an apparently normal conduction despite considerable histological damage.

NEUROMUSCULAR DISORDERS RESULTING FROM OCCUPATIONAL HAZARDS

There have been no reports of primary myopathies resulting from industrial or environmental exposure. However, myopathy as a secondary consequence of disruption of peripheral nerve and/or the neuromuscular junction has been widely described. The most prevalent and widely studied of these hazards are briefly described here.

Organophosphorus insecticides produce one or both of two peripheral toxicities. All organophosphorus agents are potent and irreversible inhibitors of the cholinesterases. The resultant accumulation of acetylcholine at the neuromuscular junction and elsewhere results in a cholinergic syndrome which is fatal if not treated with antidotes such as atropine and pralidoxime. Persistent, low level exposure to organophosphates, which occurs, for example, to workers in formulation plants, produces clinical signs similar to those seen in myasthenic patients receiving excessive therapy. There is an increased fatigability of the muscle due to depolarization blockade of the neuromuscular junction.

Some, but certainly not all, organophosphorus agents are also capable of inducing a delayed neuropathy. The clinical signs typically develop one to three weeks following a single exposure. Motor nerves appear to be somewhat selectively affected. Histologically there is overall reduction in the number of fibers in the peripheral nerves but a selective sparing of large diameter axons. This is consistent with the clinical and electrophysiological observations, which show that the muscle action potential amplitude is diminished but the maximal nerve conduction velocity is unaffected even in cases of severe limb weakness.

Clinical cases of neurotoxicity as a result of acrylamide poisoning have been relatively rare. However, acrylamide has been extensively studied as a prototype agent in the laboratory for producing a characteristic experimental peripheral neuropathy. Acrylamide produces a dying-back form of neuropathy in which the nerve fibers degenerate first at the nerve terminal and then the neuropathy extends more proximally. Large diameter axons are preferentially affected, which is consistent with the electrophysiologic and clinical data. At the onset of clinical signs, there is a slowing of the maximal motor nerve conduction velocity and an increased duration and decreased amplitude of the evoked muscle action potential. Conduction is disproportionately slowed in the distal parts of the motor nerve. As denervation progresses, there are resultant abnormalities in the EMG.

Hexacarbon neuropathy has resulted from industrial exposure in various settings including Japanese glue manufacture, a fabrics plant in Ohio, and in Italian bootmakers. Clinical signs of peripheral neuropathy are usually the result of extended low level exposure to either *n*-hexane or methyl butyl ketone. Giant axonal swellings from neurofilamentous accumulations along the nerve fiber are the primary histopathological event. Secondarily, the swellings cause thinning and retraction of myelin especially in the paranodal region. These observations are consistent with the clinical and electrophysiological observations which show a substantial slowing of the maximal motor conduction velocity in proportion to the development of clinical signs. There is not only a loss of axons but also a segmental demyelination of functioning axons. In severe cases, secondary muscle atrophy also occurs.

The neurotoxicity produced by lead is diffuse, affecting both the central and peripheral nervous systems. The clinical effects of lead on peripheral nerves are less profound than of other neuropathic agents. For instance, although segmental demyelination can be induced experimentally with lead, this has been difficult to substantiate clinically. Slowed conduction velocity has been reported in some clinical investigations of lead, but the electrophysiologic deficits do not correlate well with either the clinical effects or histopathology.

Exposure to carbon disulfide occurs mainly in the manufacture of rayon and cellophane. The histologic damage is qualitatively similar to hexacarbon toxicity, producing giant axonal swellings and segmental demyelination. Electrophysiologic measurements show a decreased maximal motor conduction velocity. Some patients also show an abnormal fatigability of muscle contraction following tetanic stimulation, indicating a disruption of the neuromuscular junction.

Physical trauma, especially vibration, is another source of neurotoxic damage. Those who use pneumatic tools and power saws, such as rock drillers and lumberjacks, are at high risk. Vibration induces a Wallerian type of degeneration and results in clinical signs of peripheral neuropathy. Electrophysiologic measurements are concurrently changed. Most commonly this is a slowing of the conduction velocity of the slower fibers in the ulner and median nerves.

REFERENCES

Albuqurque, E.S. and A.C. Oliveira (1979). Physiological studies on the ionic channel of nicotinic neuromuscular synapses. *Advances in Cytopharmacology,* Vol. 3. Raven Press, New York, p. 197.

Basmajian, J.V. (1979). *Muscles Alive, Their Functions Revealed by Electromyography.* Williams & Wilkins, Baltimore.

Brismar, T. (1985). Changes in electrical threshold in human peripheral neuropathy. *J. Neurol. Sci.* 68: 215.

Chanaud, C.M. (1987). A multiple-contract EMG recording array for mapping single muscle unit territories. *J. Neurosci. Methods* 21: 105.

Cooper, J.R., F.E. Bloom, and R.H. Roth (1982). *The Biochemical Basis of Neuropharmacology.* Oxford University Press, New York.

Dewar, A.J. (1983). *Neurotoxicity. Animals and Alternatives in Toxicology Testing.* Academic Press, New York, p. 224.

Eisen, A. (1987). Electromyography in disorders of muscle tone. *Can. J. Neurol. Sci.* 14: 501 (Suppl. 3).

Friedli, W.G. and M. Meyer (1984). Strength-duration curve: a measure for assessing sensory deficit in peripheral neuropathy. *J. Neurol. Neurosurg. Psychiatry* 47: 184.

Fuglsang-Frederiksen, A. (1988). The motor unit firing rate and the power spectrum of EMG in humans. *Electroencephalogr. Clin. Neurophysiol.* 70: 68.

Howard, B.D. and C.B. Gundersen, Jr. (1980). Effects and mechanisms of polypeptide neurotoxins that act presynaptically. *Annu. Rev. Pharmacol. Toxicol.* 20: 307.

Jorengsen, K. (1988). Electromyography and fatigue during prolonged, low-level static contraction. *Eur. J. Appl. Physiol.* 57: 316.

Korczyn, A.D., A. Kuritzky, and U. Sandback (1978). Muscle hypertrophy with neuropathy. *J. Neurol. Sci.* 36: 399.

Krarup, C. and F. Buchthal (1985). Conduction studies in peripheral nerve. *Neurobehav. Toxicol. Teratol.* 7: 319.

LeQuesne, P.M. (1978). Clinical expression of neurotoxic injury and diagnostic use of electromyography. *Environ. Health Perspect.* 26: 89.

LeQuesne, P.M. (1979). Neurological disorders due to toxic occupational hazards. *The Practitioner* 223: 40.

Maxwell, I.C., P.M. LeQuesne, J.M.K. Ekue, and J.E. Biles (1981). Effect on neuromuscular transmission of repeated administration of an organophosphorus compound, metrifonate, during treatment of children with urinary schistosomiasis. *Neurotoxicology* 2: 687.

Mayo Clinic Department of Neurology (1971). Electromyography and electric stimulation of peripheral nerves and muscles. *Clinical Examinations in Neurology.* W.B. Saunders, Philadelphia, p. 271.

Millington, W.R. and G.G. Bierkamper (1982). Chronic triethyltin exposure reduces the resting membrane potential of rat soleus muscle. *Neurobehav. Toxicol. Teratol.* 4: 255.

Mitchell, C.L., Ed. (1982). *Nervous System Toxicology.* Raven Press, New York.

Peper, K., R.J. Bradley, and F. Dreyer (1982). The acetylcholine receptor at the neuromuscular junction. *Physiol. Rev.* 62: 1271.

Seppalainen, A.M. (1975). Applications of neurophysiological methods in occupational medicine, a review. *Scand. J. Work. Environ. Health* 1: 1.

Stalberg, E., P. Hilton-Brown, B. Kolmodin-Hedman, B. Holmstedt, and K.B. Augustinsson (1978). Effect of occupational exposure to organophosphorus insecticides on neuromuscular function. *Scand. J. Work. Environ. Health* 4: 255.

Stein, R.B. (1980). *Nerve and Muscle: Membranes, Cells, and Systems.* Plenum Press, New York..

Stein, R.B. (1988). Novel uses of EMG to study normal and disordered motor control. *Can. J. Neurol. Sci.* 15: 95.

Stevens, C.F. (1985). AChRs: fivefold symmetry and the epsilon-subunit. *Trends in Neuroscience,* Vol. 8, No. 8, Aug. 1985, pp. 335–336.

Warmolts, J.R. and J.R. Mendell (1979). Open-biopsy electromyography. *Arch. Neurol.* 36: 406.

Brain Slices

Peter G. Aitken
Departments of Cell Biology and Neurobiology
Duke University Medical Center
Durham, North Carolina

INTRODUCTION

In the early part of this century, the use of brain slices was largely limited to anatomists and pathologists interested in the structure of the tissue. This was not living tissue, of course, and, in fact, pains were taken to preserve or "fix" the tissue. The earliest demonstrations that mammalian brain tissue could actually function when isolated from the body came in the 1920s from laboratories studying energy metabolism (e.g., Warburg et al., 1924), and it was another three decades before metabolic changes in response to electrical stimulation were reported (e.g., McIlwain, 1951). True electrophysiological investigations began with Li and McIlwain's (1957) demonstration of resting membrane potentials in neurons in a brain slice. Since then, and particularly during the past dozen years or so, brain slices have become an extremely popular preparation for electrophysiological and biochemical studies of the nervous system. This chapter will provide an introduction to the use of brain slices, with particular reference to toxicological research.

Today, the term, "brain slice" refers to thin (usually less than 0.8 mm) sections of tissue taken from the central nervous system (CNS) of a vertebrate animal and maintained in some semblance of normal functioning *in vitro*. The term *in vitro* is Latin for "in glass," and refers to the test tube, incubation chamber, or perfusion bath in which the brain slices are kept. Tissue from a wide variety of CNS regions has been used, including hippocampus, spinal cord, neocortex, cerebellum, and hypothalamus. It seems likely that any part of the CNS can be successfully maintained *in vitro* under the proper conditions. The great majority of slice studies use tissue taken from rats, and less frequently, guinea pigs and mice. Human brain tissue, removed surgically for the treatment of intractable epilepsy, has also been kept alive *in vitro*.

The advantages of brain slices can be summed up in two words: accessibility and control. Compared to working on the nervous system of an intact animal, a brain slice is a marvel of convenience. There are no overlying meninges or blood vessels to contend with, no bleeding to obscure the view. Tissue that normally lies deep in the brain can be studied without the need to dissect away overlying structures. A major advantage is the ability to directly visualize the laminae (layers) that are present in many parts of the nervous system. Such visualization permits extremely precise placement of recording electrodes, stimulating electrodes, and drugs. For example, the hippocampal formation has a laminar organization, as illustrated in Figure 1. The cell bodies of the principal neurons are located in distinct laminae, as are the apical and basal dendrites, and the axons. These laminae are clearly visible when hippocampal slices are viewed through a low-powered microscope. This allows application of experimental drugs to be restricted to one or more specific regions, which in turn permits a precise determination of the location(s) of drug action. The accessibility of slices also allows microknife cuts to be made, which can be used to isolate a neural population from one or more of its inputs.

The body has many regulatory mechanisms that serve to maintain the environment of the brain fairly constant. Thus, ionic milieu, temperature, oxygenation, and other potentially interesting variables are held within tight limits; any attempt by the experimenter to modify them in an intact animal results in compensatory responses by the body. This compensatory action may affect not only the variable that is being directly modified, but others as well. For example, reduction of blood oxygen will cause a compensatory increase in blood flow to the brain. These regulatory systems make it difficult or impossible to study the effects of many potentially interesting variables, in isolation, on the CNS of an intact organism.

The slice, of course, is a totally different matter. Not only is it completely exposed to our manipulations, but it is removed from the influences of the body's regulatory systems. Once the slice is in an incubation chamber, those aspects of its environment that are closely controlled in the body, such as temperature and ionic milieu, can be changed at will. This has permitted many important experiments to be done that were impossible in whole animals.

Some of the limitations of the brain slice go hand in hand with its advantages—they are, so to speak, the opposite side of the coin. Since it is isolated from the body's circulatory and regulatory systems, a slice cannot provide data on effects that would be mediated by these systems. One

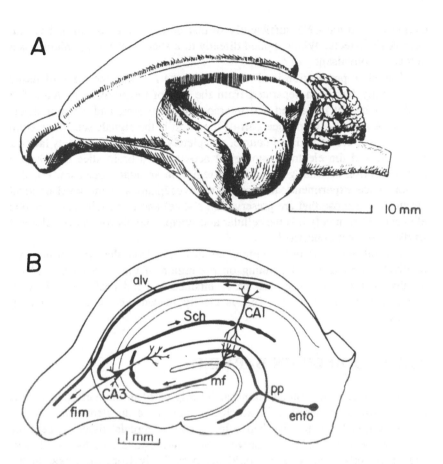

Figure 1. Anatomy of the hippocampal formation and laminar structure of the hippocampal slice. A. Lateral view of the brain with sections of neocortex removed to expose the hippocampus. The indicated section of the hippocampus is the equivalent of one *in vitro* slice, and is enlarged in B to show the major pathways and neurons present in the slice. Abrbreviations: alv, alveus; Sch, Schaffer collaterals; pp, perforant path; ento, entorhinal cortex; mf, mossy fibers; fim, fimbria; CA1 and CA3, pyramidal cell layers (Reprinted from Andersen et al., 1971).

example, would be a substance that, in whole animals, causes damage to the CNS by disrupting one of the regulatory systems that maintain the environment of the CNS, but that has no direct toxic effect on neural tissue per se. Such a substance is carbon monoxide, which causes brain damage by interfering with oxygen delivery. If applied directly to the brain slice, such a substance might appear to be harmless, resulting in a false-negative. On the other hand, there are many substances that do have a direct toxic effect on neural tissue, but which are, in the intact organism, totally or partially excluded from the CNS by the blood-brain barrier. An example of such a substance is penicillin; while not totally blocked by the barrier, penicillin is excluded from the CNS sufficiently so that normal antibiotic doses have no neurotoxic effects. When applied directly to a slice, however, penicillin is a powerful convulsant.

The slice preparation is also limited by the fact that the piece of tissue under study has been separated from the rest of the nervous system. The CNS is a highly integrated and interconnected system, and a major determinant of activity in any one part of the brain is the signals sent there from other areas. Such input is, of course, completely or almost completely lacking in slices, and the electrophysiological activity on a brain slice is at best a poor reflection of what goes on in the region in an intact organism. For this reason, slice experiments are limited to investigations at the level of local circuits (i.e., those that are preserved in a slice) and of cellular and synaptic elements. Fortunately it is the cellular and synaptic levels which toxicological studies are most concerned.

A final limitation of slice preparations results from the fact that they are relatively short lived. Depending on the region of the CNS used, and the incubation protocol, slices can be maintained in good health for 12 hours and possibly for 24 hours. This is clearly inadequate for any long term studies.

THE BLOOD-BRAIN BARRIER

The concept of a blood-brain barrier originated in the early part of this century with an experiment in which live mice were injected with a dye. It was later noted that the dye had colored all the animals' tissues except the brain, spinal cord, peripheral nerves, eyes, and testicles. Further work verified that certain compounds could not pass freely from the blood to the interstitial space of the CNS. The blood-brain barrier is believed to exist in the endothelium of capillaries in the CNS. In other parts of the body, endothelial cells are separated by relatively large gaps through which water and most molecules can pass. In the CNS, the capillary endothelial cells are closely connected by tight junctions; thus, any substance passing through the capillary wall must pass through the plasma membrane of the endothelial cells. Lipid soluble substances pass freely; these include O_2, CO_2, ethanol, and many toxic solvents. Water-soluble molecules are passed to varying degrees; electrically charged molecules are passed less freely than are nonpolar, nonionized molecules. The barrier almost totally excludes macromolecules. The exclusion of macromolecules can also result in the exclusion of certain substances that would freely pass the barrier themselves, but which in blood are bound to plasma protein.

In brain slices the circulatory system is not functioning. Substances enter and leave the interstitial space by diffusion exchange with the bathing medium. The lack of a blood-brain barrier in slices must be kept in mind, particularly when relating the neurotoxic effects of the substance *in vitro* to its effects in the whole animal. An excellent example of the problems that can arise comes from studies of kainic acid. This potent neurotoxin causes seizures and neural damage when injected into an intact organism; it also exhibits potent neurotoxicity when applied to brain tissue *in vitro*. One might at first suspect that these two findings were in fact caused by the same neurotoxic processes. On close examination, however, it turns out that kainic acid's *in vitro* effects are quite different from the effects in the whole animal; in fact, kainic acid appears to be blocked by the blood-brain barrier, and the mechanisms of its neurotoxicity in intact animals remain somewhat of a mystery.

It should be noted that there are several specific structures in the CNS in which the blood-brain barrier is absent or greatly reduced. These regions are known or suspected to have a secretory function, and include the pituitary gland, pineal body, supraoptic crest, and subfornical organ.

BRAIN SLICE TECHNIQUES

A book chapter cannot teach you how to prepare and use brain slices; rather, my intent here is to provide an introduction to the techniques used in slice experiments that will help prepare you for actual hands-on training. Even if you never actually perform experiments with slices, a basic familiarity with the procedures used will be useful when evaluating the work of others. For a more complete presentation of brain slice techniques and reviews of recent slice work, see Dingledine (1984) and Somjen et al. (1987).

Incubation Chamber

To stay healthy, brain slices must be kept in an environment that provides sufficient oxygenation, proper temperature, and bathing medium with appropriate composition. These conditions can be met by a wide variety of chamber designs, ranging from expensive and elaborate commercial models to simple, inexpensive homemade units. Almost all investigators use superfusion chambers, which immerse the slices in a continuously flowing stream of bathing medium; static bath chambers are occasionally used, but are best avoided because they provide no means other than diffusion for the removal of potentially toxic metabolic waste products from the tissue.

Superfusion chambers are two types: submersion chambers, in which the slices are totally submerged in the bathing medium; and interface chambers, in which the slices are supported at the gas-liquid interface at the surface of the bathing medium. Examples of both chamber types are shown in Figure 2. In interface chambers, the gas above the slices consists of a warmed, humidified mixture of 95% oxygen and 5% carbon dioxide. Each

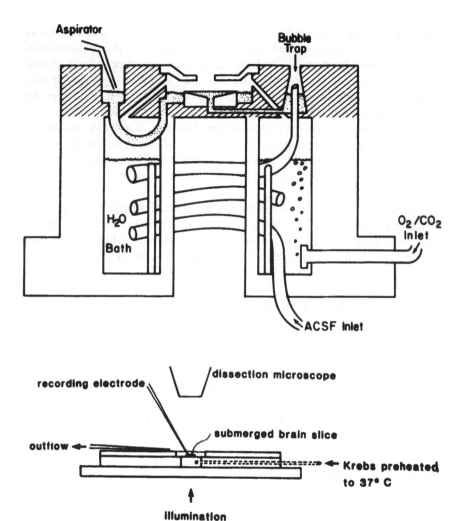

Figure 2. Diagrams of interface (top) and submersion (bottom) slice chambers. In the interface chamber, oxygenated ACSF enters the chamber and is warmed by passing through plastic tubing that is immersed in a heated water jacket. The ACSF then passes through a bubble trap, then under and around the slices to spill into a moat. The ACSF level is determined by the position of the aspirator. O_2/CO_2 is first warmed and humidified by being bubbled through the water jacket, and then directed over the slices. The chamber lid maintains the humid environment over the slices, but has a small hole for electrode placement. The submersion chamber works in a similar manner, but since the slices are totally submerged no gas phase or water jacket is needed and the maintenance of ACSF level is less critical. Heat can be provided by heating elements below the chamber or by the preheated ACSF. In some submersion chambers the flow of ACSF is from top to bottom. (Reprinted from Dingledine, 1984).

TABLE 1. Composition (in Millimoles) of Bathing
Medium Used By Various Investigators For Brain Slices

Na$^+$	K$^+$	Mg^{2+}	Ca^{2+}	Cl$^-$	HCO$_3^-$	Glucose	H$_2$PO$_4$	mOsm
150–152	3.5–6.2	1.3–2.4	1.5–2.5	132–136	24–26	4–10	1.20–1.40	302–311
143–150	5.4–6.2	1.3–2.0	2.0–2.5	127–133	26–26.2	10–11	0.90–1.25	294–307
150	6.4	1.3	2.5	134	26.0	10	1.25	304
151	5.0	2.0	2.0	133	26.0	10	1.25	307
150	6.2	1.3	2.4	134	26	10	1.20–1.24	307
150	6.2	1.3	2.4	134	26	10	1.24	307
153	5.0	1.3	2.4	136	26	10	1.20	308

Adapted from Dingledine, 1984.

type of chamber has advantages. Submersion chambers, because they bathe both surfaces of the slices with medium, permit more rapid diffusion of drugs from the bathing medium into the tissue. Interface chambers, on the other hand, provide superior oxygen delivery and therefore permit slices to be maintained at a temperature closer to normal body temperature.

Most superfusion chambers are set up in such a way that the bathing medium is passed over the slices once and then discarded. A closed-loop system is possible, in which the bathing medium is drawn from the reservoir, passed over the slices, and then returned to the reservoir. This arrangement is useful when it is desired to maximize the concentration, for later analysis, of substances released by the slice.

The Bathing Medium

To maximize the health and survival time of the slices, it is desirable to mimic as closely as possible the natural environment of the brain tissue. In the body, CNS tissue is in contact with interstitial fluid which is, for most substances, in equilibrium with cerebrospinal fluid; thus, the bathing medium for brain slices is designed to be similar to natural cerebrospinal fluid and is called, appropriately enough, artificial cerebrospinal fluid (ACSF). ACSF contains a variety of inorganic ions plus glucose, which is the primary energy substrate of the CNS. The composition of ACSF used by a number of investigators is shown in Table 1. It should be noted that these formulae almost all employ levels of potassium and calcium higher than found in natural CSF (approximately 3.5 mM potassium and 1.2 mM calcium). The reason for this is that elevated levels of potassium and calcium result in superior electrical responses and greater stability in certain kinds of electrophysiological experiments. Yet, given the well documented importance of these ions for neural function, it seems unwise to blindly depart from their normal physiological values. This is not to say that there is any one 'ideal' ACSF composition that should be used for all experiments; the formula use will depend on the problem under investigation. Modifications of ionic composition should be used only for specific purposes—for example, removal of calcium to block synaptic transmission—and with an awareness of the physiological changes that may result.

The pH of ACSF should be within the range found in natural cerebrospinal fluid (7.27 to 7.42). Such values are obtained when ACSF of standard composition is bubbled with 95% oxygen—5% carbon dioxide, as is routinely done before the ACSF is used. In the ACSF formulas given in Table 1, the pH is buffered by the bicarbonate ions in the solution; the use of Hepes, or Hepes and tris, as buffers has also been reported. While much remains to be learned about the specific effects of pH, changes in pH certainly have an effect on neural function (e.g., Balestrino and Somjen, 1988). In any case, good scientific practice would require that pH be controlled. This is particularly important when drugs or other compounds are added to the ACSF, since additions may alter the pH.

It would seem logical to keep brain slices at the normal temperature of the animal from which they are taken (38–39°C for rats and guinea pigs). Indeed, when trying to generalize the results of slice experiments to the intact animal, interpretation is always more straightforward if the *in vitro* experiments are performed at or near normal body temperature. However, it is generally the case that slice preparations survive longer and in better condition at temperatures between 30–35°C, probably because at these lower temperatures there is a more even match between oxygen demand and oxygen delivery. This is particularly true of submersion chambers, which appear to deliver oxygen to the tissue less efficiently than do interface chambers. Generally speaking, it is advisable to use the highest temperature that is consistent with the survival of healthy slices for the required length of time.

Oxygenation

Of all the tissues in the body, brain tissue is the most sensitive to oxygen deprivation. The mechanisms of oxygenation *in vitro* are quite different from those encountered in the body. In the latter case, oxygen is delivered by blood flowing through capillaries. In the CNS, the average intercapillary distance is 50 μ; thus, no neurons are more than 25 μ from their source of oxygen. In addition, the presence of hemoglobin in the blood allows it to carry significantly more oxygen per unit volume than could be carried in solution alone. In the slice, there is no blood circulation, so oxygen can only be obtained by diffusion through the surfaces of the slice from the external environment. Slice thickness is greater, often many times greater, than the average intercapillary distance, so the distance that the oxygen has to travel is correspondingly greater. Measurements of oxygen within slices has shown U-shaped profiles of oxygen partial pressure, with pO_2 highest at the two surfaces and lowest in the center of the slice (Fujii et al., 1982). With thicker slices (> 0.4 mm) there may be an anoxic core in the center of the slices. These facts must be kept in mind when planning and interpreting experiments.

For electrophysiological recordings, the ideal electrode placement depth within the slice is a compromise between being near the surface, where neurons have been damaged by slicing, and being too deep, where neural function may be compromised by oxygen deficiency.

Neural tissue is sensitive to oxygen deprivation under all conditions, and hypoxic damage is an important factor in many clinical conditions such as cardiac arrest and stroke. Because of its excellent controllability, the slice preparation has been used widely to investigate the effects of, and methods of protection against, hypoxic damage (e.g., Aitken et al., 1988; Balestrino et al., 1989).

Preparation of Slices

It will come as no surprise to anyone that CNS tissue is exceptionally fragile. The trauma of slice preparation is unavoidably severe but can be minimized with good techniques. There is a great deal of disagreement, superstition, and myth as to what constitute "proper" techniques for removal of the brain and cutting slices. One highly respected slice laboratory even speaks of a "slice goddess" who must be propitiated if good slices are to be obtained! Despite personal variations in technique, a body of shared knowledge and experience has accumulated, the basics of which I will present here.

The experimental animal must be sacrificed in a manner that prevents or minimizes pain, and that yields good slices. The two most common methods of sacrifice are decapitation and ether anesthesia; occasionally, killing by a blow to the back of the neck is used. If ether anesthesia is used, deep anesthesia appears to give superior results than does light anesthesia.

The next step is to remove the brain from the skull. Speed is important here, since brain tissue deteriorates rapidly when globally ischemic (i.e., when blood flow stops). Younger animals (rats less than 200 g, guinea pigs less than 400 g) are popular because of their thinner skulls, which make the dissection easier and faster. Once the brain has been removed, it is placed in a small beaker of chilled, oxygenated ACSF. The chilling reduces the tissue's oxygen demands and therefore lessens the damage caused by oxygen deprivation during slicing; chilling also seems to make brain tissue firmer, which facilitates dissection and slicing.

After the brain has chilled for a minute or two, a piece of the desired tissue must be removed from the brain and prepared for slicing. Specific dissection techniques will vary depending on the brain region desired and on the slicing technique to be used (more on this later). During this part of the procedure, gentleness is extremely important—more important, in fact, then speed. A good rule to follow is that contact between dissection instruments and the actual tissue that will be sliced should be minimized. The tissue should be irrigated with chilled ACSF regularly during the dissection. Any visible blood vessels should be removed, as they are considerable tougher than the tissue itself and will interfere with cutting.

There are two types of cutters, both of which use standard commercial double-edged razor blades (my favorite is Gillette "Super Blue," although others probably work just as well). The blade should be cleaned with an organic solvent, such as diethyl ether, to remove antirust oil placed on the blade by some manufacturers. Tissue choppers cut slices with a rapid, guillotine-like stroke. The piece of tissue is placed, in the proper orientation, on a layer of filter paper that is glued to the chopping block (with silicone aquarium cement) and moistened with ACSF. The blade descends, cutting off a slice which is removed from the blade with a fine sable-hair artist's brush and placed in a container of chilled ACSF. Between strokes the blade is raised and the chopping block is advanced by a distance equal to the desired slice thickness.

The other type of cutter, called a vibratome, uses a rapidly vibrating blade that is advanced slowly through the tissue. The tissue block is glued to a metal or plastic mount using a thin layer of cyanoacrylate glue. Since this glue is toxic, a larger than needed piece of tissue should be dissected free of the brain, so there will be extra tissue for attachment to the cutting block. Lateral support, if needed, can be provided by either gluing small blocks of agar around the tissue, or by pouring a viscous agar solution over the tissue and setting the agar by a quick dip in ice-cold ACSF. The mounting block is then clamped in the cutter with the tissue at the desired orientation; slices are cut and removed to a container of chilled ACSF.

A vibrating blade cutter is considerably slower and more involved to use than a chopper, but its use appears necessary to obtain viable slices from certain tissues, notably spinal cord and cerebellum. Other tissues, such as hippocampus and hypothalamus, can be successfully sliced with a chopper.

Once the slices have all been cut, they can be transferred directly to the incubation chamber where they must be allowed to "rest" for a while to recover from the trauma of slicing. The required recovery time depends to some extent on the tissue; it is 60–90 min for hippocampus. Alternatively, the slices can be placed in a holding chamber and later transferred as needed to the experimental chamber. Holding chamber designs vary widely, but they must provide the slices with essentially the same environment as the experimental chamber, with the exception that holding chambers are usually kept at room temperature to slow the deterioration of the slices. An important criticism of holding chambers is that their use makes "slice age" (i.e., time since cutting) a variable difficult to control.

Delivery of Drugs

There are four commonly used methods of delivering drugs to slices. Each has advantages and disadvantages, and the method(s) used in a given experiment will depend on the design of the slice incubation chamber and on the drug and tissue under study. In many situations, more than one method should be used to assure that the effects of the drug are not contaminated by artifacts associated with one specific delivery method.

The most straightforward method of application is to dissolve the drug in the superfusion medium. This is the only method that permits the exact extracellular drug concentration to be known. Although this method eventually results in a homogeneous distribution of the drug throughout the slice, the concentration within the slice changes relatively slowly as the drug diffuses into or out of the slice. This problem is less severe in submersion chambers than in interface chambers, since the latter expose only one surface of the slice to freely flowing ACSF. This slow equilibrium can cause problems, such as desensitization of receptors that bind the drug being applied. In addition, determination of dose-response curves can be very time consuming.

Microiontophoresis is a technique that delivers minute quantities of drug from the tip of a micropipette. Because electric current is used to eject the drug, only charged molecules can be delivered by iontophoresis. It is possible to fabricate iontophoresis pipettes with many (5–7) barrels, allowing controlled delivery of several different substances to the same location. Since the diameter of the pipette tip is typically 5–10 μ, drug action is confined to a small region of the target tissue. This has the advantage of allowing precise spatial mapping of drug responses, but can also lead to false-negatives when the pipette is positioned incorrectly. Delivery of drugs by microiontophoresis also has the advantages of relatively rapid onset and good control of the duration and termination of deliver. Disadvantages of this method include possible electrical recording artifacts caused by the iontophoresis ejection current, and the impossibility of knowing even roughly the actual concentration of the drug at its site of action. For a thorough review of the microiontophoresis technique, see Purves (1981).

The pressure pipette is another method of delivering small quantities of drug to localized regions of tissue. It differs from iontophoresis in using pressure, rather than electrical current, to eject the drug from a fine-tipped pipette. This removes the possibility of electrical recording artifacts, as well as making it possible to use uncharged drugs. However, the configuration of the pipette tip is more important, as minor changes in diameter or shape may cause significant changes in the amount of drug released at a given pressure. In addition, the use of multibarreled pipettes is more difficult. Pressure pipettes are most commonly used to deliver short "puffs" of drugs to the target tissue.

The so-called "nanodrop" technique can be used to apply small (50 nl or less) drops of drug-containing ACSF to the surface of a slice. Pipettes are used that have had their tips broken back to a diameter of 10–30 μ; the pipette is filled with the desired solution and connected, via a fluid-filled tube, to a syringe. Application of pressure causes a small drop of fluid to form at the top of the pipette, and the pipette is lowered until the drop contacts the slice. One problem with this technique is the possibility of mechanical artifacts when the drop hits the slice; this is particularly true when recording intracellularly. Local cooling effects must also be considered. On the positive side, the nanodrop technique allows nearly instantaneous drug application and, to some extent, localization of effect. Of course, this method cannot be used with submerged slices.

VALIDITY OF DATA OBTAINED FROM SLICES

It is important to know, or at least be fairly confident, that data obtained from a slice preparation are generalizable to the intact nervous system. There are two levels at which this evaluation must be done. The first level concerns the general validity of the model: does the brain region of interest, when studied *in vitro*, maintain enough of its normal electrophysiological and metabolic characteristics to provide a valid model of its function in the intact animal? For example, when the hippocampal slice preparation was first becoming popular, Schwartzkroin (1975) examined some of the electrophysiological properties of the hippocampal slice and compared them with those of the intact hippocampus and found that they were quite similar. As regards energy metabolism, however, there are some significant differences between the slice and the *in situ* brain. For example, O_2 uptake and ATP levels are approximately one third lower in the slice (see Lipton and Whittingham, 1984, for further details). The reasons for these and other metabolic differences are not entirely clear.

The second level of evaluation is related to the condition of a particular batch of slices. Even the most experienced slicer cannot be assured of having healthy slices in every experiment. A slightly careless dissection, a malfunction in the incubation chamber, or an error in mixing ACSF could result in sick or dead slices. It is very important, once the slices are cut but before experimentation is begun, to verify that the slices meet some minimum standard of "health." This is best done using electrophysiological techniques to evaluate synaptic transmission and neural function. The nature of an acceptable response depends, of course, on the specific preparation and incubation techniques being used, as well as on the experiments planned.

SLICES IN NEUROTOXICITY

Neurotoxicity research can be broadly divided into three categories:

1. Determination of whether a substance has toxic effects on the nervous system. Due to the nature of their limitations, slices are essentially useless in this type of research. The possibilities of false-positives and false-negatives, as discussed earlier, are always present. At least

for the present, the use of whole animals for neurotoxicity screening seems unavoidable. Whole animal experiments are also necessary to determine the specific location(s) of neurotoxic damage.

2. Investigations of the specific mechanisms of toxicity. Once it has been determined that a substance is toxic to the nervous system, and the location of the damage is known, brain slices *may* be useful in determining exactly how the substances exerts its effects (e.g., Aitken and Braitman, 1989). I emphasize "may" because one can certainly imagine situations in which brain slices will be of no use or, even worse, will provide misleading information. Careful planning of experiments is imperative, with close attention to the results of whole animals studies as to the location and nature of the damage, and to the possible involvement of circulatory and regulatory systems and the blood-brain barrier.

3. Use of toxic substances with known actions as tools in neurobiological research. To some, this kind of work may seem to fall outside the boundaries of neurotoxicology, but it is precisely here that brain slices have been seen, perhaps, their greatest use. Given the accessibility of a slice, the use of compounds with specific known actions has become an extremely powerful technique for unraveling the complexities of brain function. The list of substances used would almost fill a book. Examples include tetrodotoxin, which blocks regenerative sodium channels in nerve membranes; ouabain, which poisons the sodium-potassium pump; penicillin, which blocks inhibitory neural transmission mediated by gamma-amino butyric acid; and curare, which blocks nicotinic synapses.

INTERPRETATION OF DATA OBTAINED FROM SLICE PREPARATIONS

In many experiments that use a brain slice preparation to evaluate mechanisms of neurotoxicity, specific attention is directed toward excitatory synaptic transmission. The experimental protocol places an electrical stimulating electrode among presynaptic axons, and a recording electrode among postsynaptic cells. Stimulation of the presynaptic fibers at sufficient intensity causes the postsynaptic cells to fire action potentials which can be recorded as the postsynaptic response. Upon application of the substance being tested, changes in the postsynaptic response are taken as an indication of the substance's effect on synaptic transmission. Proper interpretation of this type of experiment requires an awareness of the complex steps involved in synaptic transmission. The following is an extremely abbreviated and simplified account of the possible mechanisms by which a neurotoxin could affect synaptic transmission.

On the presynaptic side, electrical stimulation evokes action potentials in the presynaptic axons. The action potentials propagate to the presynaptic terminals where they cause voltage-sensitive calcium channels to open. Calcium ions enter the terminal and trigger a chain of events that results in synaptic vesicles fusing with the cell membrane and releasing neurotransmitter into the extracellular space between the pre- and postsynaptic neurons. This chain of events offers many potential locations for neurotoxic actions, including interference with action potential generation, blockade of calcium channels, inhibition of neurotransmitter synthesis, and interference with intracellular mechanisms of exocytosis.

Postsynaptically, the released neurotransmitter binds to receptors on the surface of the postsynaptic cells. In most cases of excitatory (or inhibitory) synaptic transmission, activation of the receptors directly modifies the state of an ionophore, or ion channel, associated with the receptor. The resulting changes in membrane permeability affect the cell's membrane potential and make it more or less likely to fire action potentials. A neurotoxin could affect the postsynaptic cell by blocking the effect of the neurotransmitter on the receptors (either as a competitive or noncompetitive antagonist), by blocking the associated ion channel, by modifying intracellular second messenger systems, or by modifying the state of other ion channels not directly affected by the neurotransmitter.

SUMMARY

While brain slices have been widely used in experiments that use neurotoxins as tools for the investigation of neural function, their use in "pure" neurotoxicology research has been, to date, rather limited. The potential power of brain slice preparation would seem to insure a widening role in this field, particularly in investigations of the cellular mechanisms of acute neurotoxicity.

REFERENCES

Aitken, P.G., M. Balestrino, and G.G. Somjen (1988). NMDA antagonists: lack of protective effect against hypoxic damage in CA1 region of hippocampal slices. *Neuroscience Lett.* 89: 187–192.

Aitken, P.G. and D.J. Braitman (1989). The effects of cyanide on neural and synaptic function in hippocampal slices. *Neurotoxicology* 10: 247–256.

Anderson, P., T.V.P. Bliss, and K.K. Skrede (1971). Lamellar organization of hippocampal excitatory pathways. *Exp. Brain Res.* 13: 222–238.

Balestrino, M., P.G. Aitken, and G.G. Somjen (1989). Spreading depression like depolarization in CA1 and fascia dentata of hippocampal slices: relationship to selective vulnerability. *Brain Res.* 497: 102–107.

Balestrino, M. and G.G. Somjen (1988). Concentration of carbon dioxide, interstitial pH and synaptic transmission in hippocampal formation of the rat. *J. Physiol.* 396: 247–266.

Dingledine, R.E. (1984). *Brain Slices.* Plenum Press, New York.

Fujii, T., H. Baumgartl, and D.W. Lubbers (1982). Limiting section thickness of guinea pig olfactory cortical slices studied from tissue PO_2 values and electrical activities. *Pflügers Archives* 393: 83–87.

Li, C.L. and H. McIlwain (1957). Maintenance of resting membrane potentials in slices of mammalian cerebral cortex and other tissues in vitro. *J. Physiol.* 139: 178–190.

Lipton, P. and T.W. Whittingham (1984). Energy metabolism and brain slice function. In *Brain Slices.* R. Dingledine, Ed., Plenum Press, New York.

McIlwain, H. (1951). Metabolic response in vitro to electrical stimulation of sections of mammalian brain. *Biochem. J.* 49: 382–393.

Purves, R.D. (1981). *Microelectrode Methods of Intracellular Recording and Iontophoresis.* Academic Press, New York.

Schwartzkroin, P. (1975). Characteristics of CA1 neurons recorded intracellularly in the hippocampal in vitro slice preparation. *Brain Res.* 85: 423–436.

Somjen, G., P.G. Aitken, M. Balestrino, and S. Schiff (1987). Uses and abuses of in vitro systems in the study of the pathophysiology of the nervous system. In *Brain Slices: Fundamentals, Applications, and Implications.* A. Schurr, T.M. Teyler, and M.T. Tseng, Eds., S. Karger, Basel, pp. 89–104.

Warburg, D., K. Posener, and E. Negelein (1924). Uber den Stoffwechsel der Carcinomzelle. *Biochem. Z.* 152: 309.

Part
III

Mechanisms of Neurotoxicity

Section C. Pathological Studies

Basic Histopathological Alterations in the Central and Peripheral Nervous Systems: Classification, Identification, Approaches, and Techniques
L.W. Chang

Part

III

Mechanisms of Neurotoxicity

Section A: Pathological Basis

Basic Histopathological Alterations in the Central and Peripheral Nervous Systems: Classification, Identification, Approaches, and Techniques

Louis W. Chang
Departments of Pathology and Pharmacology/Toxicology
University of Arkansas for Medical Sciences
Little Rock, Arkansas

INTRODUCTION

Neuropathology represents one of the most important aspects of modern neurotoxicology. The basic objectives of neuropathology, morphologically speaking, are twofold: to provide information on the topography or location of the lesion (e.g., in the anterior horn of the spinal cord), and to define the nature and characteristics of the damage (e.g., neuronal swelling and vacuolation or necrosis). This neuropathology, when performed appropriately, helps to provide information on what, where, when, and sometimes even how toxic lesions occur. This information may be used as a immensely useful "indicator" or "endpoint" for a neurotoxic agent in question or be used to correlate with and to explain various other observations and data (e.g., neurophysiological, behavioral, and biochemical changes) generated.

While the basic anatomy and physiology of neurons (nerve cell bodies and their processes) are being covered in a separate chapter, this treatise intends to present only the most commonly seen histopathology (changes at light microscopic level) involving the neurons, their processes, and some of the supporting elements, such as the myelin sheaths, as occurred in toxic and other forms of cellular injuries.

A brief section to include some of the most commonly used histological techniques for the studies of neuropathology will also be presented. It is hoped that this information would be helpful to those who wish to perform basic neuropathological investigation in their own laboratories.

PATHOLOGY OF THE NEURON (NEURONOPATHY)

The basic functional unit of the nervous system is the neuron. It must be understood, however, that the nervous system is of an extremely complex nature composed of a heterogeneous group of neurons. The size and morphology of one group or type of neurons may be very different than those in another loci. The large neurons, such as those found in the anterior horn of the spinal cord, are pyramidal shaped with prominent nuclei and cytoplasmic Nissl substance. Small nerve cells, such as the granule cells in the cerebellar cortex, are round with small nuclei and minimal cytoplasm. Thus, one may expect different reactions to similar assaults, and different patterns or types of degenerative changes may be displayed by different groups of nerve cells. However, all neurons share the same characteristic in requiring a constant supply of oxygen and glucose. Under normal metabolic conditions, no neuron in the central nervous system can tolerate more than 15 min of anoxia. Moreover, since neurons in mature nervous system are incapable of cell division, replacement of neurons is not possible. Therefore, a significant loss of neurons as a result of injury may lead to permanent deficit in functions.

When neurons are injured, certain "reactions" to the injury may occur which may or may not be reversible. If the injury was proven to be too severe for recovery, eventual neuronal death may occur.

Some basic reactions of neurons to injuries may be outlined as follows.

Central Chromatolysis

This is a neuronal reaction primarily as a response to injury of its axon. It is also referred as "axonal reaction" and can be best exemplified in large neurons whose axons are partially extended into the peripheral nervous system as peripheral nerves (e.g., anterior horn motoneurons). This reaction is most prominent when the injury is proximal to the neuronal cell body. This reaction will take place approximately 48 h following injury and may reach its climax in about 2–3 weeks.

In this reaction, the neuronal cell body becomes slightly swollen (rounded) with a dissolution of the Nissl substance in the cytoplasm. This disappearance

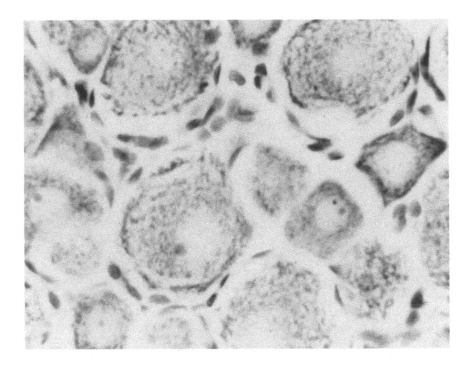

Figure 1. Large neurons from dorsal root ganglion stained with gallocyanin showing abundance of Nissl substance in the neuronal cytoplasm.

of Nissl substance begins around the nucleus (thus called "central chromatolysis") and may spread throughout the entire cell. The nucleus also moves away from the axonal hillock and acquires an eccentric position at the peripheral margin of the cell (Figures 1, 2, and 3). This type of neuronal degeneration or reaction is usually reversible depending on the severity of the injury. If the cell manages to recover, the Nissl substance will gradually reform and reaggregate, first around the nucleus and may eventually become more heavily loaded with Nissl substance (heavier staining) than normal.

Injuries to the intrinsic tracts in the central nervous system (e.g., brain stem and spinal cord) may produce similar neuronal changes (e.g., Betz cells in the motor cortex). However, these cells usually will continue to degenerate and die without recovery.

Peripheral Chromatolysis

As the term denotes, peripheral chromatolysis is, in a sense, an opposite phenomenon to central chromatolysis. It may represent a neuronal attempt to compensate or to survive a noxious assault. Nerve cells showing peripheral chromatolysis show a slight reduction in size with a depletion or disappear-

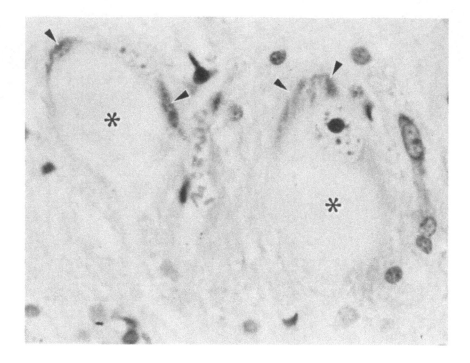

Figure 2. Central chromatolysis of the brain stem neurons induced by trimethyl lead intoxication. Note the eccentric nuclei and absence of Nissl staining in the neuronal cytoplasm (*). Remnant of Nissl substance (arrowheads) still can be seen at the periphery of the cells.

ance of Nissl granules in the periphery zone of the cytoplasm. Heavy aggregation (hyperchromatic) of Nissl substance, however, can be observed around the nucleus which remained in the center of the cell. A good example of this type of neuronal change can be observed in the anterior horn mononeurons in cases of progressive muscular atrophy.

Neuronal Atrophy

Atrophy denotes a reduction of cellular size. Simple atrophy of neurons is most commonly seen in metabolic disturbance situations which may lead to eventual degeneration and death of the neuron. It may be observed in certain vitamin deficiency states, chronic ischemia, or in systemic degeneration such as amyotrophic lateral sclerosis.

If the axons are also involved, larger fibers are more susceptible than those of smaller caliber and the distal portion of the axon is usually affected first. The nerve cell body often shrinks in size and becomes sclerotic without showing the reaction of chromatolysis. This phenomenon of involvement of the distal portion of the axon with eventual involvement of the proximal axon

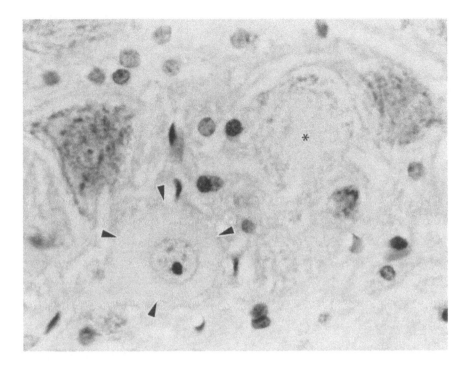

Figure 3. Chromatolysis of the brain stem neurons induced by trimethyl lead intoxication. While two neurons still showed well preserved Nissl substance, one neuron showed central depletion of Nissl substance (*) and another neuron showed total absence of Nissl substance (arrowheads).

and neuronal cell body is sometimes referred to as distal axonopathy of "dying-back" process. This kind of neuronal and axonal change is best exemplified in certain acute pheripheral neuropathy, such as in acrylamide intoxication (Blackmore and Cavanagh, 1969; Prineas, 1969; Spencer and Schaumburg, 1976).

Injuries to axons of nerve cells within the central nervous system, such as those in the dorsomedial nucleus of the thalamus (frontal lobectomy) and inferior olives (lesion in cerebellar cortex), may initially induce neuronal swelling and central chromatolysis followed by eventual atrophy of the nerve cell bodies. This type of biphasic reaction is referred to as Gudden's atrophy.

Transsynaptic atrophy or degeneration may also occur in certain neurons following damage to the axons which make synaptic contact with them. (From example: damage or lesion in the optic nerve or optic tract induces atrophy or degenerative changes of the nerve cells in the external geniculate nuclei.)

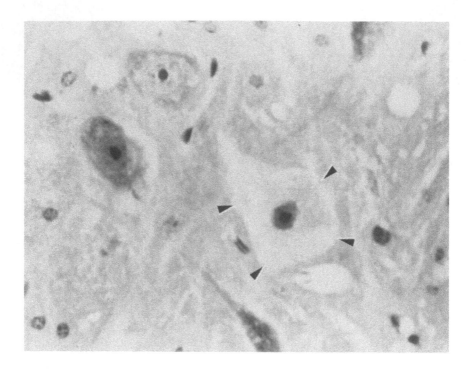

Figure 4. Neuronal edema induced by trimethyltin. The edematous neuron (arrowheads) has a distended, watery, and lacy cytoplasm with a centrally located nucleus.

Neuronal Swelling and Vacuolation

Cellular swelling represents a common and early change of cellular injury or trauma. Neuron is not an exception. Cellular swelling is the result of water accumulation in the cytoplasm of the cell yielding a distended and enlarged cell. This condition is usually referred to as edematous change and may reflect a disturbance of the electrolyte balance system (mitochondrial function, ATP-system, NA^+-pump, plasma membrane integrity, etc.) of the cell. Sodium influx into the cell is increased followed by water. The neuron involved acquires a rounded and distended appearance and the cytoplasm has an even lucid and "watery" appearance (hypochromatic). The nucleus, however, usually remains centrally located in most cases (Figure 4).

Routine paraffin sections (10 nm thick) may demonstrate a lacy or floccular cytoplasm. Thin sections (1–2 nm) from plastic embedded tissue, however, reveal tiny cytoplasmic vacuoles within these nerve cells (Figure 5). With the aid of electron microscopy, distended cisternae of the rough endoplasmic reticular (RER) can be observed. This distention of the RER is believed to be a cellular mechanism to contain the increased cytoplasmic fluid. Continued enlargement and distention of the RER cisternae will give

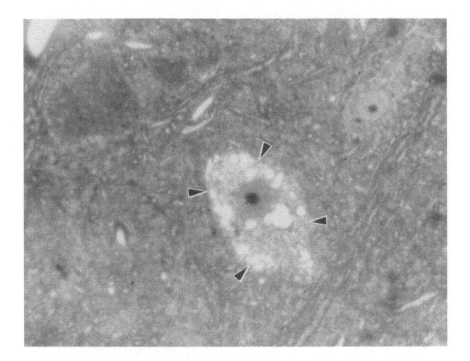

Figure 5. Neuronal vacuolation induced in an anterior horn motoneuron (arrowheads) by trimethyltin. This preparation is a thin section (1 nm) from plastic embedded tissue stained with toluidine blue. The tiny cytoplasmic vacuoles are the result of edematous distention of the endoplasmic reticular system in the cell.

rise to light microscopically observable cytoplasmic vacuoles in the neurons. Fusion of these vacuoles may give rise to the appearance of total vacuolation of the neurons (Figure 6). Such phenomenan may be termed hydropic or vacuolar degeneration reflecting intracellular edematous condition of the neuron. This type of neuronal change may be found in certain acute toxic situations such as those seen in the brainstem neurons or anterior horn cells of the spinal cord of mice under the toxic influence of trimethyltin (Chang et al., 1983, 1984).

Neurofibrillary Degeneration

This is a special form of neuronal degeneration consisting of thickening and tortuosity of fibrils within the neuronal cytoplasm. This type of degeneration is commonly seen in human patients with Alzheimer's disease, senile dementia, and Parkinsonism-dementia of Guam. Neurons in the frontal and temporal cortices as well as in the hippocampus are most frequently affected. These fibrillary changes can be best demonstrated with silver staining meth-

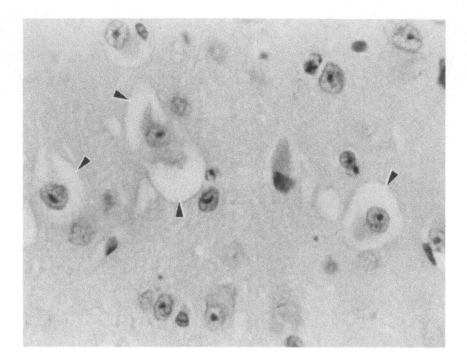

Figure 6. Neuronal vacuolation or hydropic degeneration of some olfactory neurons (arrowheads) as a result of trimethyltin intoxication. The neuronal cytoplasm is very distended and appear to be vacuolated. This change represents a severe edematous condition of the cells.

ods, such as Bielchowsky's mehtod, and are referred to as neurofibillary tangles in the neurons.

In experimental aluminum poisoning, similar, though not necessary identical fibrillary accumulation are induced in the cortical nerve cells (Klatzo et al., 1975, Wisniewski et al., 1980, Petit et al., 1980). The pathogenesis of Alzheimer's disease and aluminum intoxication are not believed to be the same; however they constitute interesting and important ground for the development of this special form of neuronal changes.

Neuronal Necrosis

Cellular necrosis (cell death) is recognized in light microscopy as pyknotic cells which possess a condensed nucleus (Pyknotic nucleus) and highly eosinophilic cytoplasm (Figure 7). Occasionally, fragmentation and breakdown of the nuclei (karyorrhexis) (Figure 8) or total dissolution and disappearance of the nuclei (karyolysis) may also occur.

Pyknotic changes of neurons are most frequently seen in acute neuronal poisoning (e.g., hippocampal neurons in trimethyltin intoxication) (Chang

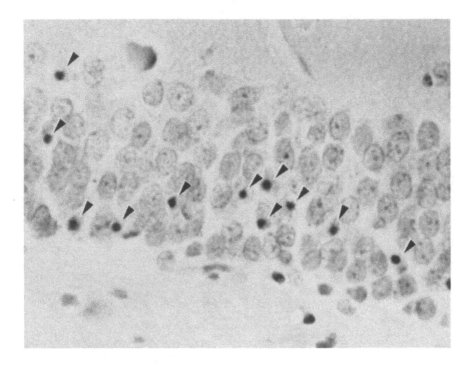

Figure 7. Trimethyltin induced neuronal death (necrosis) in the dentate granule cells of a mouse. Note the necrotic cells appear shrunken and have very densely stained pyknotic nuclei and eosinophilic cytoplasm (arrowheads).

et al., 1982) or in anoxia. It must be emphasized that the occurrence or appearance of pyknotic changes is a delayed process. There is a time lapse of 8–12 hours following cell death ("biological cell death") before such morphological change ("morphological cell death") can be detected or determined by light microscopy.

PATHOLOGY OF NEURONAL PROCESSES

The neuronal processes, including axon (myelinated or unmyelinated) and dendrites, are important functional structures of a nerve cell. Specific or selective damage to these structures by various toxicants or in certain disease states are frequent. Familiarity with these pathological changes is important for their detection and understanding of the toxic action or disease involvement which induces such changes.

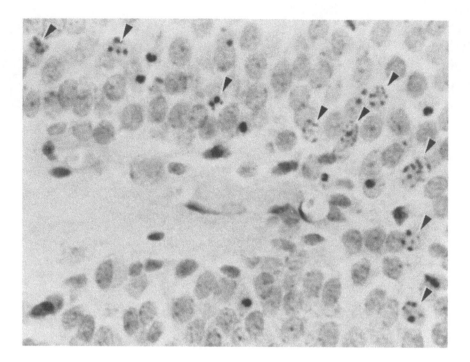

Figure 8. Neuronal necrosis induced in the dentate granule cells by trimethyltin. Besides pyknotic change of the neuclei (clumping of nuclear chromatin), fragmentation of the nuclear chromatin, giving the appearance of multiple densely stained particles (arrowheads), could also be observed. This process is termed karyorrhesis. Both pyknosis and karyorrhesis, as well as karyolysis (disappearance of nucleus) represent cell death.

Wallerian Degeneration

Wallerian degeneration refers to secondary degeneration of the axon and its myelin sheath when they are disconnected from the nerve cell body or when the nerve cell body dies from direct injury.

In this process, the axon becomes swollen, varicose, and acquires irregular shape. The associated myelin sheath also breaks down, usually beginning at the nodes of Ranvier, and forms a series of ellipsoid structures enclosing fragments of the degenerating axon (Figure 9). The myelin sheath will also eventually break into simpler lipids to be removed by macrophages. The process of Wallerian degeneration is a slow one; however, it tends to go at a rate directly proportional to the size of the fiber involved.

Axonopathies

Pathological changes primarily involving axons may be grouped together and referred to as axonopathies. Such involvement may be further classified as distal and proximal axonopathy according to the initial site of involvement.

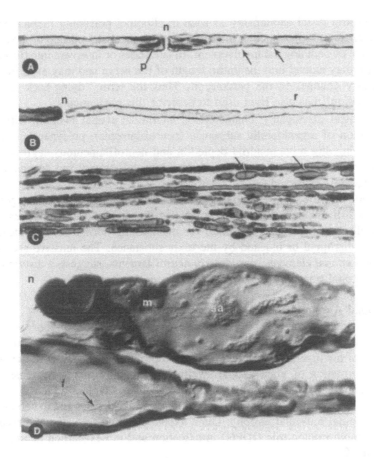

Figure 9. Nerve fiber teasing technique showing (A) normal nerve fiber with node of Ranvier (n), paranodal myelin fluting, and incisures of Schmidt-Lanterman; (B) original, thickly myelinated portion of a nerve fiber, node of Ranvier (n), and a remyelinating portion (r) of the fiber; (C) group of fibers in various stages of Wallerian degeneration with characteristic myelin ovoids (arrows); (D) nerve fibers intoxicated with *n*-hexane (m, mitochondria; sa, Schwann cell; arrow, neurofilament).

Distal Axonopathy

Distal axonopathy is characterized by degenerative changes primarily involving the distal portion of the axon. The neuronal cell body (perikaryon) is usually not directly involved and remains morphologically intact. When both the central and peripheral nervous systems are involved (e.g., motor fibers of the sciatic nerve), the condition is referred to as central-peripheral distal axonopathy. This is typical of carbon disulfide or acrylamide intoxications (Seppalainen and Haltia, 1980; LeQuesne, 1980). If the involvement is only confined to the central nervous system, this process may be referred to as central distal axonopathy as seen in clioqinol poisoning (Schaumburg and Spencer, 1980). Although in most situations the distal portion of the axon is the site of pathological involvement, in late states or in severe intoxication, changes may extend into the entire length of the nerve and may even induce secondary changes in the perikaryon. Thus the term "dying-back axonal degeneration" has also been used to describe this phenomenon.

In distal axonopathy, characteristic internodal axonal swelling with accumulation of argentophilic substance (neurofilaments), presynaptic degeneration showing mitochondrial depletion and neurofilamentous aggregation, as well as end-organ (e.g., gastrocnemius muscle) atrophy and degeneration are frequently observed.

Although the precise pathogenetic mechanism for this type of degeneration is still unclear, it has been suggested that in distal axonopathies, the metabolism or mechanism required for maintaining normal axonal integrity may be disrupted or blocked by the toxic substance. The massive accumulation of axonal elements, particularly neurofilaments, suggests a disturbance in the axonal transport or axoplasmic flow mechanism.

Proximal Axonopathy

Proximal axonopathy involves selective degeneration of the proximal axons, including the axonal hillock, the initial segment, and the proximal portion of the myelinated axon. The most important and characteristic change is the formation of giant axonal swellings containing increased density of neurofilaments. This type of degenerative change is typically seen in motor neuron diseases such as human amyotrophic lateral sclerosis and in hereditary canine spinal muscular atrophy. Giant axonal swellings are also observed in α,α-iminodipropionitrile (IDPN) intoxication and in hexacarbon neuropathy (Figure 10).

In IDPN intoxication, light microscopic examination showed large, swollen structures devoiding nuclei (thus they were referred to as "ghost cells"). Later investigation revealed that these structures were argentophilic and were enlarged portions of the proximal axons filled with massive accumulations of neurofilaments (Chou and Hartmann, 1964, 1965). The accumulation of neurofilaments in the proximal axons are found to be associated with an impairment of axonal transport of the slow component.

The basic pathology of IDPN-induced proximal axonopathy includes prominal axonal swelling, progressive atrophy of the distal axon, and secondary demyelination and remyelination attempts.

Dendritic Changes

Although axonopathies are extensively studied by neuropathologists, toxic changes of the dendrites received much less attention. This lack of investigation is most likely related to the fact that dendritic studies cannot be studied with routine histological preparations (paraffin embedded tissues) and requires much more elaborate and special techniques.

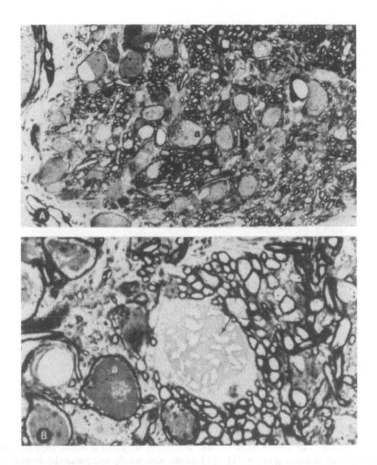

Figure 10. (A) Giant axonal swellings (a) in the gracile tract of the medulla oblongata of young adult rat with hexacarbon neuropathy. (B) Giant axonal swelling (a) in a normal but aged (ca. 2 years) rat. Note the characteristic clefts (arrow) in one swollen axon. Preparation from epony embedded section (1 nm) stained with toluidine blue.

Dendritic changes in Purkinje neurons, such as atrophy and reduction of organized secondary and tertiary branches of the dendritic aberration, have been observed in chronic mercury poisoning (Chang and Annau, 1984) and in alcoholism. This type of dendritic change is most prominently seen in developmental brains under the influence of toxic substances. As attention for developmental neurotoxicology has increased to a greater degree at the present time, more investigation on dendritic changes under this situation are warranted.

MYELINOPATHY

Myelinopathy refers to primary destruction or pathology of the myelin sheaths of the axons. Poor or underdevelopment of the myelin sheaths as a

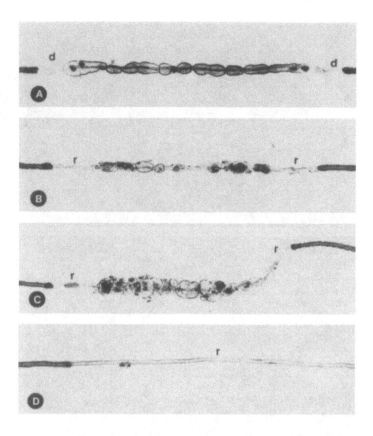

Figure 11. Nerve fiber teasing technique on tibial nerves subjected to acetyl ethyl tetramethyl tetralin intoxication showing (A) bubbling of myelin internodally and paranodal demyelination (d), (B) myelin bubbling and early paranodal remyelination (r), (C) phagocytes removing damaged myelin and paranodal remyelination, (D) advanced remyelination of a nerve fiber.

result of exposure to toxic substances during critical developmental periods of the nervous system is termed "dysmyelination." Myelin sheaths formed under this adverse condition may be thinner than those in normal animals, may have irregular internodal distances, and may show defects or degenerative changes in the myelin sheaths and node of Ranvier. This type of change can be observed in mice exposed to toxicants such as methylmercury (Chang and Annau, 1984; Chang, 1984) during early developmental lives.

Destruction of well developed myelin is referred to as demyelination. This process may involve the central axons, peripheral nerves, or both. Neurotoxic agents such as triethyltin (TET) (Watanabe, 1980; Wenger et al., 1986), lead (Krigman et al., 1980) and acetyl ethyl tetramethyl tetralin (Spencer et al., 1980a) will induce such patterns of change, respectively. Myelin bubbling, swelling, fragmentation and breakdown are frequent changes in these conditions (Figure 11). Myelin sheath destruction, frequently inter-

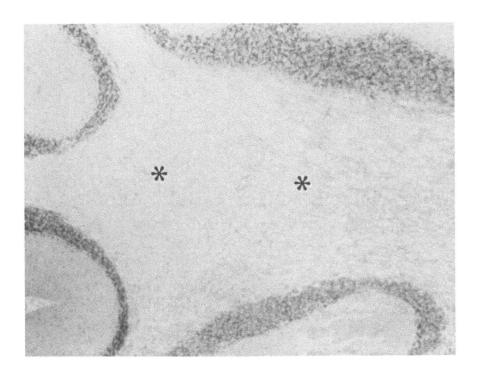

Figure 12. Swelling and distention of the white matter (*) induced by triethyltin in the cerebellar cortex. The swelling of the white matter was the result of edematous change of the myelin sheaths in the white matter.

nodal or segmental as best exemplified in lead poisoning, occurs in peripheral nerves. In TET poisoning, swelling or edematous change of the central white matter (myelin sheath) is observed (Figures 12 and 13). While routine paraffin sectioning with special staining technique (luxol fast blue) provides a good base for pathological examination of the central nervous system, nerve fiber teasing technique is more appropriate for the study of peripheral nerves.

In myelinopathy, the axons are usually spared of the damage. This should be distinguished from secondary myelin degeneration as seen in Wallerian degeneration resulting from nerve cell or axonal damage. While toxic mechanisms for the induction of myelin destruction may be different between toxic compounds, it can be safely assumed that the myelinating cells (oligodendrocytes in CNS and Schwann cells in PNS) or the metabolic system necessary to maintain the integrity of the myelin sheaths are affected by the toxicant involved. Removal of the myelin debris and lipid byproducts by macrophages or Schwann cells and attempts for remyelination (particularly in PNS) may be observed in later stages of the intoxication.

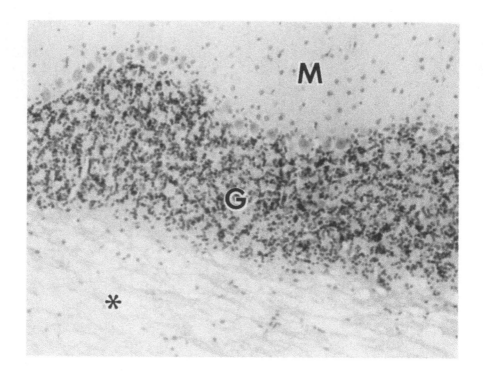

Figure 13. Cerebellar cortex, triethyltin treated. Closer examination revealed bubbling and vacuolation of the myelinated fibers in the white matter (*). G, granule cells; M, molecular layer.

APPROACHES AND TECHNIQUES FOR NEUROPATHOLOGY STUDIES

The nervous tissues, particularly those from the central nervous system, are very vulnerable to post mortem changes and other artifacteral alterations. Proper handling of the tissues should be exercised with caution to ensure preservation of the morphology. Detailed histological techniques and methodologies for pathological studies have been published previously (Chang, 1979; Spencer et al., 1980b) and may be used as valuable reference sources. Therefore, only the general approach and selected histological technique will be outlined and presented here.

General Scheme for Neuropathological Investigation

Neuropathology should be an integral part of the overall neurotoxicology investigation, designed to provide morphological information on the effects of toxic compounds on the nervous system.

Because neuropathology is by and large a morphological science, lesion or pathological changes in any given component of the nerve cell (perikaryon, axon, dendrite, etc.) must be made "visible" before such changes can be identified, located, defined, catagorized, and evaluated. As stated in 1980 by Dr. J.B. Cavanagh, the eminent British neuropathologist, "special techniques are needed to show the cellular details of damage to brain and nerve. An 'H and E,' while adequate for most general pathology, is insufficient to supply the answers to neuropathological problems. The special techniques and the special knowledge are reasons for the separation of neuropathology from its parent field."

Indeed, while H and E is an excellent general stain, it is a very nonspecific dye, staining nuclei blue and all other tissue components pink. Therefore, unless the lesions involved are of obvious or extensive nature, H and E method may fail to demonstrate changes involving the complex structures of the various tissue components of the nervous system. Special techniques are useful to selectively and specifically demonstrate the various callular and tissue components in the nervous system: axons, dendrites, Nissl patterns of the neurons, myelin sheath, etc.

Therefore a schematic approach, utilizing well established and reliable neurohistochemical methods, may be needed to help identify or confirm specific structural damages in the nervous tissues (neuronal body, axon, dendrite, myelin sheath, and peripheral nerve fiber), which may go undetected by H and E staining method. The concern is therefore not so much the lesions that are detectible by H and E method, but those which are not readily apparent by routine examination methodology (H and E staining). Recent experiences from various laboratories suggest that plastic embedded tissues yield much thinner sections (1–2 μm) which, when handled properly, can be satisfactorily stained by general histological dyes such as H and E and toluidine blue. Because of the "thinness" of the tissue sections, the resolution of the microscopic image is significantly improved (Bennet et al., 1976; Spencer and Schaumburg, 1980; Pender, 1985) and therefore be used in lieu of or in addition to the paraffin method for general screening purposes. Special histochemical staining techniques, on the other hand, should be employed as "confirmational" and/or specific detection purposes (e.g., myelin degeneration, dendritic shrinkage, neurofibrillary proliferation, etc.)

The following schemes are proposed only as "guidelines" for neuropathological investigations both on "known" neurotoxicants (i.e., the general site of injury and over all neurotoxic effect are previously known to the investigator) and on "potential" neurotoxic compounds of unknown toxic action. In both situations, the dose/pathology relationship (e.g., minimal dose which will produce a given lesion, or dose vs. severity of lesion induced) may be established.

For investigation of neurotoxicants with *known* toxic site and actions, in addition to H and E staining, selected method(s) which are specific and

General Test Scheme for "Known" Neurotoxicant

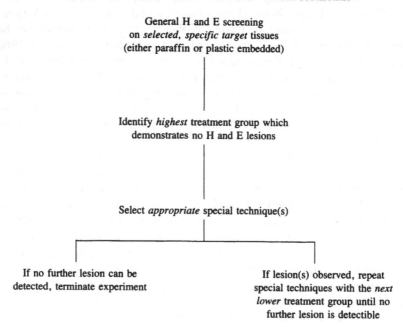

General H and E screening
on *selected, specific target* tissues
(either paraffin or plastic embedded)

Identify *highest* treatment group which
demonstrates no H and E lesions

Select *appropriate* special technique(s)

If no further lesion can be
detected, terminate experiment

If lesion(s) observed, repeat
special techniques with the *next
lower* treatment group until no
further lesion is detectible

DIAGRAM 1

pertinent for the detection of such toxic lesion(s) may be required. Such special procedure(s), including thin plastic embedded sections if so desired, should be performed only on specific tissue samples with "known" target sites for the toxicant but for one reason or another, failed to show any observable or definitive lesions by means of H and E method. Tissues from the highest treatment group showing no H and E lesion should be screened first. If lesions are detected by special technique(s), the next lower treatment groups should also be screened. If no lesion is detected, no further screening is needed (Diagram 1).

For investigation of potential neurotoxicants with *unknown* toxic site and nature, H and E method should be performed on all tissue samples before any special stains are to be inlcuded. If no clue concerning the nature of lesion is obtained from H and E stained sections, thin plastic embedded sections stained with toluidine blue, a series of special techniques could be used to ensure that no false-negative result is generated because of the relative insensitivity of the H and E method. These stains should be performed on tissues from the highest treatment group first. However, if clues on the site and nature of lesion are obtained from H and E stained sections, selected special and pertinent technique(s) may be required only on those tissues as outlined for "known" neurotoxicants (Diagram 2).

General Test Scheme for "Potential" Neurotoxicant

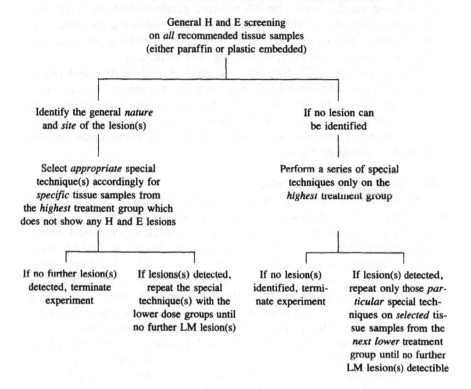

General H and E screening
on *all* recommended tissue samples
(either paraffin or plastic embedded)

Identify the general *nature*
and *site* of the lesion(s)

If no lesion can
be identified

Select *appropriate* special
technique(s) accordingly for
specific tissue samples from
the *highest* treatment group which
does not show any H and E lesions

Perform a series of special
techniques only on the
highest treatment group

If no further lesion(s)
detected, terminate
experiment

If lesions(s) detected,
repeat the special
technique(s) with the
lower dose groups until
no further LM lesion(s)

If no lesion(s)
identified, termi-
nate experiment

If lesion(s) detected,
repeat only those *par-
ticular* special tech-
niques on *selected* tis-
sue samples from the
next lower treatment
group until no further
LM lesion(s) detectible

DIAGRAM 2

It must be emphasized that while electron microscopy is an invaluable
and sensitive tool for the investigation of pathological process and intracellular events. It is too costly and too time consuming to be used as a tool for
routine pathological screening test. If the usage of electron microscopy is
justified and desired, the basic target site of toxicity (e.g., cerebellar cortex
in methylmercury poisoning) must be known. Random sampling will not
constitute any meaningful result.

General Pathological Techniques

Sacrifice of Animals and Tissue Fixation

Tissues from the nervous system, particularly those of the central nervous
system, are subjected to rapid anoxic and post mortem changes. In order to
avoid formations of artifacts, immediate fixation by appropriate fixative
solutions is required.

For basic light microscopic studies, the animal should be anesthesized and perfused via the heart with 4% buffered paraformaldehyde, 2.5% buffered glutaraldehyde, or 10% buffered formalin. Following the initial perfusion fixation, tissues must be further immersion fixed for a period of time prior to dehydration, clearing, embedding, and sectioning.

Neutral buffered 10% formalin is a good general fixative and can be used for all types of tissues for light microscopic examination. Tissues should be fixed for at least twenty-four hours before processing. For rodent central nervous system, a brief fixation (4–6 h) in Bouin's fixative prior to neutral buffered formalin fixation has proven to produce excellent tissue sections and can be considered for use.

Tissue Embedding and Staining

While the most basic and commonly used embedding medium is paraffin or paraplast, various methods have been tried for the improvement of embedding for light microscopy. The most satisfactory result was obtained with 2-hydroxy ethyl methacrylate (HEMA) or glycol methacrylate (GMA). These media are aqueous soluble monomers and will harden to plastic upon polymerization. The advantage of this technique is that it allows very thin sections (1–2 nm) to be made, thus increases the resolution of the microscopic image tremendously over the much thicker (8–10 nm) paraffin sections (Bennet et al., 1976; Pender, 1985). The disadvantages are, however, that (1) only smaller tissue samples (both surface area and thickness) can be used, (2) the time required for infiltration and processing is longer than the routine paraffin method and special microtome must be used for sectioning, (3) only a limited number of histochemical stains can be used on plastic sections, and (4) it is much more costly than the routine paraffin method. For general purposes, paraffin embedding approach is still a more practical method. Not only larger tissue sizes can be embedded by this technique, multiple tissue samples (e.g., brain, cord, ganglia) can also be embedded together in one single block to be sectioned and stained together. Since most, if not all, staining procedures for both routine and special histochemical methods were developed with paraffin embedded tissues in mind, one will also have more success on the staining results with paraffin sections.

Special Neurohistological Methods

While there are many special histological methods for the specific demonstration of various components of the nervous tissue, only the most basic, reliable, time tested, cost efficient, and easily reproducible methods are presented here. Detailed procedures for these techniques are available in standard histological technique manual (AFIP, 1968; Chang, 1979).

1. *Plastic embedding method for general lesion screening.* Plastic embedded tissues allow sectioning at a much thinner level (1–2 μm)

than those embedded in paraffin or paraplast, thus improves the resolution of the microscopic image significantly and enhances lesion detection. Commercially produced plastic embedding medium (for example, the JB-4 embedding kit by Polyscience, Inc.) is available. This type of water-soluble embedding medium allows satisfactory staining results from most general aqueous histological stains such as H and E, toluidine blue, etc. These general stains, however, are still nonspecific dyes and cannot demonstrate specific neuronal components such as dendrites, axons, myelin sheaths, etc. The routine paraffin method, by virtue of its cost efficiency and acceptable effectiveness, is still considered to be the primary basic general screening method. The plastic embedding method is a good and sensitive screening technique and can be used as an alternative method for lesion screening in the event that H and E staining from paraffin sections fails to reveal the expected or suspected lesions. Applicable on brain and cord.

2. *Gallocyanin method for neuronal Nissl substance.* This method will give a very definitive demonstration of the Nissl substance changes (rough endoplasmic reticulum and polyribosome clusters) in the nerve cell (e.g., chromatolysis of brain stem neurons in TML poisoning) (Figures 1, 2, 3). It also provides a very clear staining of the cells and their nuclei allowing a clearer view of any nuclear changes as well as cell number changes (e.g., neuronal loss in methylmercury intoxication) (Figures 14 and 15). Applicable on brain, cord, and ganglia.

3. *Bodian method for neuronal processes, especially axons.* This technique provides a selective staining on the neuronal process, especially the axons (axis cylinder) (Figures 16 and 17). It is very useful in detecting and defining axonal changes such as axonal swelling and changes in axonal contours (Figures 18 and 19) (e.g, axonal swelling in IDPN intoxication and other pathological conditions). Applicable on brain, cord, and peripheral nerves.

4. *Bielchowsky method for neurofilaments.* This technique selectively stains neurofilaments in the axonal process and nerve cell bodies (Figure 20). It is useful in detecting both axonal and nerve cell body changes related to neurofilamentous alterations (e.g., neurofibrillary tangle as seen in Alzheimer's disease). Applicable on brain, cord, and peripheral nerves.

5. *Luxol fast blue (LFB) method for myelin.* LFB staining method is very selective in staining the myelin sheath of the axons (Figures 21 and 22). It is useful in detecting alterations or loss of myelination in the nervous system as a consequence of either direct or secondary toxic influence (e.g., myelin swelling in TET intoxication). The area(s) with myelin loss will result in a negative staining. Applicable on brain, cord, and peripheral nerves.

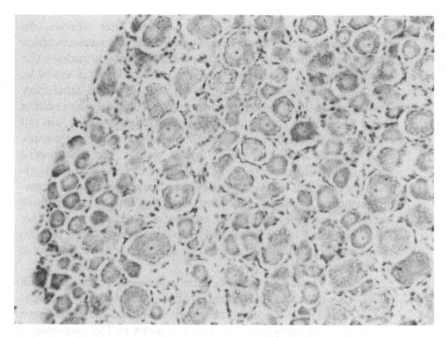

Figure 14. Gallocyanin stain, dorsal root ganglion of normal rat. The ganglion was filled with large round neurons with prominent nuclei and cytoplasmic Nissl substance.

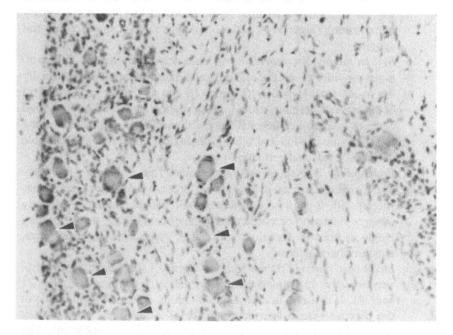

Figure 15. Gallocyanin stain, dorsal root ganglion of methyl mercury neurons intoxicated rat. Severe loss of neuron was seen. The surviving neurons (arrowheads) also appeared atrophic and lack of Nissl substance.

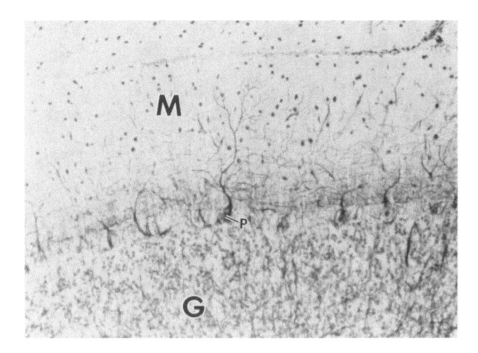

Figure 16. Bodian stain, cerebellar cortex, rat. The parallel fibers, basket cell fibers as well as some dendritic branches of the Purkinje cells (P) are demonstrated by the stain. G, granule cell layer; M, molecular layer.

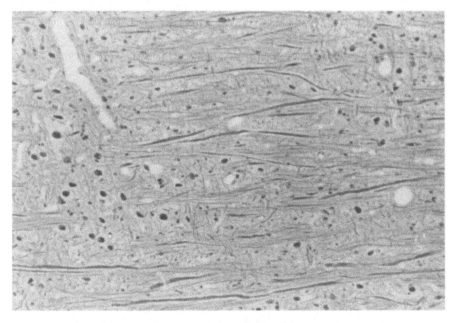

Figure 17. Bodian stain, brain stem, rat. Axons of various sizes were demonstrated.

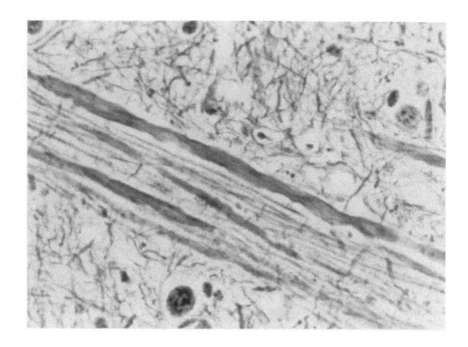

Figure 18. Bodian stain, brain stem, rat. Changes in axonal sizes and contours are observed.

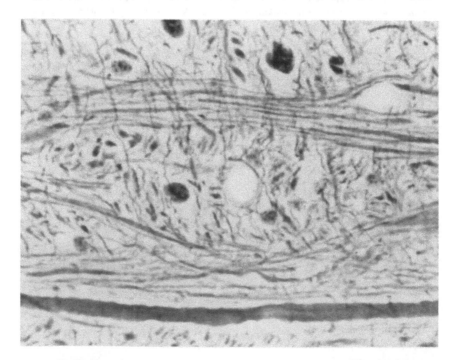

Figure 19. Bodian stain, brain stem, rat. Networks of axons and a large axon are demonstrated.

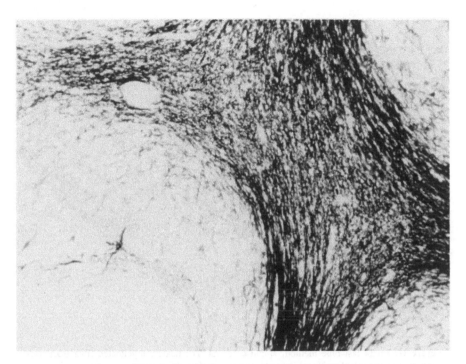

Figure 20. Bielchowsky's method, cerebellar cortex, rat. The neurofilamentous network outlining the fiber tract of the white matter was well demonstrated.

6. *Nerve fiber "teasing" technique for peripheral nerves.* Segments of peripheral nerves can be fixed, stained (osmium tetroxide), "teased," and mounted whole on a slide without sectioning. It provides a three dimensional view of the nerve fibers and is very useful in screening both axonal and myelin defects in peripheral nerves (Figure 9, 11). For peripheral nerve studies, this technique can be used as a "standard" screening method. Applicable on peripheral nerves (e.g., detecting segmental demyelination in lead poisoning).

7. *Rapid Golgi technique for dendrites.* This classical neurohistochemical method provides silver impregnation of nerve cells and their dendrites (Figure 23). It should be used only as a selective method to detect or demonstrate dendritic defects (dendritic spread, dendritic spines, etc.) in the CNS (e.g., dendritic changes in fetal methylmercury poisoning). Applicable on brain and cord.

8. *Holzer's method for astrocytic fiber.* Injuries to the central nervous system frequently induce neuronal loss and "scarring." Scarring of the CNS is accomplished by astrocytic proliferation (not unlike the scarring in other organ systems by fibroblast proliferation). The proliferation of astrocytic fibers may be visualized by means of Holzer method as darkly stained purplish blue fibers (Figure 24) (e.g., Bergmann's fiber proliferation in methylmercury intoxication). Applicable on brain and cord.

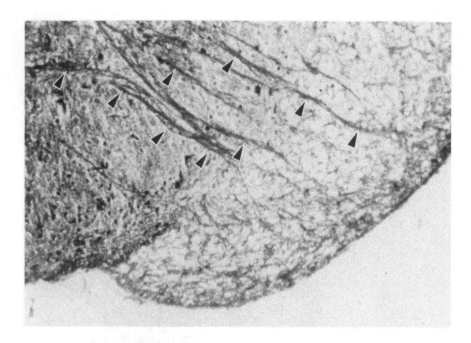

Figure 21. Luxol fast blue method, spinal cord, rat. The heavily myelinated fiber tracts (arrowheads) in the spinal cord were clearly demonstrated by this method.

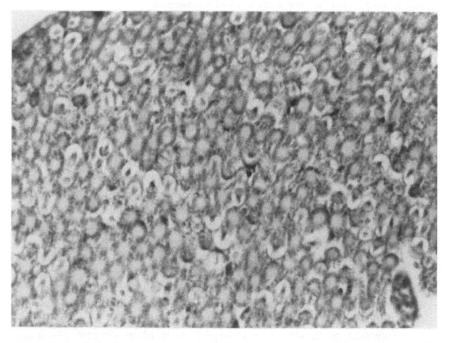

Figure 22. Luxol fast blue method, cross section of dorsal root fiber, rat. When counter stained with H and E, the center axon stained pink with the surrounding myelin sheath stained blue.

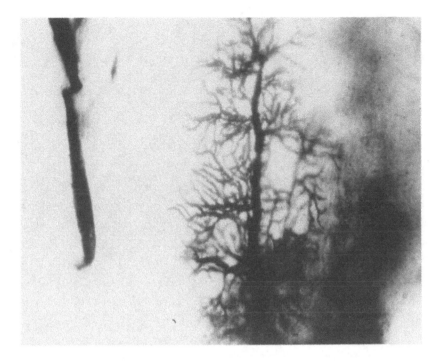

Figure 23. Rapid Golgi technique, cerebellar cortex. The elaborate dendritic aberration of a Purkinje neuron was well demonstrated.

CONCLUDING REMARKS

Neuropathology represents a unique and extremely useful tool for the study of neurotoxicology. When applied appropriately, it will provide useful and important morphological information concerning the nature, characteristic, loci, and severity of the lesion induced by a particular neurotoxicant.

It must be emphasized that the information included in the present article represents only a portion of the "morphological" aspects of neuropathology related to general toxic changes of the nervous system. The term "pathology," however, should be viewed as "the study of the basis of diseases" and morphology represents only one of the many facets of pathological investigations. Moreover, since the scope of the present article is limited only to the "histological" aspects of neuropathology at light microscopic level, many other aspects and approaches for neuropathological investigations are not included in the present treatise. This author is compelled to clarify that aside from basic histopathological techniques, as those described in the present article, there are many other approaches and techniques available for neuropathological investigations, such as neuronal and glial cultures as well as brain slice techniques for *in vitro* investigations, biochemical techniques

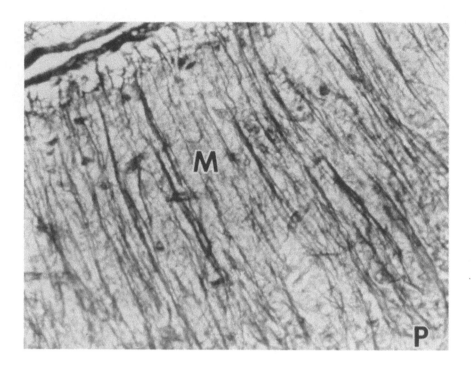

Figure 24. Holzer's method, cerebellar cortex, methyl mercury intoxicated rat. Characteristic proliferation of the astrocytic fiber (Bergmann's fibers) from the Purkinje layer (P) through the molecular layer (M) was demonstrated by this method.

for molecular and subcellular investigations (e.g., neurochemical analysis and synaptosome isolation), electron microscopic techniques for detailed cytological evaluations, nuclear magnetic resonance techniques for cellular and tissue analysis, special neurohistochemical techniques for specific evaluations, (e.g., immunofluorescence and enzyme histochemistry), and automated morphometric analysis for quantitative measurements and evaluations of pathological changes in organelles, cells, and tissues.

Although basic histopathology at light microscopic level is generally believed to be the main core of investigation in neuropathology, supplemental pathological techniques are available to create a more dynamic approach in neuropathological research. Correlated study between neuropathology, neurochemistry, neurophysiology and neurobehavioral evaluations should also be encouraged to provide a more complete portrait of the neurotoxic compound under investigation.

ACKNOWLEDGMENTS

Some of the illustrations used in this article have been previously published elsewhere. The author wishes to thank these authors and publishers for their permissions to reprint these illustrations here. The original sources for these illustrations are duly acknowledged as follows: *Figure 3:* T.J. Walsh et al., 1986, *Neurotoxicology* 7: 21–34, figure 6; *Figure 4:* L.W. Chang et al., 1984, *Environ. Res.* 34: 123–134, figure 5; *Figure 6:* L.W. Chang et al., 1983, *Neurobehav. Toxicol. Teratol.* 5: 337–350, figure 8; *Figures 7 and 8:* L.W. Chang and R.S. Dyer, 1983, *Neurobehav. Toxicol. Teratol.* 5: 443–459, figures 3 and 6; *Figures 9 and 10:* P.S. Spencer et al., 1980, in *Experimental and Clinical Neurotoxicology*, pp. 743–757, P.S. Spencer and H.H. Schaumburg, Eds., Williams & Wilkins, Baltimore, figures 50.2 and 50.5; *Figure 11:* P.S. Spencer et al., 1980, in *Experimental and Clinical Neurotoxicology*, pp. 296–308, P.S. Spencer and H.H. Schaumburg, Eds., Williams & Wilkins, Baltimore, figure 20.7; *Figures 14 and 15:* L.W. Chang et al., 1972, *J. Neuropath. Exp. Neurol.* 31: 489–501, figures 2a and c; *Figures 16, 21–24:* L.W. Chang, 1979, *A Color Atlas and Manual for Applied Histochemistry*, Charles C Thomas, Springfield, IL.

This author also wishes to pay his respects to Dr. Martin R. Krigman, a renowned neuropathologist, a dedicated scientist, and a good teacher, who passed away in 1986. This chapter is dedicated to the memory of Dr. Krigman.

REFERENCES

Armed Forces Institute of Pathology (1968). *Manual of Histologic Staining Methods.* McGraw Hill, New York.

Bennet, H.S., A.D. Wyrick, S.W. Lee, and J.H. McNeil (1976). Science and art in the preparing tissues embedded in plastic for light microscopy, with special reference to glycol methacrylate, glass knives, and simple stains. *Stain Tech.* 51: 71–97.

Blackmore, W.F. and J.B. Cavanagh (1969). Neuroaxonal dystrophy occurring in an experimental dying back process in the rat. *Brain* 92: 789–806.

Cavanagh, J.B. (1980). Foreword. In *Experimental and Clinical Neurotoxicology.* P.S. Spencer and H.H. Schaumburg, Eds., Williams & Wilkins, Baltimore, pp. v–vi.

Chang, L.W. (1979). *A Color Atlas and Manual for Applied Histochemistry.* Charles C Thomas, Springfield, IL.

Chang, L.W. (1984). Developmental toxicology of methylmercury. In *Toxicology and the Newborn.* S. Kacew and M.J. Reasor, Eds., Elsevier, Amsterdam, pp. 175–197.

Chang, L.W. and Z. Annau (1984). Developmental neuropathology and behavioral teratology of methylmercury. In *Neurobehavioral Teratology.* J. Yanai, Ed., Elsevier, Amsterdam, pp. 405–432.

Chang, L.W., T.M. Tiemeyer, G.R. Wenger, and D.E. McMillan (1982). Neuropathology of mouse hippocampus in acute trimethyltin intoxication. *Neurobehav. Toxicol. Teratol.* 4: 149–156.

Chang, L.W., T.M. Tiemeyer, G.R. Wenger, and D.E. McMillan (1983). Neuropathology of trimethyltin intoxication. III. Changes in the brainstem neurons. *Environ. Res.* 30: 399–411.

Chang, L.W., G.R. Wenger, and D.E. McMillan (1984). Neuropathology of tri-
    methyltin intoxication. IV. Changes in the spinal cord. *Environ. Res.* 34: 123–
    134.
Chou, S.M. and H.A. Hartmann (1964). Axonal lesions and waltzing syndrome after
    IDPN administration in rats. *Acta Neuropathol.* 3: 428–437.
Chou, S.M. and H.A. Hartmann (1965). Electron microscopy of focal neuroaxonal
    lesions produced by α,α-iminodipropionitrile (IDPN) in rats. *Acta Neuropathol.*
    4: 509–511.
Klatzo, I., H. Wisniewski, and E. Streicher (1975). Experimental production of
    neurofibrillary degeneration. I. Light microscopic observations. *Neuropathol. Exp.
    Neurol.* 24: 187–209.
Krigman, M.R., Bouldin, T.W. and P. Muskak (1980). Lead. In *Experimental and
    Clinical Neurotoxicology.* P.S. Spencer and H.H. Schaumburg, Eds., Williams &
    Wilkins, Baltimore, pp. 490–507.
LeQuesne, P.M. (1980). Acrylamide. In *Experimental and Clinical Neurotoxicology.*
    P.S. Spencer and H.H. Schaumburg, Eds., Williams & Wilkins, Baltimore, pp.
    309–325.
Pender, M.P. (1985). A simple method for high resolution light microscopy of nervous
    tissue. *J. Neurosci. Methods* 15: 213–218.
Petit, T.L., G.B. Biederman, and P.A. McMullen (1980). Neurofibrillary degener-
    ation, dendritic dying back, and learning-memory deficits after aluminum admin-
    istration: implications for brain aging. *Exp. Neurol.* 67: 152–163.
Prineas, J. (1969). The pathogenesis of dying-back polyneuropathies. II. An ultra-
    structural study of experimental acrylamide intoxication in the cat. *J. Neuropath.
    Exp. Neurol.* 28: 598–616.
Schaumburg, H.H. and P.S. Spencer (1980). Clioquinol. In *Experimental and Clinical
    Neurotoxicology.* P.S. Spencer and H.H. Schaumburg, Eds., Williams & Wilkins,
    Baltimore, pp. 395–406.
Seppalainen, A.A. and M. Haltia (1980). Carbon Disulfide. In *Experimental and
    Clinical Neurotoxicology.* P.S. Spencer and H.H. Schaumburg, Eds., Williams &
    Wilkins, Baltimore, pp. 356–373.
Spencer, P.S. and H.H. Schaumburg (1976). Central and Peripheral Distal Axono-
    pathy—The Pathology of Dying-back Polyneuropathies. In *Progress in Neuro-
    pathology, Vol. III.* H.M. Zimmerman, Ed., Grune & Stratton, NY, pp. 253–262.
Spencer, P.S. and H.H. Schaumburg (1980). *Experimental and Clinical Neurotoxi-
    cology.* Williams & Wilkins, Baltimore.
Spencer, P.S., M.C. Bischoff, and H.H. Schaumburg (1980a). Neuropathological
    methods for the detection of neurotoxic disease. In *Experimental and Clinical
    Neurotoxicology.* P.S. Spencer and H.H. Schaumburg, Eds., Williams & Wilkins,
    Baltimore, pp. 743–757.
Spencer, P.S., G.V. Foster, A.B. Sterman, and D. Horoupian (1980b). Acetyl ethyl
    tetramethyl tetralin. In *Experimental and Clinical Neurotoxicology.* P.S. Spencer
    and H.H. Schaumburg, Eds., Williams & Wilkins, Baltimore, pp. 296–308.
Watanabe, I. (1980). Organotins (Triethyltin). In *Experimental and Clinical Neuro-
    toxicology.* P.S. Spencer and H.H. Schaumburg, Eds., Williams & Wilkins, Bal-
    timore, pp. 545–557.
Wenger, G.R., D.E. McMillan, and L.W. Chang (1986). Effects of triethyltin or
    responding of mice under a multiple schedule of food presentation. *Toxicol. Appl.
    Pharmacol.* 8: 659–665.
Wisniewski, H., K. Iqbal and J.R. McDermott (1980). Aluminum-induced neuro-
    fibrillary changes: its relationship to senile dementia of the Alzheimer's type.
    Neurotoxicology 1: 121–135.

Part
IV

Factors Affecting Neurotoxicity

Disposition, Metabolism, and Toxicokinetics
M.B. Abou-Donia

Developmental Neurotoxicology
C.M. Kuhn and R.B. Mailman

Nutrition and Neurotoxicology
M.M. Abou-Donia

Alcohol and Neurotoxicology
M.B. Abou-Donia

Disposition, Metabolism, and Toxicokinetics

Mohamed B. Abou-Donia
Departments of Pharmacology and Neurobiology
Duke University Medical Center
Durham, North Carolina

INTRODUCTION

A significant factor affecting neurotoxicity is the various ways a neurotoxicant can enter, distribute within tissues, and leave the body. A neurotoxic chemical may enter the gastrointestinal tract through diet or drinking water; it may enter the upper airway and lungs in the form of gas, aerosol, particle, dust, or fume; or it may enter the body through the skin. Neurotoxicity occurs after neurotoxicants enter circulation and gain access to various tissues of the body. Metabolic biotransformation may produce more active or less active metabolites. These processes generally lead to more polar metabolites that eventually enhance the excretion of neurotoxic chemicals (Abou-Donia and Nomeir, 1986). The term metabolism encompasses all of these processes: absorption, tissue distribution, storage, biotransformation, and elimination. The rate of these processes that pertain to toxicity are defined as toxicokinetics (Abou-Donia, 1983).

ABSORPTION

In order for a neurotoxic chemical to reach its site of action, it must pass across various body membranes, i.e., cells of skin, cells of lung, gastrointestinal tract, erythrocyte membrane, etc.

Biological Membranes

Membranes are constituted of a phospholipid bilayer that is embedded with proteins and contains various size pores. Pores in most cells are 4 Å, endothelial cells of capillaries are 40 Å, and endothelial cells of glomerular membrane of the kidney are 100 Å (Schanker, 1962).

Transport Across Membrane

Chemicals may pass across membranes via two mechanisms: simple diffusion or specialized transport.

Simple Diffusion

Most chemicals pass across membranes by simple or passive diffusion in their unionized form. Both lipid-soluble substances and small (i.e., urea) lipid-insoluble molecules may cross body membranes by simple diffusion. This process is unsaturable (La Du et al., 1972). Some neurotoxicants (weak organic acids and bases) occur in solution in both the ionized and unionized form. While the unionized form may be lipid soluble enough to move across membranes, the ionized form is often unable to diffuse across membranes. The ratio of unionized to ionized forms depends on the association constant (PK_a) of the acid or base and the pH of the internal environment according to the Henderson-Hasselbach equation:

$$\text{For acids: } PK_a - PH = \log \frac{[\text{unionized}]}{[\text{ionized}]}$$

$$\text{For bases: } PK_a - PH = \log \frac{[\text{ionized}]}{[\text{unionized}]}$$

In addition to PK_a and pH, blood flow influences the extent of movement of chemicals across membranes. Blood flow is very important for chemicals with a high permeability constant. Thus the basic pH of the intestine favors the absorption of a basic chemical, while the acidic pH of the stomach favors absorption of an acidic chemical. Blood flow and surface area are more important than pH for the absorption of weak acids.

	pH	Blood Flow	Surface Area
Stomach	1–3	0.15 L/min	1 m^2
Intestine	5–8	1.0 L/min	200 m^2

In general, absorption of acids in the intestine is affected primarily by blood flow and surface area.

Filtration. When water flows in bulk across aqueous pores in the membrane, any solute that is small enough to pass through the pores flows with it. This process is governed by the size and shape of the molecule. An example is glomerular filtration of some chemicals in the kidney.

Special Transport. Some large, lipid-insoluble molecules and ions cross membranes through three types of special transport: active transport, facilitated diffusion, and pinocytosis.

Active Transport. This process occurs when substances move against a concentration or electrochemical gradient. It has the following characteristics:

1. can be saturated
2. selectivity for compounds of the same size
3. competitive inhibition among substances handled by the same mechanism
4. requires energy, i.e., metabolic inhibitors block transport process.

Facilitated Transport. This is similar to active transport; however, chemicals do not move against a concentration gradient. No apparent energy is required, and metabolic inhibitors do not inhibit this process. An example is the transport of a water soluble molecule, e.g., glucose through membrane down a concentration gradient (La Du et al., 1972).

Pinocytosis. In this process, the cell membrane engulfs substances within the cell. This mechanism has been postulated for transferring large molecules and particles into the cell.

Absorption of Neurotoxic Chemicals Into the Body

Absorption From the Gastrointestinal Tracts

In general, the absorption of chemicals from the gastrointestinal tract is by simple diffusion (Schanker, 1971). Almost all chemicals are absorbed at least to a small extent. Some chemicals are absorbed through specialized transport systems: sugars, amino acids, pyrimidines, and calcium and sodium ions. However, rarely may a neurotoxicant be absorbed via the active transport mechanism.

Absorption of neurotoxic chemicals takes place to some extent in the mouth. Chemicals absorbed from the buccal mucosa are not exposed to the gastrointestinal digestive juices and xenobiotic metabolizing enzymes. Also, since they are not transported to the liver directly via the hepatic portal system, they may not be rapidly metabolized, thus prolonging their neurotoxicity.

Significant absorption takes place in the stomach by diffusion. Weak organic acids are in the nonionized lipid soluble form in the stomach and are absorbed there, whereas weak organic bases are highly ionized and therefore not generally absorbed.

Most gastrointestinal tract absorption occurs in the small intestine because of its large surface area and rich blood supply. Weak organic bases are in the unionized lipid soluble form and tend to be absorbed by passive diffusion from the small intestine.

Absorption of neurotoxic chemicals from the gastrointestinal tract is affected by many factors (Brodie, 1964): neurotoxic chemicals may undergo metabolic biotransformation in the gut, e.g., intestinal flora degrade DDT to DDE; dilution increases toxicity because of more rapid absorption from intestine; food in the gastrointestinal tract may impair absorption by forming a nonabsorbable complex; normal digestion enhances absorption by increasing gastrointestinal fluid; absorption of a solid chemical will be enhanced by dissolution; newborn has poor intestinal barrier; and chemicals mixed with diet may interact with some diet components.

Absorption From the Lungs

The pulmonary epithelial lining is close to the blood because it is very thin (10 μm), possesses a large surface area (50–100 m^2), and is highly vascularized. Therefore, absorption of neurotoxic chemicals takes place here with high efficiency. Gases and aerosols with small particle size and a high lipid-to-water partition coefficient are most rapidly absorbed in lungs. Aerosol deposition in the pulmonary tract depends on the particle size, i.e., nasopharyngeal, 5 μm or larger; tracheobronchiolar, 1 to 5 μm; and alveolar, 1 μm. In whole-body inhalation studies, absorption may take place via gastrointestinal or skin absorption.

Absorption From the Skin

Absorption of neurotoxic chemicals by the skin occurs by simple diffusion. Lipid-soluble compounds penetrate the epidermis, a lipoprotein barrier, whereas the highly porous dermis is permeable to both lipid- and water-soluble chemicals (Katz and Poulsen, 1971). The absorption of neurotoxicants depends primarily upon their physical properties and, to a lesser extent, their chemical properties. Other factors that determine penetration across the skin are hydration, pH, temperature, blood supply, metabolism, and vehicle–skin interactions. Absorption is enhanced by skin abrasions.

In some studies, the chemical to be tested on the skin is applied under an occlusive dressing (Draize et al., 1944). In these events, the reversible hydration of normal skin under an occlusive dressing increases the rate of absorption of many compounds, which may exaggerate the neurotoxicity of test compounds. Skin penetration also differs with body part (Maibach et al., 1971). Thus, similar amounts of ^{14}C-labeled parathion, malathion, and carbaryl penetrate the palm and the forearm. On the other hand, follicle-rich sites, including the scalp, angle of the jaw, postauricular area, and forehead had fourfold greater penetration, whereas the scrotum allowed almost total absorption.

The rate at which test chemicals are lost to the environment is a crucial factor. The surface loss rate for a single application of most chemicals exceeds the absorption rate. Usually less than 50% of a single dose is absorbed through the skin (Feldman and Maibach, 1970). Thus, when a single dermal 50 mg/kg dose of [^{14}C]leptophos was applied to the comb of hen, 35.4% of the dose was absorbed (Abou-Donia, 1979) compared to 16.7% when the same dose was given orally (Abou-Donia, 1980). Also, chemicals applied daily at small doses are absorbed more efficiently through the skin, thus 67% of the total 10 daily dermal 0.5 mg/kg doses of EPN was absorbed (Abou-Donia et al., 1983a).

Relative Rates of Absorption

The relative rates of absorption of neurotoxic chemicals decrease in the following order: intravenous injection > inhalation > intraperitoneal injection > subcutaneous injection > oral > dermal.

DISTRIBUTION

The distribution of a neurotoxic chemical is dependent upon the relative plasma concentration, blood flow rate through organs, the rate by which the chemical penetrates cell membranes, and binding sites in plasma and various tissues. Lipid-soluble neurotoxicants tend to be distributed and localized in adipose tissue according to their lipid–water partition coefficients, e.g., the chlorinated hydrocarbon insecticides dieldrin, DDT, and DDE (Abou-Donia and Menzel, 1968) and the organophosphorus insecticides leptophos (Abou-Donia, 1976) and EPN (Abou-Donia et al., 1983 a–d). Unlike other cell membranes, the capillary membrane is permeable to foreign compounds of molecular weights of 60 kDa or less, whether lipid soluble or not. Also organic anions penetrate red blood cells much more readily than cations.

Many chemicals fail to penetrate into the brain tissue or cerebrospinal fluid as readily as into other tissues. Several membranes form the boundary between blood and brain: the blood capillary wall, the glial cells surrounding the capillary, and the membrane of the neurons or nerve cells. In addition, the brain is also protected with the blood-brain barrier, which is located at the capillary wall–glial cell region. Therefore, ionized substances and large water-soluble molecules such as proteins cannot enter the brain. The major mode of exit from brain of both lipid-soluble and polar compounds is by filtration across the arachoid villi.

BINDING

A major factor that can affect the distribution, metabolism, and elimination of a neurotoxicant, and consequently its neurotoxicity, is its affinity to binding to proteins and other macromolecules of the body. Most chemicals bind to some degree to plasma protein, with albumin being the major binding site followed by globulines. Albumin binds anions as well as cations, although with a pH 7.3, it contains a net negative charge. Neurotoxicants bound to plasma proteins become biologically inactive and cannot be filtered by the kidney. A bound chemical can displace another. Also, plasma proteins become saturated; an increase in toxicity may occur with further absorption of toxicants.

Some neurotoxic chemicals show a much greater affinity for body tissues than for plasma proteins. High concentrations of alkaloids in the liver and muscle are attributable to the affinity of these naturally occurring amines for nucleoproteins. Also, some metals and organic anions bind to plasma proteins and are then transferred from plasma to liver. Many inorganic ions (e.g., fluoride, the metal lead, and strontium) and drugs such as tetracyclin are concentrated in various tissues and organs, particularly in bones and teeth. Lipid-soluble neurotoxicants, i.e., DDT and leptophos, tend to concentrate in adipose tissue.

The placenta is a poor barrier. Chemicals cross the placenta by passive diffusion. The organophosphorus insecticide methamidophos crosses the placenta in pregnant rats (Salama et al., 1990). Viruses, e.g., rubella, bacteria, e.g., syphillis, antibodies, globuline, and erythrocytes are capable of transporting across the placenta.

EXCRETION

Neurotoxicants are excreted intact, as metabolites, or as conjugates of the parent chemical or its metabolites.

Renal Excretion

Renal excretion is the most important route of excretion of neurotoxic chemicals (Figure 1). Renal transport of neurotoxic chemicals is bidirectional: transport into renal tubules and reabsorption from renal tubules. Two mechanisms are involved in renal excretion.

1. Excretion into renal tubule
    a. *Glomerular filtration.* All water soluble chemicals with molecular masses less than 60 kDa are filtered from the blood through the highly porous glomeruli (Weiner, 1971). Compounds of

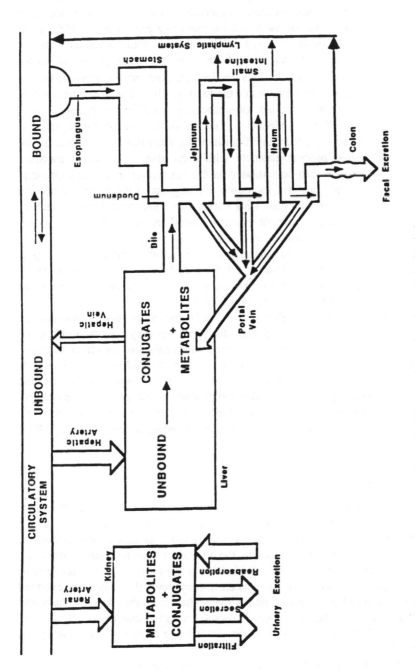

Figure 1. Pathways of excretion of neurotoxic chemicals.

high molecular weight and those bound tightly to plasma proteins do not undergo glomerular filtration. Generally neurotoxicants that are bases are excreted to a greater extent if the urine is acidic (as cationic salts), whereas acid compounds are excreted more favorably if the urine is alkaline (as anionic salts).

b. *Active secretion.* This carrier-mediated transport takes place in the proximal tubules of the kidney. There are two distinct active transport processes. one process is specific for organic anions, e.g., *P*-aminohipeurate, and the other specific for organic cations e.g., *N*-methylnicotinamide. Chemicals transported by the same process compete with each other. The active transport process can be saturated as the concentration of the compound in the plasma is increased. Conjugates of the parent compound or its metabolites with various endogenous substrates such as glycine, sulfate, or glucuronic acid are excreted via the anionic secretory process.

c. *Passive tubular diffusion.* Lipid soluble compounds are passively diffused across tubular membranes. Thus, weak acids are excreted better in alkaline urine, while weak bases are more efficiently excreted in acidic urine.

2. Passive tubular reabsorption. Water and endogenous substrates are reabsorbed from the glomerular filtrate as they pass down the renal tubule. Lipid-soluble, unionized chemicals pass in either direction in the tubule by passive diffusion. Thus, lipid-soluble compounds may be reabsorbed from the tubules by nonionic diffusion. The tubular epithelium of the distal convoluted tubule is selectively permeable or more permeable to the unionized soluble molecule than to the less lipid-soluble corresponding anion or cation (Milne et al., 1958).

Biliary Excretion

Biliary excretion is a major excretion route for neurotoxic chemicals (Figure 1). These chemicals enter the liver via the hepatic artery or through the lymph. The compound emerges from the liver, as such or as metabolites, with the bile, which passes down the bile duct into the gall bladder. At intervals, bile moves from the gall bladder by the bile duct to the duodenum. Enterohepatic circulation occurs when a compound that has been excreted via the bile into the small intestine is reabsorbed from the blood by the portal vein and then into the liver, which secretes it again into the small intestine (Gibson and Becker, 1967). Both circuits involve a gradual excretion by the feces (Williams, 1965).

Chemicals excreted into the bile have been divided into three classes according to their bile/plasma concentration ratios (Brauer, 1959). Substances of class A are those with a ratio of nearly one, e.g., Na^+, K^+, Hg^+. Class B includes substances with bile/plasma ratios between 10 and 1000, e.g., bile salts, lead, manganese, and weak acids. Class C compounds consist of those with a ratio less than one, e.g., inuline, albumin, and iron. The transport of compounds in class B across the biliary epithelium into bile is an active secretory process. Characteristically, these chemicals compete with each other for transport, which can be saturated by an excess of compound. For appreciable biliary excretion of class B compounds, they should have high molecular mass (more than 300 Da), with two or more aromatic rings and polar anionic groups. Also, highly protein-bound substances in the plasma are readily transferable directly into the secreting cells (Baker and Bradley, 1966). The delayed neurotoxicant leptophos and its metabolites are transferred from the blood to the bile (Abou-Donia, 1976, 1979, 1980). The high biliary excretion of this compound is attributed to (1) high lipid solubility, (2) high apparent volume of distribution, (3) strong and multiple site-binding to proteins, and (4) large molecular weights.

Other Routes of Excretion

Gastrointestinal Tract

Organic bases that are highly unionized at neutral pH may be transferred from the plasma to the stomach. Also, weak acids ionized at neutral pH may be partially transferred from the plasma to the lumen of the intestine. Dermally applied leptophos (Abou-Donia, 1979), EPN (Abou-Donia et al., 1983a,e) and TOCP (Nomeir and Abou-Donia, 1986) were recovered in the stomach, suggesting gastrointestinal absorption of these neurotoxicants. These chemicals are then excreted in the feces.

Exhaled Air

Many volatile neurotoxic chemicals are excreted via exhaled air, e.g., 2.5-hexanedione (Suwita et al., 1987). Also, carbon dioxide that results from exhaustive metabolic oxidation of these chemicals is excreted via this route, e.g., DDT (Abou-Donia and Menzel, 1976), leptophos (Abou-Donia, 1980).

Milk

Many neurotoxic chemicals and their metabolites are excreted in milk in either the aqueous or lipid phase. Methamidophos was excreted in the milk of treated rats (Bakry et al., 1990). Although the amount of neurotoxic chemicals excreted via the milk is minor compared to other routes of excretion, it may have a major impact on the newborn. Milk may be an important route of exposing the immature nervous system of the newborn to relatively high concentrations of neurotoxic chemicals.

Sweat

Neurotoxic chemicals are also excreted in the sweat. Some of these chemicals may cause dermatitis.

Saliva

Many chemicals are excreted in the saliva, which is usually swallowed into the gastrointestinal tract. Analysis of saliva instead of plasma may be used to determine chemical exposure.

METABOLIC BIOTRANSFORMATION

Metabolic transformation of neurotoxic chemicals is the alterations of these chemicals produced by biological environment. It results in the formation of more polar and water-soluble derivatives of these chemicals, so that they can be more readily excreted from the body (Conney, 1967).

The liver plays the major role for metabolic transformation of chemicals. The enzymes that take part in this process are present in the microsomal, soluble, and mitochondrial fractions of the cell. These enzymes are also found, to a lesser degree, in other tissues: gastrointestinal tract, kidney, lung, placenta, and blood. Also, intestinal flora may play an important role in the metabolism of neurotoxicants.

Metabolic Transformation Reactions

In biological systems, chemicals may undergo two phases of biotransformation:

Phase I. Metabolic transformation of a neurotoxicant may be restricted to a single reaction, usually oxidation, reduction, or hydrolysis. Some chemicals pass through multiple metabolic pathways that include any combination of single reactions. The result of this phase is the introduction of a polar group such as OH, CO_2H, SH, or NH_2 that results in a more polar compound.

Phase II. This phase involves the addition of polar endogenous functional groups (D-glucuronic acid, sulfuric acid, glycine, etc.) that usually render the molecule more polar, less lipid-soluble, more strongly acidic, and therefore more readily excreted. Factors that affect metabolic pathways include the nature of the chemical, dose, species, strain, sex, age, and some environmental parameters.

Types of Drug-Metabolizing Reactions

Microsomal Cytochrome P-450 Monooxygenase-Mediated Reactions

This enzyme, also known as microsomal mixed-function oxygenase (MFO), is found in the microsomes of endoplasmic reticulum (Mason, 1957).

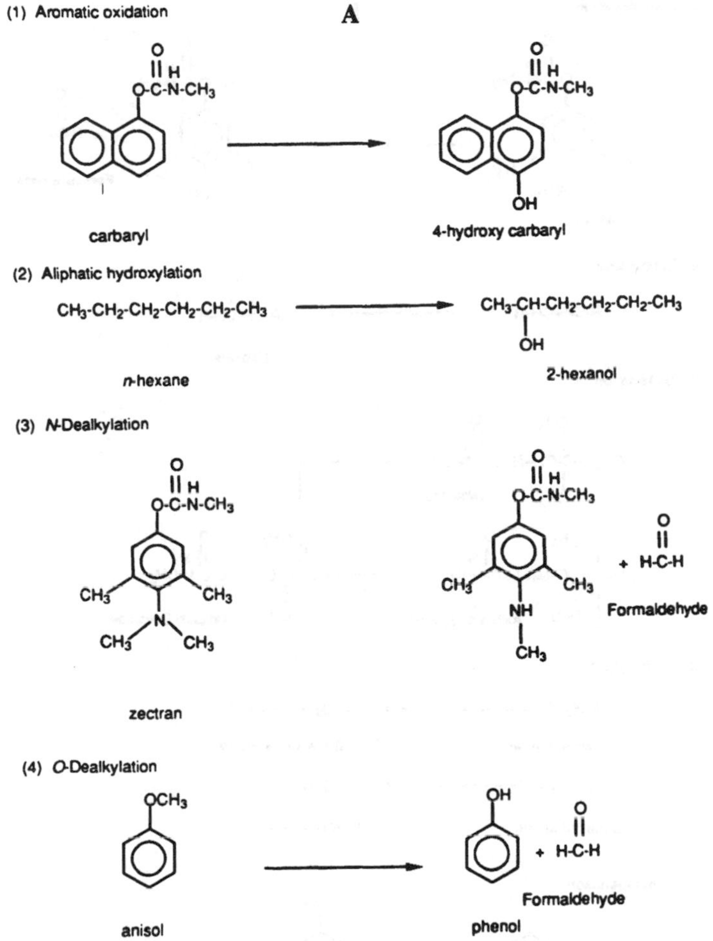

(1) Aromatic oxidation

A

carbaryl → 4-hydroxy carbaryl

(2) Aliphatic hydroxylation

$CH_3\text{-}CH_2\text{-}CH_2\text{-}CH_2\text{-}CH_2\text{-}CH_3$ → $CH_3\text{-}CH\text{-}CH_2\text{-}CH_2\text{-}CH_2\text{-}CH_3$ (OH)

n-hexane → 2-hexanol

(3) *N*-Dealkylation

zectran → + H-C-H Formaldehyde

(4) *O*-Dealkylation

anisol → phenol + H-C-H Formaldehyde

Figure 2. Oxidation reactions catalyzed by cytochrome P 450 monooxygenase. *(Continued on p. 266)*

These reactions require cytochrome P-450, nicotinamide adenine dinucleotide phosphate (reduced form, NADPH), NADPH cytochrome *c* reductase, oxygen, and magnesium. At least 20 different cytochrome P-450 isozymes have been identified (Nebert and Gonzalez, 1987).

Examples of oxidation reactions catalyzed by cytochrome P-450 monooxygenase are presented in Figure 2, A–C.

Examples of reduction reactions catalyzed by cytochrome P-450 monooxygenase are presented in Figure 3.

FAD-Containing Monooxygenase (FMO)

This enzyme has a molecular weight of 65 kDa and utilizes NADPH and could oxidize *N*-substituted amines and sulfur-containing compounds. It

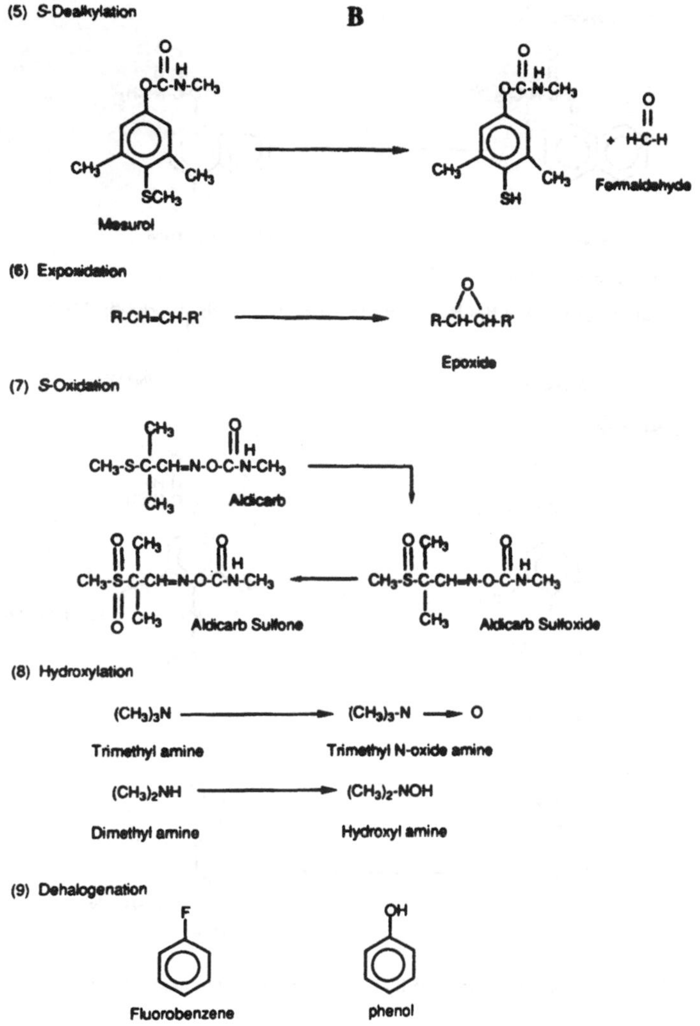

(5) *S*-Dealkylation   **B**

Mesurol   + H-C-H   Formaldehyde

(6) Expoxidation

R-CH=CH-R' ⟶ R-CH-CH-R'

Epoxide

(7) *S*-Oxidation

CH$_3$-S-C-CH=N-O-C-N-CH$_3$ Aldicarb

Aldicarb Sulfone ⟵ Aldicarb Sulfoxide

(8) Hydroxylation

(CH$_3$)$_3$N ⟶ (CH$_3$)$_3$-N ⟶ O

Trimethyl amine   Trimethyl N-oxide amine

(CH$_3$)$_2$NH ⟶ (CH$_3$)$_2$-NOH

Dimethyl amine   Hydroxyl amine

(9) Dehalogenation

Fluorobenzene   phenol

Figure 2. *(Continued on p. 267)*

is present in the nuclear envelope and in the endoplasmic reticulum of all nucleated cells, with liver containing the largest amount (Levi and Hodgson, 1989). Unlike the cytochrome P-450 enzyme system, however, FAD-containing monooxygenase is not inducible by pretreatment of animals with any chemical (Table 1).

FAD-containing monooxygenase catalyzes the oxidation of secondary and tertiary amines, hydrazines, phosphorus, and sulfur compounds outlined in Figure 4, A and B.

Both sulfur and amine substrates interact with the same catalytic site.

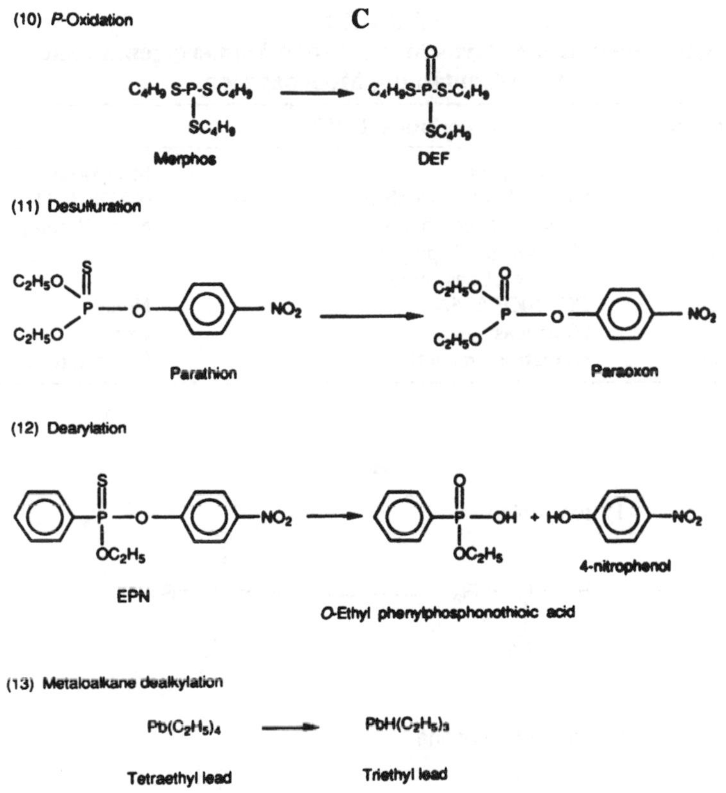

(10) *P*-Oxidation

C

$$C_4H_9 \; S\text{-}P\text{-}S \; C_4H_9 \longrightarrow C_4H_9S\text{-}P\text{-}S\text{-}C_4H_9$$

Merphos

DEF

(11) Desulfuration

Parathion

Paraoxon

(12) Dearylation

EPN

O-Ethyl phenylphosphonothioic acid

4-nitrophenol

(13) Metaloalkane dealkylation

$$Pb(C_2H_5)_4 \longrightarrow PbH(C_2H_5)_3$$

Tetraethyl lead

Triethyl lead

Figure 2. *Continued.*

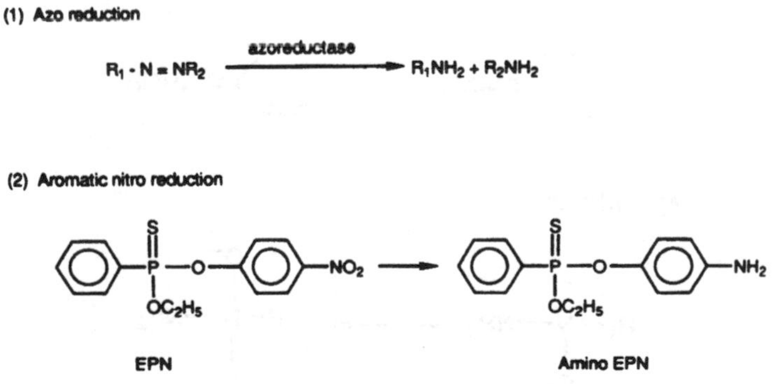

(1) Azo reduction

$$R_1 \cdot N = NR_2 \xrightarrow{\text{azoreductase}} R_1NH_2 + R_2NH_2$$

(2) Aromatic nitro reduction

EPN

Amino EPN

Figure 3. Reduction reactions catalyzed by cytochrome P-450 monooxygenase.

TABLE 1.
Characteristics of Cytochrome P-450 Monooxygenase and
FAD-Containing Monooxygenase

Characteristic	Cytochrome P-450	FMO
Location	Microsomes	Microsomes
Cofactors	NADPH, O_2, cytochrome c reductase	NADPH, O_2
Substrates	C, N, P, S compounds	N, P, S compounds
Inducers	Phenobarbitol, polyaromatic hydrocarbons, ethanol	None
Inhibitors	CO, SKF-525A	None
Isozymes	20 at least	Few
Reactions	Oxidation, reduction	Oxidation

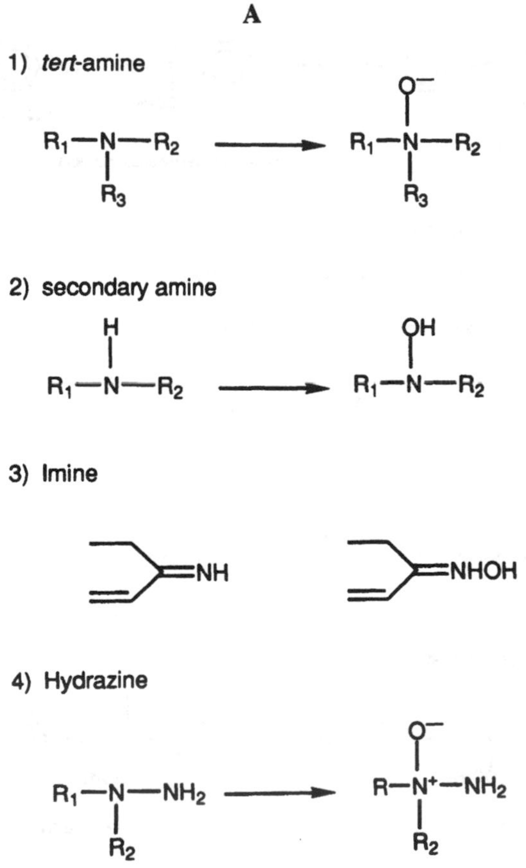

A

1) *tert*-amine

$$R_1{-}N{-}R_2 \longrightarrow R_1{-}\overset{\overset{\displaystyle O^-}{|}}{N}{-}R_2$$
with R_3 below each N

2) secondary amine

$$\overset{H}{R_1{-}N{-}R_2} \longrightarrow \overset{OH}{R_1{-}N{-}R_2}$$

3) Imine

with $=NH \longrightarrow =NHOH$

4) Hydrazine

$$R_1{-}N{-}NH_2 \longrightarrow R{-}\overset{\overset{\displaystyle O^-}{|}}{N^+}{-}NH_2$$
with R_2 below each N

Figure 4. Oxidation reactions catalyzed by FAD-containing monooxygenase. (A) nitrogen compounds; (B) sulphur compounds. *(Continued on p. 269)*

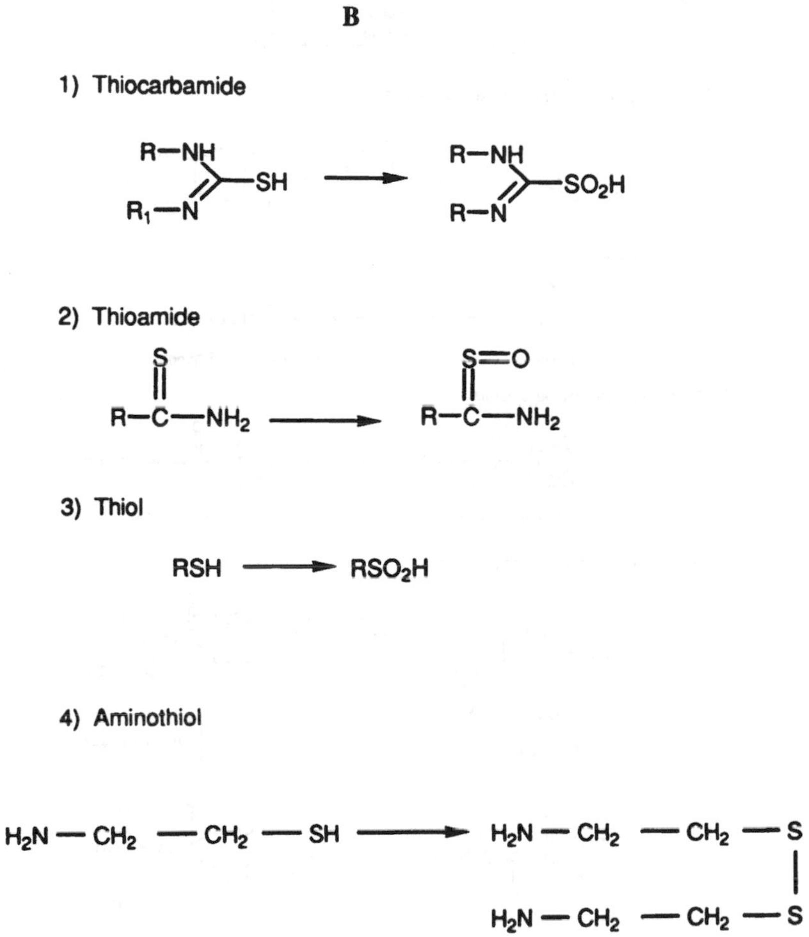

B

1) Thiocarbamide

2) Thioamide

3) Thiol

$$RSH \longrightarrow RSO_2H$$

4) Aminothiol

Figure 4. *Continued.*

Prostaglandin Endoperoxide Synthetase

Prostaglandin endoperoxide synthetase involves two enzymes: fatty acid cyclooxygenase and hydroperoxidase. Fatty and cyclooxygenase catalyzes bis-dioxygenation to produce the hydroperoxy endoperoxide PGG_2, while hydroperoxidase catalyzes the reduction of PGG_2 to PGH_2. Studies in our laboratory have shown that N-methyl carbamates that have a pair of electrons available for the last step of the reaction are themselves cooxidized (Sivarajah et al., 1981). Incubation of ram seminal vesicle microsomes, as a source of these enzymes, with arachidonic acid, in the presence of air, resulted in the demethylation of aminocarb and other mono- and dimethyl substituted carbamate insecticides. On the other hand, neither S- nor O-demethylation took

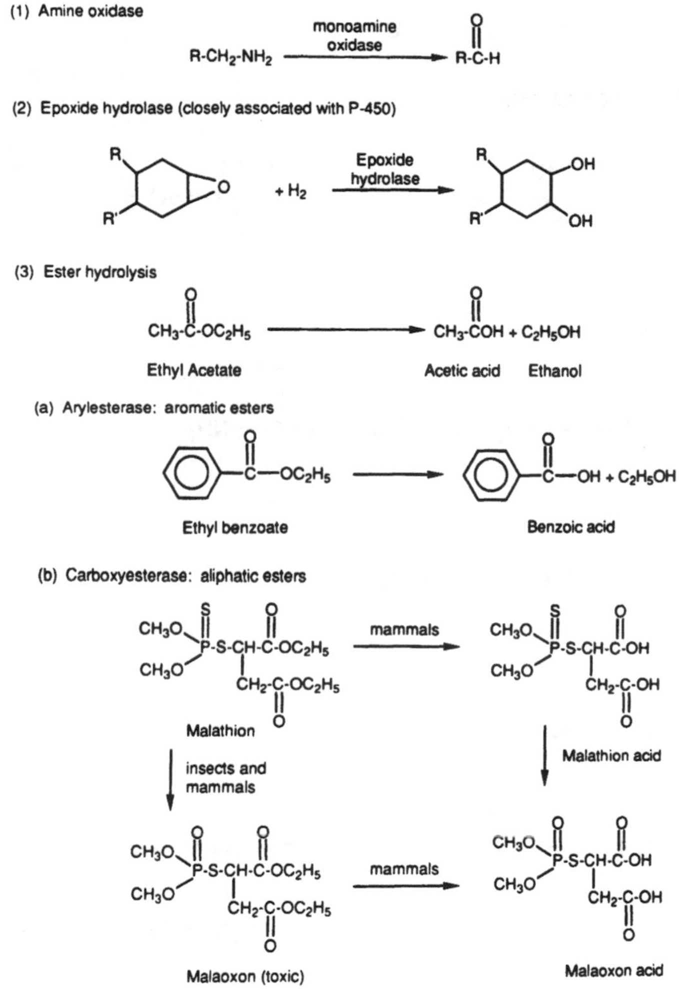

(1) Amine oxidase

$$R\text{-}CH_2\text{-}NH_2 \xrightarrow[\text{oxidase}]{\text{monoamine}} R\text{-}\overset{\overset{\displaystyle O}{\|}}{C}\text{-}H$$

(2) Epoxide hydrolase (closely associated with P-450)

(3) Ester hydrolysis

$$CH_3\text{-}\overset{\overset{\displaystyle O}{\|}}{C}\text{-}OC_2H_5 \longrightarrow CH_3\text{-}\overset{\overset{\displaystyle O}{\|}}{C}OH + C_2H_5OH$$

Ethyl Acetate   Acetic acid   Ethanol

(a) Arylesterase: aromatic esters

Ethyl benzoate   Benzoic acid

(b) Carboxyesterase: aliphatic esters

Figure 5. Other metabolic transformation reactions. *(Continued on p. 271)*

place. The rate of *N*-demethylation was similar to that produced by hepatic microsomal cytochrome P-450. Although the significance of this chemical-metabolizing enzymatic system is not yet clear, it may play an important role in tissues containing it, such as lung, kidney, and skin.

Other Reactions

Some neurotoxic chemicals may undergo metabolic transformations similar to those of normal body constituents: oxidation, reduction, deamination, and hydrolysis. The enzymes involved in these reactions are present in the soluble or mitochondrial fractions. Examples of these reactions are presented in Figure 5.

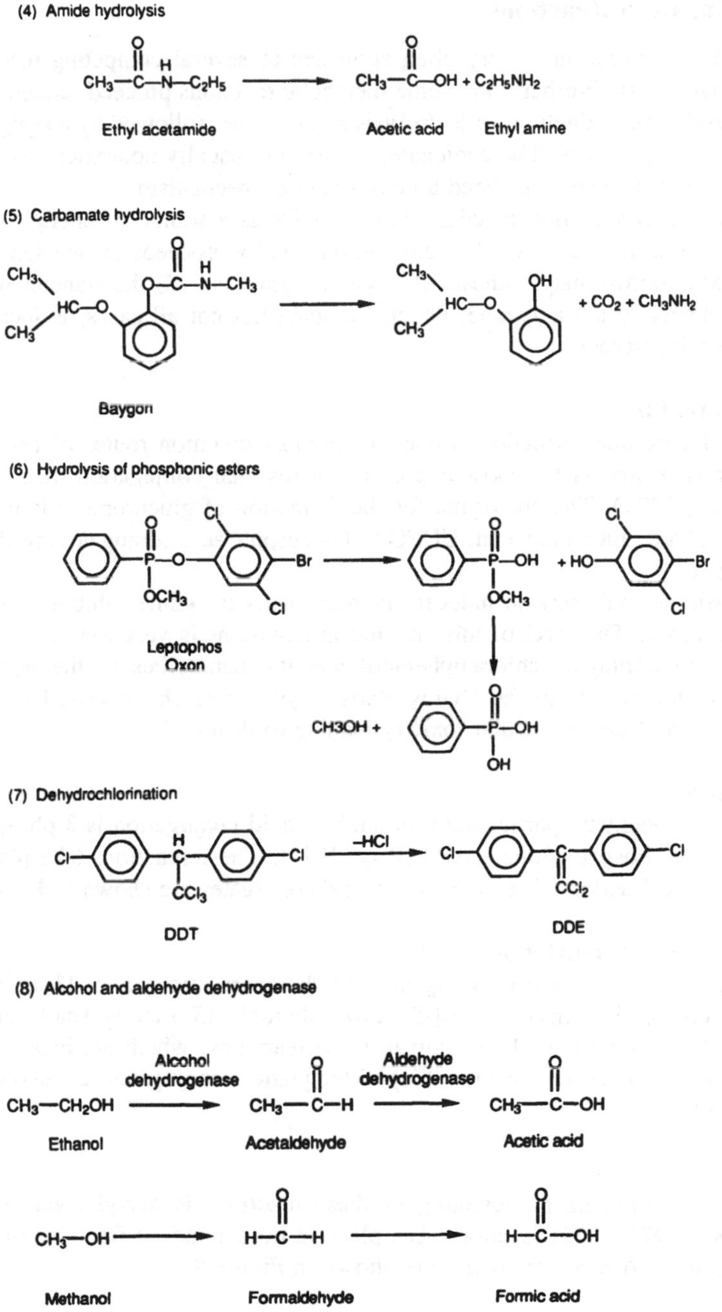

(4) Amide hydrolysis

CH$_3$—C—N—C$_2$H$_5$ $\longrightarrow$ CH$_3$—C—OH + C$_2$H$_5$NH$_2$

Ethyl acetamide   Acetic acid   Ethyl amine

(5) Carbamate hydrolysis

Baygon

(6) Hydrolysis of phosphonic esters

Leptophos
Oxon

CH3OH +

(7) Dehydrochlorination

DDT   DDE

(8) Alcohol and aldehyde dehydrogenase

CH$_3$—CH$_2$OH $\xrightarrow{\text{Alcohol dehydrogenase}}$ CH$_3$—C—H $\xrightarrow{\text{Aldehyde dehydrogenase}}$ CH$_3$—C—OH

Ethanol   Acetaldehyde   Acetic acid

CH$_3$—OH $\longrightarrow$ H—C—H $\longrightarrow$ H—C—OH

Methanol   Formaldehyde   Formic acid

Figure 5. *Continued.*

Conjugation Reactions

A neurotoxicant is very often subjected to several competing reactions simultaneously. Furthermore, some metabolic reactions proceed sequentially, and oxidation, reduction, or hydrolysis reactions are followed by conjugation of reaction products. The conjugates formed are usually nontoxic; therefore, conjugation is also considered a detoxification mechanism.

These conjugation reactions require ATP as a source of energy, coenzymes, and transferases. The conjugation usually proceeds in two steps: (1) the extramicrosomal synthesis of acylcoenzyme and (2) the transfer of the acyl moiety to the aglycone, which, in some but not all cases, is localized in the microsomes.

Glucuronides

Glucuronide formation is one of the most common routes of chemical metabolism and is the most important microsomal conjugation mechanism (Dutton, 1971). The coenzyme for the formation of glucuronides is uridine diphosphate glucuronic acid, UDPGA. Glucuronic acid conjugates are shown in Figure 6.

Microsomal enzyme inducers increase activity, while inhibitors inhibit this enzyme. The level of this enzyme in newborns is very low. In infants, failure to conjugate chloramphenicol and its metabolites to the nontoxic glucuronide results in the "Gray Baby" syndrome, characterized by with cyanosis and cardiovascular toxicity leading to death.

Sulfates

The coenzyme participating in sulfuric acid conjugation is 3-phosphadenosine-5-phosphosulfate, PAPS (Roy, 1971). These reactions take place in the soluble fraction of cells. Sulfuric acid conjugates are shown in Figure 7.

N,O-, and *S*-Methylation

The coenzyme participating in methylation reactions is 5'-[3-amino-3-carboxypropyl)methylsulfono]-5'-dioxyadenosine(*S*-adenosyl methionine). Methyl transferases (2.1.1) catalyze these reactions, which are localized in liver microsomes (Axelrod, 1971). Methylation conjugates are shown in Figure 8.

Amide Synthesis

The coenzyme participating in these reactions is acetyl coenzyme A (Weber, 1971). This reaction takes place in mitochondrial fractions of liver and kidney. Amide conjugates are shown in Figure 9.

Glutathione Conjugation

The enzyme involving the formation of glutathione conjugates is glutathione aryl transferase or glutathionekinase which form mercapturic acid

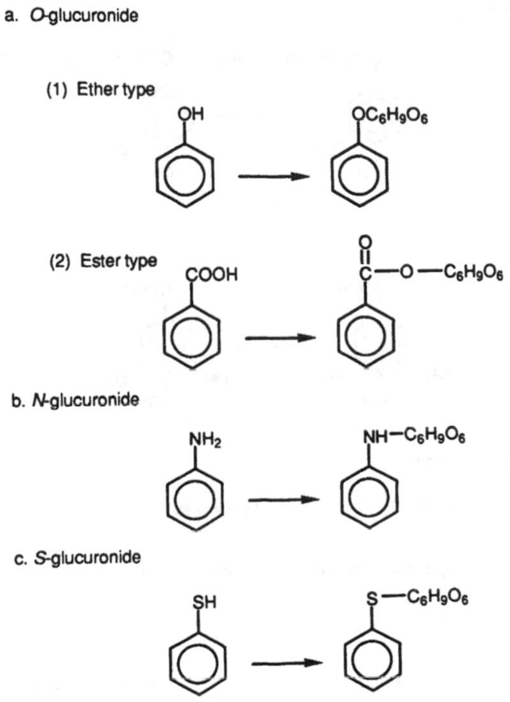
a. O-glucuronide
(1) Ether type
OH
OC6H9O6
(2) Ester type
COOH
O
C—O—C6H9O6
b. N-glucuronide
NH2
NH—C6H9O6
c. S-glucuronide
SH
S—C6H9O6

Figure 6. Glucuronic acid conjugation reactions.

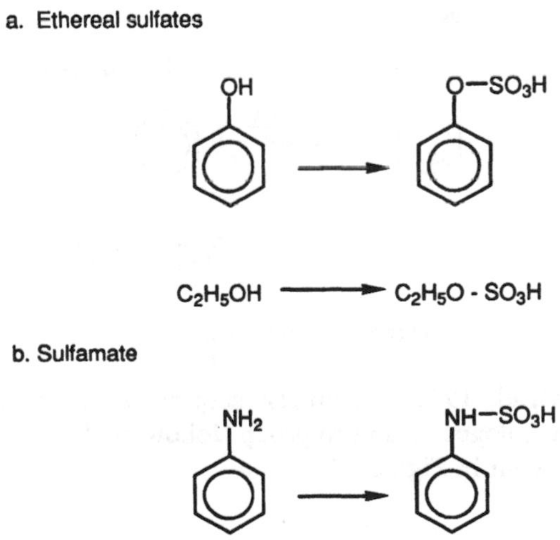
a. Ethereal sulfates
OH
O—SO3H
C2H5OH
C2H5O - SO3H
b. Sulfamate
NH2
NH—SO3H

Figure 7. Sulfuric acid conjugation reactions.

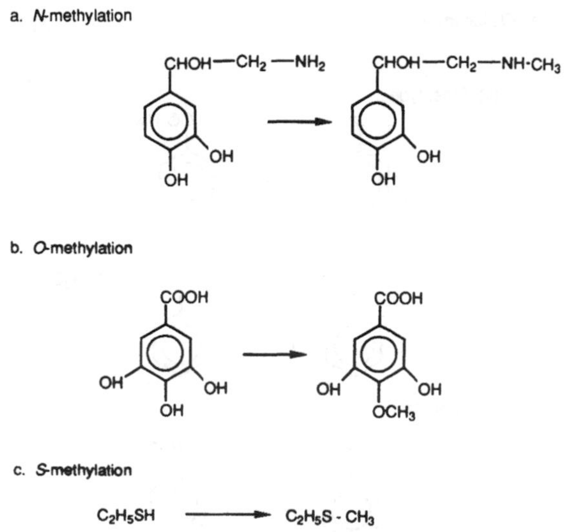

a. *N*-methylation

b. *O*-methylation

c. *S*-methylation

$$C_2H_5SH \longrightarrow C_2H_5S \cdot CH_3$$

Figure 8. Methylation conjugation reactions.

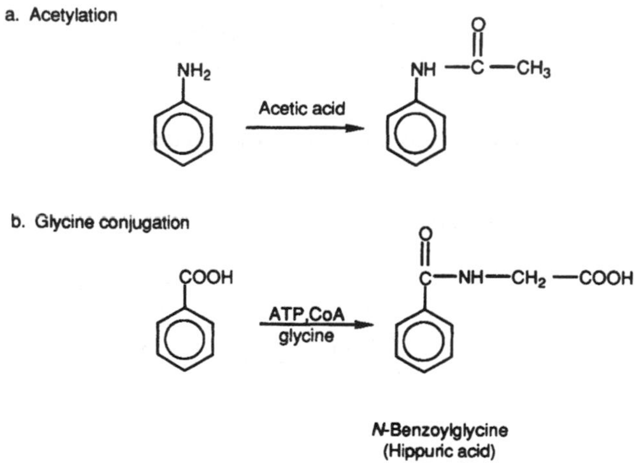

a. Acetylation

b. Glycine conjugation

N-Benzoylglycine
(Hippuric acid)

Figure 9. Amide conjugates.

derivatives (Boyland, 1971). Chemicals subject to this reaction frequently contain an active halogen or a nitro group (Jollow et al., 1974). Glutathione conjugates are shown in Figure 10.

Thiocyanate

Free cyanides or cyanide formed from nitrites or oximes form thiocyanate endogenously using sulfur transferase or Rhodanase that is found mainly in liver mitochondria.

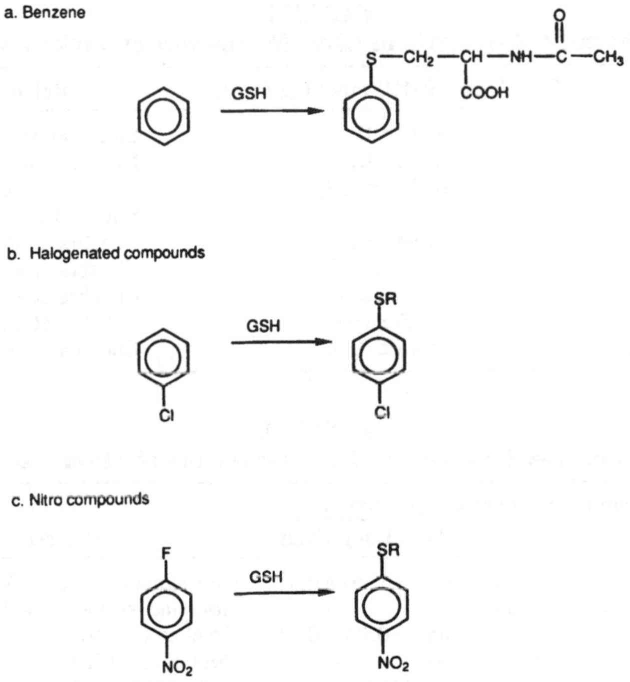

a. Benzene

b. Halogenated compounds

c. Nitro compounds

Figure 10. Glutathione conjugation reactions.

$$\text{CN}^- + \text{S}_2\text{O}_3^= \xrightarrow{\underset{\text{(rhodanase)}}{\text{sulfur transferase}}} \text{SCN}^- + \text{SO}_3$$

cyanide   thiosulfate   thiocyanate   sulfite

Glucoside Conjugation

This conjugation occurs mainly in insects and plants where it replaces glucuronide formation. Ester and ether glucosides are formed.

$$\text{ROH} + \text{UDP} - \text{glucose} \xrightarrow{\text{glucosyl transferase}}$$

$$\text{RO} - \beta\text{-glucoside} + \text{UDP}$$

Mammals do not synthesize glucoside because they lack glucosyl transferase.

Factors Affecting Metabolism

Species Variability

Various animal species metabolize neurotoxic compounds differently qualitatively and quantitatively. It is important in assessing the neurotoxicity of a compound to use an animal species that metabolizes the test chemical similar to the process in humans. Liver microsomal cytochrome P-450 content

TABLE 2.
Cytochrome P-450 Levels in Liver Microsomes of Various Species

Species	Cytochrome P-450 nmol/mg Protein	Reference
Chicken	0.13 ± 0.1	Lasker et al., 1982
Human	0.28 ± 0.12	Nelson et al., 1971
	(0.53 ± 0.33)	(Pelkonen et al., 1973)
Cat	0.34	Kato, 1966
Rat	0.84 ± 0.07	Chhabra et al., 1974
Mouse	1.1 ± 0.07	Chhabra et al., 1974
Rabbit	1.1 ± 0.32	Chhabra et al., 1974
Hamster	1.26 ± 0.07	Chhabra et al., 1974
Guinea pig	1.45 ± 0.16	Chhabra et al., 1974

TABLE 3.
Cytochrome P-450 Levels in Liver Microsomes of Three Rat Strains

Cytochrome P-450 nmol/mg Protein			
Sprague-Dawley	Fischer 344	Long-Evans	Reference
1.50 ± 0.16	1.05 ± 0.07	1.46 ± 0.17	Carrington and Abou-Donia, 1988
0.75 ± 0.16	0.59 ± 0.08	—	Gold and Widnell, 1975
0.26 ± 0.04	0.33 ± 0.05	0.04 ± 0.05	Creel et al., 1976
0.84 ± 0.03	0.67 ± 0.04	—	Dent et al., 1980

varies with animal species (Table 2) and the strain of the same species (Table 3).

Some species metabolize neurotoxic compounds at different rates. Studies on the metabolism of the delayed neurotoxicant leptophos indicate that species susceptible to delayed neurotoxicity have a higher accumulation rate coupled with a slower elimination rate of the compound. Thus orally administered leptophos, in nonsusceptible species, e.g., mice (Holmstead et al., 1973) and rats (Whitacre et al., 1976), was rapidly metabolized and excreted as polar metabolic products mainly in the urine. In hens, a sensitive species (Abou-Donia, 1976, 1979, 1980), this insecticide was persistent in the body and was eliminated at a rate 32 times slower than that of the mice, mostly as the parent compound.

Induction and Inhibition of Microsomal Cytochrome P-450 Enzymes

Some chemicals can increase the activity of microsomal cytochrome P-450 enzymes by quantitatively increasing these enzymes. Most enzyme inducers give maximum increase within 2–3 days, although some may require 2 weeks or longer. The degree of induction after reaching maximum activity may decline despite continuing treatment of the animal with the chemical. Dermally applied DEF, a neurotoxic pesticide, induced phenobarbitol-type cytochrome P-450 in hen liver microsomes (Lapadula et al., 1984).

The significance of enzyme induction in neurotoxicity is twofold. The net result of neurotoxic action will depend on the balance between activation and detoxification produced by metabolizing enzymes. If metabolism results in polar, water-soluble, nontoxic metabolites, induction will increase detoxification and excretion of the compound, thus diminishing its neurotoxicity. On the other hand, if metabolism yields a more toxic metabolite, induction will increase the neurotoxicity of a compound. An example is the metabolism of TOCP. This neurotoxic compound is bioactivated to the saligenin *o*-tolyl cyclic phosphate that is more potent as a neurotoxicant than the parent compound. It is also hydrolyzed to *o*-cresol, which undergoes further oxidation to more polar, less toxic metabolites. The balance between these activation and inactivation reactions seems to play an important role in the development of delayed neurotoxicity. In sensitive species, i.e., the cat (Nomeir and Abou-Donia, 1984, 1986) and chicken (Suwita and Abou-Donia, 1989), the active metabolite is formed to a larger extent and persists longer than in insensitive species, i.e., rat (Abou-Donia et al., 1990).

Cytochrome P-450 enzymes may be inhibited by certain chemicals, e.g., SKF-525A. This compound has been used to study the effect of enzyme inhibition on the metabolism of neurotoxicants. Methods for assaying cytochrome P-450 enzymatic activities using various substrates are listed in Table 4.

Saturation of Metabolic Pathways

Metabolic pathways may be saturated *in vivo* following the administration of large doses of neurotoxic chemicals (Jollow et al., 1974). Such saturation results in less metabolic biotransformation of the neurotoxicant and the persistence of the parent compound in various tissues thus producing severe neurotoxicity. It is very important to determine the effect of the dose size on the metabolism of a chemical to assess its neurotoxicity.

Metabolism Studies of Neurotoxic Chemicals

The overall goals of metabolism and toxicokinetic studies are to determine the amount and rate of absorption, distribution, metabolism, and excretion of the test compound. Metabolism studies are carried out on an animal species that handles the test chemical similarly to man. If such information is not available, initial studies are performed in the rat and on other nonrodent species. Animals should be acclimatized to the metabolism cage before the experiment. Experimental conditions should be standardized, e.g., light cycle, temperature, humidity, and time of feeding. The clinical condition, weight, food and water consumption of each animal should be recorded daily (La Du et al., 1972).

To determine the bioavailability of the test compound, an initial study should be carried out by administering the compound by intravenous injection

TABLE 4.
Methods for Determination of Cytochrome P-450 Enzymatic Activities

Reaction	Substrate	Reference
Aliphatic hydroxylation	Phenobarbital	Cooper and Brodie, 1965
Aliphatic, benzylic hydroxylation	Testosterone	Waxman et al., 1983
	Androsterone	Waxman et al., 1983
	Progesterone	Waxman et al., 1988
Aromatic hydrocarbon hydroxylation	Benzo(α)pyrene	Nebert and Gelboin, 1968
Aromatic hydroxylation	Aniline	Brodie and Axelrod, 1948
	Acetanilide	Guenthner et al., 1979
	Coumarin	Greenlee and Poland, 1978
	P-nitrophenol	Koop, 1986
	(R)-warfarin	Kaminsky et al., 1981
N-demethylation	N,N-dimethylaniline	Cochin and Axelrod, 1959
	Ethylmorphine	Anders and Mannering, 1966
	Benzphetamine	Hewick and Fouts, 1970
	Aminopyrine	Brodie and Axelrod, 1950
O-deethylation	Ethylumbelliferone	Ullrich and Weber, 1972
	7-Ethoxycoumarin	Greenlee and Poland, 1978
	7-Ethoxyresorufin	Burke and Mayer, 1974
O-demethylation	4-Nitrophenol	Zannoni, 1971
	7-Methoxyresorufin	Burke and Mayer, 1974
	6,7-Dimethoxycoumarin (scoparone)	Muller-Enoch and Greischel, 1988
N-hydroxylation	Aniline	Herr and Kiese, 1959
N-oxidation	N,N-Dimethyl aniline	Ziegler et al., 1973
Nitro reduction	4-Nitrophenol	Fouts and Brodie, 1957

and via the route by which humans are exposed to it. The intravenous administration will provide definite information on the kinetics of earlier phases of distribution and/or elimination. It is preferable that two doses be used: a dose equivalent to the neurotoxic dose, and another dose that is below the dose that produces signs of neurotoxicity.

TOXICOKINETICS

Toxicokinetics is the study of toxicant movement in the body. It is concerned with quantification of metabolism processes: absorption, distribution, biotransformation, and excretion. Thus, toxicokinetics assess the rates of these processes.

In toxicokinetics, the body is represented as a system of compartments, even though they have no apparent physiologic or anatomic presence. A compartment encompasses those organs, tissues, cells, and fluids that have similar rates of uptake and clearance of a chemical. The central compartment is a rapidly equilibrating compartment that may include tissues with profuse blood supply, e.g., liver and kidneys. On the other hand, peripheral compartment is a slow equilibrating compartment and may include tissues with limited blood supply, e.g., fat and bone.

One-Compartment Open-Model System

The one-compartment, open-model system is shown in Figure 11A. This model depicts the body as a single homogeneous unit.

The model assumes that the chemical equilibrates with all tissues rapidly and any change that occurs in the plasma quantitatively reflects changes occurring in tissue chemical levels (Gillette, 1974; Othman and Abou-Donia, 1988). In this model, chemical elimination occurs from the body by apparent first-order kinetics, i.e., the rate of elimination of a chemical is proportional to its concentration in the plasma at that time described in Equation (1)

$$\frac{dC}{dt} = -KC \qquad (1)$$

where C is the concentration of the chemical in the plasma at time t; K is the apparent first-order elimination rate constant for the chemical; and the negative sign indicates that drug is being lost from the plasma.

Integration of Equation 1 describes the time course of the chemical concentration in plasma as shown in Equation 2

$$C = C_0 e^{-Kt} \text{ (exponential form)} \qquad (2)$$

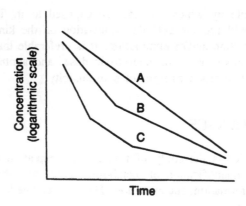

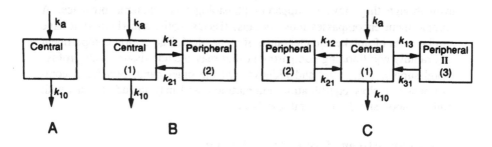

Figure 11. Plasma compound concentrations (logarithmic scale) vs. time. (A) Monoexponential, one-compartment; (B) biexponential, two-compartment; and (C) triexponential, three-compartment open model systems.

or

$$\ln C = Kt + \ln C_0 \qquad (3)$$

$$\log C = -Kt/2.303 + \log C_0 \text{ (logarithmic form)} \qquad (4)$$

where C_0 is the concentration of the chemical in the plasma at time zero. A plot of C vs. time on semilogarithmic paper yields a straight line with slope $= -K$ and an intercept $= C_0$.

Toxicokinetic Parameters

Elimination Half-Life. The elimination of half-life, $t_{1/2}$, is the time required to reduce plasma concentration to half of its original value, assuming that the chemical is eliminated by a first-order process (Gibaldi, 1982; Abou-Donia et al., 1989). Its unit is time.

$$t_{1/2} = \frac{\lin 2}{K} = \frac{0.693}{K} = \frac{0.693 \times V_d}{Cl} \qquad (5)$$

where V_d is the apparent volume of distribution, and Cl is total clearance.

Elimination half-life may be used to estimate the length of the period before multiple dosing will reach steady state; it takes 4–5 $t_{1/2}$ to reach the steady state as follows:

Steady State (%)	Number of $t^{1/2}$
50	1
75	2
87.5	3
93.75	4
96.87	5

Apparent Volume of Distribution. The apparent volume of distribution (V_d), is the apparent volume to which the chemical is distributed in the body. V_d is a proportionality constant that relates the amount of chemical in the body to its concentration in plasma and can be defined as:

$$V_d = \frac{\text{Amount of chemical in body}}{\text{concentration in plasma}} \qquad (6)$$

Following an intravenous injection, plasma concentration of the chemical declines because of elimination and distribution to tissues. Therefore, to determine V_d, it is necessary to extrapolate the curve back to the origin where no elimination has taken place at time zero. Plasma concentration obtained at time zero intercept is divided into applied dose, D_o, to obtain the volume of distribution, V_d.

$$V_d = \frac{\text{Dose}}{C_0} \qquad (7)$$

where C_0 is the hypothetical drug concentration at time zero. V_d has a volume unit of ml or l or adjusted for body weight, e.g., l/kg.

The value of V_d does not represent a real body space, but rather provides some information about the distribution of the chemical in the body. As the distribution into the tissues increases, V_d increases. If a chemical in human body is not distributed in the tissues, but its distribution is limited to plasma, extracellular fluid, or total body water, the V_d values will be approximately 40, 170, and 580 ml/kg, respectively. If the chemical is widely distributed or has a high affinity for a tissue such as fat, V_d may be more than 1000 ml/ kg. V_d also reflects the fraction of the chemical that is available for excretion. If the chemical is extensively distributed in the tissues, as indicated by large V_d, less of it will be available for excretion. If V_d is small, indicating less tissue distribution, more of it will be available for excretion.

Total Clearance. Total clearance, Cl, is the fraction of V_d that is cleared of a chemical per unit time. Cl may be determined according to Equation 8.

$$Cl = V_d \times K = \frac{0.693}{t_{1/2}} V_d \qquad (8)$$

However, Cl is usually calculated according to Equation 9.

$$Cl = \frac{Dose}{AUC} \qquad (9)$$

where AUC is the total area under the plasma concentration vs. time curve after a single dose.

Total clearance include individual organ clearances, e.g., liver (hepatic) clearance, Cl_H, and renal clearance, Cl_R. Clearance by an organ is determined by the blood flow through the organ, Q, and the extraction ratio, E, according to Equation 10.

$$Cl = Q \times E \qquad (10)$$

E is determined by the in-flowing, C_{in}, and out-flowing, C_{out}, concentration of the chemical, by Equation 11.

$$E = \frac{C_{in} - C_{out}}{C_{in}} \qquad (11)$$

The maximum value for hepatic clearance is that of liver blood flow rate, which is approximately 1.5 l/min in humans. High hepatic clearance leads to "first-pass" effect, i.e., high presystemic (biotransformation) elimination of the chemical after oral administration, e.g., propranolol. Similarly, renal clearance cannot exceed renal blood flow rate, which is approximately 650 ml/min in humans. This high Cl_R results in total tubular excretion, e.g., *P*-aminobenzoic acid. Renal clearance of a chemical may be determined according to Equation 12.

$$Cl_R = \frac{A_e,\infty}{AUC} \qquad (12)$$

where A_e,∞ is the total amount of chemical eliminated unchanged in the urine. Cl_R can be used to determine f_e, the fraction of dose of chemical eliminated through the kidneys in humans with normal renal functions as follows:

$$f_e = \frac{A_e,\infty}{Dose} = \frac{Cl_R}{Cl} \tag{13}$$

Clearance unit is volume per time, e.g., ml/min, l/h, or expressed for body weight, e.g., l/h/kg. To maximize recovery of dose, urine should be collected over at least 5 elimination half-lives to calculate f_e using this method.

The value of Cl, Cl_H, and Cl_R provide important information regarding the elimination processes of the chemical and, consequently, its toxicity. High values of Cl and Cl_H suggest high extraction ratios by the liver. This implies the chemical would have high systemic availability in patients with liver diseases. High Cl_R, e.g., 100 ml/min, suggests that the chemical will accumulate in patients with renal failure.

Bioavailability. Bioavailability or systemic availability, F, is the fraction of orally administered dose that is absorbed into the systemic circulation. It is determined as the ratio of AUCs obtained following oral and i.v. dosing according to Equation 14.

$$F = \frac{AUC_{oral} \times Dose_{i.v.}}{AUC_{i.v.} \times Dose_{oral}} \tag{14}$$

Bioavailability is a fraction and therefore has no units. Low values of F for a chemical with low Cl suggests poor absorption, while low values of F coupled with high Cl implies high first-pass effect.

Two-Compartment Open-Model System

Following the introduction of some chemicals into the central compartment, which is composed of highly perfused tissues, their concentration in these tissues declines more rapidly during the distribution phase than during the postdistribution phase. In contrast, the chemical levels in the peripheral compartment, which is composed of the poorly perfused tissues, e.g., muscle, skin, and fat, will reach a maximum and then begin to decline during the elimination phase (Gibaldi and Perrier, 1982). All poorly perfused tissues are often designated as a single "peripheral" compartment. With time, an equilibrium is attained between the tissues and fluid of the central and peripheral compartments. Chemicals pass into and out of each compartment by a first-order process, and are eliminated only from the central compartment by a first-order process resulting from metabolism and excretion.

In this model, k_{12} is the rate constant for the movement of the chemical from the central compartment 1 to the peripheral compartment 2, k_{21} is the rate constant for the movement of the chemical from the peripheral compartment back to the central compartment, and k_{10} is the rate constant of elimination (Figure 11B). The time course of chemical level in the two-compartment open-model system is described by the following equation:

$$C_t = A_0 e^{-\alpha t} + B_0 e^{-\beta t} \tag{15}$$

C_t is plasma concentration of the chemical at time t.

The rate constant for the first phase of the biexponential decline of the chemical concentration is α. After the straight line of the β-phase is extrapolated back to time zero, the extrapolated values are subtracted from the experimental values of chemical concentration and the resultant values are plotted on semilogarithmic graph paper. α may be estimated from the slope of the secondary plot according to the equation $\alpha = 2.303 \times$ slope or from the half-time according to the relationship $\alpha = 0.693/t_{1/2}$.

The rate constant for the second phase of the biexponential decline in the chemical concentration is β.

A_0 is the intercept obtained after extrapolating the secondary plot back to time zero.

B_0 is the intercept obtained by extrapolating the straight line associated with β back to time zero.

α-Phase is the rapid phase of the biphasic decline of the chemical concentration. It is described by the equation $A = A_0 e^{-\alpha t}$.

β-Phase is the slow phase of the biphasic decline of the chemical concentration. It is described by the equation $B = B_0 e^{-\beta t}$.

$$k_{21} = \frac{A\beta + B\alpha}{A + B} \tag{16}$$

$$k_{10} = \frac{\alpha\beta}{k_{21}} \tag{17}$$

$$k_{12} = (\alpha + \beta) - (k_{21} + k_{10}) \tag{18}$$

Three-Compartment Open-Model System

The three-compartment, open-model system assumes that all processes are linear and that elimination occurs from the central compartment (including plasma) which is connected to two peripheral "shallow" and "deep" compartments (Kaplan et al., 1973; Abou-Donia et al., 1989). The liver and kidneys are the major sites for metabolic transformation and excretion, respectively. These well-perfused organs are assumed to be part of the central compartment. Figure 11C simulates plasma concentration curve representing a three-compartment system following rapid intravenous administration of a chemical. The chemical is first rapidly distributed and equilibrated to well perfused tissues, then slowly to tissues constituting the "shallow" compartment, and finally more slowly to the "deep" compartment. Assuming all transfer processes are first-order, the plasma concentration curve appears triphasic, and is described by the following triexponential equation

$$C_p = Pe^{-\pi t} + Ae^{-\alpha t} + B^{-\beta t} \tag{19}$$

where C_p is the chemical concentration in the central compartment at time t, and P, π, A, α, B, and β are constants. The values of the three hybrid rate constants, π, α, and β, reflect a rapid initial distribution of the chemical and a slower apparent elimination rate of the chemical from the body. The individual rate constants, k_{12}, k_{21}, and k_{13}, k_{31}, reflect the rate of distribution into and out of the shallow and deep peripheral compartments, respectively.

Calculation of Kinetic Parameters

The rate constants k_{12}, k_{21}, k_{13}, k_{31}, and k_{10} may be determined as follows:

$$P = \frac{X_0 (k_{21} - \pi)(k_{31} - \pi)}{V_c (\pi - \alpha) (\pi - \beta)} \tag{20}$$

$$A = \frac{X_0 (k_{21} - \alpha)(\alpha - k_{31})}{V_c (\pi - \alpha) (\alpha - \beta)} \tag{21}$$

$$B = \frac{X_0 (k_{21} - \beta)(k_{31} - \beta)}{V_c (\alpha - \beta) (\pi - \beta)} \tag{22}$$

$$C_0 = P + A + B \tag{23}$$

$$C_0 = \frac{X_0}{V_c} \tag{24}$$

$$V_c = \frac{X_0}{P + A + B} \tag{25}$$

$$a = 1 \tag{26}$$

$$b = -\frac{\pi B + \pi A + \beta P + \beta A + \alpha P + \alpha B}{P + A + B} \tag{27}$$

$$c = \frac{\alpha\pi B + \pi\beta A + \alpha\beta P}{P + A + B} \tag{28}$$

$$k_{21} = 1/2 \, (-b + \sqrt{b^2 - 4C}) \tag{29}$$

$$k_{31} = 1/2 \, (-b - \sqrt{b^2 - 4C}) \tag{30}$$

$$k_{10} = \frac{\alpha\beta\pi}{k_{21}K_{31}} \tag{31}$$

$$k_{12} = \frac{(\alpha\beta + \alpha\pi + \beta\pi) - K_{21}(\alpha + \beta + \pi) - k_{10}k_{31} + k_{21}^2}{k_{31} - k_{21}} \tag{32}$$

$$k_{13} = \alpha + \beta + \pi - (k_{10} + k_{12} + k_{21} + k_{31})$$

$$X_0 = \text{The intravenous dose} \tag{33}$$

REFERENCES

Abou-Donia, M.B. (1976). Pharmacokinetics of a neurotoxic oral dose of leptophos in hens. *Arch. Toxicol.* 36: 103–110.

Abou-Donia, M.B. (1979). Pharmacokinetics and metabolism of a topically-applied dose of O-bromo-2,5-dichlorophenyl O-methyl phenylphosphonothioate in hens. *Toxicol. Appl. Pharmacol.* 51: 311–328.

Abou-Donia, M.B. (1980). Metabolism and pharmacokinetics of a single oral dose of O-4-bromo-2,5-dichlorophenyl O-methyl phenylphosphonothioate (leptophos) in hens. *Toxicol. Appl. Pharmacol.* 55: 131–145.

Abou-Donia, M.B. (1983). Toxicokinetics and metabolism of delayed neurotoxic organophosphorus esters. *Neurotoxicology* 4: 113–130.

Abou-Donia, M.B. and D.B. Menzel (1968). The *in vivo* metabolism of DDT, DDD and DDE in the chick by embryonic injection and dietary ingestion. *Biochem. Pharmacol.* 17: 2143–2161.

Abou-Donia, M.B. and D.B. Menzel (1976). The degradation of ring-labeled ^{15}C-DDT to $^{14}CO_2$ in the rat. *Experientia* 32: 500–501.

Abou-Donia, M.B. and A.A. Nomeir (1986). The role of pharmacokinetics and metabolism in species sensitivity of neurotoxic agents. *Fundam. Appl. Toxicol.* 6: 190–207.

Abou-Donia, M.B., H.M. Abdel-Kader, and S.A. Abou-Donia (1983a). Tissue distribution, elimination, and metabolism of O-ethyl O-4-nitrophenylphenylphosphonothioate in hens following daily dermal doses. *J. Am. Coll. Toxicol.* 2: 391–404.

Abou-Donia, M.B., Y.M. Hernandez, N.S. Ahmed, and S.A. Abou-Donia (1983b). Distribution and metabolism of O-ethyl O-4-nitrophenyl phenylphosphonothioate after a single oral dose in one-week old chicks. *Arch. Toxicol.* 54: 83–96.

Abou-Donia, M.B., C.G. Kinnes, K.M. Abdo, and T.D. Bjornsson (1983c). Physiologic disposition and metabolism of *O,O*-diethyl *O*-4-nitrophenyl phosphonothioate in male cats following a single dermal administration. *Drug Metab. Dispos.* 11: 31–36.

Abou-Donia, M.B., B.L. Reichert, and M.A. Ashry (1983d). The absorption, distribution, excretion, and metabolism of a single oral dose of *O*-ethyl *O*-4-nitrophenyl phenylphosphonothioate in hens. *Toxicol. Appl. Pharmacol.* 70: 18–28.

Abou-Donia, M.B., K. Sivarajah, and S.A. Abou-Donia (1983e). Disposition, elimination, and metabolism of *O*-ethyl *O*-4-nitrophenyl phenylphosphonothioate after subchronic dermal application in male cats. *Toxicology* 26: 93–111.

Abou-Donia, M.B., M.A. Othman, and P. Obih (1989). Interspecies comparison of pharmacokinetic profile and bioavailability of ($\pm$)gossypol in male Fischer-344 rats and male B6C3 F mice. *Toxicology* 55: 37–51.

Abou-Donia, M.B., A.A. Nomeir, J.H. Bower, and H.A. Makkawy (1990). Absorption, distribution, excretion, and metabolism of a single oral dose of [^{14}C]tri-*o*-cresyl phosphate (TOCP) in the male rat. *Toxicology* 65: 61–74.

Anders, M.W. and G.J. Mannering (1966). Inhibition of drug metabolism. I. Kinetics of the inhibition of the *N*-demethylation of ethylmorphine by 2 diethylaminoethyl 2,2-diphenylvalerate HCl (SKF 525A) and related compounds. *Mol. Pharmacol.* 2: 319–327.

Axelrod, J. (1971). Methyltransferase enzymes in the metabolism of physiologically compounds and drugs. In *Concepts in Biochemical Pharmacology*, Vol. 2. B.B. Brodie and J.R. Gillette, Eds., Springer-Verlag, Berlin, pp. 609–619.

Baker, K.J. and S.E. Bradley (1966). Binding of sulfobromophthalein (BSP) sodium by plasma albumin. Its role in hepatic BSP extraction. *J. Clin. Invest.* 45: 281.

Bakry, N.M., A.K. Salama, H.A. Aly and M.B. Abou-Donia (1990). Milk transfer, distribution, and metabolism of a single oral dose of [^{14}CH$_3$S] methamidophos in Sprague Dawley rats. *The Toxicologist* 10: 346.

Boyland, E. (1971). Mercapturic acid conjugation. In *Concepts in Biochemical Pharmacology*, Vol. 2. B.B. Brodie and J.R. Gillette, Eds., Springer-Verlag, Berlin, pp. 584–608.

Brauer, R.W. (1959). Mechanisms of bile secretion. *J. Am. Med. Assoc.* 169: 1462–1466.

Brodie, B.B. (1964). Physiochemical factors in drug absorption. In *Absorption and Distribution of Drugs*. T.B. Binns, Ed., Williams & Wilkins, Baltimore, pp. 16–48.

Brodie, B.B. and J. Axelrod (1948). The estimation of acetanilide and its metabolis products, aniline, *N*-acetyl-*p*-aminophenol and *p*-aminophenol (free and total conjugates) in biological fluids and tissues. *J. Pharmacol. Exp. Ther.* 94: 22–28.

Brodie, B.B. and J. Axelrod (1950). The fate of aminopyrine (Pyramidon) in man and methods for the estimation of aminopyrine and its metabolites in biological material. *J. Pharmacol. Exp. Ther.* 99: 171–184.

Burke, M.B. and R.T. Mayer (1974). Ethoxyresorufin: direct fluorimetric assay of a microsomal O-dealkylation which is preferentially inducible by 3-methylcholanthrene. *Drug Metab. Dispos.* 2: 583–588.

Carrington, C.D. and M.B. Abou-Donia (1988). Variation between three strains of rats: inhibition of neurotoxic esterase and acetylcholinesterase by tri-o-cresyl phosphate. *J. Toxicol. Environ. Health* 25: 259–268.

Chhabra, R.S., R.J. Pohl and J.R. Fouts (1974). A comparative study of xenobiotic-metabolizing enzymes in liver and intestine of various animal species. *Drug Metab. Dispos.* 2: 443–447.

Cochin, J. and J. Axelrod (1959). Biochemical and pharmacological changes in the rat following chronic administration of morphine, nalorphine and normorphine. *J. Pharmacol. Exp. Ther.* 125: 105–110.

Conney, A. H. (1967). Pharmacological implication of microsomal enzyme induction. *Pharmacol. Rev.* 19: 317–366.

Cooper, J.R. and B.B. Brodie (1965). Enzymatic oxidation of pentobarbital and thiopental. *J. Pharmacol. Exp. Ther.* 120: 75–87.

Creel, D., D.E. Shearer, and P.F. Hall (1976). Differences in cytochrome P-450 of various strains of rats following chronic administration of pentobarbital. *Pharmacol. Biochem. Behav.* 5: 705–707.

Dent, J.G., M.E. Schnell, and J. Lasker (1980). Constitutive and induced hepatic microsomal cytochrome P-450 monooxygenase activities in male Fischer-344 and CD rats. A comparative study. *Toxicol. Appl. Pharmacol.* 52: 45–53.

Draize, J.H., G. Woodard, and H.O. Calvery (1944). Methods for the study of irritation and toxicity of substances applied topically to the skin and mucous membrane. *J. Pharmacol. Exp. Ther.* 82: 377–390.

Dutton, G.J. (1971). Glucuronide-forming enzymes. In *Concepts in Biochemical Pharmacology,* Vol. 2. B.B. Brodie and J.R. Gillette, Eds., Springer-Verlag, Berlin, pp. 378–400.

Feldman, R.J. and H.I. Maibach (1970). Absorption of some organic compounds through the skin of man. *J. Invest. Dermatol.* 54: 399–404.

Fouts, J.R. and B.B. Brodie (1957). The enzymatic reduction of chloramphenicol, *p*-nitrobenzoic acid and other aromatic nitro compounds in mammals. *J. Pharmacol. Exp. Ther.* 119: 197–207.

Gibaldi, M. and D. Perrier (1982). *Pharmacokinetics,* 2nd ed. Marcel Dekker, New York.

Gibson, J.E. and B.A. Becker (1967). Demonstration of enhanced lethality of drugs in hypoexcretory animals. *J. Pharm. Sci.* 56: 1503–1505.

Gillette, J.R. (1974). The importance of tissue distribution in pharmacokinetics. In *Pharmacology and Pharmacokinetics.* T. Toerell, R.L. Dedrick, and P.G. Condliffe, Eds., Plenum, New York, pp. 209–231.

Gold, G. and C.D. Widnell (1975). Response of NADPH cytochrome c reductase and cytochrome P-450 in hepatic microsomes to treatment with phenobarbital-differences in rat strains. *Biochem. Pharmacol.* 24: 2105–2106.

Greenlee, W.F. and A. Poland (1978). An improved assay of 7-ethoxycoumarin O-deethylase activity: induction of hepatic enzyme activity in C57BL/6J and DBA/2J mice by phenobarbital, 3-methylcholanthrene and 2,3,7,8 tetrachlorodibenzo-P-dioxin. *J. Pharmacol. Exp. Ther.* 205: 596–605.

Guenthner, T.M., M. Negishi, and D.W. Nebert (1979). Separation of acetanilide and its hydroxylation metabolites and quantitative determination of "acetanilide 4-hydroxylase activity" by high-pressure liquid chromatography. *Anal. Biochem.* 96: 201–207.

Herr, F. and M. Kiese (1959). Determination of nitrosobenzol in the blood. *Arch. Exp. Pathol. Pharmakol.* 235: 351–353 (in German).

Hewick, D.S. and J.R. Fouts (1970). Effects of storage on hepatic microsomal cytochromes and substrate difference spectra. *Biochem. Pharmacol.* 19: 457–472.

Holmstead, R.L., T.R. Fukuto and R.B. March (1973). The metabolism of *O*-(4-bromo-2,5-dichlorophenyl) *O*-methyl phenylphosphonothioate (leptophos) in white mice and on cotton plants. *Arch. Environ. Contam. Toxicol.* 1: 133–147.

Jollow, P.J., J.R. Mitchell, N. Zampoglione, and J.R. Gillette (1974). Bromobenzene-induced liver necrosis: protective role of glutathione and evidence for 3,4-bromobenzene oxide as the hepatotoxic metabolite. *Pharmacology* 11: 151–169.

Kaminsky, L.S., M.J. Fasco and F.P. Guengerich (1981). Production and application of antibodies to rat liver cytochrome P-450. *Methods Enzymol.* 74: 262–272.

Kaplan, S.P., M.L. Jack, K. Alexander, and R.E. Weinfeld (1973). Pharmacokinetic profile of Diazepan in man following single intravenous oral and chronic oral administration. *J. Pharm. Sci.* 62: 1789–1796.

Kato, R. (1966). Possible role of P-450 in the oxidation of drugs in liver microsomes. *J. Biochem. Tokyo* 59: 574–583.

Katz, M. and B.J. Poulsen (1971). Absorption of drugs through the skin. In *Concepts in Biochemical Pharmacology*, Vol. 1. B.B. Brodie and J.R. Gillette, Eds., Springer-Verlag, Berlin, pp. 103–174.

Koop, D.R. (1986). Hydroxylation of p-nitrophenol by rabbit ethanol-inducible cytochrome P-450 isozyme 31. *Mol. Pharmacol.* 29: 399–404.

La Du, B.N., G.H. Mandel, and E.L. Way (1972). Fundamental of drug metabolism and drug disposition. Williams & Wilkins, Baltimore.

Levi, P.E. and E. Hudgson (1989). Monooxygenations: interactions and expression of toxicology. In *Insecticide Action from Molecule to Organ*. T. Narahashi and J.E. Chambers, Eds., Plenum Press, New York, pp. 233–244.

Lapadula, D.M., Carrington, C.D. and M.B. Abou-Donia (1984). Induction of hepatic microsomal cytochrome P-450 and inhibition of brain, liver, and plasma esterases by an acute dose of S,S,S-tri-n-butyl phosphorotrithioate (DEF) in the adult hen. *Toxicol. Appl. Pharmacol.* 73: 300–310.

Lasker, J.M., D.G. Graham, and M.B. Abou-Donia (1982). Differential metabolism of O-4-nitrophenyl phenylphosphonothioate by rat and chicken hepatic microsomes. *Biochem. Pharmacol.* 31: 1961–1967.

Maibach, H.I., Feldman, R.J., T.H. Milby and W.F. Serat (1971). Regional variation in percutaneous penetration in man. *Arch. Environ. Health* 23: 208–211.

Mason, H.S. (1957). Mechanism of oxygen metabolism. *Adv. Enzymol.* 19: 79–233.

Milne, M.D., B.N. Schribner, and M.A. Craford (1958). Non-toxic diffusion and excretion of weak acids and bases. *Am. S. Med.* 24: 709–729.

Muller-Enoch, D. and A. Greischel (1988). Differentiation of cytochrome P-450 activities with scoparone as substrate. Arzneim. Forsch./Drug Res. 38: 1520–1522.

Nebert, D.W. and H.V. Gelboin (1968). Substrate inducible microsomal aryl hydroxylase. II. Cellular responses during enzymic induction. *J. Biol. Chem.* 243: 6250–6261.

Nebert, D.W. and F.J. Gonzalez (1987). P-450 genes. Structure, evolution, and regulation. *Annu. Rev. Biochem.* 56: 945–993.

Nelson, E.B., P.P. Raj, K.J. Belfi, and B.S.S. Masters (1971). Oxidative drug metabolism in human liver microsomes. *J. Pharmacol. Exp. Ther.* 178: 580–588.

Nomeir, A.A. and M.B. Abou-Donia (1984). Disposition of tri-o-cresyl phosphate (TOCP) and its metabolites in various tissues of the male cat following a single dermal application of [^{14}C]TOCP. *Drug Metab. Dispos.* 12: 705–711.

Nomeir, A.A. and M.B. Abou-Donia (1986). Studies on the metabolism of the neurotoxic tri-*o*-cresyl phosphate. Synthesis and identification by infrared, protron magnetic resonance, and mass spectrometry of five of its metabolites. *Toxicology* 38: 1–13.

Othman, M.A. and M.B. Abou-Donia (1988). Pharmacokinetic profile of (±)-gossypol in male Sprague-Dawley rats following single intravenous and oral and subchronic oral administration. *Proc. Soc. Exp. Biol. Med.* 188: 17–22.

Pelkonen, O., P. Jouppila, and N.T. Karki (1973). Attempts to induce drug metabolism in human fetal liver and placenta by the administration of phenobarbital to mothers. *Arch. Int. Pharmacodyn. Ther.* 202: 288–297.

Roy, A.B. (1971). Sulfate conjugation enzymes. In *Concepts in Biochemical Concepts*, Vol. 2. B.B. Brodie and J.R. Gillette, Eds., Springer-Verlag, Berlin, pp. 536–563.

Salama, A.K., N.B. Bakry, H.A. Aly, and M.B. Abou-Donia (1990). Placental transfer, disposition, and metabolism of a single oral dose of [$^{14}CH_3S$] methamidophos in Sprague Dawley rat. *The Toxicologist* 10: 346.

Schanker, L.S. (1962). Passage of drugs across body membranes. *Pharmacol. Rev.* 14: 501–530.

Schanker, L.S. (1971). Absorption of drugs from the gastrointestinal tract. In *Concepts in Biochemical Pharmacology*, Vol. 1. B.B. Brodie and J.R. Gillette, Eds., Springer-Verlag, Berlin, pp. 9–24.

Sivarajah, K., J.M. Lasker, T.E. Eling and M.B. Abou-Donia (1981). Metabolism of N-alkyl compounds during the biosynthesis of prostaglandins. *Mol. Pharmacol.* 21: 133–141.

Suwita, E., A.A. Nomeir and M.B. Abou-Donia (1987). Disposition, pharmacokinetics, and metabolism of a dermal dose of [^{14}C]2,5-hexanedione in hens. *Drug Metab. Dispos.* 15: 779–785.

Ullrich, V. and P. Weber (1972). The *O*-dealkylation of 7-ethoxycoumarin by liver microsomes: a direct fluorimetric test. *Z. Physiol. Chem.* 353: 1171–1177.

Waxman, D.J., C. Attisano, F.P. Guengerich, and D.P. Lapenson (1988). Human liver microsomal steroid metabolism: identification of the major microsomal steroid hormone 6a-hydroxylase cytochrome P-450 enzyme. *Arch. Biochem. Biophys.* 263: 424–436.

Waxman, D.J., A. Ko, and C. Walsh (1983). Regioselectivity and stereoselectivity of androgen hydroxylations catalyzed by cytochrome P-450 isozymes purified from phenobarbital-induced rat liver. *J. Biol. Chem.* 258: 11937–11947.

Weber, W.W. (1971). Acetylating, deacetylating and amino acid-conjugating enzymes. In *Concepts in Biochemical Pharmacology*, Vol. 2. B.B. Brodie, and J.R. Gillette, Eds., Springer-Verlag, Berlin, pp. 564–583.

Weiner, I.M. (1971). Excretion of drugs by the kidney. In *Concepts in Biochemical Pharmacology*, Vol. 1. B.B. Brodie and J.R. Gillette, Eds., Springer-Verlag, Berlin, pp. 328–353.

Whitacre, D.M., M. Badie, B.A. Schwemmer, and L.I. Diaz (1976). Metabolism of ^{14}C-leptophos and ^{14}C-4-bromo-2,5-dichlorophenol in rats: a multiple dosing study. *Bull. Environ. Contam. Toxicol.* 16: 689–696.

Williams, R.T. (1965). The influence of enterohepatic circulation on toxicity of drugs. *Ann. N.Y. Acad. Sci.* 123: 110–124.

Zannoni, V.G. (1971). Experiments illustrating drug distribution and excretion. In *Fundamentals of Drug Metabolism and Drug Disposition*. B.N. La Du, H.G. Mandel, and E.L. Way, Eds., Williams & Wilkins, Baltimore, pp. 583–590.

Ziegler, D.M., E.M. McKee, and L.L. Poulsen (1973). Microsomal flavoprotein-catalyzed *N*-oxidation of arylamines. *Drug Metab. Dispos.* 1: 314–320.

Developmental Neurotoxicology

Cynthia M. Kuhn
Department of Pharmacology
Duke University Medical Center
Durham, North Carolina

Richard B. Mailman
Departments of Psychiatry and Pharmacology
University of North Carolina
Chapel Hill, North Carolina

INTRODUCTION

The development of the mammalian central nervous system is highly complex: during ontogenesis of the CNS, very specialized morphological and biochemical changes occur that may be profoundly altered by changes in the physical or chemical milieu. Although this chapter will focus almost exclusively on changes in the chemical environment (i.e., exposure to xenobiotics), it is important to realize that changes in the physical environment (e.g., contact with the mother or siblings) may also have profound effects. These latter changes, although initiated by physical forces, may involve physiological mechanisms. These may then serve as the antecedent chemical stimulus for alterations in CNS development.

The developing nervous system generally is more susceptible to damage by neurotoxic agents that the adult nervous system. There are several reasons for this, the primary one being that the developmental process itself involves a number of biochemical and morphologic events which do not occur in the adult brain. Therefore, chemicals may perturb the developing nervous system by affecting these developmental processes, above and beyond any actions they may have on the adult nervous system. Agents may cause no detectable effect on the adult nervous system, or they may alter both the adult and the developing nervous system. The distribution of chemicals into the developing brain also differs somewhat from that of the adult brain. In general, many compounds that are excluded from the adult brain by the "blood-brain barrier" may enter the developing CNS. Another point to consider is that some chemicals having defined actions on specific neuronal functions or neurotransmitter systems can have different, often longer lasting, effects on the same systems if they are delivered to the developing brain. In contrast to these points, it is also true that developing animals can exhibit greater "plasticity" than adults; that is, the ability to recover from damage. For example, after surgical sectioning of a nerve, the young animal shows more sprouting and, in turn, greater recovery than the adult. In the following chapter, we will review the characteristics of nervous system development responsible for its sensitivity to neurotoxic injury and will discuss the effects of several agents which are known to disrupt nervous system development. These agents do not comprise an all-inclusive list by any means but are simply representative of the types of damage which do occur.

Effects on Developmental Processes

Development of the nervous system involves a precise sequence of events which must occur in orderly fashion for the establishment of the exact anatomical and biochemical relationships required for proper function. (See Purves and Lichtman [1985] for an excellent review.) These include, in sequence:

1. proliferation of cells, followed by migration, aggregation, and cytodifferentiation
2. axonal outgrowth and synaptogenesis
3. "pruning" of inappropriate connections and cell death.

The first phase is particularly critical, as nerve cells (neurons) do not continue dividing as do many other cell types. Instead, they undergo a final division which must occur during a narrow temporal window. Therefore, the developing CNS is much more susceptible to agents that may cause even transient impairment of cell division. The timing of these events is particularly crucial. For each group of neurons there is a "critical period" during which the death of a single or several cells may have extreme consequences for the differentiating nervous system. For most cell groups, this period is during the first and second trimester of gestation, during the period of active organogenesis. Like most developing systems, the CNS shows different susceptibility to damage at different stages of development. During the early part of the first trimester, before implantation in the uterus, most toxic agents are lethal, and so do not cause specific nervous system defects. During the period from implantation through mid-gestation, active organogenesis for the central nervous system occurs, and this phase of development is very sensitive to disruption by exogenous forces. Toxicant exposure during this phase of development frequently is associated with major CNS abnormalities like microcephaly, spina bifida, and other major structural defects. Finally, during late gestation and the early neonatal period (which for humans extends to several years of age), the process of synaptogenesis and myelination predominates. Xenobiotic exposure during this time generally has more selective and subtle behavioral effects or effects specific for the brain regions which mature late.

The effects of neurotoxic agents are also influenced by the caudal to rostral sequence of CNS development. Neural systems in the spinal cord mature first, followed in sequence by the brainstem, midbrain, cortex, and finally, the hippocampus and cerebellum. The latter two areas are unusual in that a significant amount of neurogenesis occurs postnatally in several mammalian species and so these areas provide extremely useful experimental models for studying nervous system development and its alteration by neurotoxins.

Within a circumscribed area, specific cell types differentiate in a definite sequence, generally with motor neurons preceding sensory neurons and large, outflow neurons preceding small, local interneurons. Finally, differentiation of neurons within a given brain area precedes the formation of glial cells. Therefore, exposure during certain "windows" of development may profoundly influence the particular brain region or cell population undergoing rapid differentiation at that time, while sparing regions that are relatively more mature and regions that are not yet developed. For example, in the rat, the cerebellum matures mainly postnatally and so exposure to agents after birth can completely disrupt normal cerebellar ontogeny without affecting the maturation of other areas. In contrast, the developing spinal cord is relatively spared following postnatal exposure to neurotoxic substances as the major part of its differentiation occurs prenatally.

The specific biochemical process disrupted by a neurotoxicant will determine the period of brain development that represents a "critical period" and the areas affected. Each of the processes listed above (differentiation, migration, etc.) involves a different series of biochemical processes. For example, cell differentiation is extremely dependent upon DNA replication. Myelination, which is one of the final events, requires synthesis of specific glycolipids, while establishment of appropriate synaptic contacts relies both on availability of appropriate "trophic" agents as well as synthesis of neurotransmitters and their receptors.

Each of the above processes occurs in a caudal to rostral sequence. Therefore, the effects of neurotoxic agents will be determined both by the particular process disrupted and the period during which the developing CNS is exposed. To continue the analogy developed above with the cerebellum, administration of drugs which impair cell differentiation during postnatal life does not decrease the number of Purkinje cells, as these undergo final division before birth, but the same treatment will completely eliminate the population of granule, basket and stellate cells which appear postnatally. Similarly, exposure of early fetal brain to agents which disrupt myelination will affect only those brain areas (spinal cord and brain stem) in which brain development is far enough advanced for significant myelination to occur.

Finally, another critical characteristic of CNS maturation which renders it extremely sensitive to disruption by exogenous agents is the marked degree of interaction among different cellular elements that is required for normal development. Again, in the developing cerebellum, the morphology and pattern of synaptic connections established by the three cell types which mature postnatally (basket, stellate, and granule) is determined in part by the morphology of the large output neurons (the Purkinje cells) which develop first. Therefore, administration of an agent during the early period of time which disrupts the normal ontogeny of these cell types can completely alter the cytoarchitectural organization of the cerebellum, despite the absence of the drug during differentiation of the other cell types. This effect has been demonstrated to contribute, in part, to the disruption of cerebellum development associated with early hypo- and hyperthyroidism (Nicholson and Altman, 1972a,b).

Pharmacokinetic Determinants of Developmental Neurotoxicology

Several pharmacokinetic considerations that are unique to the fetus and early neonate contribute to the disruption of CNS development by xenobiotics. The first is the enhanced movement of drugs into developing brain. In the normal adult brain, three membranes divide the intracellular space from the systemic circulation: the capillary membrane, which contains "tight junctions" between cells, the basement membrane, and a glial cell layer which covers 90% of the capillary surface area. These three membranes comprise the "blood-brain barrier" which provides a significant diffusion barrier to most large and/or polar molecules (Dobbing, 1961). The only exception are the vital nutrients which pass the blood-brain barrier by special transport mechanisms. Such systems have been demonstrated for hexoses, carboxylic acids, amino acids (separate ones for neutral, basic and acidic amino acids), amines, and inorganic ions (Rapaport, 1976). In the developing brain these cell layers are thinner, and one (the glial covering) is not complete until later, during CNS development. Therefore, there is less impediment to the movement of substances into the brain (Johnson et al., 1975, for review of barrier ontogeny). Furthermore, nonpolar, lipid-soluble molecules, including many potent neurotoxic agents penetrate the blood-brain barrier easily even in the adult brain. Such agents tend to concentrate in the developing CNS, because lipid content in the remainder of the developing fetus is low. Finally, the specific transport systems cited earlier may facilitate the passage of toxicants into the brain (Woodbury, 1977). Though these systems are usually highly specific in nature, potential toxicants with structures analogous to physiological substrates may be transported into the brain, thus bypassing the blood-brain barrier. Finally, another factor to consider is the presence of structures outside the blood-brain barrier (known as supraependymal structures) that are bathed in cerebrospinal fluid and can be exposed to molecules that will not penetrate into deeper brain tissue. These are particularly vulnerable to toxicants.

There are other pharmacokinetic factors that may be of particular significance to potential neurotoxicity of the mammalian CNS. For example, if toxicant exposure occurs prenatally, a primary determinant of effects on the developing organism is the ability of the compound in question to pass into the fetal circulation. This is determined by the blood-placental barrier, which in many ways resembles the blood-brain barrier discussed above. Although somewhat less selective in the variety of molecules which are excluded, it often serves to reduce the severity of insult from many compounds. In mammals, a primary source of toxicants postnatally may be the mother's milk. Although it is clear that some environmental contaminants such as DDT or PCBs may become somewhat concentrated in mother's milk, the final concentrations reached are usually not of significant concern (Briggs et al., 1983; Knowles, 1965).

An additional consideration may be the presence of therapeutic or illicit drugs in the maternal circulation and the consequences for the nursing infant. In general, this concern has led to the suggestion that all drugs are contraindicated in the pregnant or lactating female unless absolutely necessary. However, rigorous studies have shown that the risk from specific agents is small. As an example, the tricyclic anti-depressants are believed to act by inhibiting the uptake of biogenic amine neurotransmitters (i.e., norepinephrine and serotonin). Clearly, if significant inhibition of such processes took place *in vivo* during late neural development, it is highly likely that significant perturbation of the nervous system would occur. However, well-controlled studies with specific drugs have shown that this class of compounds is usually not concentrated in mother's milk (Knowles, 1965). Therefore, the total dose to an infant obtaining nutrition solely from nursing would never be sufficient to cause concentrations adequate to inhibit the biogenic amine uptake systems. Thus, it would seem that available data are consistent with the safety of many of this class of drugs in the lactating mother. Conversely, a similar evaluation of risk cannot be made regarding the consequences of use of these drugs prenatally. In the latter case, great caution is still the most conservative approach. As these and later examples make clear, pharmacokinetic considerations should be intrinsic to all aspects of toxicology.

Types of Defects

The spectrum of CNS defects induced by exposure to a neurotoxicant during fetal or early postnatal life includes:

1. morphologic abnormalities ranging from microcephaly, spina bifida, and anencephaly to disruption of specific brain areas
2. marked functional defects resulting from disruption of specific processes (such as myelination) which affect the entire developing nervous system
3. permanent alterations of specific neural systems, with consequent changes in specific behaviors or physiological functions dependent on these systems
4. subtle behavioral abnormalities that cannot readily be associated with morphologic defects.

The type of defect caused by a toxic agent will be determined by the period during gestation that the CNS is exposed, the specific mechanism by which the agent acts, and the levels of exposure. Generally, marked disruption of cell differentiation during the early embryonic stage results in major defects, while exposure to neurotoxic agents during the fetal stage (second and third trimester in humans) causes defects ranging from microcephaly to more specific defects in multiple neuronal systems. Exposure during the perinatal period tends to disrupt synaptogenesis, formation of local circuit neurons, and myelination.

In most general terms, agents which dramatically impair cell division, like drugs used in cancer chemotherapy, will disrupt differentiation of a large population of cells and produce the first type of defect. Agents with general effects (e.g., disruption of the broad class of sulfhydryl enzymes by heavy metals) will affect several specific biochemical processes and produce multiple defects. It should be emphasized that frequently such agents cause marked perturbation in other developmental events and the CNS damage is part of a constellation of developmental defects. This is the case for antimetabolites, heavy metals, alcohol, and nicotine.

More specific effects on developing CNS result from exposure to agents which affect a specific population of neurons. Such effects have been demonstrated for a number of drugs which act on the adult CNS including sedative-hypnotics, neuroleptics, stimulants, opioids, and certain agents with toxic effects on specific neurotransmitter systems (Hutchings, 1983; Vernadakis, 1982; Hans et al., 1984; Tucker, 1985; Mueller et al., 1980). Some actions result simply from biochemical effects identical to those observed in adults (i.e., sedation for barbiturates, analgesia for opioids). For example, it has been suggested that suppression of respiratory movements by CNS depressants contributes to the neurotoxic effects of these agents. Similarly, developing animals develop tolerance to drugs like opiates which induce tolerance in adults.

However, in the developing nervous system, a frequent outcome of chronic exposure to drugs is a series of counteradaptations in other neuronal systems which persist far longer than in adults. This is thought to be one important action of neurotransmitter-specific agents. Finally, drugs which interfere with specific neurotransmitter systems can have considerably different effects on the developing nervous system if the neuronal system involved has trophic effects on other neuronal systems. For example, effects of drugs which impair serotonergic neurotransmission are thought to act on the developing nervous system by impairing trophic actions of serotonin (Lauder and Krebs, 1978).

The effects produced by developmental neurotoxins depend not only on the mechanism of action and period during which the agent is administered, but also on the dose to which the animal is exposed. For example, high doses of vitamin A are overtly teratogenic and associated with marked CNS deficits including microcephaly (Giroud and Martinet, 1955; Kalter and Warkany, 1961). However, administration of smaller doses results in subtle behavioral changes. For example, a specific type of hyperactivity results that is associated with selective damage to certain cell groups in the hippocampus (Hutchings and Gaston, 1974). Similar results have been observed following administration of a number of agents including heavy metals, cancer chemotherapeutic agents, and alcohol.

The effects listed above have been mentioned with the assumption that it is possible to determine specific mechanisms for developmental neurotoxicity of xenobiotics. In many cases, this assumption cannot be made, and it is vital to understand the limitations of different experimental strategies and methodologies for studying nervous system development. While the study of teratology is predicated upon identification of morphologic abnormalities in development, identification of such abnormalities represents only one possible strategy for studying CNS development. Ultimately, all effects on CNS development are reflected in disrupted behavioral development, and it has been proposed that evaluation of behavioral effects of neurotoxic agents, "behavioral teratology," is the most sensitive mechanism for identifying developmental neurotoxicants (Hutchings, 1983). The rationale for this hypothesis is that a minor perturbation in a very small population of cells which would be difficult to detect by morphologic or biochemical determinations can have dramatic behavioral consequences. However, while evaluation of behavior is a sensitive indicator of disrupted CNS development, it is extremely nonspecific. Most behavioral tests (i.e., conditioning, learning, locomotor activity, as well as the complex intellectual functions that are evaluated in humans) can be perturbed by almost an infinite number of different mechanisms. More specific studies of precise biochemical mechanisms mediating neurotoxic injuries represent the future direction of this area; for example, detection of changes in DNA synthesis in specific brain areas or changes in specific neurotransmitters. However, one important caution in interpretation of biochemical studies of developmental neurotoxins is that any changes observed might well reflect events that are adaptive changes in response to some other injury. For example, destruction of catecholamine nerve terminals causes a number of adaptations in the postsynaptic cell which alter its sensitivity to neurotransmitter. However, these adaptive events might be the mediators of behavioral changes and so represent an important area of investigation themselves.

MECHANISMS FOR DISRUPTION OF FUNCTION

As mentioned briefly above, there are a wide variety of biochemical effects which can ultimately disrupt nervous system development and behavioral ontogeny. Furthermore, a number of agents have multiple effects. Therefore, an exhaustive review of the processes involved would be necessarily confusing and incomplete. Instead, we will mention some specific examples of postulated mechanisms for xenobiotic action on developing CNS.

Inhibition of DNA Synthesis or Other Aspects of Cell Division and Differentiation

A host of interventions may cause effects of this type. This may range from agents deliberately given because of their effects on such processes; for example, cancer chemotherapeutic drugs (cytosine arabinoside, 5-azacytidine, ethylnitrosourea), or gamma or X-irradiation (Pfaffenroth et al., 1974). Other compounds (such as alcohol, vitamin A, methylazoxymethanol) also inhibit the process of DNA synthesis in a broad range of cells. In addition, even some relatively specific drugs (such as the dopaminergic antagonist haloperidol) may affect specific populations of cells in which specific target sites (i.e., dopamine receptors) are present (Patel and Lewis, 1982). The defects associated with inhibited cell replication generally involve disappearance of a specific population of cells and a resultant disruption of synaptic contacts among cells normally communicating with these neurons (Langman and Shimada, 1971; Butcher et al., 1983; Hall and Das, 1978; Hicks and D'Amato, 1966; Koyama et al., 1970).

Inhibition of Specific Biochemical Processes (Myelination, Protein Synthesis, etc.)

Heavy metals, alcohol, and other agents have been shown to retard normal developmental increases in protein and nucleic acid synthesis, increases in glycoproteins specific for the myelin sheath surrounding neurons, and a number of enzymes associated with energy metabolism (Konat, 1984;

Grundt and Neskovic, 1980; Grundt et al., 1980; Menon and Lopez, 1985; Sarafien et al., 1984; Syversen, 1977). These effects can cause defects ranging from gross microcephaly if animals are exposed to extremely high levels during early gestation to subtle behavioral abnormalities if animals are exposed to low doses, especially late in development. The role of specific processes in the neuropathology associated with developmental exposure to these agents is completely unknown with agents like these which have multiple actions.

Disruption of Synthesis of Neurotransmitter, Receptors or Second-Messenger Generation

In general, administration of agents with specific effects on some neurotransmitter systems can lead to changes in the targeted neuronal systems and cells communicating with them. This can often be thought of as a "resetting" of the sensitivity of the system following chronic stimulation or blockade during the time that synaptic function is first established. In general, these effects resemble the similar adaptations that occur in adults, although the effects are permanent rather than transitory. Thus, administration of dopaminergic antagonists during postnatal life in the rat may cause a permanent change in sensitivity of dopaminergic regulation of locomotor activity as well as other functions, while such changes are reversible in adults (Werboff and Havelena, 1962; Spear et al., 1980). There is also a significant body of data demonstrating that opiates may cause similar effects on targeted opioid systems.

Interference with Normal Trophic Actions of Specific Neurons

Certain neurotransmitters or other substances in brain serve a trophic function in developing brain, and so disruption of neurotransmitter functions has unique effects in the developing animal which are related to interruption of trophic activity. For example, in adult brain, reserpine causes behavioral depression and a number of other effects by depletion of catecholamines from storage sites in vesicles. However, during gestation, a similar depletion of catecholamines is thought to inhibit cell division. A similar trophic role for the neurotransmitter serotonin have been proposed, and drugs which interfere with serotonergic function have been similarly reported to influence cell division.

SPECIFIC EXAMPLES OF DEVELOPMENTAL NEUROTOXICITY

Toxic Metals

Trace elements are ubiquitous as cofactors in biological systems. Because calcium is known to have numerous regulatory roles in the CNS, it is unremarkable that other elements that are polyvalent cations and are able to form coordination compounds may interfere with normal physiological function when present at sufficient levels to compete or interact with the endogenous metals. "Heavy metals," so named because they are usually of greater atomic weight than the physiologically necessary elements, have been documented to have many CNS actions. It has become more customary to now refer to the agents as "toxic metals." It should be recognized that even essential metals may cause toxicity if their concentrations becomes too great.

Mercury, by its ability to bind to organic molecules (especially at sulfhydryl groups), has been used both experimentally and clinically as a gastrointestinal helminthicide and as a topical skin agent. Its therapeutic use is a result of low penetration of inorganic mercury through the skin, gut, and blood-brain barriers. However, there have been many episodes of human illness caused by intake of organomercury compounds. Minemata disease (organomercuric poisoning) is so named because the first widely recognized and accurately identified incidence resulted from consumption of fish that had been exposed to dimethyl and methyl mercury formed by biomethylation of inorganic mercury in sediments of the waters near Minemata, Japan. Poisoning resulted because methyl mercuries penetrate biological membranes nearly two orders of magnitude more effectively than inorganic forms of mercury. Thus, the toxicants could accumulate in fish and also pass into the human CNS. This episode resulted in more than 100 cases of illness and dozens of deaths. The symptoms of Minemata disease are a compendium of CNS disturbances, including loss of cognitive functioning, palsied movements and other permanent mental disorders (Kurland et al., 1960).

The young and the unborn were markedly more sensitive to the effects of the mercurials (Mottet et al., 1985; Sanstead, 1986). One reason for the extreme sensitivity of developing brain for damage by mercury is that mercury is concentrated highly in the fetal tissues (Yang et al., 1972; Null et al., 1973). The defects noted included mental retardation, impairment of gait, speech and visual ability, delayed cognitive developmental and flaccid as well as spastic paralysis. In these cases, brains on autopsy were observed to be atrophic with thinning of the cerebral cortex and widespread damage and neuronal death (Takeuchi, 1977). The developing cerebral cortex and cerebellum seemed to be particularly susceptible to the effects of prenatal mercury exposure. Disorientation, disrupted synaptogenesis and impaired development of Purkinje cell dendritic processes suggest that migratory disturbance as well as impaired growth contributed to the deficits (Chang and Annau, 1984). In addition, poor myelination was observed.

Interestingly, another case of mercury poisoning occurred in Iraq, which resulted from a shorter exposure to higher doses of mercury and resulted in a somewhat different developmental neuropathology (Amin-Zaki et al., 1967). Recent attempts to model Minemata disease in animals by pre- and postnatal administration of methyl mercury have produced similar defects in cytoarchitectural development including microcephaly, disrupted cortical and cerebellar synaptogenesis, as well as associated biochemical defects including impaired myelination, decreased protein and nucleic acid synthesis. As in the human disease, the effects are dose and time related. Administration of low doses can produce effects on behavior (impaired learning, slowed reflex development) and extremely sensitive biochemical markers like ornithine decarboxylase activity that are not associated with overt neuropathology, while higher doses produced more widespread deficits (Slotkin et al., 1985).

The molecular site of action of methyl mercury is unknown, and it is likely that numerous loci play a role in toxicity. Changes in numerous enzymes involved in energy metabolism as well as protein, nucleic acid, and myelin synthesis have been reported. These include succinic dehydrogenase, DPN diaphorase, cytochrome oxidase, ATPase (Grundt et al., 1974, 1980; Grundt and Neskovic, 1980; Omata et al., 1978; Sarafien et al., 1984; Syversen, 1977). Interestingly, impaired glucose production could possibly contribute to methyl mercury neurotoxicity in developing brain since the CNS is extremely sensitive to blood glucose. The importance of dose and period of exposure cannot be emphasized too much. For example, it has been shown that postnatal administration of methyl mercury to rats is associated with significant increases in DNA and RNA content of the cortex and cerebellum but not the brainstem, an effect attributed to repair in later-developing areas (Slotkin et al., 1985). Furthermore, the broad spectrum of enzymes affected by mercury provide an additional explanation for the dose sensitivity to mercury: such effects could be explained by either a differential sensitivity of different cell types (i.e., different ability to repair damage or by differential sensitivity of various enzyme systems).

Subacute lead toxicity has also recently become an important clinical question in developmental neurotoxicology. Exposure to high levels of lead is known to cause permanent alterations of the central and peripheral nervous system in both humans and laboratory animals. It is less certain how lead affects the CNS when lower levels of exposure (detected in humans via measurement of blood lead and alterations in protoporphyrin biosynthesis occurs. Pediatric studies suggest that lead burdens lower than those causing clinical symptoms are correlated with both behavioral and cognitive deficits (Winder and Kitcher, 1984). The high incidence of lead exposure of infants and juveniles in urban areas, certain industrial populations, and, until recently, populations in areas located near major automotive thoroughfares makes the potential effect of these lead-induced CNS alterations very important.

Recent research has found that exposure of rodents to lead during postnatal development may cause a series of subtle but consistent alterations in behavioral development including alterations in the ontogeny of basal loco-motor activity as well as the ability to learn specific tasks. Anatomical changes resembling those seen following mercury exposure have been reported, including altered dendritic growth and synaptogenesis in the cerebellum. Work from the authors' laboratory has recently shown that single, small doses of lead, given only during discrete times of postnatal development of the rat, can also cause permanent changes in certain centrally mediated pharmacological responses (Mailman et al., 1979; Hanin et al., 1984). Since this happens in the absence of any detectable anatomical or neurochemical abnormality, it suggests long term but subtle changes in CNS function.

The identity of the specific biochemical mechanisms by which lead or other agents without selected biochemical actions alter behavioral development have not been identified. For example, administration of both triethyltin and mercury to developing animals causes significant impairment of myelination (Konat et al., 1984), brain DNA and RNA synthesis and protein synthesis and also induces changes in a number of neurotransmitter systems including GABA, dopamine, and serotonin (Slotkin et al., 1985; Taylor and DiStefano, 1976). The diversity of such effects suggests that the marked inhibition of energy metabolism caused by heavy metals and disruption of numerous enzyme activities involved in synthesis of specific cerebral constituents both contribute to these effects. The time of exposure is also a key variable as shown by the fact that rats exposed to lead after critical development periods do not show the same neural changes seen following earlier exposure.

Methylazoxymethanol: A Model of Inhibited Cell Division

Cycasin (methylazoxymethanol-β-d-glucoside) is a naturally-occurring alkylating agent that is produced by the cycad. Ingestion of this plant by animals or humans has been associated with hepatotoxicity, carcinogenicity, teratogenesis and neurotoxicity which have been linked to the active agent methylazoxymethanol (MAM) (Jones et al., 1972). This agent is an alkylating agent which is probably metabolized to diazomethane, which subsequently forms active methyl groups which methylate nucleotide bases and impair DNA and RNA synthesis.

Administration of MAM during fetal life causes a microcephaly which is associated with extensive necrosis in the area normally occupied by differentiating cells and seems to result from the disruption of cell replication (Jones et al., 1972; Haddad and Rabe, 1980). As predicted, administration restricted to postnatal life has selected effects on areas which mature postnatally. Marked cerebellar lesions characterize the neuropathology observed following postnatal administration to experimental animals, with measurable but less dramatic effects on the hippocampus and olfactory bulb, two other regions which undergo a significant part of their development postnatally. The cerebellar pathology associated with postnatal MAM administration provides a good example of the spectrum of effects which result from inhibition of cell division during differentiation of a specific population of cells (in this case, the cerebellar granule cells). Behaviorally, the animals show fairly specific neurologic signs related to disruption of motor function. Anatomical findings are characterized by decreased cerebellar mass and a diminished number of cells that can be attributed to the deficit in granule cells. In contrast, postnatal MAM effects on developing spinal cord are characterized by fairly specific decreases in indices of glia and myelin formation, again consistent with the mainly prenatal differentiation of this region, which is therefore spared (Jones et al., 1972; Hirono and Hayashi, 1969).

Among the effects of MAM which are most instructive for an understanding of developmental neurotoxin action are the effects on cells which had undergone division prenatally and so were fully differentiated and not directly affected by MAM. For example, the organization of cerebellar Purkinje cells is markedly disrupted; cells are not located deep in the cerebellum as usual but, instead, are scattered randomly through more superficial layers of the cerebellum. In addition, a marked increase in the volume of Purkinje cell dendrites covered by astrocytic processes rather than synaptic contacts with other cells was noted (Hirono and Hayashi, 1969; Jones et al., 1972). This "glial overgrowth" is a typical response to the loss of innervation. The later-maturing glial cells grow in and occupy space that has been left open by the absence of normal synaptic contacts with the parallel fibers of the granule cells.

Alcohol

Alcohol represents one of the most commonly used, yet poorly understood, agents which produce marked abnormalities in CNS development. The effects of this drug are considerably different from and far more devastating on the developing than the adult nervous system. "Fetal alcohol syndrome" is a disorder characterized by a pattern of prenatal and postnatal growth deficiency, developmental delay or mental deficiency, microcephaly and fine motor dysfunction (Rosett and Warner, 1984; Jones, 1988). Children with the most severe form of this disorder typically have characteristic facial abnormalities including short palpebral fissures, micropthalmia, midfacial hypoplasia and epicanthal folds. Although the occurrence of mental retardation in the children of chronic alcoholic mothers was understood as early as the 17th century, the characteristic complex of symptoms was not studied scientifically until the last 30 years.

The effects of alcohol typify the response described above for agents affecting CNS development in association with other congenital abnormalities, most frequently growth retardation, facial abnormalities, and with significant dose- and duration-dependence of the effect. The full-blown syndrome with grossly retarded development is usually associated with other congenital abnormalities. However, hyperactivity, sleep disturbances, language dysfunction, and altered evoked potentials in the EEG have been observed in the absence of somatic dysmorphogenesis. Generally, less severe deficits are observed with decreasing levels of alcohol intake. However, the increased sophistication of behavioral testing available for high-risk children has identified abnormalities including decreased habituation and low arousal during the neonatal period and poor school performance following as little as 2–2.5 ounces of alcohol per day (Landesman-Dwyer et al., 1977; Streissguth, 1983). Similar behavioral deficits have been observed in animal studies of FAS with frequent reports of delayed maturation of behavioral motor and cognitive abilities and associated hyperactivity (Abel, 1984; Randall, 1982).

Some alcohol effects may result from the suppression of normal neural activity in the developing CNS while others may reflect more primary actions on specific maturational events. For example, certain behavioral abnormalities observed in newborn infants of alcohol-addicted mothers improve with age and might reflect neonatal "withdrawal" from chronic alcohol. These effects are not actually "teratogenic," but instead reflect actions on the developing CNS similar to those produced in the adult CNS. For example, the marked sleep disturbances observed during the neonatal period improve with age and might reflect CNS "hyperactivity" following alcohol withdrawal. However, sedation itself can have significant development consequences as neural activity is thought to be important in the establishment of appropriate synaptic connections during the period of synaptogenesis in many neural systems.

Alcohol is also a model for neurotoxicant action by secondary effects on maternal physiology and behavior. It is well known that the disrupted maternal behavior typical of the chronic alcoholic is associated with significant developmental delays. Furthermore, the poor nutritional status and endocrine abnormalities and other physiologic abnormalities in the mother all potentially contribute to FAS. In both humans and animal models, levels of alcohol ingestion associated with significant abnormalities generally reflect up to 30–40% of daily calorie intake. Levels of nutrient restriction this great are known to have significant and not dissimilar effects on developing brain.

Alcohol almost certainly impairs brain development by multiple actions. Marked disruption of cell migration in the cerebral cortex, decreased cell size and decreased dendritic arborization and synaptogenesis have been observed in brains of severely affected infants (Pennington and Kalmus, 1987; West et al., 1987). Investigation of specific mechanisms of alcohol effects on the developing nervous system is an area of active research interest and many deficits have been described. Administration of alcohol to pregnant rats is known to impair the normal ontogeny of neurotransmitter systems (including catecholamines and serotonin) as well as to significantly impair myelination and synthesis of other lipid components including gangliosides and glycoproteins (Slotkin et al., 1980; Druse et al., 1981; Shoemaker et al., 1983). As marked decreases in cell number have been observed, it is likely that alcohol inhibits cell replication and/or accelerates the process of cell death which is a normal part of synaptogenesis.

Neuroleptics and Stimulants: Drugs Causing Behavioral Teratogenesis by Actions on Specific Neurotransmitter Systems

For several clinically relevant reasons, there has been considerable interest in the effects of drugs which affect dopaminergic neurotransmission. An irreversible condition caused by long term use of such drugs even in adults is "tardive dyskinesia," a bizarre and abnormal motor movement principally oral-facial in nature. It is hypothesized that these irreversible changes result from long term adaptation in dopaminergic function. Since many mentally retarded and behaviorally disturbed children are treated chronically with dopaminergic antagonists, the potential for lasting damage to brain dopaminergic systems following chronic administration of such drugs to a child must be considered.

These pressing clinical questions motivated the first studies in what now is termed the area of "behavioral teratology" through the work of Werboff and others who studied the effects of the dopaminergic antagonist chlorpromazine on behavioral development. However, a number of other drugs which affect dopaminergic systems as agonists (rather than antagonists) also are likely to affect the developing nervous system. This may occur in a therapeutic context (amphetamine and methylphenidate are commonly used to treat hyperactivity in children) or through rapidly increasing illicit use of amphetamine and cocaine. Despite the significant human population that is exposed to dopaminergic agonists or antagonists during development, there has been relatively little investigation of the effects of these drugs on central nervous system development. However, the studies which have been completed suggest that these agents represent a significant risk to CNS development, particularly because the damage is subtle and not easily ascribed to drug treatment. While none of these agents is teratogenetic at clinically relevant doses, significant changes in behavioral development have been observed in animal studies (Mueller et al., 1980). Furthermore, clinical studies suggest that the development of permanent changes in dopamine receptor sensitivity which occasionally develop in adults during chronic treatment with these agents is much more common during treatment of children. The consequences of perinatal exposure to cocaine present one dramatic exception to this pattern. The explosion of cocaine use in the United States in the late seventies and eighties has resulted in a rapid rise in births of babies exposed to cocaine. The most dramatic effects reflect actions on maternal physiology (uterine vasoconstriction for example) which result in increased incidence of stillbirth, abruptio placentae and premature birth (Chasnoff and Griffith, 1989). Cocaine also has direct effects on the developing fetus leading to strokes *in utero,* some mild congenital malformations, as well as effects on the normal ontogeny of CNS functions, including but not restricted to dopaminergic neuron development (Dow-Edwards, 1989; Spear et al., 1989; Cohen et al., 1989).

Administration of a number of dopaminergic antagonists including chlorpromazine, haloperidol, penfluridol and pimozide during development is associated with significantly impaired performance in learning tasks which persist long after drug withdrawal (Rosengarten and Friedhoff, 1979; Cagiano et al., 1988).

These behavioral deficits are accompanied by specific deficits in brain dopamine systems as assessed biochemically. Similarly, exposing developing rats to the indirectly acting dopamine agonist amphetamine also disrupts performance in locomotor tasks and decreases dopamine and norepinephrine content (Tonge, 1973). Unfortunately, the effects of different dopaminergic agents have not been compared in the same study using identical behavioral tasks and biochemical assessments. Therefore, it is impossible to describe the specific deficits in dopamine neuron function associated with perinatal blockade and stimulation. However, both types of agents are associated with behavioral deficits which seem to be associated with specific changes in catecholamine neurons. Therefore, these changes reflect the subtle effects which can be detected most easily with behavioral measures.

It is timely to emphasize again that subtle effects on specific neuronal systems in the brain can impair CNS function enough to significantly disrupt behavior. Teratology often has emphasized morphologic abnormality to the exclusion of changes which can be equally devastating. However, the effects of dopaminergic agonists and antagonists reviewed above suggest that subtle biochemical effects on specific neural population can have permanent effects on behavior. The presence of uncontrollable choreoathetoid movements as characteristic of tardive dyskinesia in neuroleptic-treated children are every bit as disabling as are the limb-bud deficits experienced by the offspring of women treated with thalidomide.

Opioids and Other Drugs of Abuse: Perinatal Dependence

The effects of opioids on the developing nervous system reflect the opposite end of the spectrum from the effects of teratogens which totally inhibit cell formation. It is well known that infants born to opioid-addicted women are small for gestational age and show signs of narcotic withdrawal at birth. These signs include hyperflexia, tremor, irritability, disturbed sleep, and excessive high-pitched crying (Hans et al., 1984; Zagon and McLaughlin, 1984). These signs often are slow to appear and can persist for up to six months after birth. It is thought that most behavioral disturbances of opiate-addicted neonates reflect the increased CNS arousal associated with withdrawal. Therefore, the most dramatic effect of opioids on the developing brain, as assessed behaviorally, is an effect which resembles that observed in adults. Withdrawal symptoms also are observed in infants exposed to barbiturates, phenytoin, benzodiazepines, and alcohol, although the behavioral signs are different from those of opiate-exposed infants.

While neonatal withdrawal is not typically classified as a ''neurotoxic'' event, a growing appreciation of the role of infant behavior in subsequent neurological development suggests that this type of behavioral disruption can have devastating consequences. For example, it has been shown that "mother-infant" bonding is critical to the behavioral development of the human infant and that this process can be significantly disrupted with measurable consequences on neonate cognitive development in situations in which either the mother or the infant are behaviorally impaired. In the home of the opioid addict both situations usually exist. Furthermore, recent animal studies suggest that perinatal opioid withdrawal elicits marked physiologic effects which can be lethal in certain animal models and definitely can have long term consequences on subtle behavioral organization (Lichtblau and Sparber, 1984). Therefore, the behavioral consequences of neonatal opioid withdrawal can have significant effects on nervous system development, although the mechanism by which these changes occur is completely unknown.

There are few human or animal studies of behavioral function in adult animals that were addicted to opioids during the perinatal period. Those studies which exist suggest that perinatal opioid addiction is associated with fairly subtle behavioral effects, particularly increased reactivity and/or decreased ability to habituate to sensory stimuli (Hans et al., 1984). While these effects contrast with the marked mental retardation associated with alcohol use during pregnancy, the impaired ability to concentrate and mild impairment in fine motor control are sufficient to significantly interfere with function in a school environment.

Until the discovery of the endogenous opioid systems, the mechanism by which perinatal opioid addiction could influence behavioral development was completely unknown and was most often attributed to generalized effects on protein synthesis, hypoxia, and other physiological effects. However, some very recent provocative studies suggest that endogenous opioids exist in the germinal cell layers of developing brain and that excessive stimulation of these systems can be associated with slowed differentiation of certain cell populations. For example, the opioid peptide enkephalin is present in developing cerebellum but is absent from this brain region in adulthood. Studies demonstrating significant effects of endogenous opioid peptides on replication support the hypothesis that endogenous opioid systems play a role in modulating brain growth (Vertes et al., 1982; Zagon and McLaughlin, 1983; Zagon, 1987).

Sedative-Hypnotics

There is considerable controversy about whether sedative-hypnotic agents affect CNS development. This issue has considerable clinical relevance because of the large population of epileptic women who must be maintained on such medication through pregnancy as well as a significant population of children who receive such medication chronically during a period of active synaptogenesis and myelination. In the past, such agents as barbiturates, phenytoin and other sedatives were thought to be fairly safe during pregnancy. However, our increasing sophistication about CNS "teratogenesis" has led to a number of studies suggesting that these agents are not benign (Vorhees et al., 1988).

Several problems have been associated with prenatal exposure to sedative-hypnotics. First, exposure to several specific agents, including phenytoin and trimethadione, has been associated with a small but significant incidence of congenital abnormalities including slight mental retardation. More commonly, infants are tolerant to these agents at birth and undergo a period of withdrawal during the neonatal period associated with hyperactivity and other transient behavioral disruptions (Hutchings, 1983).

An understanding of possible mechanisms by which sedative-hypnotic agents act have provided considerable insight into the effects of these agents on developing brain. Recent studies suggest that many of these agents act on the GABA-benzodiazepine-chloride channel complex. Identification of a specific site of action for these agents has led to more focused behavioral studies which demonstrate that specific behavioral tasks which might be modulated by activity of this complex can be disrupted by prenatal exposure to benzodiazepines, barbiturates or other sedative-hypnotics. For example, perinatal sedative-hypnotic administration has been suggested to alter seizure threshold, change behavior in a test for anxiety, and alter certain characteristics of the startle responses, all effects in which endogenous "anxiogenic" systems are involved (Kellogg, 1980, 1983a,b). Studies underway in several laboratories are focusing on specific biochemical mechanisms mediating these changes, especially changes in benzodiazepine receptor binding and the interaction of developing benzodiazepine neurons with other neural systems, especially catecholaminergic systems.

Experimental Neurotoxicants in Development

There are a number of drugs which have been developed for research to selectively destroy specific cell populations. These include 6-hydroxydopamine (6-OHDA), which destroys catecholamine terminals, 5,7-dihydroxytryptamine (5,7-DHT), which destroys serotonergic terminals, ibotenic and kainic acids, which destroy cell bodies and AF-64A, which lesions acetylcholine neurons. Studies of the effects of these agents in developing nervous system have demonstrated the considerable difference in the sensitivity of immature and mature neural tissue to neurotoxins. In some cases, for example 5,7-DHT, treatment of the immature animal is far more devastating than treatment of the adult. 5,7-DHT destroys serotonergic nerve terminals in the adult animal but kills the entire neuron including the cell body in the immature animal. However, the converse can also be the case: following 6-OHDA treatment of neonatal rat pups, nerve terminals are destroyed as in adults but axonal sprouting from the remaining cell bodies occurs in pups, which can actually result in the inappropriate hyperinnervation of brain regions (e.g., the cerebellum) that are closest to the remaining cell bodies (Mueller et al., 1980). While the former paradigm provides an interesting model for permanent disruption of CNS development, the latter provides a model for studying CNS adaptation and has proven extremely useful for studying adaptation of dopaminergic and other neural systems following these lesions.

It is also important to note that new toxicants may become of specific developmental interest. Several years ago, Langston and co-workers (1983) found that 1-methyl-4-phenyl-2,3,4,6-tetrahydropyridine (MPTP, an undesired contaminant of the "designer drug" synthesis of a synthetic opiate) caused pseudoparkinsonism in humans. Apparently MPTP is metabolized by monoamine oxidase B to 1-methyl-4-phenylpyridinium (MPP+). Active uptake into dopamine and adrenergic neurons brings MPP+ into cells where it preferentially destroys dopaminergic neurons. While it has been hypothesized that environmental compounds of similar structure may be responsible for idiopathic Parkinson's disease, the question of whether similar drugs may affect development (a more sensitive process) has not been posed.

Nutrition, Hormone Levels, and Environmental Stimuli

As has been mentioned in the preceding pages, many of the purported effects of neurotoxins on the developing nervous system do not result from actions on a specific neuronal population or target enzyme but instead result from indirect actions on maternal physiology and behavior or fetal physiology. The developing nervous system is affected drastically by malnutrition, perturbed endocrine status or by altered behavioral interactions with the caretaker. In fact, hypo- or hyperthyroidism or hypercortisolism elicit what are almost classical effects on cell growth, synaptogenesis, neurotransmitter ontogeny and behavioral development as discussed earlier. Although it is beyond the scope of this chapter to cover this area extensively, it has been shown that many of the drug treatments used in animal models disrupt nutritional or endocrine status sufficiently to account for at least some of the observed effects. For example, many of the effects of alcohol resemble those of malnutrition, a common concomitant of alcohol abuse (Krigman and Hogan, 1976). Similarly, maternal behavior is significantly affected by maternal ingestion of alcohol, opioids, and sedative-hypnotics, in some cases to an extent that offspring development is affected. Disruption of maternal and fetal endocrine status accompanies alcohol, opioid, sedative-hypnotic, and probably mercury exposure. Even if such changes are not sufficient to significantly impair CNS development, any such change in physiologic status will affect the ability of the organism to respond to insult and so magnify the effects of xenobiotics.

Childhood Hyperactivity: Neurotoxicants and Food Colors

In humans, it is estimated that as many as 1–10% of all children suffer from the syndrome formerly called minimal brain dysfunction but now referred to as attention deficit disorder (ADD). These children exhibit both behavioral and cognitive difficulties, and the etiology is often not known. Attempts have been made to correlate the occurrence of ADD with exposure to certain toxicants.

One correlation that has been proposed is that some children with these learning and behavioral disabilities have increased blood lead concentrations. These reports spurred studies on animal models, and some investigators found that exposure of rats or mice to lead during developmental periods caused behavioral alterations that they proposed were similar to ADD. These same animals appeared to have alterations in catecholamine and acetylcholine neuronal systems in the brain, though mechanisms for these changes were not found. It has also been found that levels of lead exposure during development that cause no observable pathological changes cause permanent alterations in certain pharmacological responses known to be dependent on dopamine neurons in the brain. If further studies can associate these animal studies with human exposure, it may indicate that presently tolerated levels of lead exposure will have to be reevaluated. It is interesting to note that adult animals similarly exposed to equivalent lead levels do not seem to be perturbed similarly, if at all.

Another recent controversy related to ADD was the "Feingold hypothesis" (named after the California pediatrician who proposed it in 1973). This proposal links specific dietary factors, through unknown mechanisms, to significant alterations in brain function, resulting in clinically observable symptoms. The agents implicated by Feingold included specific food colors and natural salicylates. Proponents of this hypothesis suggest that elimination of the offensive agents through dietary manipulation can markedly improve many children. One synthetic color, FD&C Red #3, was even reported to have biochemical effects on dopamine systems that might explain the clinical speculation. However, more careful biochemical studies demonstrated that the actions were not specific for dopamine systems. Moreover, pharmacokinetic studies indicated that even at very high levels of intake it was extremely unlikely that enough color would enter the CNS to cause effects, an hypothesis borne out by behavioral studies in rats and humans. Thus, although the link of environmental chemicals to ADD is still open, these examples underline the importance of pharmacokinetics noted earlier in this chapter.

SUMMARY AND CONCLUSIONS

It is clear that many compounds can cause neurotoxic changes in the developing animal. Even if one knows what the proximate neurotoxicant is and how it is distributed, it is often difficult to know how to go about determining what aspect of development (especially at the molecular level) is perturbed. Two broad classes of actions can be considered. Does the neurotoxicant have a relatively specific effect (e.g., it could be targeted at a specific protein and result in a specific perturbation in an enzyme activity or a particular ion channel), or are the effects more generalized (e.g., it could cause generalized membrane perturbations which would secondarily affect a broad class of membrane bound enzyme activity, membrane ion channels, etc.)?

Another essential distinction to be made is whether observed neurochemical perturbations are directly related to an action of the toxicant or whether they are part of some compensatory response. Because the nervous system has considerable plasticity, compensatory responses presumably can be set in motion rapidly. For example, compensatory effects as neuronal sprouting or compensatory increases in development and activity of certain neuronal pathways. It is useful to consider the case of the neurotoxicant 6-hydroxydopamine. The specific lesion caused by this toxicant is a destruction of catecholamine terminals and, in some cases, neuronal death (Johnson et al., 1975). However, it also is well documented that after certain treatment paradigms of the developing animal one may see large compensatory increases in innervation of either the "lesioned" transmitter or of modulating neuronal systems. Moreover, a variety of changes in receptor characteristics may be observed, not all of which were directly predictable on the basis of the lesion alone. Thus, interpretation of the neurotoxic effects of 6-hydroxydopamine is greatly dependent on the dose of the toxicant, which parameters were being measured, and when after exposure the assays were conducted.

REFERENCES

Abel, E.L. (1984). *Fetal Alcohol Syndrome and Fetal Alcohol Effects*. Plenum Press, New York.

Amin-Zaki, L., S. Elhassani, M.A. Majeed, T.W. Clarkson, R.A. Doherty, and M.R. Greenwood (1967). Intra-uterine methylmercury poisoning in Iraq. *Pediatrics* 54: 587–595.

Briggs, G.G., T.W. Bodendorfer, R.K. Freedman, and S.Y. Jaffee (1983). In *Drugs in Pregnancy and Lactation*. Williams & Wilkins, Baltimore.

Butcher, R.E., W.J. Scott, K. Kazmaier, and E.J. Ritter (1983). Postnatal effects in rats of prenatal treatment with hydroxyurea. *Teratology* 7: 161–165.

Cagiano, R., R.J. Barfield, N.R. White, E.T. Pleim, M. Weinstein, and V. Cuomo (1988). Subtle behavioral changes produced in rat pups by in utero exposure to haloperidol. *Eur. J. Pharmacol.* 157: 45.

Chang, L.W. and Z. Annau (1984). Developmental neuropathology and behavioral teratology of methylmercury. In *Neurobehavioral Teratology*. J. Yanai, Ed., Elsevier, pp. 405–432.

Chasnoff, I.J. and D.R. Griffith (1989). Cocaine: clinical studies of pregnancy and the newborn. *Ann. N.Y. Acad. Sci.* 562: 260–266.

Cohen, M.E., E.K. Anday, and D.S. Leitner (1989). Effects of in-utero cocaine exposure on sensorineural reactivity. *Ann. N.Y. Acad. Sci.* 562: 344–347.

Dobbing, J. (1961). The blood brain barrier. *Physiol. Rev.* 41: 130–188.

Dow-Edwards, D.L. (1989). Long-term neurochemical and neurobehavioral consequences of cocaine use during pregnancy. *Ann. N.Y. Acad. Sci.* 562: 280–289.

Druse, M.J., C.S. Waddell, and R.G. Haas (1981). Maternal ethanol consumption during the third trimester of pregnancy: synaptic plasma membrane glycoproteins and gangliosides in offspring. *Substance and Alcohol Actions/Misuse* 2: 359–368.

Giroud, A. and M. Martinet (1955). Malformations diverses du foetus de rat suivant les stades d'administration de Vitamin A en excès. *C.R. Seances Soc. Biol. Filiales* 149: 1088–1090.

Grundt, I. and N.M. Neskovic (1980). Comparison of the inhibition by methylmercury and triethyllead of galactolipid accumulation in rat brain. *Environ. Res.* 23: 282–291.

Grundt, I., H. Offner, G. Konat, and J. Clausen (1974). The effect of methylmercury chloride and triethyllead chloride on sulphate incorporation into sulphatides of the cerebellum slices during myelination. *Environ. Physiol. Biochem.* 4: 166.

Grundt, I.K., E. Stensland, and T.C.M. Syversen (1980). Changes in fatty acid composition of myelin cerebrosides after treatment of developing rat with methylmercury and diethylmercury. *J. Lipid Res.* 21: 162–168.

Haddad, R. and A. Rabe (1980). Use of the ferret in experimental neuroteratology: cerebral, cerebellar and retinal dysplasias induced by methylazoxymethanol acetate. In *Advances in the Study of Birth Defects. Vol. 4. Neural and Behavioral Teratology.* T.V.N. Persaud, Ed., MTP Press, Lancaster, England, pp. 45–62.

Hall, B.H. and G.D. Das (1978). N-ethyl-N-nitrosourea induced teratogenesis of brain in the rat. *J. Neurol. Sci.* 39: 111–122.

Hanin, I., M.R. Krigman, and R.B. Mailman (1984). Central neurotransmitter of organotin compounds: trials, tribulations and observations. *Neurotoxicology* 5: 267–278.

Hans, S.L., J. Marcus, R.J. Jeremy, and J.G. Auerbach (1984). Neurobehavioral development of children exposed in utero to opioid drugs. In *Neurobehavioral Teratology.* J. Yanai, Ed., Elsevier, pp. 249–273.

Hicks, S.P. and C.J. D'Amato (1966). Effects of ionizing radiation on mammalian development. *Adv. Teratology* 1: 195–250.

Hirono, I. and K. Hayashi (1969). Induction of a cerebellar disorder with cycasin in newborn mice and hamsters. *Proc. Soc. Exp. Biol. Med.* 131: 593–599.

Hutchings, D.E. (1983). Behavioral teratology. *Handb. Exp. Pharmacol.* 65: 207–234.

Hutchings, D.E. and J. Gaston (1974). The effects of vitamin A excess administered during the midfetal period on learning and development in rat offspring. *Dev. Psychobiol.* 7: 225–233.

Johnson, C.E. (1989). Ontogeny and phylogeny of the blood-brain barrier. In *Implications of the Blood Brain Barrier and its Manipulation,* Vol. 2, Edward A. Neuwelt, Ed., Plenum Press, New York, pp. 157–198.

Johnson, G., T. Malmfors, and C. Sachs (1975). 6-Hydroxydopamine as a Denervation Tool in Catecholamine Research. North-Holland, Amsterdam.

Jones, C., E. Gardner, and M. Yang (1972). The pathogenesis of methylazoxymeth-
anol-induced cerebellar hypoplasia: ultrastructural observations. *Am. J. Path.* 66:
3a.

Jones, D.G. (1988). Influence of ethanol on neuronal and synaptic maturation in the
central nervous system—morphological investigations. *Prog. Neurobiol.* 31: 171–
197.

Jones, M., O. Mickelsen, and M. Yang (1973). Methylazoxymethanol neurotoxicity.
In *Progress in Neuropathology, Vol. 11.* H.M. Zimmerman, Ed., Grune & Strat-
ton, pp. 91–114.

Kalter, H. and J. Warkany (1961). Experimental production of congenital malformations in strains of inbred mice by maternal treatment with hypervitaminosis.
Am. J. Pathol. 38: 1–21.

Kellogg, C., D. Tervo, J. Ison, T. Parisi, and R.K. Miller (1980). Prenatal exposure
to diazepam alters behavioral and development in rats. *Science* 207: 205–206.

Kellogg, C.J., R. Chisholm, R.K. Millert, and J.R. Ison (1983a). Neural and behavioral consequences of prenatal exposure to diazepam. *Monogr. Neural Sci.* 9:
119–129.

Kellogg, C., J.R. Ison, and R.K. Miller (1983b). Prenatal diazepam exposure; effects
on auditory temporal resolution in rats. *Psychopharmacology* 79: 332–337.

Knowles, J.A. (1965). Excretion of drugs in milk: a review. *J. Pediatr.* 66: 1068–
1082.

Konat, G. (1984). Triethyllead and cerebral development: an overview. *Neurotoxicology* 5: 87–96.

Koyama, T., J. Handa, and S. Matsumoto (1970). Methylnitrosourea-induced malformations of brain in SD-JCL mice. *Arch. Neurol.* 22: 342–347.

Krigman, M.R. and E.L. Hogan (1976). Undernutrition in the developing rat: effect
on myelination. *Brain Res.* 107: 239–255.

Kurland, L.T., S.N. Faro, and H. Siedler (1960). Minamata disease. The outbreak
of a neurologic disorder in Minamata, Japan and its relationship to the ingestion
of seafood contaminated by mercuric compounds. *World Neurol.* 1: 370–395.

Landesman-Dwyer, S., L.S. Keller, and A.P. Steissguth (1977). Naturalistic observation of high and low risk newborns. *Alcohol* 2: 177–178.

Langman, J. and M. Shimada (1971). Cerebral cortex of the mouse after prenatal
chemical insult. *Am. J. Anat.* 132: 355–363.

Langston, J.W., P. Ballard, J.W. Tetrud, and I. Irwin (1983). Chronic Parkinsonism
in humans due to a product of meperidine analog synthesis. *Science* 219: 979–
980.

Lauder, J.M. and H. Krebs (1978). Serotonin as a differentiation signal in early
neurogenesis. *Dev. Neurosci.* 1: 15–30.

Lichtblau, L. and S.B. Sparber (1984). Opioids and development; a perspective on
experimental models and methods. *Neurobehav. Toxicol. Teratol.* 6: 3–8.

Mailman, R.B., G.R. Breese, M.R. Krigman, P. Mushak, and R.A. Mueller (1979).
Lead exposure during infancy permanently increases lithium-induced polydipsia.
Technical comments rebuttal to above paper. *Science* 205: 726.

Menon, N.K. and R.R. Lopez (1985). The effects of mild congenital methylmercury
intoxication on the metabolism of 3-hydroxybutyrate and glucose in the brains of
suckling rats. *Neurotoxicology* 6: 55–62.

Mottet, N.K., C.M. Shaw, and T.M. Burbacher (1985). Health risks from increase
in methylmercury exposure. *Environ. Health Perspect.* 63: 133–140.

Mueller, R.A., R.B. Mailman, and G.R. Breese (1980). Behavioral and monoaminergic consequences of exposure to neurotoxins during development. In *Biogenic Amines in Development*. H. Parvez and S. Parvez, Eds., Elsevier/North Holland, Amsterdam.

Nicholson, J.L. and J. Altman (1972a). The effects of early hypo- and hyperthyroidism on the development of rat cerebellar cortex. I. Cell proliferation and differentiation. *Brain Res*. 44: 13–23.

Nicholson, J.L. and J. Altman (1972b). The effects of early hypo- and hyperthyroidism on the development of the rat cerebellar cortex. II. Synaptogenesis in the molecular layer. *Brain Res*. 44: 25–36.

Null, D.H., P.S. Gartside, and E. Wei (1973). Methylmercury accumulation in brains of pregnant, nonpregnant and fetal rats. *Life Sci*. 12: 65–72.

Omata, S., H. Sakimura, H. Tsubaki, and H. Sugano 91978). In vivo effect of methylmercury on protein synthesis in brain and liver of the rat. *Toxicol. Appl. Pharmacol*. 44: 367–378.

Patel, A.J. and P.D. Lewis (1982). Effect on cell proliferation of pharmacological agents acting on the central nervous system. In *Mechanisms of Actions of Neurotoxic Substances*. K.N. Prasad and A. Pernadakis, Eds., Raven Press, New York, pp. 181–218.

Pennington, S. and G. Kalmus (1987). Brain growth during ethanol-induced hypoplasia. *Drug Alcohol Depend*. 20: 279–286.

Pfaffenroth, M.J., G.D. Das, and J.P. McAllister (1974). Teratologic effects of ethylnitrosurea on brain development in the rat. *Teratology* 9: 305.

Purves, D. and J. Lichtman (1985). *Principles of Neural Development*. Sinauer Associates, Sutherland, MA.

Randall, C. (1982). Alcohol as a teratogen in animals. In *Biomedical Processes and Consequences of Alcohol Use*. Alcohol and Health Monogr. 2, Department of Health and Human Services, pp. 291–335.

Rapaport, S.E. (1976). *The Blood Brain Barrier in Physiology and Medicine*. Raven Press, New York.

Rosengarten, H. and A.J. Friedhoff (1979). Enduring changes in dopamine receptor cells of pups from drug administration to pregnant and nursing dams. *Science* 203: 1133–1135.

Rosett, H.L. and L. Warner (1984). Alcohol and the fetus: a clinical perspective. Oxford University Press, New York.

Sanstead, H.H. (1986). A brief history of the influence of trace elements on brain function. *Am. J. Clin. Nutr*. 43: 293–298.

Sarafien, T.A., M.K. Cheung, and M.A. Verity (1984). In vitro methylmercury inhibition of protein synthesis in neonatal cerebellar perikarya. *Neuropath. Appl. Neurobiol*. 10: 85–100.

Shoemaker, W.J., G. Baetge, R. Azad, V. Sapin, and F.E. Bloom (1983). Effects of prenatal alcohol exposure on amine and peptide neurotransmitter systems. *Monogr. Neural Sci*. 9: 130–139.

Slotkin, T.A., S. Pachman, R.J. Kavlock, J. Bartolome (1985). Early biochemical detection of adverse effects of a neurobehavioral teratogen: influence of prenatal methylmercury exposure on ornithine decarboxylase in brain and other tissues of fetal and neonatal rat. *Teratology* 32: 195–202.

Slotkin, T.A. and P.V. Thadani (1980). Neurochemical teratology of drugs and abuse. In *Advances in the Study of Birth Defects. Vol. 4. Neural and Behavioral Teratology.* T.V.N. Persaud, Ed., MTP Press, Lancaster, England, pp. 199–234.

Spear, L.P., I.A. Shalaby, and J. Brick (1980). Chronic administration of haloperidol during development: behavioral and psychopharmacological effects. *Psychopharmacology* 70: 47–58.

Spear, L.P., C.L. Kristein, and N.A. Frambes (1989). Cocaine effects on the developing central nervous system: behavioral, psychopharmacological, and neurochemical studies. *Ann. N.Y. Acad. Sci.* 562: 290–307.

Streissguth, A. (1977). Maternal drinking and the outcome of pregnancy: implications for child mental health. *Am. J. Orthopsychiatry* 47: 422–431.

Syversen, T.L.M. (1977). In vitro methylmercury inhibition of protein synthesis in isolated cerebral and cerebellar neurons. *Neuropath. Appl. Neurobiol.* 3: 225–233.

Takeuchi, T. (1977). Pathology of fetal Minamata disease. *Pediatrics* 6: 69–87.

Taylor, L.L. and V. DiStefano (1976). Effects of methyl mercury on brain biogenic amines in the developing rat pup. *Toxicol. Appl. Pharmacol.* 38: 489–497.

Tonge, S.R. (1973). Some persistent effects of psychotropic drugs on noradrenaline metabolism in discrete parts of the rat brain. *Br. J. Pharmacol.* 48: 364–365.

Tucker, J.C. (1985). Benzodiazepines and the developing rat: a critical review. *Neurosci. Biobehav. Rev.* 9: 101–111.

Vernadakis, A. (1982). Neurotoxic effects of phenytoin: the developing organism. In *Mechanisms of Actions of Neurotoxic Substances.* K.N. Prasad and A. Vernadakis, Eds., Raven Press, New York, pp. 155–178.

Vertes, Z., G. McLegh, M. Vertes, and S. Kovacs (1982). Effect of naloxone and D-met2-Pro5 enkephalinamide treatment on the DNA synthesis in the developing rat brain. *Life Sci.* 31: 119–126.

Vorhees, C.V., D.R. Minck, and H.K. Berry (1988). Anticonvulsants and brain development. *Prog. Brain Res.* 73: 229–244.

Wedeen, R.P. (1979). Lead enhancement of lithium-induced polydipsia. *Science* 205: 725–726.

West, J.R., C.R. Goodlett, and S.J. Kelly (1987). Alcohol and brain development. *National Institute on Drug Abuse Research Monogr. Ser.* 78: 45–60.

Werboff, J. and J. Havelena (1962). Abstract behavioral effects of tranquilizers administered to the gravid rat. *Exp. Neurol.* 6: 262–269.

Winder, C. and I. Kitcher (1987). Lead neurotoxicity: a review of the biochemical, neurochemical and drug-induced behavioral evidence. *Prog. Neurobiol.* 22: 59–87.

Woodbury, D.M. (1977). Maturation of the blood-brain and blood-CSF barriers. In *Drugs and the Developing Brain.* A. Vernadakis, Ed., Plenum Press, New York, pp. 259–280.

Yang, M.G., K.S. Krawford, J.D. Garcia, J.H. Wang, and K.Y. Lei (1972). Deposition of mercury in fetal and maternal brain. *Proc. Soc. Exp. Biol. Med.* 141: 1004–1007.

Zagon, I.S. and P.J. McLaughlin (1984). An overview of the neurobehavioral sequelae of perinatal opioid exposure. In *Neurobehavioral Teratology.* J. Yanai, Ed., Elsevier, pp. 197–234.

Zagon, I.S. and P.J. McLaughlin (1983). Increased brain size and cellular content in infant rats treated with an opiate antagonist. *Science* 221: 1179–1180.

Zagon, I.S. (1987). Endogenous opioids, opioid receptors and neuronal development. *NIDA Research Monogr.* 78: 72–94.

Zagon, I.S., R.E. Rhodes and P.J. McLaughlin (1985). Distribution of enkephalin immunoreactivity in germinative cells of developing rat cerebellum. *Science* 27: 1049–1051.

Nutrition and Neurotoxicology

Martha M. Abou-Donia
Department of Clinical Neurosciences
Burroughs Wellcome Company
Research Triangle Park, North Carolina

INTRODUCTION

The interaction between nutritional status and the ultimate expression of neurotoxicity is complex and can be modified by a number of factors. Certain nutritional elements, such as pyridoxine, can be neurotoxic in excess, whereas a deficiency of another nutritional element, such as thiamine, can also influence neuronal functioning. Alternatively, a marginal nutritional status may become inadequate when a toxicant, such as alcohol, is ingested. The ultimate expression of the interaction between nutritional status and toxicants is reflected in the overall functioning of the nervous system. Furthermore, some foods that are routinely ingested in certain cultures contain neurotoxicants. The developmental stage at which toxic levels of a nutritional factor are present can influence the expression of neurotoxicity. As described in detail below, high levels of phenylalanine can be toxic in certain infants because of an enzyme deficiency and can result in severe mental retardation. However, if dietary phenylalanine is restricted until the child is approximately five years old, the diet can be less restrictive after that age without affecting mental development. The precise nature of nutritional interactions and neurotoxicants is difficult to delineate since one factor rarely changes alone.

The basic nutritional elements are carbohydrates, fats, proteins, vitamins, and minerals (as shown in Table 1).

TABLE 1.
Basic Nutritional Needs

1. CARBOHYDRATES	Riboflavin (B_2)
2. PROTEINS	Niacin
Essential Amino Acids	Pyridoxine (B_6)
Histidine	Pantothenic Acid
Isoleucine	Biotin
Leucine	Folic Acid
Lysine	Cobalamine (B_{12})
Methionine (Cysteine)[a]	Ascorbic Acid (C)
Phenylalanine (Tyrosine)[a]	b. Fat-Soluble
Threonine	Retinol (A_1)
Tryptophan	Cholecalciferol (D_3)
Valine	α-Tocopherol (E)
3. LIPIDS	Menaquinone-h (K_2)
Essential Fatty Acids	5. MINERALS
Linoleic	Calcium
Linolenic	Phosphorus
Arachidonic	Magnesium
4. VITAMINS	Iron
a. Water-Soluble	Zinc
Thiamine (B_1)	Iodine

[a] Amino acid in parentheses is not considered essential because it can be derived from preceding amino acid.

It is important to understand that although certain guidelines (i.e., recommended daily allowances, RDA) have been established to quantitate the requirements of man for these basic nutritional elements and to ascertain whether a particular diet is nutritionally adequate, there are limitations to the applicability of these recommended daily allowances to the specific diet of an individual. The value of these recommended daily allowances is bound by the methods that were used to determine them; the subpopulation that was used as subjects; and, in most cases, the failure to examine the effects of physiological status on nutritional need. In general, these RDAs were derived either from short term balance studies, from the levels of an exogenous nutrient required to produce saturation of the blood or a tissue, or by determining the minimal amount of an exogenous nutrient required to prevent the expression of a particular deficiency state. Thus, the value established as an RDA for a specific nutrient does not necessarily represent the level of that nutrient required to maintain optimal nutritional health. Additionally, these studies have been done primarily in young men and the recommended allowances for other subpopulations (i.e., women, children, elderly) have been approximated from the values estimated for men. Moreover, the recommended daily allowances fail to account for interactions between certain nutrients and naturally occurring food constituents (i.e., iron and oxalates) that may modify the availability of the nutrient or for the potential increase or decrease in the requirement for certain essential elements due to disease, stress, activity level, or the administration of drugs and/or alcohol.

BASIC NUTRITIONAL ELEMENTS

Carbohydrates

Although carbohydrates are primarily thought of as a source of energy, they serve several other essential functions in meeting basic nutritional needs. Carbohydrates spare other essential elements. When carbohydrate levels are low, the bodily requirement for energy is met by metabolizing protein and lipids for energy. However, the generation of energy from lipids in the presence of low levels of carbohydrates results in the formation of ketone bodies (i.e., acetoacetic acid, acetone, β-hydroxybutyric acid) which produce various physiological effects including appetite suppression, dehydration, and ultimately, coma. Carbohydrates in the diet, therefore, can function as an antiketogenetic factor. Although protein can be used to supply energy through gluconeogenesis, this use of protein would reduce the pool of essential amino acids that are available. Thus, the incorporation of carbohydrates in the diet for energy spares the protein for more essential functions. Additionally, carbohydrates can provide the carbon backbone for synthesizing more complex molecules. Carbohydrates are the major source of energy for the brain. The energy requirement of the adult brain is estimated to be 4.6 g glucose/h or approximately 400 kcal/day. Although lactate, pyruvate, fumarate, acetate, and glyceraldehyde can be metabolized in the brain, these compounds are not an important source of energy for the brain since their levels are low in the circulating blood, and they do not pass the blood-brain barrier.

The two sources of glucose for the brain are the uptake of circulating glucose and glycogen stores. In adults 90% of glucose is metabolized by the tricarboxylic (TCA) cycle to generate adenosine triphosphate (ATP). Glucose is also metabolized by the hexose monophosphate shunt to generate the NADPH that is required to synthesize the lipid utilized in myelination and the ribose molecules needed for nucleic acid synthesis. During periods of oxygen deprivation, the brain uses energy from the phosphate bonds of ATP and phosphocreatine. This pool of energy-rich phosphate bonds will, however, maintain the brain for only minutes. Since brain glycogen stores are extremely small, they are reserved as an emergency energy source. During periods of starvation, the brain can take up ketone bodies derived from peripheral fatty acid metabolism as a source of energy. Since the normal level of ketone bodies in the blood is approximately 1 mg/ml, this is not an important source of energy for the brain except during periods of starvation. Ultimately, the conversion of protein to glucose peripherally via gluconeogenesis will provide the brain with energy.

Although there is no established recommended daily allowance for carbohydrates, 50–100 g (200–400 kcal) of carbohydrates daily appear to be required to prevent ketosis and the catabolism of peripheral proteins to supply the brain with energy.

Proteins and Amino Acids

In addition to providing a source of protein for growth and maintenance and a supply of energy, dietary proteins also serve as a source of the essential amino acids that humans cannot synthesize. These essential amino acids are required for such diverse functions as the precursors for the synthesis of certain neurotransmitters (tyrosine converted to epinephrine, norepinephrine, and dopamine; tryptophan converted to serotonin), the synthesis of purine and pyrimidine bases, and the synthesis of specialized proteins such as transport proteins (i.e., albumin), protein hormones (i.e., insulin), tissue proteins (i.e., collagen), and enzymes.

The essential amino acids in humans include histidine, isoleucine, leucine, lysine, methionine, phenylalanine, threonine, tryptophan, and valine. Clinically, protein deficiency is most easily recognized where there is a relatively high intake of calories from carbohydrates and a relatively low intake of protein in the diet. Protein malnutrition in the postweaning years is referred to as kwashiorkor. The symptoms of kwashiorkor are subnormal weight; mental apathy; an abnormal electrocardiogram; edema of the hands, feet, legs, and face; variations in the amount and color of the hair; changes in the texture and pigmentation of the skin; fatty liver; anemia; and symptoms of vitamin A deficiency. Marasmus is another protein deficient state and is characterized by deficits in both protein and calories.

The recommended daily allowance for protein varies from 56 g/day in adult men to 46 g/day in adult women. This requirement has been estimated to increase in women who are pregnant or lactating. Additionally, the requirement for protein varies with certain disease states and their severity. Fevers, fractures, burns, and surgical trauma cause extensive loss of protein that must be replaced by increases in the dietary intake of protein. On the other hand, protein intake must be restricted in conditions such as acute liver failure and uremia where the capacity to metabolize and excrete nitrogenous compounds is limited.

Protein in the diet is digested by pepsin secreted into the stomach and by the proteolytic enzymes (i.e., carboxypepitase, chymotrypsin, trypsin) released from the pancreas and mucosal cells lining the lumen of the small intestine. The free amino acids and small peptides released by this proteolytic action are absorbed by the mucosal cells of the small intestine. The small peptides are cleaved into free amino acids within the mucosal cell by peptide hydrolases. The free amino acids pass into the portal vein and are transported to the liver. The liver is responsible for monitoring and regulating the levels of circulating amino acids. The enzymes responsible for the metabolism of various essential and nonessential amino acids are induced in the liver when the dietary uptake increases beyond physiological needs. In the absence of normal liver function (i.e., cirrhosis), plasma amino acid levels are not closely regulated, and the brain can be exposed to toxic levels of certain amino acids. The toxic effects of phenylalanine have been shown clearly in children born with a disorder of phenylalanine metabolism known as phenylketonuria or hyperphenylalaninemia. Children with this metabolic error develop severe mental retardation. The mechanism for the toxic effect of phenylalanine on the brain is unknown. Classical phenylketonuria is caused by the inability to convert phenylalanine to tyrosine due to absence of the enzyme phenylalanine hydroxylase in the liver. The conversion of phenylalanine to tyrosine also requires the participation of a cofactor, tetrahydrobiopterin, and the regeneration of this cofactor by dihydropteridine reductase.

As early as 1934, Folling suggested that an inherited defect in the metabolism of phenylalanine was responsible for the development of the mental retardation. However, it was not until 1949 that the first patient was treated with a low phenylalanine diet and showed some improvement in development and behavior. It has since been shown that there is an inverse correlation between the age at which the low phenylalanine diet is started and the ultimate I.Q. of a child. The first screening tests for phenylketonuria were introduced in 1957 and were based on a positive urinary color test utilizing ferric chloride. The ferric chloride test was of limited value since it could not detect the disease early enough to prevent significant mental retardation. However, the introduction of the Guthrie test in 1963 has made it possible to detect increases in plasma phenylalanine levels within the first week of life and to introduce a phenylalanine restricted diet early enough to prevent mental deterioration.

The incidence of phenylketonuria is approximately 1 in 20,000 live births. Over the years, approximately 3% of phenylketonuric children have failed to respond to a phenylalanine restricted diet and have developed mental and physical retardation. In the last decade, it has become apparent that a deficiency of either the cofactor, tetrahydrobiopterin, required by phenylalanine hydroxylase or of the enzyme dihydropteridine reductase, required to regenerate the cofactor, could also result in the development of an atypical phenylketonuria (or hyperphenylalaninemia). Diagnosis of an atypical hyperphenylalaninemia is based on a differential response to a phenylalanine load test. After a phenylalanine load in a patient with a deficiency of the enzyme dihydropteridine reductase, the plasma level of the cofactor, tetrahydrobiopterin, will increase. However, patients with a deficiency in cofactor levels due to a defect in its synthesis are unable to respond to a phenylalanine load with increased plasma cofactor levels. Treatment of atypical hyperphenylalaninemias is still experimental and is based on both the restriction of dietary phenylalanine as well as the addition of L-dopa and 5-hydroxytryptophan to replace the potential neurotransmitter deficit. Carbidopa is also administered to minimize the peripheral utilization of L-dopa and 5-hydroxytryptophan. Since tetrahydrobiopterin is also a cofactor for tyrosine and tryptophan hydroxylases, a potential deficit in dopamine, norepinephrine, epinephrine, and serotonin synthesis could exist.

Lipids

Dietary fats and lipids are a rich source of energy and provide 9 kcal/g, over twice the energy yield of both carbohydrates (4 kcal/g) and proteins (4 kcal/g). Additionally, fat stored in the body can provide a compact form of reserve energy. Dietary lipids are also required to provide a pool of the essential fatty acids that man cannot synthesize. These essential fatty acids include linoleic acid, linolenic acid, and arachidonic acid. These essential fatty acids are required for the regulation of cholesterol metabolism, as components of the phospholipids in cell membranes and organelles, and as precursors of the prostaglandins and leukotrienes. Deficiency symptoms have been observed in a number of species and include decreased fertility, decreased growth, loss of hair, eczematous dermatitis, and changes in cell membrane function and enzyme activity. It is difficult to produce an essential fatty acid deficiency in adults because of the large stores of essential fatty acids in the body. However, a deficiency of essential fatty acids has been observed in individuals on total parenteral nutrition for prolonged periods of time.

There is no recommended daily allowances for lipids but it has been suggested that a minimum of 1% of the total calories in the diet should consist of essential fatty acids.

Vitamins

There have been three eras in the medical use of vitamins. Initially, vitamins were used for the prevention and treatment of a deficiency syndrome, such as scurvy and pellagra. More recently, vitamins have been used for the treatment of vitamin responsive inborn errors of metabolism such as the use of pharmacological doses of pyridoxine to treat infantile convulsions. The use of vitamins in this manner is based on the theory that genetic disorders can lead to a ten- to thousand-fold increase in a vitamin requirement. This increased requirement may arise from an impairment in the absorption, transport, or conversion of the vitamin to an active form (i.e., deficiency of cobalamin reductase), a defect in binding to a genetically abnormal enzyme (i.e., vitamin C responsive lysyl hydroxylase deficiency) or a decreased half-life of a genetically altered enzyme (i.e., pyridoxine responsive cystathionine β-synthase deficiency). There are 25 vitamin responsive disorders involving 8 of the 9 water-soluble vitamins and one of the 4 fat-soluble vitamins. Most recently, the medical use of vitamins has centered around the controversial use of megavitamin doses in the treatment of major mental illnesses. The theoretical basis of orthomolecular psychiatry involves the delivery of megadoses of the vitamins to the tissues so that apoenzymes will be saturated with their cofactors and the catalytic rates of the reactions will be maximized by mass action. Although there is no experimental evidence to support this theory and much evidence of feedback regulation to suggest that it would not work, orthomolecular psychiatry and the use of megavitamin doses has been readily popularized by the press. The mass media exposure of this theory and the ready availability of over-the-counter vitamin preparations have combined to create the possibility of vitamin toxicity.

Water-Soluble Vitamins

Thiamine (B₁). Thiamine deficiency was first characterized by Eijkman in 1907 in chickens fed polished rice who developed a polyneuritis that resembled beriberi, a human disease that had been endemic to the Orient for 4000 years. Beriberi is characterized by anorexia, constipation, lethargy, depression, Korsakoff's psychosis with confabulation, and losses of memory and cognitive function. The symptoms also include palpitations, tachycardia, epigastric pain, systolic hypotension, venous distention, and peripheral cyanosis. Similar symptoms have been observed in chronic alcoholics.

Thiamine functions as the cofactor thiamine pyrophosphate in the oxidative decarboxylation of pyruvate and α-ketoglutarate and the transketolation of xylose-5-PO₅ in the hexose monophosphate shunt for the generation of NADPH.

The recommended daily allowance for thiamine is 0.5 mg/1000 kcal. There is no known toxicity to thiamine since the maximum quantity that can be absorbed from an oral dose is 5 mg. However, repeated parenteral administration can result in vasodilation, nausea, tachycardia and dyspnea. Death due to depression of the respiratory center in animals has been observed following intravenous injection of thiamine in large doses (i.e., 125 mg/kg in mice, 250 mg/kg in rats).

Riboflavin (B₂). Riboflavin is a constituent of two coenzymes: flavin 5-monophosphate (FMN) and flavin 5-pyrophosphate (flavin adenine dinucleotide, FAD). The flavin coenzymes serve as cofactors to catalyze the transfer of electrons for various dehydrogenases, including succinate dehydrogenase, dihydrolipoyl dehydrogenase, and glyceraldehyde-3-phosphate dehydrogenase. A deficiency of riboflavin is characterized by cheilosis, angular stomatitis, glossitis, corneal vascularization, seborrheic dermatitis, and scrotal dermatitis.

The recommended daily allowance for riboflavin is 0.6 mg/1000 kcal. The relatively low solubility of riboflavin may account for its lack of toxicity.

Niacin. Nicotinamide is the physiological form of niacin and is a constituent of two coenzymes: nicotinamide adenine dinucleotide (NAD) and nicotinamide adenine dinucleotide phosphate (NADP). There are over 150 reactions that utilize NADH or NADPH. These cofactors are involved as coenzymes in electron transport, in cellular respiration, and in oxidative/reduction reactions.

Classical niacin deficiency first appeared in 1720 with the introduction of maize planting in Europe. The classical deficiency disease is pellagra, which occurs when the diet is heavily dependent on corn. Corn is deficient in tryptophan and contains only low levels of bound niacin. In the early 1900s the disease was thought to be caused by a tryptophan deficiency. It has since been shown that tryptophan can be converted to nicotinamide and serves as a precursor for the vitamin. The conversion ratio is approximately 60 mg tryptophan to yield 1 mg niacin. Pellagra is an easily diagnosed disease and was very common in the southern United States until the 1930s. Pellagra is characterized by the four "Ds": dermatitis (a butterfly-shaped dermatitis on the face), diarrhea, dementia, and ultimately, death.

The recommended daily allowance for niacin is 6.6 niacin equivalents/ 1000 kcal with a minimum of 13 equivalents (i.e., 60 mg tryptophan = 1 mg of niacin). Nicotinic acid can also be converted to the active vitamin form nicotinamide.

Ingestion of large doses of nicotinic acid can produce flushing and reduce the levels of cholesterol, beta-lipoproteins, and triglycerides in blood. Prolonged ingestion of nicotinic acid can lead to gastrointestinal irritation and liver damage. Reports of niacin toxicity have appeared in the literature coincident with its use in the treatment of schizophrenia as part of the regimen of orthomolecular psychiatry. Advocates of orthomolecular psychiatry maintain that some major mental illnesses are due to metabolic disorders and can be successfully treated with megadoses of vitamins. Doses as high as 50 g daily of niacin have been recommended for treatment of mental disorders. Large doses of niacin have been associated with flushing, pruritis, hyperkeratosis, heartburn, nausea, vomiting, diarrhea, low blood pressure, tachycardia, and syncope.

Pyridoxine (B_6). The vitamin pyridoxine occurs naturally in three forms: pyridoxol, pyridoxal, and pyridoxamine. The active form of the enzyme is pyridoxal-5-phosphate. There are sixty reactions that are dependent on pyridoxal-5-phosphate as a cofactor, including all transaminase reactions, deamination and dehydration of serine to produce pyruvate, desulfhydration of cysteine, decarboxylation of dihydroxyphenylalanine (dopa) to dopamine and 5-hydroxy tryptophan to serotonin, phospholipid synthesis (sphingosine), and the synthesis of the porphyrin nucleus for the heme of hemoglobin and the cytochromes. A deficiency of pyridoxine is characterized by abnormal electroencephalogram, insomnia, confusion, nervousness, depression, seborrhea, cheilosis, glossitis, and stomatitis.

The recommended daily allowance for pyridoxine is 2 mg/day and the requirement parallels the protein intake. The availability of pyridoxine can be affected by certain drug interactions. Drugs that are considered to be pyridoxine antagonists and increase the pyridoxine requirement are penicillamine, isoniazide, semicarbazide, and cycloserine.

Pyridoxine has received wide acceptance in the lay and medical communities. Under certain conditions, however, it can be toxic. This vitamin is believed to be helpful as a component of body building programs and as a treatment for premenstrual syndrome, carpal tunnel syndrome, schizophrenia, autism, hyperkensia, and peripheral neuropathies associated with isoniazide and hydralazine therapy. A recent report documents 7 patients who consumed large doses of pyridoxine (2–6 g daily) over 4 to 40 months and developed symptoms of sensory neuropathy. The clinical profile that developed over time included an unstable gait, numb feet, numbness and clumsiness in the hands, and finally, perioral numbness. Improvement occurred in all cases on removal of pyridoxine supplements and the results of extensive clinical evaluation indicated that megavitamin doses of pyridoxine were the causative agent. The mechanism of pyridoxine neurotoxicity is unknown.

Pyridoxine is a pyridine and, as a family of compounds, pyridines are neurotoxic. Pyridine neurotoxicity is limited to the periphery since the blood-brain barrier prevents access to the central nervous system. Peripheral nerves, such as the cell bodies of the peripheral sensory fibers located in the dorsal ganglia, can be exposed to high concentrations of pyridines since there is unlimited gastrointestinal transport of pyridines and no blood-brain barrier to protect them. The toxicity of pyridoxine may also be complicated by the fact that it must be converted to pyridoxal phosphate, the active form of the vitamin. The administration of a large dose of pyridoxine could exceed the capacity for conversion and result in the inactive form of the vitamin competing with the active form for binding to the apoenzyme, leading to the development of a pyridoxal phosphate deficiency. A third explanation for the toxicity of megadoses of pyridoxine may be the presence of an unknown impurity in the over-the-counter preparations of pyridoxine that are available. These vitamin preparations are required to be only 98% pure. The consumption of vitamin doses at 2500 times the recommended daily allowance could result in the ingestion of significant quantities of toxic impurities.

Pantothenic Acid. Pantothenic acid is an integral part of coenzyme A and, as such, is a part of acetyl CoA and acyl CoA. Coenzyme A functions in the energy metabolism of carbohydrates and fatty acids. Coenzyme A is involved in many reactions including the oxidation of fatty acids, pyruvate α-ketoglutarate, and acetaldehyde. As acyl CoA, pantothenic acid functions in the synthesis and degradation of fatty acids. Pantothenic acid is also involved in the synthesis of cholesterol, citrate, acetoacetate, porphyrins, and sterols. As acetyl CoA, it functions in the synthesis of acetylcholine. A deficiency of pantothenic acid is rarely observed except in cases of gross malnutrition such as chronic alcoholics. Anecdotal accounts from prisoners in World War II Japanese prisoner of war camps describe their deficiency symptoms as including "burning feet," nausea, vomiting, abdominal cramps, malaise, insomnia, personality changes, muscle fatigue, and cramps. In experimental animals, pantothenic acid deficiency is characterized by degeneration of the peripheral nerve myelin and lesions of the spinal cord.

There is no recommended daily allowance for pantothenic acid but a daily intake of 5–10 mg is considered adequate. When pantothenic acid has been administered in doses as large as 10–20 mg, diarrhea has occurred.

Biotin. Biotin functions as a cofactor in carboxylation and decarboxylation reactions (acetyl CoA to malonyl CoA to fatty acids; pyruvate to oxaloacetate). The best source of biotin is egg yolk. Egg white contains avidin, a natural antivitamin for biotin. A deficiency of biotin is uncommon unless the diet contains a large amount of raw egg white. Biotin deficiency is characterized by glossitis, gray mucosa and skin pallor, depression, dermatitis, anorexia, nausea, vomiting, anemia, and hypercholesterolemia.

There is no recommended daily allowance for biotin; however, 150–300 μg/day is estimated to be adequate.

Folic Acid. Folic acid functions in one-carbon transfer and regeneration in a manner analogous to the two-carbon transfers involving coenzyme A. The active form of the enzyme is tetrahydrofolic acid. Reactions utilizing folic acid include the formation serine from glycine, methionine from homocysteine, the heme moiety of hemoglobin, and the biosynthesis of purines and pyrimidines such as thymidine. Folic acid deficiency arrests cell replication, leading to a nutritional megaloblastic anemia which is pancytopenic. This pancytopenia involves all bone marrow cells and is characterized by macroovalocytes and hypersegmented polymorphonuclear neutrophils. Symptoms of folic acid deficiency include weakness, stomatitis, glossitis, diarrhea, malabsorption, and weight loss.

The recommended daily allowance of folic acid for adults is 0.4 mg. The amount of folic acid required by an individual can be influenced by chronic blood loss or the inadequate utilization of folic acid in the presence of a vitamin B_{12} deficiency. Vitamin B_{12} is required for the removal of the methyl group from N^5N^{10} methylene tetrahydrofolate. The presence of a vitamin B_{12} deficiency essentially creates a folic acid trap. The therapeutic dose of folic acid commonly given to adults is 5 mg/kg. This dose is sufficient to correct a megaloblastic anemia that results from a true vitamin B_{12} deficiency. The treatment of this megaloblastic anemia resulting from a B_{12} deficient state with excess amounts of folic acid may mask the underlying vitamin B_{12}-induced progressive degeneration of nerves in the spinal cord that can lead to severe neurological damage.

Cobalamin (Vitamin B_{12}). Vitamin B_{12} is required for red blood cell formation, nerve function and growth, and functions biochemically in the removal of the methyl group from folate and, therefore, in the maintenance of folate levels. It is also required for the transfer of alkyl groups; for the methylation of homocysteine to methionine and, therefore, protein synthesis. It also functions as a reducing agent for sulfhydryl compounds, such as glutathione, and has an unknown role in myelin synthesis. Dietary deficiency of vitamin B_{12} is normally only a problem for vegetarians since plants do not contain B_{12}. The main cause of a B_{12} deficiency is a failure of glands in the stomach to secrete an *intrinsic factor* that is essential for the gastrointestinal absorption of B_{12}. A vitamin B_{12} deficiency is characterized by severe chronic atrophic gastritis with a secondary anemia (pernicious anemia) and inadequate myelin synthesis. The inadequate myelin synthesis and subsequent nerve damage may be exhibited in an individual as paresthesias, loss of proprioception, ataxia, loss of stretch reflexes, and spasticity. A deficiency in the secretion of the intrinsic factor and the subsequent malabsorption of B_{12} may be the result of a genetic predisposition, a gastrointestinal disease and/or surgery, or chronic aspirin ingestion. B_{12} deficiencies are very slow to develop because of an efficient enterohepatic recirculation with losses of less than 1.5 µg/day. It has been calculated to require 3–6 years to develop a B_{12} deficiency even on a vegetarian diet.

The recommended daily intake of B_{12} for adults is 3 µg.

Ascorbic Acid (Vitamin C). A deficiency of ascorbic acid is one of the oldest recorded nutritional deficiency states, having been described in an ancient Egyptian papyrus ca. 1500 B.C. Long sea voyages have been associated classically with the development of a hemorrhagic disease known as scurvy. The introduction of fresh fruit and vegetables into the diet of British sailors in 1747 represented the first known cure for a nutritional disease. Interestingly, although scurvy was the first nutritional disease with a known cure, Scott and his group of Antarctic explorers died from this disease as recently as 1912.

Vitamin C functions as a coenzyme in hydrogen ion transfers and oxidation/reduction reactions such as dopamine β-hydroxylation, the hydroxylation of proline and lysine for collagen, synthesis of glucocorticoids in the adrenal cortex, and facilitation of the absorption of dietary iron. Most animals can synthesize ascorbic acid from UDP-glucuronic acid in their livers, but exceptions include man, monkeys, guinea pigs, and Indian fruit bats. Vitamin C deficiency is slow to develop because of a physiological adaption that increases absorption in gastrointestinal tract and reabsorption in the kidneys. It takes approximately four months to develop an ascorbic acid deficiency.

Vitamin C is required to build and maintain the bone matrix containing collagen and connective tissue. The classic lesion of scurvy is the development of capillary fragility and pinpoint petechial hemorrhages due to impaired collagen synthesis. Vitamin C deficiency is also characterized by swollen and bleeding gums, dryness of the mouth and eyes, loss of hair, loose teeth, anemia, edema, and death.

The recommended daily allowance for ascorbic acid is 45 mg. Ingestion of megadoses of vitamin C in the range of 5–15 g/day to prevent the common cold can be associated with the development of nausea and diarrhea. A large daily intake of ascorbic acid may be associated with increased urinary excretion of oxalic acid and calcium and decreased sodium excretion. Formation of oxalate or urate crystals may be enhanced by these factors. In men there have been reports of an increased incidence of uric acid and cystine stones in the urinary tract. Men may also develop hemosiderosis because ascorbic acid facilitates the absorption of dietary iron. Less than 10% of dietary iron is absorbed in the presence of normal dietary levels of ascorbic acid. Ascorbic acid may also interfere with anticoagulant effects of heparin and dicumarol.

Fat-Soluble Vitamins

Vitamin A (Retinol). Vitamin A (retinol) functions in vision as the chromophore (rhodopsin) in the photoreceptors of rods and cones; regulates the biosynthesis of glycoproteins which are common constituents of cell membranes and as such as involved in the maintenance of cell membrane integrity and function; is involved in the synthesis of squalene, a precursor of sterols such as cholesterol and ubiquinone, coenzyme Q which can function in electron transfer in the TCA cycle. Vitamin A also has an integral role in the formation and maintenance of epithelial tissue. Absorption of vitamin A is facilitated by bile acids and fats and is transported in blood by a binding protein. The majority of the body's retinol is stored in the liver. Plant carotenoids can serve as a source of vitamin A. The conversion of carotenoids to retinol requires thyroxine. A complication of hypothyroidism can be night blindness. A deficiency of vitamin A is characterized by night blindness and keratomalacia. The keratomalacia can lead to xerophthalmia, corneal ulceration, secondary infection, and perforation and prolapse of the iris.

The recommended daily allowance of vitamin A is 1 mg or 5,000 I.U. Toxic tissue levels of vitamin A can develop because it is fat soluble and excess amounts of the vitamin are stored in the body rather than being excreted like the water-soluble vitamins. Acute toxicity in adults requires the ingestion of 2–5 million I.U. Signs of acute toxicity may develop in 6–8 h and include severe headaches, dizziness, drowsiness, and nausea with vomiting. Symptoms may continue for several weeks. Chronic toxicity is characterized by fatigue; malaise; headaches, insomnia; restlessness; abdominal discomfort; constipation; bone and joint pain; night sweats; brittle hair and nails; irregular menses; dry, scaly skin; peripheral edema; and exophthalmia.

Vitamin D (Cholecalciferol). Vitamin D is active with calcitonin and parathyroid hormone in the regulation of serum calcium levels; it increases the absorption of dietary calcium, stimulates the synthesis of calcium binding proteins and mobilizes calcium from the bone. Vitamin D is synthesized from steroid precursors present in the skin. The conversion of dehydrocholesterol to cholecalciferol in the skin depends on ultraviolet light intensity and skin color. Melanin blocks ultraviolet light; lighter skins with their lower levels of melanin require less light exposure to produce adequate levels of vitamin D. In addition, milk in the United States is routinely supplemented with vitamin D.

A deficiency of vitamin D is characterized by hypocalcemia and defective calcification of the bones. In vitamin D deficient children a condition known as rickets develops; the condition in adults is known as osteomalacia.

No recommended allowance of vitamin D has been established for adults since individual requirements are influenced by lifestyle and subsequent exposure to sunlight. However, 400 I.U. are recommended for children and women who are pregnant or lactating. With the American fortification program for milk, cereal, and vitamin supplementation, approximately 10% of American children consume greater than 1000 I.U./day and are at risk of hypervitaminosis. Symptoms of acute vitamin D toxicity take several days to develop and include anorexia, nausea, vomiting, diarrhea, headache, polyuria, and polydipsia. In chronic cases of toxicity, calcification of soft tissues, such as lung and kidney, can occur.

Vitamin E (α-Tocopherol). Vitamin E functions as an antioxidant to protect polyunsaturated fatty acids, vitamin C, and cellular phospholipids. Vitamin E is believed to prevent the initiation and propagation of lipid peroxidation. Vitamin E also seems to be involved through an unknown mechanism in the preservation of red blood cell integrity and of muscle tissue structure and function. The recommended daily allowance for vitamin E is 12 and 15 mg for women and men, respectively. Ingestion of large doses of vitamin E in excess of 2–4 g for prolonged periods has been associated with disruption of gonadal function, creatinuria, chapping, cheilosis, angular stomatitis, gastrointestinal disturbances, and muscle weakness.

Vitamin K (Menaquinone). Vitamin K is required for the synthesis of coagulation factors: II (prothrombin), VII, IX, and X. Vitamin K deficiency rarely occurs since it is synthesized by intestinal bacteria. A deficiency state may occur in patients who have a malabsorption syndrome, such as sprue or celiac disease; who are receiving prolonged therapy with broad spectrum antibiotics; who have biliary disease; or who are receiving anticoagulants like dicumarol. Due to its structural similarity, dicumarol is an antimetabolite of vitamin K.

No recommended daily allowance has been established for vitamin K because of its active synthesis by the gut microflora. Hemolytic anemia, hyperbilirubinemia, and kernicterus have occurred in infants after intravenous doses of vitamin K.

Minerals

Minerals that are required in doses greater than 100 mg/kg/day are considered major metals. Trace metals are those minerals that are required in amounts less than a few milligrams per day. Recommended daily allowances have been established for only six minerals: calcium, iodine, iron, magnesium, phosphorous, and zinc.

Minerals participate in many physiologically important activities. They can function as cofactors in metabolic reactions (iron, cobalt), as components of organic compounds (hemoglobin, myoglobin, cytochromes), as ions affecting muscle contractibility (calcium, magnesium), and as minerals in bone structure (calcium, phosphorus, magnesium, fluoride). A detailed discussion of the toxicologic actions of metals is included in Part V, *Neurotoxic Agents*.

Calcium

Although 99% of the total body calcium is present in bones, calcium has a number of other critical physiological functions. Calcium is responsible for maintaining nerve and muscle excitability, muscle contractibility, and normal myocardial function. It also functions in the blood coagulation cascade, as well as serving as a cofactor for enzymes such as adenosine triphosphatase. Circulating calcium is regulated by an intricate interaction between parathyroid hormone, calcitonin, and vitamin D. A calcium deficiency can lead to rickets or osteomalacia depending on whether it occurs in children or adults, respectively.

The recommended daily allowance for calcium is 800 mg.

Phosphorus

Phosphorus, like calcium, is of major importance in bone structure but is also a metabolically important element in the synthesis of phospholipids, glyceraldehyde-3 phosphate, pyridoxal-phosphate, and thiamine pyrophosphate. In addition, phosphorylation is associated with the activation of numerous proteins and energy storage is associated with these phosphorylated compounds (ATP, GTP, UDP).

Phosphorus deficiency is associated with weakness, anorexia, malaise, and bone pain with demineralization of the bones. An intake of phosphorus equal to that of calcium is recommended.

Iron

Iron is an integral part of hemoglobin (which contains 85% of the total body iron), myoglobin, cytochromes, and various metalloenzymes including catalase, peroxidase, and tyrosine and phenylalanine hydroxylases. A deficiency of iron causes a hypochromic, microcytic anemia and can be accompanied by weakness and fatigue.

The recommended daily allowance of iron is 10 mg per day. Excessive dietary intake of iron can result in hemochromatosis. Alcoholics and individuals with chronic liver disease are at risk of developing iron toxicity. The use of iron cooking containers can also lead to excessive iron consumption.

Iodine

Iodine is required to synthesize the hormones thyroxine and triiodothyronine. The thyroid hormones influence all cell metabolism with the exception of the brain. As mentioned above, thyroxine is also required in the conversion of carotenoids to retinol and as such, plays an important role in vision. Thyroxine is also important for normal growth and development; it influences the rate of bone growth, affects gastrointestinal motility, and sets the basal metabolic rate for all cells except brain cells. A congenital deficiency of iodine can result in severe mental retardation. Iodine deficiency in adults can lead to the development of goiters.

The recommended daily allowance of iodine is 100 and 140 µg for women and men, respectively. Diets high in seaweed which contain excessive amounts of iodide can lead to the development of iodide goiter or myxedema.

Magnesium

Like calcium and phosphorus, magnesium is an integral part of the bones. Additionally, magnesium is a cofactor for various enzymes and maintains normal neuromuscular function. A deficiency of magnesium results in impaired neuromuscular functioning, tremors, convulsions, and behavioral disturbances.

Magnesium toxicity may develop when individuals with renal insufficiency are treated with drugs containing magnesium. Infusion of magnesium causes paralysis of skeletal muscles in humans and has been reported to cause respiratory depression, coma, and death in experimental animals.

The recommended daily allowance is 350 mg.

Zinc

Zinc functions as an integral part of such diverse enzymes as carbonic anhydrase, alcohol dehydrogenase, and carboxypeptidase. Zinc is also involved in RNA synthesis. Zinc deficiency is characterized by anemia, hepatosplenomegaly, hypogonadism, hyperpigmentation, and dwarfism.

The recommended daily allowance is 15 mg. Cases of zinc toxicity have been reported in individuals consuming water from galvanized pipes over prolonged periods. Symptoms include irritability, stiffness and pain in the neck and back muscles, loss of appetite, and nausea.

NATURALLY OCCURRING NEUROTOXICANTS

Lathyrism

Lathyrism is a neurological disease caused by the consumption of seeds from *Lathyrus sativus*, *L. cicera*, or *L. clymenum* and is largely restricted to India. During droughts and period of famine, lathyrus plants continue to grow and can become a major food source. The first symptoms of lathyrism include a feeling of heaviness in the legs, muscle tremors, a jerky gait, and a loss of sensory perception. Although several toxic substances have been isolated from *L. sativus* and *L. odoratus,* the toxic agent responsible for human lathyrism is unknown.

Oxalates

Oxalates occur in high concentrations in spinach, Swiss chard, beet tops, lamb's quarters, poke, and rhubarb. The most obvious effects of toxic levels of oxalates is on calcium metabolism. The occurrence of convulsions has been associated with the ingestion of raw rhubarb leaves and stalks.

Pyridoxine Antagonist

Linseed meal contains approximately 100 ppm 1-(n-γ-L-glutamyl) amino-D-proline. The 1-amino-D-proline forms a stable derivative with pyridoxal phosphate, creating a pyridoxine deficiency.

Thiaminase

Bracken fern contains a thiaminase, and when it is ingested by monogastric animals, the animals exhibit symptoms of thiamine deficiency within a few weeks. Fiddlehead greens, which are the sprouts of young ferns, are sometimes consumed by humans, but there have been no reported cases of thiamine deficiency. Thiaminases are present in a number of other foods consumed by humans including blackberries, black currants, red beets, Brussels sprouts and red cabbage. However, the occurrence of these thiaminases is primarily of theoretical interest.

Cycad

Consumption of cycad (palm-like tree) nuts has been reported to be associated with a greater than normal incidence of amyotrophic lateral sclerosis. In Guam, cycad nuts provide an emergency food source, since these are among the few plants that survive droughts and typhoons. The people of Guam have tried to minimize the toxicity of the cycad nut by soaking the kernels in water for a week or more prior to consumption. Cycad neurotoxicity has been reported in Australian cattle consuming the cycad leaves.

Muscarine

Muscarine is an important neurotoxicant occurring in the mushroom, *Amanita muscaria.* Most of the effects of muscarine are associated with stimulation of the parasympathetic ganglionic nerves. Muscarine poisoning is associated with cholinergic signs such as rapid onset of excessive salivation and lacrimation, contracted pupils, nausea, cramps, diarrhea, dizziness, and confusion. Atropine is the specific antidote for muscarine poisoning.

Paralytic Shellfish Poisoning

Toxic levels of paralytic shellfish poison occurs in mussels, clams, scallops, and oysters feeding in areas of growing dinoflagellates. The symptoms of paralytic shellfish poisoning include numbness of the lips and fingertips and, in some cases, death due to respiratory paralysis.

Tyramine

Aged cheese and wine contain large amounts of tyramine, a potent vasopressor substance, that is normally metabolized by monoamine oxidases. In patients receiving monoamine oxidase inhibitors to treat depression, toxic levels of tyramine can develop after the ingestion of aged cheese or wine. When the metabolism of the tyramine is blocked by the monoamine oxidase inhibitors, the tyramine can produce severe hypertension.

REFERENCES

DiPalma, J.R. and D.M. Ritchie (1977). Vitamin toxicity. *Annu. Rev. Pharmacol. Toxicol.* 17: 133–148.

Goodhart, R.S. and M.E. Shils (1980). *Modern Nutrition in Health and Disease,* 6th edition. Lea and Febiger, Philadelphia, PA.

Guttler, F. (1980). Hyperphenylalaninemia: diagnosis and classification of the various types of phenylalanine hydroxylase deficiency in childhood. *Acta Paediatr. Scand. Suppl.* 280: 1–80.

Hayes, K.C. and D.M. Hegsted (1973). Toxicity of vitamins. In *Toxicants Occurring Naturally in Foods,* National Academy of Sciences, Washington, D.C., pp. 235–253.

Lipton, M.A., R.A. Mailman, and C.B. Nemeroff (1979). Vitamins, megavitamin therapy and the nervous system. In *Nutrition and the Brain,* R.J. Wurtman and J.J. Wurtman, Eds., Raven Press, New York, pp. 183–264.

Scriver, C.R. and C.L. Clow (1980). Phenylketonuria and other phenylalanine hydroxylation mutants in man. *Annu. Rev. Genet.,* 14: 179–202.

Shoden, R.J. and W.S. Griffin (1980). *Fundamentals of Clinical Nutrition.* McGraw-Hill, New York.

Williams, R.D. (1974). *Essentials of Nutrition and Diet Therapy.* C.V. Mosby, St. Louis, MO.

Alcohol and Neurotoxicology

Mohamed B. Abou-Donia
Departments of Pharmacology and Neurobiology
Duke University Medical Center
Durham, North Carolina

INTRODUCTION

Ethanol or ethyl alcohol has numerous and varied applications. It is used as a beverage, and as a solvent for perfumes, aftershaves, and cologne. Cough preparations may contain up to 20% ethanol and mouth washes and some rubbing alcohols contain ethanol as well. Distilled spirits (80–100 proof) typically contain 40 to 50% ethanol by volume; wines (20–40 proof) range from 10–20%, averaging 12%; beer (4–12 proof) contains 2 to 6%; mouth washes up to 75%; and colognes from 40–60%.

In the United States alcohol abuse represents a major social problem. An estimated 10 million persons are alcohol-dependent and approximately 200,000 deaths a year are alcohol-related including 70% of fire deaths, 67% of drownings, 67% of murders, 35% of suicides, and a large number of deaths from liver failure (Lewis, 1980).

DISPOSITION AND METABOLISM OF ETHANOL

Absorption
Approximately 80% of ingested ethanol is absorbed by the small intestine, and the remainder is absorbed from the stomach. Absorption of ethanol from the gastrointestinal tract may be delayed by ethanol concentration, e.g.,

beer slows absorption; gastric contents, e.g., high protein and large volumes decrease absorption; and gastrointestinal motility. These factors delay peak blood level of ethanol. In adults, absorption of 80% of ingested ethanol occurs within 30–60 min; food may delay absorption for 4–6 hours. Absorption of ethanol follows zero-order kinetics (Wilkinson, 1980).

Tissue Distribution

Because it is soluble in water and lipids, ethanol is distributed into body water and penetrates the blood-brain barrier and placenta. The volume of distribution (V_d) of ethanol is 0.6 L/kg body weight in men and 0.7 L/kg in children. Women have a smaller V_d because they have more fat and less water than men (Wilkinson, 1980).

Elimination

Only 5–20% of an ethanol dose is excreted as the parent compound in urine, sweat, and breath (Bogusz et al., 1977). An average-sized man can metabolize and eliminate 10 ml of ethanol per hour which is equivalent to the ethanol present in a 12-oz. can of beer (4.5% by volume), a 4-oz. glass of wine (12%) or a scant jigger (1.2 g) of whiskey (45%). Thus, if one drinks a six pack of beer (60 ml ethanol) in one hour, ethanol is eliminated according to zero-order kinetics where the rate of elimination remains constant and is independent of its concentration in the body as follows:

	Hour					
	1	2	3	4	5	6
Ethanol eliminated (ml)	10	10	10	10	10	10
Ethanol remaining (ml)	50	40	30	20	10	0
Ethanol eliminated (%)	17	20	25	33	50	100
(% of total ethanol)	(10/60)	(10/50)	(10/40)	(10/30)	(10/20)	(10/10)

Metabolic Biotransformation

The major metabolism of ethanol takes place in the liver (Lieber, 1984). The brain's ability to metabolize ethanol is very slight. Ethanol is metabolized to carbon dioxide and water by three main enzymatic systems present in hepatocytes.

Alcohol Dehydrogenase (ADH)

This enzyme is present in the cytosol (soluble fraction) of liver homogenate. ADH has a low K_m for ethanol (approximately 1.0 mM) and is responsible for 80–85% of its metabolism. This enzyme is not inducible by ethanol or other factors.

$$C_2H_5OH + NAD^+ \xrightleftharpoons[Zn^{2+}]{Alcohol\ dehydrogenase} \underset{Acetaldehyde}{CH_3-\overset{O}{\overset{\|}{C}}-H} + NADH + H^+$$

Ethanol

Acetaldehyde formation is the rate-limiting step.

$$\underset{Acetaldehyde}{CH_3-\overset{O}{\overset{\|}{C}}-H} + H_2O + NAD^+ \xrightleftharpoons[dehydrogenase]{Acetaldehyde} \underset{Acetic\ acid}{CH_3-\overset{O}{\overset{\|}{C}}-OH} + NADH + H^+$$

Both reactions result in the decrease in the NAD/NADH ratio, which may affect normal enzymatic reactions dependent on these cofactors. Acetate is metabolized to carbon dioxide and water through the tricarboxylic acid cycle. Thus alcohol supplies calories, without supplying vitamins or amino acids, depressing the appetite and resulting in nutritional deficiency.

Microsomal Ethanol-Oxidizing System (MEOS)

MEOS is present in microsomal fractions of liver and can oxidize ethanol in the presence of O_2 and NADPH. It has a high K_m for ethanol in the range of 10 mM and is responsible for 10–15% of its metabolism. MEOS is distinct from cytochrome P-450. This system is inhibited by carbon monoxide, but unlike cytochrome P-450 is not inhibited by SKF 525A or by pyrazole. Chronic ingestion of ethanol in rats increased MEOS activity, hepatic microsomal proteins, smooth endoplasmic reticulum, and cytochrome P-450.

Peroxidase-Catalase System

This enzymatic system, present in peroxisomes in the liver, plays a minor role in oxidizing ethanol *in vivo*.

ACUTE TOXICITY OF ETHANOL

Dose Level

The lethal dose of ethanol is 5–8 g/kg of body weight for adults and 3 g/kg for children. The legal blood level for intoxication (drunkenness) is 100 ml ethanol/100 ml blood, which may be expressed as follows: 0.02 M, 1.0 mg ethanol/mg blood, or 0.1%. Death results from respiratory depression at ethanol concentrations exceeding 500 ml/100 ml blood.

Ethanol Action on the Central Nervous System

Ethanol produces its acute effects by depressing the central nervous system (Davis and Lipson, 1986). Its initial effect is a selective depression of the reticular activating system. This action results from ethanol interference with Na^+ transport not at the synapse but at the cell membrane. Ethanol affects the frontal lobes, resulting in alteration of thought and mood; the occipital lobe, causing vision changes; and the cerebellum, producing incoordination.

Signs of Acute Intoxication

Reduced attention, concentration, motor coordination, and reaction time result at blood concentrations of ethanol that have no effect on visual and verbal memory (Taylor and Hudson, 1977). Initially, ethanol causes exhilaration, but progresses to loss of restraint, behavioral abnormalities, slurred speech, ataxia, irritability, drowsiness, and finally coma. These central nervous system symptoms may be accompanied by flushed face, dilated pupils, excessive sweating, and gastrointestinal distress. Acute ingestion of ethanol causes arrhythmias and increases both systolic and diastolic pressures.

Chronic ingestion of ethanol (alcoholism) results in serious, sometimes fatal side effects.

Metabolism Disorders

These disorders are related to decreased NAD/NADH ratio, resulting in alterations in the cellular reduced state. Changes include

1. acidoses resulting from lactate formation instead of pyruvate
2. increased serum uric acid
3. accumulation of hepatic triglycerides
4. decreased gluconeogenesis
5. increased collagen disposition
6. depression of protein synthesis.

Nutritional Deficiency

Chronic alcoholism causes a deficiency in vitamin B_1, B_6, and B_{12} and the trace element Zn with resultant alcoholic beriberi and Wernicke's encephalopathy.

Lung Cancer

A prospective epidemiologic study suggests a significant positive correlation between ethanol consumption and the incidence of lung cancer (Pollack, 1984).

Lungs

Chronic uptake of ethanol produces aspiration, pneumonia, and tuberculosis.

Zieve's Syndrome

Prolonged consumption of ethanol results in Zieve's syndrome, a disorder characterized by jaundice, hyperlipidemia, and hemolytic anemia.

Nervous System

Ethanol is a central nervous system depressant. It causes decreased motor skills and cognition, peripheral motor/sensory neuropathy, and Wernicke-Korsakoff syndrome.

Endocrine System

Chronic uptake of ethanol decreases release of testosterone and oxytocin and increases secretion of cortisol, aldosterone, and insulin.

Cardiovascular System

Chronic ethanol toxicity results in cardiomyopathy and cardiac beriberi (Rich et al., 1985).

Gastrointestinal Tract

Alcoholism results in esophageal varicose, gastritis stomach, ulcerations and atrophy, small intestinal malabsorption, motility changes, pancreatitis, fatty liver, hepatitis, cirrhosis, and hepatic failure.

Muscle

Acute and chronic myopathy and myoglobinuria result from chronic ethanol uptake.

Hematologic Abnormalities

Chronic consumption of ethanol causes suppression of bone marrow, and megaloblastic hematopoiesis accompanied by increased mean corpuscular volume. Hyperfermia, interference with iron utilization, and thrombocytopenia are also produced.

Bone

Chronic ethanol use decreases density of vertebral callous bone.

Wernicke-Korsakoff Syndrome

Ethanol reduces the active gastrointestinal transport of thiamine (vitamin B$_1$) and decreases thiamine activation and storage in the liver (Reuler et al.,

1985). The resulting thiamine deficiency decreases levels of the coenzyme thiamine pyrophosphate, disrupting intermediate carbohydrate metabolism, and also changes membrane structure of the nervous system, a condition known as Wernicke's encephalopathy. This condition is characterized by ocular abnormalities, ataxia, vestibular paresis, global convulsion, hypotension, hypothermia, and coma. Ocular changes (usually reversible) include diplopia, blurred vision, mystagmus, and lateral gaze palsy. Ataxia results from a polyneuropathy and central dysfunction. Tremor often occurs.

The majority of individuals with Wernicke's encephalopathy (80%) develop Korsakoff's psychosis. This condition is characterized by anterograde and retrograde amnesia. Anterograde amnesia is the inability to assimilate new formation and retrograde amnesia is the inability to recall information.

Fetal Alcohol Syndrome

The safe level of ingested ethanol during pregnancy has not been determined. Numerous studies have suggested that in large doses, i.e., 400 g daily, ethanol is both teratogenic and fetotoxic (Streissguth et al., 1980). Babies born to mothers who consume large quantities of ethanol during pregnancy not only experience withdrawal after delivery, but also may suffer permanent retardation. Many infants with birth abnormalities have been born to alcoholic mothers. These children are characterized by specific facial abnormalities known as "fetal alcohol syndrome" (Clarren and Smith, 1978). Children with this syndrome exhibit severe growth deficiencies and are intellectually handicapped. The infants are small and show muscular incoordination. Their facial features are characterized by shortened palpebral fissures, epicanthal folds, broadened nasal bridge, upturned nose, and thinned upper lips.

Genetic Factors

Studies in families indicate that genetic factors control alcohol drinking behavior (Vesell et al., 1971). Ethanol is metabolized in some individuals to 2,3-butanediol resulting from condensation of two molecules of acetaldehyde, an ethanol oxidation metabolite. The presence of atypical liver alcohol dehydrogenase is related both to interracial and interindividual differences, e.g., 85% of Japanese and 5–20% of Europeans possess this atypical enzyme. Individuals with high alcohol dehydrogenase levels may be genetically predisposed to an inborn disulfiram effect resulting from excess production and accumulation of acetaldehyde in the blood (Wolff, 1972).

Acetaldehyde decreases protein synthesis, interferes with microtubule formation, impairs mitochondrial functions, and causes flushing (a reaction to disulfiram). Chronic acetaldehyde levels may cause peroxidation of the cellular membrane. High concentrations of acetaldehyde result from the oxidation of ethanol with alternate microsomal pathway, i.e., MEOS.

TOLERANCE TO ETHANOL

Drug tolerance is the decrease of its effect following a period of administration of that drug. Drug tolerance may be dispositional or pharmacodynamic (functional).

Dispositional Tolerance

In dispositional tolerance less of the drug reaches the site of action, leading to reduced effect (Lieber and DeCarli, 1970). Chronic ethanol use results in an increase of its metabolism, shorter half-life in the blood and less ethanol reaching the brain. There is little evidence that ethanol or any other chemical induces alcohol dehydrogenase activity. However, liver microsomal MEOS and cytochrome P-450 activities are increased following prolonged consumption of ethanol. In general, this type of tolerance is not significant and accounts for only a 30–50% decrease in ethanol effects.

Pharmacodynamic Tolerance

In functional tolerance there is a decrease in the sensitivity of the brain to a given concentration of ethanol (LeBlanc et al., 1973). This type of tolerance may be considered an adaptive mechanism.

Cross-Tolerance

Cross-tolerance takes place when exposure to one chemical reduces the sensitivity to another. Metabolic cross-tolerance is common because the liver microsomal cytochrome P-450 system responsible for the metabolism of many drugs may be induced by administration of many chemicals. Chronic uptake of ethanol induces the activity of MEOS and liver microsomal cytochrome P-450. This activity produces metabolic cross-tolerance to other chemicals metabolized by these enzymes, e.g., barbiturates, which are metabolized by MEOS. Enzyme induction is not an important mechanism for cross-tolerance of ethanol by other chemicals because MEOS plays a minor role in ethanol metabolism and alcohol dehydrogenase, the major enzyme for ethanol metabolism, is not known to be induced.

An effect opposite of cross-tolerance may result if two drugs have a common metabolic system. Concurrent use of both chemicals may block their metabolism by competing for the same enzyme, resulting in an increase in the effects of both drugs. An example of this interaction occurs between ethanol and acetaminophen (Sato et al., 1981). Acetaminophen is metabolized to a potent hepatotoxicant which is blocked by ethanol competing for the same enzyme sites. This results in less liver damage induced by acetaminophen in the presence of ethanol. In contrast, chronic consumption of ethanol induces microsomal enzymes and enhances the metabolic activation of acetaminophen and its hepatotoxicity.

Functional cross-tolerance may also develop between ethanol and other drugs such as halothane. This may be related to a change in membrane lipids, leading to less chemical crossing these membranes and less present at the site of action. This may be considered a molecular form of dispositional tolerance.

Physical Dependence

Chronic use of ethanol results in a marked physical dependence that leads to actual physical illness after termination of drinking (Goldstein, 1974). Physical dependence may develop rapidly, sometimes following a single dose of ethanol. Benzodiazepines are the drugs of choice for treatment of alcohol withdrawal reactions (Kaim et al., 1969). These drugs have replaced barbiturates and paraldehyde which may have dangerous side effects.

Disulfiram (Antabuse)

Disulfiram or tetraethylthiuram is used in treatment of alcohol abuse. Disulfiram taken daily does not have adverse effects. If ethanol is ingested several hours after taking disulfiram, the person develops nausea, vomiting, flushing, palpitation, heart rate increases, and headache (Asmussen et al., 1948). Later stages of ingestion are characterized by dyspnea, hyperventilation, nausea, vomiting, and a sharp fall in blood pressure. The mechanism of disulfiram action is the inhibition of aldehyde dehydrogenase and the accumulation of acetaldehyde in the blood.

$$
\begin{array}{ccccc}
CH_3\text{–}CH_2 & & S & S & CH_2\text{–}CH_3 \\
& \diagdown & \parallel & \parallel & \diagup \\
& & N\text{–}C\text{–}S\text{–}S\text{–}C\text{–}N & & \\
& \diagup & & & \diagdown \\
CH_3\text{–}CH_2 & & & & CH_2\text{–}CH_3 \\
\end{array}
$$

Disulfiram

Disulfiram is usually administered in a dosage of 1–2 g the first day, decreased gradually to 0.5 g in four days, then to 0.25 g as a maintenance dose.

Inhibitors of Aldehyde Dehydrogenase

Drugs or chemicals that inhibit aldehyde dehydrogenase in the presence of ethanol produce disulfiram-like reactions resulting from the accumulation of acetaldehyde. These include drugs such as sulfonamides, metronidazole, and hypoglycemia agents; mushrooms (*Coprinus atramentariuis*); industrial chemicals (amides and oximes for example), pesticides including carbamates, dithiocarbamates, and thiram derivatives (Hills and Venable, 1982).

Central Nervous System (CNS) Drugs

Simultaneous ingestion of ethanol and sedative-hypnotics results in CNS depression. Also, concomitant consumption of ethanol and aspirin tablets leads to prolonged bleeding time (Deykin et al., 1982). Impaired performance induced by ethanol is not significantly improved by either caffeine or dextroamphetamines.

MECHANISMS OF ETHANOL TOXICITY TO THE CENTRAL NERVOUS SYSTEM

Ethanol affects the central nervous system more rapidly and severely than any other organ or system in the body. It disturbs CNS functions and causes emotional and psychic disorders. Because of its lipid solubility, ethanol acts on biological membranes. It changes the lipid components of cell membranes into a more fluid state that may result in their disruption. Ethanol also increases membrane resistance and charge ion fluxes, resulting in alteration in the action potential. Changes of membrane proteins which include ion channels, carriers, membrane-bound enzymes, etc., may underlie ethanol action on synthesis, utilization, metabolism, and release of neurotransmitters. Studies of molecular mechanisms in ethanol action on the central nervous system are conducted to elucidate symptoms of ethanol dependence syndrome, including tolerance, withdrawal, craving, and positive reinforcement.

Animal Models for Ethanol Neurotoxicity

To study alcoholism in experimental animals, various methods are used to force chronic ingestion of ethanol, i.e., alcoholization in animals. These methods include

1. depriving animals of food prior to alcohol consumption
2. making ethanol more palatable by adding saccharine, glucose, etc.
3. gastric intubation
4. intravenous infusion, and
5. exposure to ethanol vapor.

Although these methods are useful in investigating the toxicological actions of chronic uptake of ethanol, they do not address the development of ethanol consumption motivation, tolerance, or physical dependence. Alcohol motivation studies are carried out on noninbred rats by evaluating their consumption rate of 15% ethanol solution in water when having a choice between ethanol and water. These studies showed that 52% of the animal population consumed less than 20 ml/kg of 15% ethanol solution/day (Burov and Kampov-Polevoi 1985). The rest, 48%, had alcohol motivation sufficient to induce consumption of more than 20 ml/kg/day of the ethanol solution. About half of this group exhibited intense alcohol motivation and consumed more than 40 ml/kg/day. The remaining half had a daily intake of 20 to 40 ml/kg/day of 15% ethanol solution.

Effect on Monoamine Oxidase Activity

Monoamine oxidase (MAO, EC 1.4.3.4) mainly occurs in the outer mitochondrial membrane. There are two forms of this enzyme: MAO-A and MAO-B (Fowler et al., 1982). The form MAO-A is inhibited by the acetylenic inhibitor clorgyline. In humans, serotonin (5-HT) and norepinephrine (NE) are substrates mainly for the A-form, while dopamine (DA) is an equally preferred substrate for both forms of the enzyme. Furthermore, MAO-B is inhibited by 1-deprenyl. While both forms are present in the brain, only the B-form occurs in human platelets, which is similar to MAO-B in the brain. Recent studies have shown their MAO-A inhibition in rat brain results in significant changes in the levels of 5-HT and DA.

Effect of Chronic Ingestion of Ethanol on Brain MAO Activity

Low MAO activity was reported in the hypothalamus and caudates of alcoholics (Oreland et al., 1983). No changes were found in the cortex gyrus unguilli, or hippocampus. Similar results were found for platelet MAO activity. These results are consistent with the hypothesis that low platelet activity is associated with a compromised brain monoaminergic system. Experimental studies with rats, either by adding ethanol to drinking water for nine months or following voluntary consumption, revealed no significant effect on either form of the enzyme in the brain.

Relationship Between Platelet MAO-B Activity and Alcoholism

Human studies have suggested platelet MAO-B activity as a biological marker for high risk paradigm. Low platelet MAO-B individuals exhibited instability, and psychiatric disturbances. They experienced more convictions, experimented more with illegal drugs, and reported more familial mental health problems than subjects with high platelet MAO-B (Revely et al., 1983). These problems are especially evident with respect to depression, alcoholism, and attempted suicide.

Numerous studies have shown that humans with low platelet MAO activity are sensation seekers (Wiberg et al., 1977). These subjects are usually cigarette smokers and are often chronic alcohol users. They experiment with glue and cannabis (marijuana), and have tried a variety of illegal drugs

(Stillman et al., 1978). Alcoholics usually have low MAO activities in their brains. During abstinence, however, there is a transient increase in platelet MAO activity that returns to its previous low level after the abstinence phase. Relatives of alcoholics have low levels of platelet MAO activity as well. Platelet MAO activity is genetically controlled. Certain personality traits are common in individuals with low platelet MAO activity and with chronic consumption of ethanol, e.g., sensation seeking, impulsiveness, and aggressiveness. Family studies have shown an increased incidence of alcoholism in individuals related to alcoholics; however, the relative roles of genetics and environment are not clear.

Biosynthesis of Catecholamines

Catecholamines are biosynthesized in the brain, chromaffin cells, sympathetic nerves, and sympathetic ganglia (Figure 1). Their precursor amino acid is tyrosine that is normally present in circulation in a concentration of 5 to 8 $\times$ 10^{-5} M. L-Tyrosine is hydroxylated to dopa via tyrosine hydroxylase. Aromatic amino acid decarboxylase catalyzes the decarboxylation of tyrosine and dopa to tyramine and dopamine, respectively. These two compounds are subsequently hydroxylated via dopamine-β-oxidase to octopamine and norepinephrine (NE), respectively. NE is also formed by the hydroxylation of octopamine via catechol-forming enzyme which also catalyzes the hydroxylation synephrine to epinephrine. Synephrine results from the methylation of octopamine using phenylethanolamine-N-methyl transferase that also catalyzes the formation of epinephrine from epinine NE and its subsequent methylation to N-methyl epinephrine. Dopamine is methylated by N-methyl transferase to form epinine which is then oxidized to epinephrine using dopamine-β-oxidase.

Effect of Chronic Ethanol Ingestion on
CNS Norepinephrine (NE) Metabolism

Animal studies have reported increased turnover of the central nervous system NE after acute or chronic alcohol ingestion. Depletion of the dorsal noradrenergic pathways was also reported in animals after heavy consumption of ethanol (Liljequist, 1979). Tolerance seems to be related as well to the central adrenergic activity, since more intoxication was produced following the depletion of NE in animals.

Studies on the effect of ethanol consumption on the central nervous system noradrenergic metabolism in living individuals were conducted by measuring the metabolites of NE in the lumbar cerebrospinal fluid (CSF). Studies have determined that the concentration of 3-methoxy-4-hydroxyphenylethyleneglycol (MOPEG or HMPG) is a good reflection of the activity of the central noradrenergic system (Chase et al., 1973). An increase in tissue MOPEG concentration resulted from an elevated release of NE due to increased functional activity of the central NE neurons. The level of MOPEG

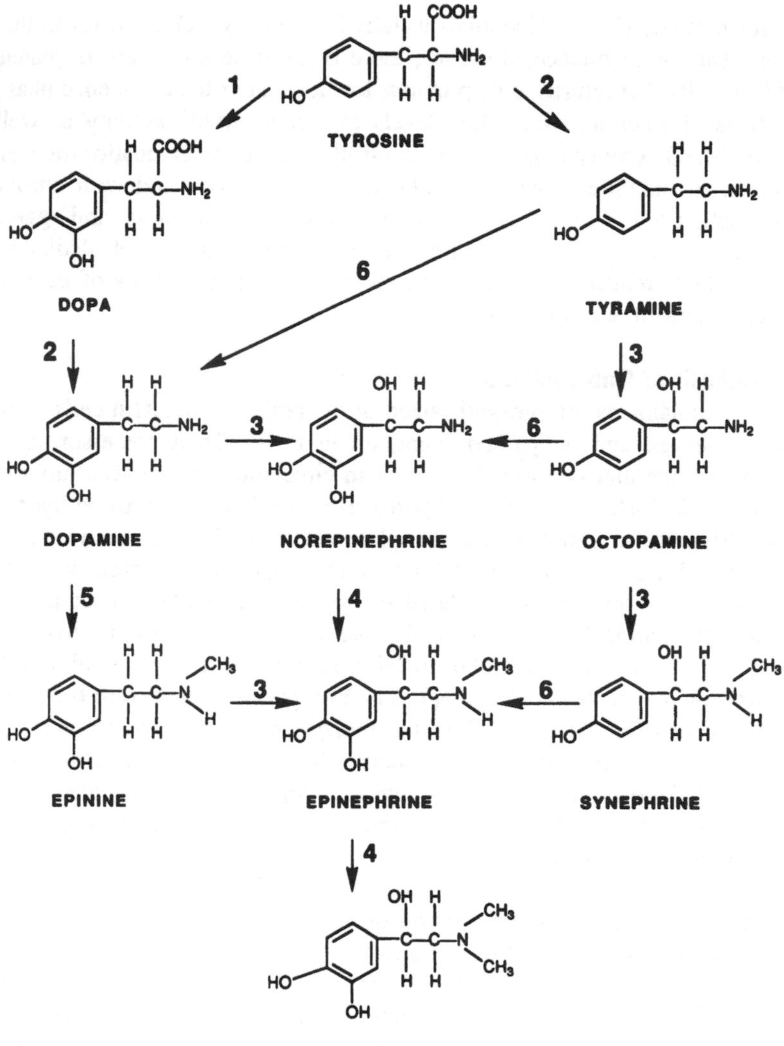

Figure 1. Biosynthetic pathways of catecholamines. Enzymes involved in these reactions are: 1, tyrosine hydroxylase; 2, aromatic amino acid decarboxylase; 3, dopamine-β-oxidase; 4, phenylethanolamine-*N*-methyl transferase; 5, nonspecific *N*-methyl transferase in lung, folate-dependent *N*-methyl transferase in brain; 6, catechol-forming enzyme.

was measured in the CSF of alcoholics and healthy controls during alcohol consumption, acute withdrawal, and after abstinence of one, three, and six weeks (Borg et al., 1983). High levels of MOPEG in CSF were measured during intoxication. These levels were lower during withdrawal than during intoxication, but were still significantly higher than those in healthy controls.

Similar results were obtained in healthy volunteers given alcohol. These results support the hypothesis that ethanol intake increases central noradrenergic turnover. It is known that the alpha-agonist Clonidine, which diminishes noradrenergic turnover, is effective in treating acute withdrawal in alcoholics (Björkqvist, 1975).

Effect of Ethanol on Catecholamines

Ethanol behavioral effects have been linked to its action on the release and destruction of the brain's neurotransmitter norepinephrine. It has been hypothesized that prolonged ethanol uptake causes excess release of NE in the hypothalamus and midbrain, leading to the phase of psychic, vegetative, and motor irritation (Anokhina and Kogan, 1975). This phase is followed by diminished NE and dopamine (DA) concentrations, resulting in suppression of the activity of those brain regions and leading to the phase of psychic and motor inhibition. By contrast, consumption of a moderate ethanol dose increases release of NE, improving certain functions of CNS for a short period. However, the NE released is rapidly metabolized resulting in decreased NE in the CNS. This temporary improvement leads to more drinking resulting in more psychic alcohol dependence.

When ethanol consumption is terminated during abstinence, catecholamine destruction becomes normal. Catecholamine synthesis remains intensified, resulting in an increase in DA content in the CNS. The current concepts assume that the development of psychotic states is related to disturbed metabolism of DA in the CNS. Ethanol ingestion intensifies the circulation of DA and its acid metabolites, i.e., HUA and DOPAC (Karoum et al., 1976).

When blood DA content was determined in alcoholics, the following increases were evident, as compared to normal subjects: 48% in patients with no abstinence symptoms, 108% in alcohol abstinence, 114% in predelirium, and 400% in alcohol delirium.

Alcohol craving was accompanied by an elevated DA content in the blood of alcoholics during remission. Concurrently, there was a decrease in DA excretion in urine. These results suggest the presence of some factors preventing DA elimination and contributing to its increased blood and CNS concentration. The role of increased DA concentration in alcoholics was studied using low doses of DA receptor stimulators such as apomorphine and bromocriptine. These chemicals regulate DA circulation primarily by affecting the presynaptic receptors. Administration of apomorphine decreased DA content in patient's blood, and arrested abstinence states (Balsara et al., 1982). Apomorphine positively treated alcohol cravings and most somatoneurological and psychopathological symptoms of abstinence. Apomorphine normalized DA concentration in the hypothalamus and midbrain of rats following chronic uptake of ethanol. Also, bromocriptine decreased alcohol consumption in experimental animals accompanied by normalization of the DA content in midbrain and hypothalamus (Bitran and Bustos, 1982). In some patients with alcohol abstinence syndrome, effective treatment was achieved by using charcoal to actively absorb DA, while the concentration of other neurotransmitters in the blood remained unchanged after hemosorption. These results show a direct correlation between DA concentration (and subsequent brain dopaminergic functions), and alcohol craving, psychic and physical alcohol dependence.

Studies on the effect of ethanol on enzyme system regulating DA may shed some light on the mechanism by which ethanol affects DA concentration. These enzymes are dopamine-β-hydroxylase (DBH) which affects norepinephrine anabolism and monoamine oxidase (MAO), the major enzyme responsible for catecholamine degradation. In alcohol intoxication there is an increase in DBH activity in blood and brain, while MAO levels decrease.

Effect of Ethanol on Opiate System

Patients with alcohol delirium exhibited an increase in β-endorphin content in the CSF and the blood. This increase may result in effects such an analgesia, poor appetite, irritability, and loss of self-control following high uptake of ethanol (Anokhina et al., 1981). These endorphinergic actions may explain the hallucinations and severe convulsions seen in alcohol delirium stage. These studies suggest that the brain's opiate system is involved in the pathogenesis of alcoholism.

Chronic ingestion of ethanol causes a significant decrease of metenkaphalin content in the rat midbrain and striatum. On the other hand, acute administration of ethanol caused an activation of enkephalinase A in the rat midbrain and hypothalamus greater than in the case of chronic uptake of ethanol (Schultz et al., 1980).

Since the activation of enkephalinergic system produces a stimulating effect on the synthesis and release of dopamine in the brain, it is reasonable to expect that the development of resistance to ethanol in opiate systems to the occurrence of ethanol resistance in the dopaminergic system following chronic alcohol uptake. These studies suggest that chronic use of ethanol results in changes in the brain's opiate system.

Effect of Ethanol on the Dopaminergic System

Several reports have suggested the involvement of brain dopaminergic system in the acute effects of ethanol, tolerance, dependence, and withdrawal syndrome.

Acute Effect of Ethanol

Ethanol has dual effects on the central nervous system: stimulation and depression. Low doses of ethanol produce behavioral stimulation, while large doses have a sedative action. These effects have been demonstrated both in experimental animals and in humans.

Low doses produce behavioral stimulation and euphoria in humans. In animals low ethanol doses result in locomotor stimulation, Y-maze running, avoidance acquisitions and electric self-stimulation. Ethanol-induced behavioral stimulation results from actions in the catecholaminergic system. This suggestion is supported by the finding that low doses of ethanol in rats and mice were antagonized by pretreatment with α-methyltyrosine (α-MT), a specific inhibitor of catecholamine synthesis. Also, small doses of L-dopa, which do not affect motor activity in control animals, partially antagonized the effect of ethanol. Similarly, α-MT antagonized the stimulatory effect of small doses of ethanol in humans (Engel and Carlsson, 1977).

The relative role of noradrenergic and dopaminergic systems in the mechanism of ethanol-induced stimulation of the CNS is not quite understood. Both the α-adrenergic receptor antagonist phenoxybenzamine and the DA receptor agonist pimozide antagonize locomotor stimulation induced by ethanol in rats (Engel and Liljequist, 1983). Furthermore, clonidine, an inhibitor of noradrenergic neurons by an action on autoreceptors, antagonized ethanol-induced motor stimulation in mice (Ahlenius et al., 1973). Similar action was produced by small doses of apomorphine, which inhibits dopaminergic neuronal activity by acting on DA autoreceptors (DiChiara et al., 1976).

Increased GABAergic activity has been credited for the depression and sedation produced by large doses of ethanol. This suggestion was supported by the finding that muscimol, a GABA mimetic drug, increased duration of sleeping time induced by large doses of ethanol. Also, both picrotoxin and bicuculline (specific GABA antagonists) reversed ethanol-induced sleep (Mohler and Okada, 1977).

Several studies have been carried out to investigate the role that DA plays in the acute effects of ethanol. While the activity of dopaminergic neurons is increased by low doses of ethanol, large doses decrease their activity (Walker et al., 1980). Similarly, while low doses of ethanol increase the synthesis of DA, large doses decreased it. Although the mechanism by which low ethanol levels increase DA activity is not known, it was suggested that it may result from the removal of the GABAergic inhibitory control on DA neurons. This suggestion is consistent with the reduction of GABAergic activity by ethanol. Thus GABAergic agonists would be expected not only to stimulate ethanol action on motor activity, but also enhance DA synthesis (Cott et al., 1976). This is in agreement with the finding that low doses of ethanol decreased GABA levels in all parts of the brain except the pons. It is also in harmony with the decrease in GABA synthesis in the caudate nucleus after low ethanol doses.

In contrast, large doses of ethanol resulted in an increase in GABA concentration in the brain (Sytinsky et al., 1973). This increase may explain the inhibitory effect of high ethanol doses on DA neurons.

Chronic Effect of Ethanol

Chronic ethanol treatment alters synaptic and postsynaptic dopaminergic functions. In ethanol-dependent rats, there is a massive release of DA, but ethanol does not stimulate DA synthesis (Fadda et al., 1980). In rats, dependence on ethanol develops rapidly following three to nine days of uptake. Withdrawal signs are characterized by CNS hyperexcitability, including tremors, hyperactivity, hyperirritability, spasticity, and convulsive seizures. The severity of these signs depends on the dose, frequency, and duration of ethanol administered.

Chronic ethanol treatment does not activate DA synthesis in the brain, unlike acute ethanol administration. These results suggest that both acute and chronic ethanol may increase the release of DA from stores; however, only the acute administration of ethanol increases DA synthesis, which compensates for its increased release and results in no change in DA levels.

Tolerance to ethanol results after it develops tolerance to the stimulant effect on DA synthesis, while the DA-releasing action becomes more intense and long-lasting (Hoffmann et al., 1980).

It has been proposed that ethanol abstinence syndrome is mediated by a release of DA in the brain. This hypothesis is in agreement with the attenuation of ethanol withdrawal convulsions in mice with systemic L-dopa and intracerebrally injected DA and the increase convulsions with haloperidol, a DA receptor blocker (Blum et al., 1976).

Effect of Ethanol on Cerebrovascular and Cardiovascular System

Morbidity and mortality are significantly higher among chronic users of alcohol; almost 50% of the increased deaths result from circulatory diseases (Ashley, 1982). Chronic ingestion of alcohol is related to many cardiovascular diseases: high blood pressure, cardiac arrhythmias, ischemic heart disease, angina, cardiomyopathy (degeneration of heart muscle), atherosclerosis, cerebrovascular diseases, strokes, and vascular-related headaches.

Some epidemiologic studies have suggested that moderate daily consumption of alcohol may be beneficial to the coronary circulation (Turner et al., 1981). Several mechanisms are possible to explain the protective action of small amounts of alcohol against ischemic diseases and cardiac arrhythmias.

1. Increased high-density lipoproteins (HDL) by ethanol resulting in anti-atherogenic effect.
2. Decreased blood coagulation by ethanol.
3. Ethanol-induced coronary vasodilators actions.
4. Prevention of coronary spasms, which are involved in the development of ischemic heart disease and angina, either by an effect on calcium entry or by vasodilation.

Alcohol intoxication in humans produced 30% increase in cerebral blood flow (CBF) and 30% decrease in cerebral oxygen consumption ($CMRO_2$). However, both CBF and $CMRO_2$ were severely depressed for several weeks in patients with Wernicke-Korsakoff syndrome (Berglund, 1981). Chronic alcoholics usually exhibit significant reduction in blood flow to the gray and white matter as well as to the frontal and parietal regions of the brain. These results are consistent with the confusion characteristic of Korsakoff's psychosis. They are also in agreement with blackouts and functional neuronal deficits seen in alcoholics (Berglund, 1981).

It is noteworthy that heavy alcohol ingestion predisposes some individuals to strokes and sudden death. Also, young middle-aged men, unlike women, seem to be very sensitive to alcohol-induced cerebral infarctions. Ethanol action on cerebral circulation seems to result in hypoxia (Goldman et al., 1983). This is consistent with the hallucinations often observed in cases of Korsakoff's syndrome.

Recent studies have shown that ethanol has spasmogenic actions on cerebral blood vessels. These effects may be related to high incidence of hypertension in chronic alcoholics, high incidence of cerebrovascular accidents and sudden death following excess drinking (Hillbom et al., 1983).

Effect of Ethanol on Thiamine Metabolism

Humans are unable to synthesize thiamine, hence it has to be supplied in the diet. Thiamine is synthesized by plants. It is involved in carbohydrate metabolism and consequently calorie intake. Its daily requirement is 120 μg in humans and it was estimated in rats to be 4 mg/kg/day (Rogers, 1979). The rate at which thiamine is lost from brain may be as high as 1 nmole/g brain tissue/day. Thiamine occurs in equilibrium with its three phosphorylated states: thiamine monophosphate, thiamine diphosphate, and thiamine triphosphate as shown in Figure 2.

Thiamine Transport

Transport Across Biological Membranes. Thiamine crosses biological membranes mostly by carrier mediated transport. This transport system is characterized by: (1) stereospecificity of carrier proteins; (2) saturation; (3) competition with other related chemicals for transport; (4) active transport, i.e., inhibition by metabolic inhibitors; and (5) sodium-dependent.

Transport Across the Gastrointestinal Wall. Thiamine is most absorbed in the upper small intestine where intestinal alkaline phosphatase catalyzes the dephosphorylation of phosphorylated thiamine (Schaller and Holler, 1975). This absorption proceeds against a concentration gradient.

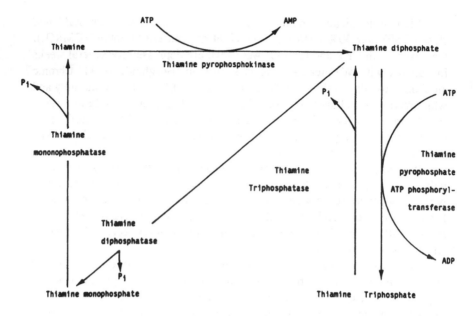

Figure 2. Phosphorylation and dephosphorylation reactions of thiamine.

Transport Into the Central Nervous System. The transport of thiamine into brain cells is a carrier-mediated process (Sharma and Quastel, 1965). Similarly, thiamine crosses the blood-brain barrier through a carrier-mediated process that is saturable with a minor nonsaturable component. The uptake of thiamine by cells occurs as an energy-dependent process of active transport that moves against concentration gradient.

The Significance of Thiamine in the Central Nervous System

Although the brain accounts for only 2% of the total body weight, it uses approximately 20% of the body's total oxygen requirement. Thiamine-dependent metabolism of glucose is the only source of energy in the central nervous system. This is consistent with the vulnerability of the brain to thiamine deficiency that results in neurological deficits characterized by tremor and such symptoms may result from damage to both the central (Prickett, 1934) and peripheral (Swank, 1940) nervous systems. Thiamine depletion in the CNS results in intracellular edema of glial cells and neuronal necrosis of the lateral vestibular nucleus and the brain stem. Neurological syndromes related to thiamine deficiency include Wernicke's encephalopathy, Leigh's disease, and cerebrocortical necrosis.

Wernicke's encephalopathy, first described in 1881, is characterized by lesions in the periventricular gray matter, degeneration of neurons, and proliferation of microglia, astrocytes, and small blood vessels. Wernicke's disease is usually accompanied by Korsakoff's psychosis, first defined in 1887, and characterized by hallucinations, disorientation in space and time, and loss of recent memory.

Leigh's disease, first described in 1951, is characterized by necrotic lesions, mostly in the gray matter of the CNS. These lesions are similar to those seen in Wernicke's encephalopathy except that in Leigh's disease the mammillary bodies remain normal.

Thiamine Deficiencies

Animals fed some types of raw fish and ferns containing thiaminase may develop thiamine deficiency. Thiaminase breaks down thiamine at the methylene bridge resulting in decreased thiamine in the body. These animals develop Chastek paralysis, a pattern of typical neurological deficits associated with thiamine deficiency (Green and Evans, 1940). Another thiamine-deficient disease related to the presence of thiaminase in the rumin of ruminant animals or in nonruminant animals following the administration of amprolium (an antimetabolite of thiamine) has been described. This disease is known as polioencephalomalacia or cerebrocortical necrosis and is characterized by necroses of the cerebrocortical neurons accompanied by perineuronal and pericapillary edema (Markson et al., 1972). This disease is completely different from Wernicke's encephalopathy. Thus ingestion or administration of thiamine antimetabolites may produce thiamine deficiency. These chemicals may competitively inhibit thiamine at its transport, phosphorylation, or coenzyme activity. Chronic ingestion of ethanol contributes to thiamine deficiency in a number of mechanisms.

1. Decreased dietary intake of thiamine, since alcoholic beverages are devoid of thiamine.
2. Reduced absorption of thiamine from the gastrointestinal tract by:
   a. Ethanol inhibition of Na^+-K^+ ATPase-dependent transport of thiamine.
   b. Change of gastric emptying.
   c. Folate deficiency.
   d. Induction of gastrointestinal changes, i.e., chronic pancreatitis, cirrhosis, portal hypertension.
3. Diminished conversion of thiamine to its active form, i.e., thiamine pyrophosphate (TPP).
4. Reduced liver storage of thiamine because of fatty infiltration, hepatocellular lesions, and cirrhosis.
5. Increased metabolic breakdown of thiamine.
6. Enhanced excretion of thiamine.

Biochemical Consequences of Thiamine Deficiency in the CNS

In the central nervous system, thiamine functions as a coenzyme for the following enzymatic reactions:

$$\text{pyruvate} + \text{CoA} + \text{NAD}^+ \xrightarrow{\text{pyruvate}\atop\text{dehydrogenase}}$$

$$\text{acetyl CoA} + CO_2 + \text{NHDH} + H^+$$

$$\alpha\text{-ketoglutarate} + \text{CoA} + \text{NAD}^+ \xrightarrow{\alpha\text{-ketoglutarate}\atop\text{dehydrogenase}}$$

$$\text{succinyl CoA} + CO_2 + \text{NADH} + H^+$$

$$\begin{array}{cc} O & O \\ \| & \| \\ \end{array}$$
$$R\text{–}C\text{–}C\text{–}OH + \text{CoA} + \text{NAD}^+ \xrightarrow{\alpha\text{-ketoacid}\atop\text{decarboxylase}}$$

$$\begin{array}{c} O \\ \| \\ \end{array}$$
$$R\text{–}C\text{–}S\text{–}CoA + CO_2 + \text{NADH} + H^+$$

$$\text{xyluse-5-phosphate} + \text{erythrose-4-phosphate} \xrightarrow{\text{transketolase}}$$

$$\text{fructose-6-phosphate} + \text{glyceraldehyde-3-phosphate}$$

Early studies have demonstrated an increase in pyruvate concentration in blood of thiamine deficient animals (Thompson and Johnson, 1935). Thus, it was hypothesized that thiamine deficiency impairs pyruvate oxidation, which might be responsible for brain lesions. Other studies reported that transketolase rather than pyruvate dehydrogenase in rat brain was more sensitive to inhibition by thiamine deficiency. However, several follow-up studies have demonstrated that the decrease in thiamine pyrophosphate-dependent enzymes in the central nervous system is not the primary cause for pathologic lesions seen in thiamine-deficient animals.

Effect of Ethanol Upon Thiamine Transport

Chronic ingestion of ethanol reduces thiamine absorption from the gastrointestinal tract in humans and animals (Thomson and Leavy, 1972). Low concentrations of thiamine are actively absorbed while higher concentrations are absorbed passively. Active transport of thiamine was blocked by ATPase inhibitors, e.g., ouabain. These results were confirmed by the finding that

Na$^+$/K$^+$ ATPase activity was reduced in basolateral enterocyte membrane by ethanol which was also correlated with thiamine reduction (Hoyumpa et al., 1974). These results are also consistent with extremely low intake of thiamine in alcoholics.

Thiamine elimination from the body takes place via urine and perspiration since it is not reabsorbed by the tubules (Pearson, 1967). Thus by increasing blood concentration of thiamine, the rate of its uptake by brain, both by active and passive transport, is increased. This may treat ethanol effects such as Leigh's disease.

REFERENCES

Ahlenius, S., A. Carlsson, G. Engel, T.H. Svensson, and P. Sadersten (1973). Antagonism by α-methyltyrosine of the ethanol-induced stimulation and euphoria. *Man. Clin. Pharm. Ther.* 14: 586–591.

Anokhina, I.P. and B.M. Kogan (1975). The role of alteration of brain catecholaminergic system in the pathogenesis of alcoholism. *Korsakow J. Neuropathol. Psychiatry,* 75: 1874–1883 (in Russian).

Anokhina, I.P. and L.F. Panchenko (1981). Effect of acute and chronic ethanol exposure on the functioning state of rat brain opiate receptors. In *Alcoholism (Clinical, Therapeutical, Pathogenic, Forensic, and Psychiatric Aspects).* G.V. Morozov, Ed., Moscow, pp. 52–56 (in Russian).

Ashley, M.C. (1982). Alcohol consumption, ischemic heart disease and cerebrovascular disease. *J. Stud. Alcohol* 43: 869–887.

Asmussen, E., J. Hold, and V. Larsen (1948). The pharmacological action of acetaldehyde on the human organism. *Acta Pharmacol.* 4: 311–320.

Balsara, S., T.R. Barat, V.P. Gada, and A.G. Chandorkar (1982). Small dose of apomorphine induce catalepsy and antagonize methamphetamic stereotypy in rats. *Psychopharmacology* 78: 192–194.

Berglund, M. (1981). Cerebral blood flow in chronic alcoholics. *Alcoholism: Clin. Exp. Res.* 5: 293–303.

Bitran, M. and G. Bustos (1982). On the mechanism of presynaptic autoreceptor-mediated inhibition of transmitter synthesis in dopaminergic nerve terminals. *Biochem. Pharmacol.* 31: 2851–2860.

Björkqvist, S.E. (1975). Clonidine in alcohol withdrawal. *Acta Psychiatr. Scand.* 52: 256–263.

Blum, K., J.D. Eubanks, J.E. Wallace, H.A. Schwertner, and W.W. Morgan (1976). Suppression of ethanol withdrawal by DA. *Experientia* 32: 493–495.

Bogusz, M., J. Pach, and W. Stasko (1977). Comparative studies on the rate of ethanol elimination in acute poisoning and in controlled studies. *J. Forensic Sci.* 22: 446–451.

Borg, S., H. Kvande, and G. Sedvall (1983). Monoamine metabolites in lumbar cerebrospinal fluid and urine in alcoholic patients before and after treatment with Disulfiran. *Alcohol Alcoholism* 18: 61–65.

Burov, Yu. V. and A.B. Kampov-Polevoi (1985). Peculiarities of forming experimental alcoholism in non-inbred rats. In *Progress in Alcohol Research*, Vol. 1, S. Parvez, Y. Burov, H. Parvez, and E. Burns, Eds., VNU Science Press, Utrecht, The Netherlands, pp. 69–81.

Chase, T.N., E.K. Gordon, and L.K.-Y. Ng (1973). Norepinephrine metabolism in the central nervous system of man. Studies using 3-methoxy-4-hydroxyphenylethylene glycol levels in cerebrospinal fluid. *J. Neurochem.* 21: 581.

Clarren, S.K. and S.W. Smith (1978). The fetal alcohol syndrome. *N. Engl. J. Med.* 298: 1063–1067.

Cott, J., A. Carlsson, J. Engel, and M. Lindquist (1976). Suppression of ethanol-induced locomotor stimulation by GABA-like drugs. *Naunyn-Schmiedeberg's Arch. Pharmacol.* 295: 203–209.

Davis, A.R. and A.H. Lipson (1986). Central nervous system depression and high blood ethanol levels. *Lancet* 1: 566.

Deykin, D.D., P. Janson, and L. McMahon (1982). Ethanol potentiation of aspirin induced prolongation of the bleeding time. *N. Engl. J. Med.* 306: 852–854.

DiChiara, G., M.L. Proceddu, L. Vargiu, A. Argiolus, and G.L. Gessa (1976). Evidence for dopamine receptors mediating sedation in the mouse brain. *Nature (London)*, 264: 564–566.

Engel, J. and Carlsson, A. (1977). In *Current Development of Psychopharmacology*, Vol. 4. W.B. Essman and L. Valzelli (Eds). Spectrum, New York, pp. 1–32.

Engel, J. and Liljequist, S. (1976). The effect of long-term treatment on the sensitivity of the dopamine receptors in the nucleus accumbens. *Psychopharmacology* 49: 253–257.

Fadda, F., A. Argiolas, M.R. Melis, G. Serra, and G.L. Gessa (1980). Differential effect of acute and chronic ethanol on dopamine metabolism in frontal cortex, caudate nucleus, and substantia nigra. *Life Sci.* 29: 979–986.

Fowler, C.J., K.F. Tipton, A.V.P. MacKay, and M.B.H. Youdim (1982). Human platelet monoamine oxidase—a useful enzyme to study psychiatric disorders? *Neuroscience* 7: 1577–1594.

Goldstein, D.B. (1974). Rates of onset and decay of alcohol physical dependence in mice. *J. Pharmacol. Exp. Ther.* 190: 377–383.

Goldman, H., L.A. Sapirstein, S. Murphy, and J. Moore (1983). Alcohol and regional blood flow in brains of rats. *Proc. Soc. Exp. Biol. Med.* 144: 983–988.

Green, R.G. and C.A. Evans (1940). A deficiency disease of foxes. *Science* 92: 154–155.

Hillbom, M., M. Kaste, and V. Rasi (1983). Can ethanol intoxication affect hemocoagulation to increase the risk of brain infarction in young adults? *Neurology* 33: 381–384.

Hills, B.W. and H.C. Venable (1982). The interaction of ethyl alcohol and industrial chemicals. *Am. J. Ind. Med.* 3: 321–333.

Hoffman, P.L., S. Urwyler, and B. Tabakoff (1980). Changes in opiate receptor function in ethanol-treated rats. *Alcoholism: Clin. Exp. Res.* 4: 218.

Hoyumpa, A.M., H.M. Middelton, F.A. Wilson, and S. Scheuker (1974). Dual system of thiamine transport: characteristics and the effect of ethanol. *Gastroenterology* 66: 714.

Kaim, S.C., C.J. Klett, and B. Rothfeld (1969). Treatment of the acute alcohol withdrawal state: a comparison of four drugs. *Am. J. Psychiatry* 125: 1640–1646.

Karoum, F., R.J. Wyatt, and E. Majchrowicz (1976). Brain concentration of biogenic amine metabolites in acutely treated and ethanol dependent rats. *Br. J. Pharmacol.* 56: 403–411.

LeBlanc, A.E., R.J. Gibbins, and H. Kalant (1973). Behavioral augmentation of tolerance to ethanol in the rat. *Psychopharmacologia* 30: 117–122.

Lewis, D.C. (1980). Diagnosis and management of the alcoholic patient. *R.I. Med. J.* 63: 27–31.

Lieber, C.S. (1984). Metabolism and metabolic effects of alcohol. *Med. Clin. North Am.* 68: 3–31.

Lieber, C.S. and L.M. DeCarli (1970). Hepatic microsomal ethanol-oxidizing system. In vitro characteristics and adaptive properties in vivo. *J. Biol. Chem.* 245: 2505–2512.

Liljequist, S. (1979). Behavioral and biochemical effects of chronic ethanol administration. Thesis, Department of Pharmacology, University of Göteborg, Göteborg, Sweden.

Markson, L.M., G. Lewis, S. Terlecki, E.E. Edwin, and S.E. Ford (1972). The aetiology of cerebrocortical necrosis: the effects of administering antimetabolites of thiamine to preruminant calves. *Br. Vet. J.* 128: 488–499.

Mohler, H. and T. Okada (1977). GABA-receptor binding with H(+) bicucullin-metiodite in rat CNS. *Nature (London)*, 276: 65–67.

Oreland, L., C.G. Gottfries, K. Kiianmaa, B. Wiberg, and B. Winblad (1983). The activity of monoamine oxidase-α and -β in brains from chronic alcoholics. *J. Neurol Transm.* 56: 73–83.

Pearson, W.N. (1967). Blood and urinary vitamin levels as potential indices of body store. *Am. J. Clin. Nutr.* 20: 514–525.

Pollack, W.S. (1984). Prospective study of alcohol consumption and cancer. *N. Engl. J. Med.* 310: 617–621.

Prickett, C.O. (1934). The effect of a deficiency of vitamin B_1 upon the central and peripheral nervous system of the rat. *Am. J. Physiol.* 107: 459–470.

Rueler, J.B., D.E. Girard, and T.G. Cooney (1985). Wernicke's encephalopathy. *N. Engl. J. Med.* 312: 1035–1039.

Revely, M.A., A.M. Revely, C.A. Cliffort, and R.M. Murray (1983). Genetics of platelet MAO activity in discordant schizophrenic and normal twins. *Br. J. Psychiatry* 142: 560–565.

Rich, E.C., C. Siebold, and B. Champion (1985). Alcohol related acute atrial fibrillation. A case-controlled study and review of 40 patients. *Arch. Intern. Med.* 145: 830–833.

Rogers, A.E. (1979). Nutrition. In *The Laboratory Rat*, Vol. 1. H.J. Baker, J.R. Lindsey, and S.H. Weisbroth, Eds., Academic Press, New York, pp. 123–146.

Sato, C., Y. Matsuda, and C.S. Lieber (1981). Increased hepatotoxicity of acetaminophen after chronic ethanol consumption in the rat. *Gastroenterology* 80: 140–148.

Schaller, K. and H. Holler (1975). Thiamine absorption in the rat. II. Intestinal alkaline phosphatase activity and thiamine absorption from rat small intestine *in vitro* and *in vivo*. *Int. J. Vitam. Nutr. Res.* 45: 30–38.

Schultz, R., M. Wuster, T. Duka, and A. Herz (1980). Acute and chronic ethanol treatment changes endorphin levels in brain and pituitary. *Psychopharmacology* 68: 221–229.

Sharma, S.K. and J.H. Quastel (1965). Transport and metabolism of thiamine in rat brain cortex *in vitro*. *Biochem. J.* 94: 790–800.

Stillman, R.C., R.J. Wyatt, D.L. Murphy, and F.P. Rauscher (1978). Low platelet monoamine oxidase activity and chronic marijuana use. *Life Sci.* 23: 1577–1582.

Streissguth, A.P., S. Landesman-Dwyer, J.C. Martin, and D.W. Smith (1980). Teratogenic effects of alcohol in humans and laboratory animals. *Science* 209: 353–361.

Swank, R.L. (1940). Avian thiamine deficiency. A correlation of the pathology and clinical behavior. *J. Exp. Med.* 71: 683–702.

Sytinsky, I.A., B.M. Guzicov, M.J. Gomanko, L.A. Safegina, and N.M. Soboleva (1973). Alcohol and gamma-amino-butyric acid of brain. *Vofr. Biokhim. Mozga Acad. Sci. Armen. SSR*, p. 331.

Taylor, H.L. and R.P. Hudson (1977). Acute ethanol poisoning: a two year study of deaths in North Carolina. *J. Forensic Sci.* 22: 639–653.

Thompson, R.H.S. and K.E. Johnson (1935). Blood pyruvate in vitamin B_1 deficiency. *Biochem. J.* 29: 694–700.

Thomson, A.D. and C.M. Leavy (1972). Observations on the mechanism of thiamine hydrochloride absorption in man. *Clin. Sci.* 43: 153–163.

Turner, T.B., V.L. Bennett, and H. Hernandez (1981). The beneficial side of moderate alcohol use. *Johns Hopkins Med. J.* 148: 53–63.

Vesell, E., J.G. Page, and G.T. Passananti (1971). Genetic and environmental factors affecting ethanol metabolism in man. *Clin. Pharmacol. Ther.* 12: 192–201.

Walker, D.W., D.E. Barnes, S. Zornetzer, B. Hunter, and P. Kubanis (1980). Neuronal loss in hippocampus induced by prolonged ethanol consumption in rats. *Science* 209: 711–713.

Wiberg, O., C.-G. Gottfries, and L. Oreland (1977). Low platelet monoamine oxidase activity in human alcoholics. *Med. Biol.* 55: 181–186.

Wilkinson, P.K. (1980). Pharmacokinetics of ethanol: a review. *Alcoholism: Clin. Exp. Res.* 4: 6–21.

Wolff, P.H. (1972). Ethnic differences in alcohol sensitivity. *Science* 175: 449–450.

Part

V

Neurotoxic Agents

Metals
M.B. Abou-Donia

Solvents
M.B. Abou-Donia

Gases and Vapors
M.B. Abou-Donia

Pesticides
M.B. Abou-Donia

Drugs of Abuse
M.B. Abou-Donia

Naturally Occurring Toxins
M.B. Abou-Donia

Metals

Mohamed B. Abou-Donia
Departments of Pharmacology and Neurobiology
Duke University Medical Center
Durham, North Carolina

INTRODUCTION

Sources of Metals

Toxic metal sources include natural occurrence in soil and water, cookware, burning of coal and petroleum, pesticides, drugs, and industrial chemicals.

Action of Metals

Metals exert toxic actions by reacting with one or more chemical groups essential for normal physiologic functions, such as hydroxyl, carboxyl, aldehyde, carbonyl, sulfhydryl, disulfide, amine, and imine groups. At the molecular level, some metals interact with proteins, leading to denaturation, precipitation, or enzyme inhibition. Other metals bind to nucleic acids, e.g., DNA, resulting in mutation or carcinogenicity. At the subcellular level, some metals decrease the stability of lysosomal membranes and disrupt cell functions by releasing various hydrolases. At the cellular level, some metals may affect the permeability of cell membranes and disturb energy metabolism.

Factors Affecting Metal Toxicity

The toxicity of metals is dependent upon: (1) the physical form of the metal (ionic, colloidal, or radiochemical); (2) the solubility, stability, and reactivity of the metal in the body; and (3) the ability of the metal to bind with ligands of biological macromolecules and the stability of these metal-bound metals.

Chelating Agents

Metal toxicity is antagonized by a group of chemicals called chelating agents. Effective chelating agents have the following properties:

1. High water solubility
2. Stability to metabolic biotransformation
3. Ability to penetrate the sites of metal storage
4. Chelate-metal complexes are less toxic than free metallic ion
5. Retain chelating activity at the pH of the body fluids (chelators are less stable at low pH)
6. Form five- or six-membered chelating rings, since they are the most stable
7. A low affinity to Ca^{2+}, since Ca^{2+} in plasma is susceptible to chelation
8. Metal affinity to chelating agents varies with the ligand atoms. Thus, lead and mercury have greater affinities for sulfur and nitrogen than for oxygen, and calcium has opposite properties
9. Ready excretion of the chelates.

Dimercaprol (BAL)

Its chemical nomenclature is 2,3-dimercaptopropanol, also known as BAL or British Antilewsite.

Dimercaprol antagonizes the biological actions of metals that form mercaptides with essential sulfhydryl groups, such as arsenic and mercury. Poisoning by selenites, which oxidize sulfhydryl enzymes, is not influenced by dimercaprol.

Dimercaprol should be administered as soon as possible after exposure to heavy metals, because it is more effective in preventing their inhibition of sulfhydryl enzymes than in reactivating them. It is essential to maintain a complex that consists of one metal atom and two molecules of dimercaprol. Such a complex is water soluble and more stable than the 1:1 complex, which is not soluble in water. Dimercaprol is given intramuscularly as 10% solution in oil, since it cannot be administered orally. It is a toxic chemical and results in severe side effects including nausea, vomiting, headache, salivation, sweating, and abdominal pain.

Penicillamine

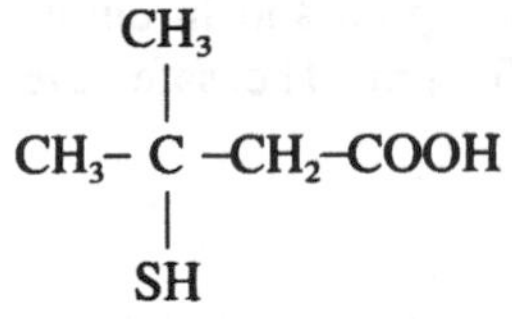

$$CH_3$$
$$|$$
$$CH_3- C -CH_2-COOH$$
$$|$$
$$SH$$

D-β,β'-dimethylcysteine

Penicillamine is prepared by the hydrolytic degradation of penicillin. It effectively chelates copper, mercury, zinc, and lead and enhances their excretion in the urine. It is relatively nontoxic. It is administered orally, 1–4 g/day in four divided doses.

Ca,Na$_2$-edetate

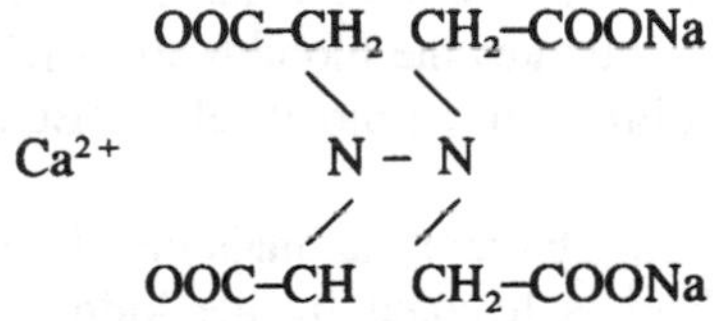

$$OOC-CH_2 \quad CH_2-COONa$$
$$\diagdown \quad \diagdown$$
$$Ca^{2+} \qquad N - N$$
$$\diagup \quad \diagup$$
$$OOC-CH \quad CH_2-COONa$$

Calcium disodium ethylenediamine tetraacetic acid is more effective than BAL in chelating lead. This compound is also less toxic than BAL. It should be given intravenously at a concentration that does not exceed 3%. Orally administered Ca,Na$_2$-edetate is not well absorbed, i.e., less than 1% is absorbed. Furthermore, if an excessive amount of lead is present in the gastrointestinal tract, its chelate complex with Ca,Na$_2$-edetate is well absorbed.

Deferoxamine

Deferoxamine is isolated from siderochrome pigment from a streptomycete. It chelates iron and forms a very stable complex. It will remove iron from transferrin and from ferritin, but not from hemoglobin or cytochromes, to which iron is bound more tightly.

HEAVY METALS

Lead

Acute Toxicity

The acute lethal dose of lead acetate is 10 g and of the less water soluble lead carbonate is 30 g. However, ingesting as little as 2 mg/day of these lead salts for weeks causes chronic lead poisoning.

The threshold limit value (TLV) of lead in inhaled air is 0.2 mg/m^3 and its tolerance in food is 7 ppm and in citrus fruits is 1 ppm. The TLV for tetraethyl lead is 0.075 ppm. The safe level of lead in drinking water is below 0.1 ppm.

Disposition

Ingested inorganic lead is more efficiently absorbed from the gastrointestinal tract of children (50%) than adults (8%). Calcium and ferrous ion deficiency enhance the absorption of inorganic lead. Inorganic lead is also absorbed via the respiratory tract. The extent of this absorption depends on the particle size and chemical form. At high concentrations, lead is also absorbed through the skin.

Absorbed inorganic lead initially distributes itself in blood, liver, and kidney. Prolonged exposure to inorganic lead results in the deposition of 95% of the body burden in the bone (Manton, 1985). Lead readily crosses the placenta. It also penetrates the blood-brain barrier to the central nervous system in children. It binds to red blood cells (RBC) with an RBC to plasma ratio of 16:1.

Lead is excreted in laboratory animals mostly via the bile. On the other hand, in humans, lead is handled by the kidney similar to Ca^{2+} and is eliminated in the urine. The half-life of lead in humans is about 1 month in soft tissues and 10 years for skeleton. In humans, 75% of the lead is excreted in urine, 8% in feces and 4% in sweat.

Action

Effect on Membranes. Lead toxicity is accompanied with damage to many membranes. In rats, lead results in erythrocyte fragility and ultrastructural damage to mitochondria (Hoffman, 1972). It crosses the blood-brain barrier; and it causes breakdown of active transport systems in cerebral capillaries.

Lead induces damage in myelin-containing membranes. It causes peripheral nerve damage in adult humans and animals. This results in peripheral neuropathy characterized by segmental demyelination and a decrease in nerve conduction velocity.

In the central nervous system, organic and inorganic leads decrease the myelination process. Lead also damages the protective barrier of nerve tissues resulting in an increased susceptibility to viral infection and changed drug absorption as secondary effects.

Lead damage to membranes may be related to its effect on oxidative processes. Protective enzymes such as glutathione peroxidase and superoxide dismutase have been affected by lead compounds. Lead acetate increased glutathione peroxidase levels in neonatal rats. Triethyl lead chloride did not affect this enzyme in adult rats but increased the level of superoxide dismutase. This effect was most prominent in the hippocampus, which is especially sensitive to lead and tin toxicity. Occupational exposure to inorganic lead reduced levels of Na^+/K^+-ATPase in human erythrocytes. Trialkyl lead chloride inhibited both oxidative phosphorylation and nerve conduction by allowing chloride ion to enter the neuron. This explains the higher toxicity of triethyl lead compounds compared to trimethyl lead compounds.

Effect on Energy. Organic and inorganic lead accumulate in mitochondria and inhibit oxidative phosphorylation *in vivo* and *in vitro* (Bondy et al., 1988). This effect may be related to lead damage to mitochondrial membranes or its effect on calcium hemostasis in nerve cells. Lead neurotoxicity has also been attributed, at least in part, to its vascular damage in the brain, which compromises both glucose and oxygen supplies.

Calcium. Dietary calcium retards lead absorption from the gastrointestinal tract (Bondy et al., 1988) suggesting that calcium and lead may have a common absorption mechanism. In the nervous system, lead displaces Ca^{2+} in the brain resulting in changes in its deposition.

Mitochondria actively remove Ca^{2+} from the cytosol through an energy-dependent process. Lead is concentrated in the mitochondria, which results in decreased uptake of Ca^{2+} from the cytosol to the mitochondria. This may result in a high lethal level of Ca^{2+} in the cell. Lead directly blocks the Ca^{2+} pumps. Increased levels of intracellular free Ca^{2+} may explain the effect of lead on the Ca^{2+}-dependent synaptic release of neurotransmitters. This action may result in the inhibitory effect of lead on the cholinergic system and the activation of the catecholaminergic system. Lead may compete with presynaptic Ca^{2+}, which reduces cholinergic activity. Also, it may substitute Ca^{2+} and act as an agonist resulting in dopaminergic hyperactivity. Since humans in the U.S. with no overexposure to lead have lead levels of 10^{-7} M, which is the same as the intracellular concentration of Ca^{2+}, this level may adversely interfere with Ca^2/calmodulin-dependent biological functions such as Ca^2/calmodulin-dependent protein kinase phosphorylation. The effect of lead on Ca^{2+} may contribute to its neurotoxicity to both the central and peripheral nervous systems.

δ-Aminolevulinic Acid Dehydrogenase (ALAD). Inorganic lead inhibits ALAD, a zinc-dependent enzyme, both *in vivo* and *in vitro*. Plasma levels below 1 nmol lead per unit result in significant depression of ALAD concentrations in erythrocytes. The inhibitory action of organic lead derivatives of ALAD parallels their toxicity. Triethyl lead is a very potent inhibitor to ALAD, while the less toxic diethyl and tetraethyl lead derivatives are weak inhibitors of the enzyme. ALAD is a critical enzyme in heme biosynthesis. Inhibition of ALAD results in an increased ALA synthetase and an elevated ALA level in plasma. Lead results in reduced heme synthesis.

Effect on Hippocampus. Lead exhibits regioselective toxicity in the brain. The hippocampus is particularly sensitive to damage induced by organic and inorganic lead derivatives. There is evidence that lead toxicity is exacerbated by zinc deficiency. However, not all toxic effects of lead are related to interference with zinc-containing enzymes.

Trimethyl and triethyl lead chloride increase hippocampal concentration of superoxide dismutase, suggesting oxidation damage. Superoxide dismutase is a copper- and zinc-requiring enzyme.

Symptoms

Acute Toxicity. Symptoms of acute lead poisoning are characterized by acute gastroenteritis, with burning of the pharynx, vomiting and diarrhea. In severe poisoning, shock develops and death follows.

Chronic Toxicity. Central nervous system symptoms result primarily from chronic exposure to tetraethyl lead. Early chronic lead signs are characterized by facial pallor, anemia with basophilic stippling of the erythrocytes, lead line of the gum, colic, and lead palsy.

1. *Facial pallor.* This is one of the earliest and most consistent signs of chronic lead poisoning. The pallor may result from vasospasm.
2. *Lead anemia.* A mild anemia is always present in lead toxicity. Red blood cell count is 3.5–4 million and hemoglobin concentration is 65–80%. Lead anemia results from the inhibition of delta-amino levulinic acid (ALA) dehydrogenase, which impairs the synthesis of heme from protoporphyrin and of porphobilinogen from ALA. Lead may cause anemia by increasing fragility of the red cell membrane. Low concentrations of lead cause RBC membranes to be more permeable to K^+ and decrease Na^+-K^+-ATPase leading to hemolysis.
3. *Retinal stippling.* During lead exposure, an early sign of lead poisoning is the retinal stippling about the optic disc, with gray lead particles. It disappears upon removal of exposure.
4. *The "lead-line."* This sign appears occasionally, i.e., 2–50% of the patients develop it. It consists of a subepithelial deposit of granules, i.e., lead sulfide, at the gingival margins. It is not known how it is made.
5. *Colic.* Lead-induced colic is characterized by spasmodic pain that involves the large and small intestine, ureter, uterus, and blood vessels. Lead seems to act on smooth muscle, and the resultant spasm is temporarily relieved with atropine.
6. *Lead palsy.* Less than 10% of lead-poisoning cases show lead palsy. Tremor, numbness, hyperesthesia, fibrillation, or cramps may precede muscle weakness. Usually there is unilateral paralysis of the extensor muscles of the wrist, which is known as "wrist drop." Sometimes "foot drop" may be involved. The paralysis results from the degeneration of the nerve and atrophy of the muscle. Slow recovery that ensues cessation of exposure may be complete.

7. *Lead encephalopathy.* This occurs frequently in young children (Lin-fu, 1982). It is characterized by change of personality, restlessness, dullness, fatigue, in mild poisoning. In extreme cases there is convulsion, delirium, or coma. Mortality may reach 50% of severe encephalopathy cases.

Absorption of tetraethyl lead may lead to acute anemia, headache, blurring of vision, aphasia, and severe elevation of cerebrospinal fluid pressure.

8. *Reproductive toxicity.* Lead exposure leads to male and female reproductive toxicity, miscarriages, and degenerate offspring. Abortion may result due to atrophy or spasm in the uterine muscle.
9. *Hair.* Chronic lead exposure may cause loss of hair.
10. *Urinary changes.* Individuals exposed to lead excrete lead in the urine and as much as 500 µg of porphyrin per day. ALA concentrations in the urine are also increased especially in children.

Histopathological Lesions

1. *Bone marrow.* Marrow shows hyperplasia of leukoblasts and erythroblasts with a decrease in fat cells.
2. *Gastrointestinal tract.* Ulcerative or hemorrhagic changes may accompany lead colic.
3. *Nervous system.* Experimental animals develop peripheral neuritis, degeneration of the anterior horn cells, and meningoencephalitis. Schwann cells proliferate and degenerate leading to demyelination.

Cause of Death

Acute Lead Poisoning. Gastroenteritis and subsequent shock are the cause of death in acute lead poisoning.

Chronic Lead Poisoning. Malnutrition; intercurrent infection; failure of liver function; failure of respiration, renal, and liver function; and encephalopathy are the causes of death due to chronic lead poisoning.

Treatment

Acute Lead Poisoning. When lead is ingested, magnesium or sodium sulfate in large amounts of water is administered. This changes unabsorbed lead salts to the highly insoluble lead sulfate and enhances its excretion through the intestine.

Chronic Poisoning

1. Calcium lactate or calcium gluconate (1/2–1 teaspoonful) will enhance removal of lead from blood. Similar results are also obtained by potassium or sodium citrate or disodium phosphate.

2. A more effective chelating agent than citrate is calcium disodium ethylendiamine tetraacetate (calcium edathamil, calcium disodium edetate), 1–10 g i.v. daily (Osterloh and Becker, 1986). (Trisodium salt combines with calcium and produces calcium deficiency.) Within the first hour, 50% of circulating lead is excreted, and 90% is excreted in seven hours. This drug removes circulating lead from blood and tissues but not from bone.
3. Dimercaprol in a dose of 4 mg/day i.m. every 4–6 hours is used. Larger doses cause too rapid an excretion of lead, leading to a danger of renal tubular necrosis.
4. Penicillamine is used in oral doses of 0.3–0.5 g 3–5 times/day to accelerate the elimination of lead (Marcus, 1982). This drug has serious side effects.
5. Folic atropine sulfate is used in 0.5–1 mg doses orally.
6. Vomiting is controlled by chlorpromazine injection, 50–100 mg.
7. Lead in bones is gradually removed by the use of iodides. Calcium disodium edetate has been used with some success.

Mercury

Sources
Mercury containing compounds are classified as:

1. elemental mercury
2. mercurous (Hg^+) and mercuric (Hg^{2+}) salts, or
3. organic aliphatic (CH_3Hg^+) or aromatic mercury compounds.

Disposition
Inhaled elemental mercury vapor is almost completely absorbed, with about 75% retained (Joselow et al., 1972). Ingested elemental mercury is not readily absorbed and is relatively harmless. If delayed in the gastrointestinal tract, elemental mercury may be oxidized to mercuric oxide. Only about 15% of ingested inorganic mercury may be absorbed from the gastrointestinal tract. Inorganic mercury salts are able to cross the placenta. Organic mercury compounds are readily absorbed from the gastrointestinal tract. Methyl mercury undergoes complete absorption, while the absorption of phenyl mercury ranges from 80–100%. Methyl mercury crosses into placenta and concentrates in the fetus. It is also found in the mother's milk. Methyl mercury crosses biological membranes such as red blood cell membrane and binds to hemoglobin. Both elemental mercury and methyl mercury readily cross the blood-brain barrier into the nervous system.

Approximately 7% of inhaled elemental mercury vapor is exhaled. Its half-life in humans is about 58 days (Jaffe et al., 1983).

Inorganic mercury salts deposit mostly in the kidneys and in the liver, spleen, bone marrow, red blood cells, intestine, skin, and respiratory mucosa (Sketris and Gray, 1982). These compounds are excreted primarily through the kidney via glomerular filtration and tubular secretion and to a much lesser extent through the gastrointestinal tract. The half-life of inorganic mercury is about 40 days.

Methyl mercury accumulates primarily in the central nervous system and in the liver (Gage, 1964). It undergoes biotransformation and conjugation (acetylation or conjugation) in the liver and is eliminated primarily via the bile but is reabsorbed from the gastrointestinal tract. It is excreted to a lesser degree through the urine. Mercury is excreted in sweat and salivary secretions. Sweat has higher mercury concentrations than in urine, while saliva concentration is two-fifths of that in the blood.

Action

The primary target of biochemical action of methyl mercury has not been determined (Joselow et al., 1972). The putative site, however, contains sulfhydryl (SH) groups that are critical for biochemical and/or physiological functions. Mercurials react strongly with sulfhydryl group. Recent studies have demonstrated that Hg^{2+} inhibits tubular polymerization that apparently resulted from its binding to sulfhydryl groups in tubulin. These results are in agreement with previous findings that SH reagents inhibit tubulin polymerization stoichiometry by blocking SH groups of tubulin. Histopathologic changes that are induced in the brain following exposure to methyl mercury are

1. reduction of RNA and protein synthesis
2. disintegration of ribosomes and rough endoplasmic reticulum
3. irreversible damage to parenchymatous cells
4. inhibition of DNA synthesis, chromosome disturbances, and nuclear damage
5. breakdown of biological membrane
6. disintegration of mitochondria.

Symptoms

Mercury Vapor

1. *Acute exposure.* Following acute exposure to mercury vapor, the patient may exhibit the following symptoms: cough, dyspnea, tightness of the chest, chills, fever, weakness, salivation, nausea, vomiting, diarrhea, and a metallic taste in the mouth. If these symptoms do not subside rapidly, they may progress to emphysema, necrotizing bronchiotitis, pulmonary edema, and interstitial fibrosis of the lungs.

2. *Chronic exposure.* The allowable threshold limit value for an 8-hour day and 40-hour working week (TLV-TWA) is only 0.05 mg/m^3. Repeated exposure to levels of 100 mg/m^3 produces symptoms of mercury poisoning. Chronic exposure to elemental mercury results in neurotoxicity, since it crosses the blood-brain barrier into the brain where it is slowly metabolized to mercuric oxide. Chronic exposure to metal mercury produces the following symptoms: tremors, anxiety, emotional instability, irritability, depression, forgetfulness, insomnia, excessive salivation, nausea, vomiting, diarrhea, constriction of visual fields, fatigue, muscular weakness, anorexia, headache, vertigo, peripheral neuritis, proteinurea, polyurea, anuria, and chronic pneumonitis which may be accompanied by low grade fever and dermatitis.

Inorganic Mercury Salts

1. *Acute toxicity.* Ingestion of 1/2–4 g of mercuric chloride may be fatal to adults. The following symptoms may be evident following the intake of a toxic dose of an inorganic mercury salt: severe nausea, vomiting, hematemesis, abdominal pain, a metallic taste, bloody diarrhea, colitis, and necrosis of intestinal mucosa (Hilmy et al., 1976). These symptoms may be followed by dehydration and cardiovascular collapse. As a result of excreting mercury through glomerular filtration and tubular secretion, the urine may contain red blood cells and protein. Also, acute tubular necrosis may develop accompanied by oliguria and uremia.

2. *Chronic toxicity.* Following chronic exposure to inorganic mercury salts, these symptoms may develop: tremors, erethism, ataxia, slurring of speech, excessive salivation, loosening of teeth, anxiety, and mental deterioration. Colitis, chronic renal failure, and dementia were seen following years of mercurous chloride use. Also, acrodynia (pink disease), which is characterized by redness of the palms and soles, edema of the hands and feet, skin rashes, diaphoresis, tachycardia, hypertension, photophobia, irritability, anorexia, insomnia, and constipation or diarrhea, developed in children exposed to mercurous chloride in teething lotions and diaper powders and from heating of paint containing mercurial fungicide.

Organic Mercury Compounds

1. *Aliphatic mercury compounds.* Exposure to aliphatic mercury compounds results in central nervous system symptoms characterized by ataxia; constricted visual fields; paresthesias; neurosthemia; loss of libido; hearing loss; memory loss; depression; mental deterioration; loss of emotional stability; incoordination; involuntary movements including static tremor, chorea, asthetosis and myoclonus; paralysis; and coma, which may result in death (Adams et al., 1983). These

symptoms are typical of methyl mercury poisoning. In addition to these symptoms, ethyl mercury may produce gastrointestinal symptoms and renal damage. Alkyl mercury compound may also cause eye, skin, and mucous membrane irritation and dermatitis.

2. *Aromatic mercury compounds.* Acrodynia may be produced from chronic exposure to vapors of mercurial fungicides in paint and to repeated injection of gamma globulin containing sodium ethylmercurithiosalicylate (merthiolate) as a bacteriostatic agent. Symptoms of inorganic mercury may be evident following exposure to phenyl mercury.

Treatment

Elemental mercury ingestion is nontoxic and does not require any treatment. Ingested organic or inorganic mercury compounds are first treated by gastric lavage or emesis. Activated charcoal is helpful in treating inorganic mercury. BAL is used for antidotal treatment of acute poisoning of inorganic mercury compounds and aryl organic mercury compound, e.g., phenylmercuric acetate, which is converted to inorganic mercury in the body. BAL is ineffective in reversing chronic neurologic deficits of mercury compounds (Campbell et al., 1986). d-Penicillamine is useful in treatment of inorganic mercury exposure and elemental mercury inhalation. This compound reverses sulfhydryl binding to mercury in blood resulting in increased excretion of mercury.

Arsenic

Sources

These chemicals are classified into three groups:

1. Organic or arsenate compounds, which are the pentavalent As^{5+} form, e.g., alkyl arsenates.
2. Inorganic or arsenite compounds, which are the trivalent As^{3+} form, e.g., arsenic trioxide, lead arsenate, sodium arsenite, arsenic acid, and arsenious acid.
3. Arsine gas (AsH_3), is a colorless nonirritating gas, that is formed by the action of acids on arsenic.

Approximately 80% of arsenic compounds are used as insecticides, rodenticides, and herbicides. Other uses include glassware, marine paints, alloy components, and pigment production. Arsine gas is used in the semiconductor industry to produce microchips.

Microorganisms in the environment convert arsenic to the dimethylarsenic form, which accumulates in fish. Arsenic compounds are also present as contaminants of well water in "moonshine" whiskey (Gerhard et al., 1980).

Action

In general the pentavalent compounds are less toxic than the trivalent derivatives. Also, water-insoluble salts, e.g., arsenic trioxide and lead arsenate, are less toxic than the water-soluble compounds, e.g., sodium arsenite, arsenic acid, and arsenious acid. Arsine gas is the most toxic form; the TLV-TWA is 0.05 ppm (Parish et al., 1979). Severe poisoning in children may result from less than 1 mg/kg, while 2 mg/kg may be fatal.

Arsenite (As^{3+}) compounds are lipid soluble and may be absorbed by the skin. Thus toxicity may result following ingestion, inhalation, or skin contamination.

Within 24 hours of entry, arsenic distributes to liver, lungs, intestinal wall, and spleen, where it binds to sulfhydryl groups of tissue proteins (Watson et al., 1981). It replaces phosphorus in bone where it may be stored for years. Only a small amount penetrates the blood-brain barrier. Most of the absorbed dose is eliminated via the kidney within four days. Minor amounts are excreted in the feces and sweat. Although the milk contains only insignificant amounts of the absorbed arsenic compounds, these chemicals can cross the placenta and cause harm to the fetus.

Arsenic binds to the sulfhydryl cofactor, dihydrolipoate, which inhibits the oxidation of pyruvate and succinate (Schoulmeester and White, 1980). Another less important action is the inhibition of oxidative phosphorylation by phosphorus replacement.

Symptoms

In acute poisoning, within 30 min to 2 hr of ingestion, severe gastrointestinal symptoms begin, e.g., vomiting, watery bloody diarrhea, severe abdominal pain, and burning esophageal pain. Salivation and garlic odor may occur (Watson et al., 1981). Vasodilation and myocardial depression may be evident as cold clammy skin, cyanosis, weakness, and lightheadedness leading to shock. Cerebral edema leads to headache, lethargy, delirium, coma, and convulsions. A stocking-glove distal sensory peripheral neuropathy, resulting from resorption of myelin and axonal degeneration, may be followed by muscle atrophy (Hine et al., 1977). Later stages of poisoning include jaundice and renal failure. Death results from circulatory failure within 24 hr to 4 days.

Chronic exposure results in nonspecific symptoms such as diarrhea, abdominal pain, hyperpigmentation, and hyperkeratosis. This is followed by a symmetrical sensory neuropathy. Late changes include a Wernicke-like encephalopathy, gangrene of the extremities, anemia, leukopenia, squamous cell carcinoma of the skin, lung cancer, and nasal septal perforations and possibly hepatic angiosarcoma.

Arsine gas toxicity is characterized by a 20-hour delay period, followed by abdominal pain, hemolysis, and renal failure (Kleinfeld, 1980). These symptoms are followed by ones similar to those of arsenic compounds.

Treatment

BAL is used as an antidotal treatment of arsenic toxicity except for arsine gas. Also, penicillamine provides additional therapy with fewer adverse effects than BAL. Treatment of arsine is mostly supportive care such as administration of oxygen, intravenous administration of fluids, and cardiac monitoring. Care should be exercised to treat hyperkalemia if it develops.

Cadmium

Sources

Cadmium does not form organometallic compounds. Its sources in the environment are coal and fossil fuels as well as superphosphate fertilizers. Occupational exposure takes place in lead and zinc smelters and workplace for the manufacturing pigments and cadmium/nickel batteries in electroplating industries.

Action

The main toxicity target for cadmium is the kidney (Perry et al., 1976). Approximately 10% of the dose is absorbed by the lung and gastrointestinal tract. Metallothionein, a small molecular weight plasma protein, binds to cadmium (Buel, 1975). The elimination half-life of cadmium ranges from 16–33 years. About one-half of the body burden is stored in the kidney and to a much lesser extent in the liver.

Symptoms

Acute inhalation of cadmium metal fumes results in fever, headache, dyspnea, pleuritic chest pain, conjunctivitis, sore throat, and cough within 4–12 hr of exposure. The symptoms may progress to pulmonary edema and death.

Acute ingestion of cadmium results in gastroenteritis, characterized by vomiting, diarrhea, and abdominal pain and metabolic acidosis, pulmonary edema, respiratory arrest, and finally, death (Buckler et al., 1986).

Chronic exposure to cadmium causes emphysema and renal dysfunction characterized by proteinurea, glucosuria, and aminoaciduria (DeSilva and Donnan, 1981). In Japan, environmental contamination resulted in a disease called itai-itai (ouch-ouch) that was characterized by arthralgias and osteomalacia in women with calcium and vitamin D deficiency.

Treatment

Following ingestion, the use of ipecac/lavage is indicated. An antidotal treatment is $CaNa_2$ EDTA to chelate cadmium. BAL is not indicated since it may increase nephrotoxicity.

NON-ESSENTIAL METALS

Aluminum

Sources
Aluminum hydroxide and aluminum phosphate are sources of aluminum poisoning.

Action
Accumulation of aluminum in bone may prevent incorporation of Ca^{2+} into osteoid, resulting in osteomalacia. Excess Ca^{2+} leads to an increase in its concentration in the blood (Wills and Savory, 1982). Elevated Ca^{2+} inhibits the release of the parathyroid hormone by the parathyroid glands. Renal failure and anemia may result from aluminum poisoning. A normal serum aluminum level is 10 $\mu g/ml$, the potential toxic level is 50 $\mu g/ml$, and symptoms appear at the 200 $\mu g/ml$ level.

Symptoms
Symptoms of aluminum toxicity include hypercalcemia, anemia, and vitamin D refractory osteodystrophy. Dialysis encephalopathy ensues and is characterized by dementia, facial seizures, mixed dysarthia in apraxia of speech, asterixis, tremulousness, and myoclonus. Symptoms include bone pain and proximal myopathy. These symptoms may develop over months to years of chronic exposure (Longstretch et al., 1985).

Some workers in the aluminum industry developed severe encephalopathy, spinocerebellar degeneration, incoordination, tremor, and cognitive deficits. The contribution of aluminum is undetermined since bone aluminum levels were normal.

Aluminum and Alzheimer's Disease
In 1973 aluminum involvement in Alzheimer's disease was proposed because of the elevated aluminum levels in brains at autopsy. These autopies, however, did not show increased aluminum concentrations in hair, serum, or cerebrospinal fluid. It was postulated that aluminum may change the blood-brain barrier (Banks and Kastin, 1983). Further studies showed elevated levels of aluminosilicates in the central region of senile plaque cores. In other diseases that do not produce dementia, e.g., hepatic and renal failure, metastatic carcinoma, and amyotropic lateral sclerosis, there is an increased brain aluminum concentration. Other factors may be involved in the etiology of Alzheimer's disease such as viruses or autoimmune reaction.

Treatment
Dialysis encephalopathy and osteomalacia are treated with deferoxamine with good results. Ca-Na-edetate is less effective than deferoxamine.

Antimony (Sb)

Sources
Antimony oxides and antimony-containing alloys are sources of toxicity. Sibine gas (SbH_3) is produced when Sb alloys are treated with acids.

Action
It binds to sulfhydryl groups of many enzymes.

Symptoms
Antimony oxide produces pneumoconiosis, severe pulmonary edema, cutaneous burns, antimony spots, and cardiomyopathy. Stibine gas toxicity is characterized by hemolytic anemia, myoglobinuria, renal failure, weakness, profuse vomiting, nausea, headache, abdominal and low back pain, and hematuria.

Treatment
Antidotal treatment with BAL. Dialysis may be performed.

Tin (Sn)

Sources
Organotin chemicals are used extensively in industry and agriculture. Although their toxicity has been recognized for many years, organotins are of considerable importance because of their industrial and environmental hazards and because of their use as probes to study selectively nervous system functions.

Action

Humans. The following signs and symptoms of trialkyltin have been reported in humans (Barnes and Stoner, 1959; Rey et al., 1984).

Triethyltin (TET)	Trimethyltin (TMT)
Headaches	Headaches
Abdominal pain	Generalized pain
Visual disturbances	Visual disturbances
Vertigo	Disorientation
Weight loss	Loss of appetite
Hypothermia	Memory deficits
Paralysis	Sleep disturbances
Papilloedema	Loss of libido
	Intermittent depression
	Attacks of rage

Of the 217 persons reportedly affected in an outbreak of organotin poisoning in France in the 1950s, 100 died.

Animals. TET and TMT produce qualitatively similar signs of acute toxicity in experimental animals. Signs of poisoning include muscle weakness, tremors, and convulsions followed by death within 2–5 days of administration.

Several studies have reported TMT-induced lesions in rat hippocampus. These distributions of the changes were reported in the following areas: $CA_{1,2}$ and CA_{3c}, but not $CA_{3a,b}$; CA_{3c} not $CA_{3a,b}$; CA_{3a}, CA_{3b}, and CA_{3c} but relative sparing of $CA_{1,2}$. Severity of lesion depended on the regional distribution. Thus in the ventral and dorsolateral areas, the fascia dentata granule cells were most involved, while in dorsomedial areas CA_{3a} and CA_{3b}, pyramidal cells were most affected. $CA_{1,2}$ cells of the dorsal hippocampus were most affected at the temporal pole.

The injection of TMT to neonatal rat pups produced selective destruction of hippocampal neurons that depended on the neonatal age. Lesions in the Ammon's horn pyramidal neurons, and CA_3 region in particular, occurred in mature and intact granule cells and their fibers. Pathological lesion appearance in the developing hippocampus depended on the neonatal age, with the most sensitive age period being between 9 and 15 days postnatal. Growth of these animals was markedly retarded. Their brains were significantly smaller than untreated animals.

Biochemical Changes. In both *in vivo* and *in vitro* studies TMT was shown to inhibit the uptake of GABA (Doctor et al., 1982a,b). It was suggested that TMT-induced blockage of GABA uptake could result in a prolongation of the postsynaptic inhibitory action of released GABA. Prolonged increased extracellular concentration of GABA due to TMT-inhibition of GABA uptake may result in depletion of GABA in presynaptic vesicles. Recent studies have demonstrated dose-dependent changes in the uptake of endogenous glutamic and GABA in synaptosomes isolated from mature rat hippocampus (Valdes et al., 1983). Chang (1986) proposed that TMT may produce its pathological lesions by acting on the inhibitory systems of the hippocampus, which leads to a hyperexcitatory state of the dentate granule cells and hyperstimulatory damage to their target cells (CA_3 pyramidal neurons). It is also possible that TMT may have an effect on the neuronal ions, e.g., Ca^{2+} or Cl^-.

Disposition

Trimethyltin undergoes biomethylation of tin by estuarine microorganisms. TMT has been identified in fresh, saline, and estuarine surface waters. It binds to rat liver mitochondria and inhibiting of oxidative phosphorylation.

The disposition of TMT at 1 h after treatment of mice, in order of decreasing mean tissue concentration, was liver > testis > kidneys > lungs > blood > brain > skeletal muscle > adipose tissue. The elimination of

TMT from these tissues did not follow first-order kinetics, suggesting the redistribution of TMT. TMT deposited more slowly in the brain, skeletal muscle, and adipose tissue, where peak TMT levels were reached at 6, 10, and 16 h after administration, respectively (Cook et al., 1984a,b). In the other tissues, the maximum TMT concentration was observed by 16 h following administration.

The total brain concentration of TMT associated with the CNS effects at early stages of intoxication in mice (i.e., sedation, antinociception, decreased responsiveness to induced seizures) was 1.1–1.3 μg/g wet brain tissue. The mean concentration of TMT in the brain associated with delayed excitability at 16 h was 1.53 μg/g of wet tissue.

TMT binds to rat hemoglobin but not to the hemoglobin of the gerbil and hamster. Also, the blood level of TMT in mice was found to be 2% of that of a similar dose in rats. The low hemoglobin TMT-binding capacity has been suggested as an explanation for the high sensitivity of these species to TMT neurotoxicity compared to rats.

Treatment

Management of tin poisoning is supportive and preventative. BAL does not chelate organic metals well.

Thalium

Sources

Thalium was used in the treatment of veneral disease, ringworm, gout, dysentery, and tuberculosis. Its use as an rodenticide has been banned because of its severe toxicity. Recent uses include manufacturing of imitation jewelry pigments, low-temperature thermometers, semiconductors, scintillation counters, and optical lenses.

Action

Thalium inhibits sulfhydryl-containing enzymes such as succinic dehydrogenase, protease, and monoamine oxidase in the corpus striatum. Thalium exchanges with potassium and interferes with oxidative phosphorylation by inhibiting Na^+/K^+ ATPase. It is stored in axonal mitochondria, resulting in their degeneration (Lovejoy, 1982).

Symptoms

The average lethal dose for adults is 1 g of thalium salts. Poisoning reaches a maximum in 2–3 weeks followed by improvement or death. Initial symptoms are nausea, vomiting, and diarrhea. These symptoms are followed by central nervous system effects such as disorientation, lethargy, thirst, insomnia, psychosis, convulsions, coma, and cerebral edema with central respiratory failure (Stevens, 1978). Other symptoms include peripheral neuropathy, tachycardia, hypertension, fever, salivation, and sweating. Cardiovascular symptoms are dysrhythmias, hypotension, bradycardia, optic neuritis and ophthalmoplegias may occur. Thalium toxicity may produce blue-gray lines on gums and dark circles around hair root. Also, acne may occur.

Treatment
Ingested thalium is decontaminated using emesis, lavage, or activated charcoal. Chelating agents are not effective. Potassium chloride diuresis may be used to increase thalium excretion.

Nickel

Sources
Most nickel toxicity is occupational.

Action
Nickel compounds are potent allergens and coronary vasoconstrictors.

Symptoms
Nickel and nickel compounds are carcinogenic for the pulmonary tract. Occupationally, skin contact produces dermatitis known as "Nickel Itch." Inhalation of nickel carbonyl leads to acute chemical pneumonitis or a delayed pulmonary edema that may be fatal (Leach et al., 1985).

Phosphorus

Sources
Yellow phosphorus is used as a rodenticide, in matches, and in the manufacturing of acetyl cellulose, bronze alloy, fertilizer and semiconductors. The rough surface of matches had red phosphorus, phosphorus sesquisulfide, potassium chlorate, and glue. The tips of safety matches contain antimony sulfide and potassium chlorate, which is the major toxic chemical causing mucosal irritation and methemoglobinemia. Phosphine gas, PH_3, is involved in the production of acetylene gas. It also evolves from reacting aluminum phosphide grain fungicide or zinc phosphide rodenticide with water. This gas has a garlic odor.

Action
Yellow phosphorus uncouples oxidative phosphorylation. Target organs include the gastrointestinal tract, liver, kidney, and bone.

Symptoms
Initial symptoms of yellow phosphorus poisoning are vomiting and abdominal pain, which progress to diarrhea and bleeding. Central nervous

system effects are lethargy, restlessness, hypotension, shock, and coma. Hepatorenal toxicity, bone degeneration, and osteoporosis, especially the mandible "phossy jaw," also ensue. Hepatotoxicity includes liver enlargement, jaundice, and increased hepatic amino-transferase levels.

Phosphine gas causes headache, fatigue, nausea, vomiting, cough, dyspnea, paresthesias, ataxia, tremor, weakness, and diplopia. Other symptoms include jaundice, centrilobular necrosis of the liver, congestive heart failure with pulmonary edema, and fatal myocardial necrosis.

Treatment

Emesis and lavage are indicated. Lavage is administered with 1:5,000 potassium permanganate. There is no antidotal treatment. Symptomatic treatment is indicated.

Potassium Permanganate

Sources

Diluted potassium permanganate solutions are used to treat dermatitis or as a mild antiseptic.

Action

Potassium permanganate is a powerful oxidizing agent. It causes methemoglobinemia and produces necrosis similar to acid action.

Symptoms

Ingestion causes burning of the mouth, chest pain, abdominal pain, nausea, and vomiting (Cowie and Escreet, 1981). Nervous system damage is manifested as paresthesias, sweating, loss of coordination, visual impairment, and muscle fasiculation. A delayed effect is a Parkinsonian syndrome. Hypermanganemia may develop.

Treatment

Treatment is supportive and symptomatic. Methemoglobinemia is treated with methylene blue. Ingestion requires dilution with milk or water, but no emesis is indicated. Eyes and skin are irrigated with water or saline. Neither dimercaprol (BAL) nor $CaNa_2$-EDTA increases manganese excretion.

Lithium

Sources

Lithium carbonate is used as the medication of choice to treat manic depression. Occupational lithium overexposure is rare, although it is used as a nuclear reactor coolant, in alkaline storage batteries, and in alloys.

Action

The major target for lithium toxicity is the central nervous system. It replaces body cations, resulting in changes in ion exchange and related normal cellular processes (Singer and Rotanberg, 1973).

Mild toxicity results when blood levels of lithium reach 1.5 mEq/L and results in hypotension and peripheral vascular collapse at 4.0 mEq/L (Marcus, 1983).

Symptoms

Lithium carbonate is almost completely absorbed from the gastrointestinal tract. Its volume of distribution is the same as the total body water (0.6 L/kg) suggesting low tissue distribution. It enters the liver and kidney rapidly and is entirely eliminated by glomerular filtration. It diffuses slowly to the brain, where it may persist in high concentrations.

Most toxicity symptoms of lithium result from effects on the nervous system. Early signs are nausea, vomiting, and diarrhea. This is followed by encephalopathy, i.e., disorientation, and poor memory. These symptoms may be accompanied by other features, e.g., blurred vision, muscular weakness, drowsiness, dizziness, vertigo, confusion, slurred speech, blackouts, and fasciculations. The clinical condition may worsen and victims may exhibit myoclonic twitch, myoclonic movement of entire limb, choreoathetoid movement, urinary and fecal incontinence, seizure, and coma. Cardiovascular toxicity results from hypokalemia due to displacement of potassium by intracellular lithium leading as cardiac arrhythmias. Hypotension is a secondary effect to coma. Lithium impairs the ability of the kidney for concentration and acidification. This leads to a transient natriuresis, nephrogenic, diabetes insipidus, and partial distal renal tubular acidosis. It may also disrupt antidiuretic hormone-induced water flow (Bejar, 1985).

Treatment

Following ingestion of lithium carbonate, emesis or gastric lavage is administered (El Mallakh, 1984). Charcoal is not useful in absorbing lithium. No specific antidotal treatment is available. Supportive and symptomatic treatments are indicated.

Barium

Sources

Barium compounds include barium sulfate, which is water insoluble, and the acid-soluble barium salts, i.e., acetate, carbonate, chloride, hydroxide, nitrate, and sulfide.

Action

The initial action of barium is the stimulation of striated, cardiac, and smooth muscle and depression of serum potassium. The ensuing direct depolarization and neuromuscular blockade of barium results in muscle weakness.

Symptoms

Barium sulfate is generally nontoxic, while the acid-soluble salts are highly toxic when ingested.

Ingestion of acid salts causes gastrointestinal irritation, muscle twitching, progressive flaccid paralysis, and severe hypokalemia and hypertension. Other symptoms that may ensue are respiratory failure, renal failure, and occasional cardiac dysrhythmias (Phelan et al., 1984).

Treatment

Decontamination using sodium sulfate as cathartic is the best treatment. Large doses of potassium may be administered to treat the hypokalemia.

Beryllium

Sources

It is used for alloys used in space exploration, military equipment, and nuclear reactors.

Action

The lung is the target organ for beryllium toxicity (Hooper, 1981).

Symptoms

Acute beryllium exposure results in chemical pneumonitis and pulmonary edema. It causes skin sensitization, irritation, skin granulomas, and ulcers. Chronic exposure affects lymph nodes, spleen, liver, myocardium, kidney, and bone. Hypercalcemia may develop.

Treatment

In test animals, aurin tricarboxylic acid protected against lethal doses of beryllium (King et al., 1964).

Bismuth

Sources

Bismuth is used as bismuth subsalicylate (Pepto-Bismol®). There are incidence resulting from industrial exposure to bismuth.

Symptoms

Bismuth causes a reversible encephalopathy that reverses spontaneously in 2–10 weeks (Hasking and Duggan, 1982). Chronic ingestion results in black spots in the mucosa and gums (bismuth lines) and salivation.

Treatment

d-Penicillamine may be used as a chelating agent.

Gold

Sources
  Gold sodium thiomalate is used to treat rheumatoid arthritis.

Symptoms and Treatment
  Overdosing with gold sodium thiomalate results in encephalopathy (Guilliford et al., 1985). The following symptoms also occur: interstitial pneumonitis, dermatitis, hepatitis, nephritis, stomatistis, and bone marrow suppression resulting in anemia, leukopenia, and thrombocytopenia (nephrotic). BAL is used as an antidote.

ESSENTIAL METALS

Chromium

Sources
  Toxicity to chromium is mostly occupational. Trivalent chromium compounds are relatively nontoxic compared to compounds with a valence of 6^+ (Langard and Vigander, 1983). Trivalent chromium ion (Cr^{3+}) is essential for glucose metabolism.

Action
  Chromium compounds are both skin and pulmonary sensitizers.

Symptoms
  Acute ingestion of chromium compounds results in acute gastroenteritis, hepatic necrosis, bleeding, and acute tubular necrosis with renal failure. Symptoms may also include nausea, vomiting, dark red urine, and loose reddish stool. Chronic exposure to chromium compounds produces a generalized irritation of the conjunctiva and mucous membranes, nasal perforations, and contact dermatitis. These symptoms are known as Blackjack disease and result from exposure to chromium in green felt by card players (Fisher, 1976). Epidemiological studies have suggested increased incidence of lung cancer in workers exposed to chromium compounds.

Treatment
  Treatment is symptomatic (Bader, 1986).

Copper

Sources
  Copper is an essential trace element that is important for heme synthesis and iron absorption.

Action

Copper ions increase lysosomal membrane fragility and cause the release of acid hydrolases with subsequent cell degeneration. Copper ions increase cellular permeability of erythrocytes leading to lysis, or cause agglutination of erythrocytes. Copper inhibits glutathione reductase in erythrocytes and loss of intracellular energy metabolism.

Symptoms

Copper ingestion causes a metallic taste, nausea, vomiting, epigastric burning, and vomiting of greenish-blue materials, but diarrhea is unlikely (Walsh et al., 1977). Hemoglobinurea and hemolysis may appear as may liver centrilobular necrosis. The effect on kidney results in anuria, oliguria, and hematuria. Nervous system effects are characterized by lethargy, refractory hypotension, and coma. Inhalation of copper dust causes ulceration and perforation of the nasal septum and pharyngeal congestion (Jantsch et al., 1984).

Wilson's Disease

This is a genetic disorder of copper metabolism in humans that is an autosomal recessive trait (Pfeiffer and Camo, 1988). Copper accumulates in the liver, cornea, and brain. This results in hepatic cirrhosis, necrosis, and sclerosis of the corpus striatum and trauma in the brain.

Treatment

Since copper toxicity results in vomiting, there is no need for emesis/lavage (Cole and Lirenman, 1978). Activated charcoal may be used. Chelating agents, e.g., intravenous $CaNa_2$-EDTA or intramuscular BAL are the antidotal treatment for copper poisoning. Also, d-penicillamine may be administered orally.

Iron

Sources

The large majority of iron poisoning is from ferrous sulfate mostly by children ingesting iron-containing medications (Doolin and Drueck, 1980). The nontoxic dose is less than 10–20 mg of elemental iron ingested/kg body weight (e.g., three tablets of ferrous sulfate). The lethal dose is more than 180 mg of elemental iron ingested per kilogram of body weight (e.g., 30 tablets of ferrous sulfate).

Action

Systemic blood vessels are damaged by free-circulating iron. This effect is aggravated by the release of ferritin, a vasodilator, and perhaps the release of serotonin and histamine. Ferrous ions disrupt mitochondrial membranes and the tricarboxylic acid cycle as the result of its catalyzing lipid peroxidation. Iron also produces severe acidosis. This results from the iron action on oxidative enzymes, release of hydrogen ions, the formation of ferric hydroxide, and the accumulation of lactic acid from anaerobolic metabolism. The absorbed ferrous form is oxidized to the ferric state and binds to the storage protein ferritin. Subsequently, iron is dissociated from ferritin to the globulin transferrin in the plasma and then to the blood-forming sites. Absorbed iron is poorly excreted. Elimination of iron seems to be limited to blood loss.

Symptoms

Iron ingestion causes severe hemorrhage resulting from its corrosive action on the stomach, followed by intestinal erosions and ulceration. Vomiting and diarrhea may occur (Robotham et al., 1974). Hepatic failure may result from the accumulation of iron in the Kupffer cells and hepatocytes followed by hypoprothrombineuria and hypoglycemia. Central nervous system effects such as lethargy, convulsion, and coma may also occur. Cardiovascular effects, e.g., pallor, tachycardia, and hypotension may be evident in severe poisoning (Robotham and Leitman, 1980).

Treatment

Decontamination of the gastrointestinal tract by emesis or gastric lavage is indicated. This is followed by the administration of biocarbonate solution (1–4%) to form the relatively nonsoluble ferrous carbonate. In severe cases, the addition of 5–10 g of deferoxamine is indicated (Proudfoot et al., 1986). Activated charcoal binds to the iron-deferoxamine complex although it does not effectively bind iron.

Magnesium

Sources

Magnesium poisoning is most likely to result from the ingestion of large doses of magnesium sulfate (Epsom salt) or magnesium citrate (AuCamp et al., 1981).

Action

Magnesium is the second most common cation in the cell. It is essential for the action of Ca^{2+}-Mg^{2+} ATPase and other essential enzymes, e.g., CaM kinase II. The major targets are the nervous system and the cardiovascular system (Mordes et al., 1975). Approximately 30% of the ingested dose is absorbed by the gastrointestinal tract. Magnesium ions are filtered by the glomerulus, 95% of which is reabsorbed by the proximal tubule. Both magnesium and calcium ions have similar resorption mechanisms. Excretion of Mg^{2+} is increased in the presence of hypercalcemia and high sodium intake. Magnesium toxicity results when it is not effectively excreted by the kidney.

Symptoms

Hypermagnesemia is more likely to occur in the presence of intestinal or renal disease (Graber et al., 1981). The major symptoms are paralysis resulting from action on the neuromuscular junction and hypotension, bradycardia, and heart block resulting from an effect on the cardiac system. Symptoms also include central nervous system depression, respiratory depression, decreased bowel movement, muscle paralysis, and hyperflexia. Electrocardiographic changes are characterized by prolonged QRS, QT, and PR intervals, peaked T waves, interventricular conduction defects, atrioventricular block, and finally asystole; these symptoms are similar to those seen in hyperkalemia. Increased magnesium ion concentration may result in hypocalcemia (Morton, 1985). A common consequence of magnesium poisoning is acute or chronic renal failure. Uncommon consequences of magnesium poisoning include hypothyroidism, viral hepatitis, pituitary dwarfism, acute diabetic ketoacidosis, and Addison's disease (Graber et al., 1981).

Hypermagnesemia may be accompanied by sudden respiratory arrest. Both hypercalcemia and hypocalcemia may develop (Collinson and Burroughs, 1986). Other electrolytes, i.e., K^+, Ca^{2+} and PO_4^{+3}, may be affected.

Treatment

Emesis or lavage is indicated within 4 hr of ingestion. Activated charcoal is not useful. Magnesium ions may be removed by hemodialysis. Antidotal treatment is carried out with calcium, which may act by replacing magnesium from cell membrane. The adult i.v. dose of calcium gluconate is 10 ml of 10% solution, which may be repeated once.

Manganese

Sources

Manganese ore, manganese fume, and manganese oxide are sources.

Action

Manganese exerts its major toxic action on the nervous system (Hine and Pasi, 1975).

Symptoms

Manganese produces Parkinson-like symptoms characterized by coarse intention tremor, loss of facial expression, muscle rigidity, gait disturbance, and ataxia (Cook et al., 1974). Neuropathologic lesions are present in the globus pallidus and striatum regions of the basal ganglia, unlike Parkinson's disease, where lesions are seen in the substantia nigra. Neurologic deficits improve upon withdrawal.

It also causes "manganese madness" syndrome, a progressive psychiatric disturbance that is characterized by apathy, confusion, bizarre behavior, visual hallucinations, increased lability, impotence, decreased libido, difficulty with speech, and loss of balance. Less prominent symptoms include anorexia, paresthesias, memory loss, anxiety, lumbosacral pain, incontinence, and cramps.

Treatment

L-Dopa may treat some of the neurologic dysfunctions. CaNa$_2$-EDTA may be used as a chelating agent.

Cobalt

Cobalt is an essential trace element and functions as a component of vitamin B$_{12}$. Chronic administration of this element may result in goiter and diminished thyroid activity.

Chronic exposure to the metal produces a disease known as "hard-metal pneumoconiosis" characterized with hypersensitivity reaction, malaise, cough, and wheezing (Finkel, 1983). The use of cobalt as a defoaming agent resulted in "beer drinker's" cardiomyopathy and polycythemia, i.e., increased density of red blood cells (Kennedy et al., 1981).

Potassium

Sources

Potassium chloride is used orally or intravenously for treatment of dietary potassium deficiency, treatment of dysrhythmias of cardiotonic glycoside toxicity resulting by loss of potassium, use as a salt in sodium chloride restricted diet, by myasthenia gravis patients to help restore muscular strength, and as a treatment in thalium poisoning.

Action

Reduced renal excretion of potassium results in hyperkalemia, like magnesium toxicity. Ingested potassium is well absorbed by the gastrointestinal tract. All of the absorbed potassium is secreted by the distal renal tubule.

Symptoms

Hyperkalemia is characterized by neuromuscular effects characterized by weakness, paresthesias, hyporeflexia, paralysis and cardiac diseases, i.e., bradycardia, decreased conduction, hypotension, ventricular fibrillation, and asystole (Hoyt, 1986).

Treatment

Intravenous injection of 5–10 ml of 10% calcium gluconate to antagonize the cardiac and neuromuscular effects of hyperkalemia is indicated. Sodium bicarbonate is administered intravenously to move potassium into the cells rapidly. The administration of glucose causes an intracellular shift of potassium. The cation exchange resin kayexalate is used to remove potassium. Dialysis effectively removes potassium.

Selenium

Sources
Selenium is used in the following industries: electronics, glass, ceramics, steel, and pigments manufacturing. It also has medicinal uses such as anti-dandruff shampoos, dietary supplements, and treatment of cystic fibrosis. Selenium has the following forms: elemental selenium, selenium salts, and selenious acid, which is the most toxic form.

Action
Selenium is an essential trace metal (Burke and Lane, 1983). It functions as a cofactor for glutathione peroxidase that destroys hydrogen peroxide, thus reducing tissue peroxide.

Symptoms
Toxic doses of ingested elemental selenium results in nausea, vomiting, nail changes, fatigue, and irritability. Selenium dust leads to respiratory tract irritation characterized by nasal discharge, loss of smell, and cough (Civil and McDonald, 1978).

Sodium selenate is less toxic than sodium selenite, which produces nausea, vomiting, abdominal pain, and tremor that resolves in 24 hr. Vitamin C reduces selenite to the elemental form. Selenium dioxide decomposes in water to selenious acid and produces apnea and asystole. Chronic selenium toxicity is similar to arsenic toxicity. It causes hair loss, white horizontal streaking on fingernails, fatigue, nausea, vomiting, garlic odor or breath, and metallic taste. Ingestion of selenious acid usually causes death. Signs of toxicity are respiratory depression, hypotension, and death (Carter, 1966). Inhalation of hydrogen selenide produces upper respiratory tract irritation and wheezing that is followed by dyspnea, and reduced expiratory flow rates. Residual effects may remain for years.

Treatment
There is no antidotal treatment of selenium toxicity. Treatment is symptomatic and supportive, i.e., intravenous infusion, supplemental oxygen, and ventilation as needed.

Zinc

Sources
Zinc chloride is used in dry batteries, oil refining, and dentists' cement. Zinc oxide is used in pharmaceuticals, white pigment, electroplating, and galvanizing (Finkel, 1983).

Action

Zinc is an essential trace element. It functions as a cofactor for some enzymes, e.g., alcohol dehydrogenase, carbonic anhydrase, carboxypeptides. Zinc toxicity results from impairment of lymphocyte and neutrophil function and reduces serum level of high-density lipoprotein cholesterol (Potter, 1981). It also causes irritation of the alimentary canal.

Symptoms

Exposure to zinc fumes results in "metal fume fever" that initially resembles a flue and is characterized by fatigue, chills, fever, cough, dyspnea, leukocytosis, thirst, metallic taste, and salivation (Murphy, 1984). Tolerance may be developed in factory exposure but is often lost over the weekend resulting in "Monday Morning Fever." Metal fume fever may also result from exposure to fumes of copper, magnesium aluminum, antimony, iron, manganese, or nickel. Ingestion of zinc compounds results in abdominal pain, hematesis, lethargy, slight ataxia, and difficulty in writing.

Treatment

Emesis or gastric lavage are indicated. $CaNa_2$-EDTA is used as an antidote. N-acetylcysteine may be used.

REFERENCES

AuCamp, A.K., S.M. Van Auchtenberg, and E. Theron (1981). Potential hazards of magnesium sulfate administration. *Lancet* 2: 1057.

Adams, C.R., D.K. Ziegler, and J.T. Lin (1983). Mercury intoxication simulating amyotrophic lateral sclerosis. *JAMA* 250: 642–643.

Bader, T.F. (1986). Acute renal failure after chronic acid injection. *West. J. Med.* 144: 608–609.

Banks, W.A. and A.J. Kastin (1983). Aluminum increases permeability of the blood-brain barrier to labelled DSIP and beta-endophrin: possible implications for senile and dialysis dementia. *Lancet* 2: 1227–1229.

Barnes, J.M. and H.B. Stoner (1959). The toxicology of tin compounds. *Pharmacol. Rev.* 11: 211–231.

Bejar, J.M. (1985). Cerebellar degeneration due to acute lithium toxicity. *Clin. Neuropharmacol.* 8: 379–381.

Bondy, S.C. (1988). The neurotoxicity of organic and inorganic lead. In *Metal Neurotoxicity*. S.C. Bondy and K.N. Prasod, Eds., CRC Press, Boca Raton, FL, pp. 2–17.

Buckler, H.M., W.D.F. Smith, and W.D.W. Rees (1986). Self poisoning with oral cadmium chloride. *Br. Med. J.* 1559–1560.

Buel, G. (1975). Some biochemical aspects of cadmium toxicology. *J. Occup. Med.* 17: 189–195.

Burk, R.F. and J.M. Lane (1983). Modification of chemical toxicity by selenium deficiency. *Fundam. Appl. Toxicol.* 3: 218–221.

Campbell, J.R., T.W. Clarkson, and M.D. Omar (1986). The therapeutic use of 2,3-dimercaptopropane-1-sulfonate in two cases of inorganic mercury poisoning. *JAMA* 256: 3127–3130.

Carter, R.F. (1966). Acute selenium poisoning. *Med. J. Aust.* 1: 525–528.

Civil, I.D.S. and M.J.A. McDonald (1978). Acute selenium poisoning. Case report. *N.Z. Med. J.* 87: 354–356.

Cole, D.E.C. and D.S. Lirenman (1978). Role of albumin enriched peritoneal dialysate in acute copper poisoning. *J. Pediatr.* 92: 955–957.

Collinson, P.O. and A.K. Burroughs (1986). Severe hypermagnesemia due to magnesium sulphate enemas in patients with hepatic coma. *Br. Med. J.* 293: 1013–1014.

Cook, D.G., S. Fahn, and K.A. Brait (1974). Chronic manganese intoxication. *Arch. Neurol.* 30: 59–64.

Cook, L., K.E. Stine, and L.W. Reiter (1984a). Tin distribution in adult rat tissues after exposure to trimethyltin and triethyltin. *Toxicol. Appl. Pharmacol.* 76: 344–348.

Cook, L.L., S.M. Heath, and J.P. O'Callaghan (1984b). Distribution of tin in brain subcellular fractions following the administration of trimethyltin and triethyltin to the rat. *Toxicol. Appl. Pharmacol.* 73: 564–568.

Cowie, R.L. and B.C. Escreet (1981). Potassium permanganate toxicity. *S. Afr. Med. J.* 60: 304.

DeSilva, P.E. and M.B. Donnan (1981). Chronic cadmium poisoning in a pigment manufacturing plant. *Br. J. Ind. Med.* 38: 76–86.

Doctor, S.V., L.G. Costa, D.A. Kendall, and S.D. Murphy (1982a). Trimethyltin inhibits uptake of neurotransmitters into mouse forebrain synaptosomes. *Toxicology* 25: 215–221.

Doctor, S.V., L.G. Costa and Murphy, S.D. (1982b). Effect of trimethyltin on chemically-induced seizures. *Toxicology Lett.* 13: 217–223.

Doolin, E.J. and C. Drueck (1980). Fatal iron intoxication in an adult. *J. Trauma* 20: 518–522.

El Mallakh, R.S. (1984). Treatment of acute lithium toxicity. *Vet. Hum. Toxicol.* 26: 31–35.

Finkel, A.J., Ed. (1983). *Hamilton and Hardy's Industrial Toxicology, 4th ed.* John Wright, PSG, Boston, p. 53–142.

Fisher, A.A. (1976). "Blackjack diseases" and other chromate puzzles. *Cutis* 18: 21–36.

Gage, J.C. (1964). Distribution and excretion of methyl and phenyl mercury salts. *Br. J. Ind. Med.* 21: 197–202.

Gerhard, R.E., E.A. Crecelius, and J.B. Hudson (1980). Moonshine related arsenic poisoning. *Arch. Intern. Med.* 140: 211–213.

Graber, T.W., A.S. Yee, and F.J. Baker (1981). Magnesium: physiology, clinical disorders and therapy. *Ann. Emerg. Med.* 10: 49–57.

Guilliford, M.C., N.P. Archard, and W. Vant Hoff (1985). Gold encephalopathy. *Br. Med. J.* 290: 1744.

Hasking, G. and J. Duggan (1982). Encephalopathy from bismuth subsalicylate. *Med. J. Aust.* 2: 167.

Hilmy, M.I., S.A. Rahim, and A.H. Abbas (1976). Normal and lethal mercury levels in humans. *Toxicology* 5: 155–159.

Hine, C.H. and A. Pasi (1975). Manganese intoxication. *West. J. Med.* 123: 101–107.

Hine, C.H., S.S. Pinto, and K.W. Nelson (1977). Medical problems associated with arsenic exposure. *J. Occup. Med.* 19: 391–396.

Hoffman, N.E., R.A. Trejo, N.R. DiLuzio, and J. Lamberty (1972). Ultrastructural alterations of liver and spleen following acute lead administration in rats. *Exp. Mol. Pathol.* 17: 159.

Hooper, W.F. (1981). Acute beryllium lung disease. *N.C. Med. J.* 42: 551–553.

Hoyt, R.E. (1986). Hyperkalemia due to salt substitutes. *JAMA* 256: 1726.

Jaffe, K.M., D.B. Shurtleff, and W.O. Robertson (1983). Survival after acute mercury vapor poisoning. Role of supportive care. *Am. J. Dis. Child.* 137: 749–751.

Jantsch, W., K. Kulig, and B.H. Rumack (1984). Massive copper sulfate ingestion resulting in hepatotoxicity. *Clin. Toxicol.* 22: 585–588.

Joselow, M.M., D.B. Louria, and A.A. Browder (1972). Mercuralism: environmental and occupational aspects. *Ann. Intern. Med.* 76: 119–130.

Kennedy, A., J.D. Dornan, and R. King (1981). Fatal myocardial disease associated with industrial exposure to cobalt. *Lancet* 1: 412–414.

King, M.E., A.M. Shefner, and R. Ehrlich (1964). Effectiveness of aurin tricarboxylic acid as an antidote for beryllium poisoning. *Ind. Med. Surg.* 33: 566–569.

Kleinfeld, M.J. (1980). Arsine poisoning. *J. Occup. Med.* 22: 820–821.

Langard, S. and T. Vigander (1983). Occurrence of lung cancer in workers producing chromium pigments. *Br. J. Ind. Med.* 40: 71–74.

Leach, C.N., Jr., J.V. Linden, and S.M. Hopfer, et al. (1985). Nickel concentrations in serum of patients with acute myocardial infarction or unstable angina pectoris. *Clin. Chem.* 31: 556–560.

Lin-Fu, J.S. (1982). Children and lead. *N. Engl. J. Med.* 307: 615–617.

Longstretch, W.T., L. Rosenstock, and N.J. Heyer (1985). Potroom palsy? Neurologic disorders in three aluminum smelter workers. *Arch. Intern. Med.* 45: 1972–1975.

Lovejoy, F.H. (1982). Thalium. *Clin. Toxicol. Rev.* 4(5): 1–2.

Marcus, S.M. (1982). Experience with d-Penicillamine in treating lead poisoning. *Vet. Hum. Toxicol.* 24: 18–20.

Marcus, S.M. (1983). Hydrogen sulfide. *Clin. Toxicol. Rev.* 5(4): 1–2.

Mordes, J.P., R. Swartz, and R.A. Arky (1975). Extreme hypermagnesemia as a cause of refractory hypotension. *Ann. Intern. Med.* 83: 657–658.

Morton, A.R. (1985). Severe hypermagnesemia after magnesium sulphate enemas. *Br. Med. J.* 291: 516.

Murphy, S.M., R.T. Owen, and P.J. Tyrer (1984). Withdrawal symptoms after six weeks treatment with diazepam. *Lancet* 2: 1389.

Osterloh, J. and C.E. Becker (1986). Pharmacokinetics of CaNa$_2$ EDTA and chelation of lead in renal failure. *Clin. Pharmacol. Ther.* 40: 686–693.

Parish, G.G., R. Glass, and R. Kimbrough (1979). Acute arsine poisoning in two workers cleaning a clogged drain. *Arch. Environ. Health* 34: 224–227.

Perry, H.M., C.S. Thind, and A.B. Perry (1976). The biology of cadmium. *Med. Clin. North Am.* 60: 759–768.

Pfeiffer, C.C. and B. Camo (1988). Wilson's disease. *Arch. Neurol.* 45: 247.

Phelan, D.M., S.R. Hagley, and M.D. Guerin (1984). Is hypokalemia the cause of paralysis in barium poisoning? *Br. Med. J.* 289: 882.

Potter, J.L. (1981). Acute zinc chloride ingestion in a young child. *Ann. Emerg. Med.* 10: 267–269.

Proudfoot, A.T., D. Simpson, and E.H. Dyson (1986). Management of iron poisoning. *Med. Toxicol.* 1: 83–100.

Rey, C.H., H.J. Reinecke, and R. Besser (1984). Methyltin intoxication in six men: toxicologic and clinical aspects. *Vet. Hum. Toxicol.* 26: 121–122.

Robotham, J.L. and P.S. Leitman (1980). Acute iron poisoning. A review. *Am. J. Dis. Child.* 134: 875–879.

Robotham, J.L., R.F. Troxler, and P.S. Lertman (1974). Iron poisoning: another emergy crisis. *Lancet* 2: 664–665.

Schoulmeester, W.L. and D.R. White (1980). Arsenic poisoning. *South. Med. J.* 73: 198–208.

Singer, I. and D. Rotanberg (1973). Mechanisms of lithium action. *N. Engl. J. Med.* 289: 254–260.

Sketris, I.S. and J.D. Gray (1982). Mercury poisoning. *Clin. Toxicol. Consult.* 5: 10–17.

Stevens, W.J. (1978). Thalium intoxication caused by a homeopathic preparation. *Toxicol. Eur. Res.* 1: 317–320.

Valdes, J.J., C.F. Mactutus, R.M. Santos-Anderson, R. Dawson, Jr., and Z. Annau (1983). Selective neurochemical and histological lesions in rat hippocampus following chronic trimethyltin exposure. *Neurobehav. Toxicol. Teratol.* 5: 357–361.

Walsh, M., F.J. Crosson, and M. Bayley (1977). Acute copper intoxication. *Am. J. Dis. Child.* 131: 149–151.

Watson, W.A., J.C. Veltri, and T.J. Metcalf (1981). Acute arsenic exposure treated with oral d-Penicillamine. *Vet. Hum. Toxicol.* 23: 164–165.

Wills, M.R. and J. Savory (1982). Aluminum poisoning: Dialysis encephalopathy, osteomalacia and anemia. *Lancet* 2: 29–34.

Solvents

Mohamed B. Abou-Donia
Departments of Pharmacology and Neurobiology
Duke University Medical Center
Durham, North Carolina

INTRODUCTION

Threshold Limit Value (TLV)

Solvents evaporate in the environment at rates dependent on their vapor pressures. The health risks and hazard of solvents are determined by their TLV values.

TLV refers to an airborne concentration of a substance. It represents the concentration that causes no adverse effects to workers after repeated exposure. Intended for use in the practice of industrial hygiene, TLV is determined by one or more of the following sources: (1) industrial experiences, (2) experimental investigations, and (3) human studies.

There are three categories of threshold limit values.

1. *The Threshold Limit Value-Time Weighted Average* (TLV-TWA) is the time-weighted average concentration for a normal eight-hour workday and a forty-hour work week, to which workers may be exposed repeatedly without adverse effect.
2. *Threshold Limit Value-Short Term Exposure Limit* (TLV-STEL) is the concentration to which workers may be exposed continuously for a short period of time (15 min) without suffering from irritation, chronic or irreversible tissue damage, or narcosis. STELs are recommended only when toxic effects have been reported from high,

short term exposures in either humans or animals. Exposures at the STEL should not be longer than 15 min and should be at least 60 min between successive exposures.

3. *Threshold Limit Value-Ceiling* (TLV-C) is the concentration that should not be exceeded, even instantaneously.

Threshold Limit Values for Mixtures

When two or more hazardous substances are present, their combined TLV should be determined. In the absence of information to the contrary, the effects of the combined chemicals should be considered as additive.

$$\text{TLV of mixture} = \frac{1}{\dfrac{f_a}{TLV_a} + \dfrac{f_b}{TLV_b} + \dfrac{f_c}{TLV_c} + \cdots \dfrac{f_n}{TLV_n}}$$

f = the percent composition (by weight) of the substance. TLV must be calculated in mg/mg^3.

$$\text{Substance concentration in ppm} = \frac{mg/m^3 \times 24.45}{MW}$$

PPM = parts per million; 24.45 is gas constant at 25°C and 760 mm of mercury barometric pressure; MW = molecular weight of the substance; m = meter.

A mixture contains by weight the following solvents: 50% heptane, 30% 1,1,1-trichloroethane, and 20% tetrachloroethylene.

Heptane: 50%   TLV = 400 ppm, MW = 100.21

Convert ppm to mg/m^3.

$$mg/m^3 = \frac{ppm \times MW}{24.45} = \frac{400 \times 100.21}{24.45} = 1,639$$

1,1,1-trichloroethane: 30%   TLV = 350 ppm, MW = 133.42

$$mg/m^3 = \frac{350 \times 133.41}{24.45} = 1,910$$

Tetrachloroethylene: 20%   TLV = 50 ppm, MW = 165.83

$$mg/m^3 = \frac{165.83 \times 50}{24.45} = 339$$

$$\text{TLV of mixture} = \cfrac{1}{\cfrac{0.5}{1,639} + \cfrac{0.3}{1,910} + \cfrac{0.2}{339}}$$

$$= \frac{1}{0.00031 + 0.00016 + 0.0006}$$

$$= \frac{1}{0.00107}$$

$$= 935 \text{ mg/m}^3$$

Of this mixture, 50% or $935 \times 0.5 = 468$ mg/m³ is heptane, 30% or $935 \times 0.3 = 281$ mg/m³ is 1,1,1-trichloroethane, 20% or $935 \times 0.2 = 187$ mg/m³ is tetrachloroethylene. To convert mg/m³ to ppm:

$$\text{Heptane} = \frac{468 \times 24.45}{100.32} = 114 \text{ ppm}$$

$$\text{1,1,1-trichloroethane} = \frac{281 \times 24.45}{133.41} = 51 \text{ ppm}$$

$$\text{Tetrachloroethylene} = \frac{187 \times 24.45}{165.83} = 28 \text{ ppm}$$

$$\text{TLV of mixture} = 114 + 51 + 28 = 193 \text{ ppm or } 935 \text{ mg/m}^3$$

ALIPHATIC HYDROCARBONS

Aliphatic hydrocarbons are long-chain hydrocarbon-containing chemicals or alicyclic compounds prepared from crude petroleum oil. This group is subdivided into three subgroups:

1. Alkanes (paraffins). These compounds are saturated hydrocarbons with the formula C_nH_{2n+2}.
2. Alkenes (olefins). These compounds contain one or more double bonds with the formula C_nH_{2n}.
3. Alkynes (acetylenes). These compounds contain one or more triple bonds with the formula C_nH_{2n-2}.

Gaseous Aliphatic Hydrocarbons

Chemistry
Compounds containing four or less carbon atoms are gases, i.e., methane (CH_4), ethane (CH_3-CH_3), propane ($CH_3-CH_2-CH_3$), butane ($CH_3-CH_2-CH_2-CH_3$), 1,3-butadiene ($CH_2=CH-CH=CH_2$), and acetylene ($HC\equiv CH$). Methane is the major constituent of natural gas (up to 85%) with ethane, propane, butane, and nitrogen making up the balance. 1,3-Butadiene is used in the manufacturing of synthetic rubber, rocket fuel, plastics, and resins.

Action
These compounds (especially methane and ethane) are simple asphyxiants that cause toxicity by displacing oxygen (Aviado and Belej, 1974). No systemic or irritant effect has been attributed to these compounds. Low concentrations, e.g., 15% of propane, *n*-butane, and isobutane produced myocardial sensitization and dysrhythmias in experimental animals. Propane, butane, and acetylene have low central nervous system depressant effects. Commercial grades of acetylene contain impurities such as arsine, hydrogen sulfide, carbon disulfide, and carbon monoxide; these impurities cause the worst toxic effects of exposure to commercial acetylene.

Liquid Aliphatic Hydrocarbons

Long-Chain Compounds
Hydrocarbon compounds containing 5–16 carbons are liquid at room temperature, with C_5-C_8 compounds being very volatile solvents.

Action.
1. *Acute toxicity.* These aliphatic hydrocarbons are central nervous system depressants at high concentrations. Aspiration of these chemicals produces severe chemical pneumonitis. They may also irritate mucous membrane and respiratory tract. Liquid aliphatic hydrocarbons cause drying and defatting of the skin followed by dermatitis.
2. *Gamma diketone-induced neurotoxicity (GDIN).* *n*-Hexane and other liquid aliphatic hydrocarbons that undergo biotransformation into gamma diketone produce central-peripheral distal axonopathy. In the early 1960s in Japan, workers exposed to *n*-hexane in the laminating industry developed sensor-motor polyneuropathy. This was followed by similar reports in Europe, and the U.S. Subsequently, neuropathological damage was demonstrated in exposed rats and in nerve biopsies from patients with industrial or glue-sniffer's neuropathy (Abou-Donia and Lapadula, 1988, 1989).
   A complicating factor in some *n*-hexane neurotoxicity cases was

the exposure of individuals who developed neuropathy to other solvents such as methyl isobutyl ketone (MiBk), methyl ethyl ketone (MEK), and acetone (all are nonneurotoxicant), ethyl acetate, xylene, methanol (produces retinal and optic nerve damage), toluene (implicated in causing neurobehavioral effects in humans), isopropyl alcohol, *n*-heptane (neurotoxicant) and *n*-octane.

MnBK was first implicated in an outbreak of peripheral neuropathy at a polyvinyl fabric printing factory in Columbus, Ohio. Only aliphatic hydrocarbons with 1,4 (gamma) spacing of the carbonyl groups are neurotoxic. Therefore 2,5-heptanedione and 3,6-octanedione are neurotoxic. In contrast, the following ketones are not neurotoxic: 3-heptanone; 3,5-heptanedione; 2,4-hexanedione; 2,3-hexanedione; 1,6-hexancdione; 1,4-butanediol; gluteraldehyde; and acetone.

Hexane is available as a 100% *n*-hexane and as a technical- or commercial-grade hexane containing 45–86% *n*-hexane with the remaining constituents being isohexanes, cyclopentanes, benzene, toluene, xylene, acetone, and chlorinated hydrocarbons. *n*-Hexane is a straight 6-carbon and 14-hydrogen chemical. It is an unreactive chemical and is stable to treatment with concentrated sulfuric acid, boiling nitric acid, and molten sodium hydroxide. *n*-Hexane undergoes metabolic biotransformation to 2-hexanol, which is subsequently oxidized to 2,5-hexanediol, 2-hydroxy-5-hexanone, 2-hexanone (MnBK), 2,5-hexanediol (2,5-HCOH), and 2,5-hexanedione (2,5-HD) (Abdel Rahman et al., 1976). All of these metabolites of *n*-hexane are more neurotoxic than the parent compound. Neurotoxic potency of *n*-hexane and some of its metabolites are in ascending order: *n*-hexane < MnBK < 2,5-HDOH < 2,5-HD (Abo et al., 1982; Abou-Donia et al., 1982). The metabolite 2,5-HD is believed to be the neurotoxic agent responsible for *n*-hexane's and its metabolites' neurotoxicity. Accumulation of the 10-nm neurofilaments is a pathognomonic feature of this neuropathy (Spencer and Schaumberg, 1975). The presence of neurofilamentous masses in the axons above the nodes of Ranvier results in the formation of giant axonal swelling. Several hypotheses have been proposed to account for the pathogenesis of gamma-diketone-induced neurotoxicity.

a. *Inhibition of glycolysis.* 2,5-HD inhibits glycolytic enzymes phosphofructokinase and glyceraldehyde-3-phosphate dehydrogenase (Sabri et al., 1979). This has led to the hypothesis that this compound interferes with axonal transport through the inhibition of glycolysis and energy production. This hypothesis is unlikely, however, because of the extremely high concentrations of 2,5-HD needed for *in vivo* inhibition of glycolysis.

b. *Pyrrole formation.* 2,5-HD was proposed to react with lysine amino groups of neurofilaments to form pyrrole which would interfere with their interaction with other proteins (DeCaprio et al., 1982). Neurotoxic potency correlated well with ease of formation of pyrrole adducts and accumulation of neurofilaments produced by analogs of 2,5-HD. 3,5-Dimethyl-2,5-hexanedione (DMHD) that is 10–20 times more neurotoxic than 2,5-HD, also formed pyrrole more easier than 2,5-HD and its neuropathy was more proximal. Furthermore, 3-methyl-2,5-hexanedione resulted in an accumulation of neurofilaments in the middle of the axon, between the location of that by 2,5-HD and DMHD.

c. *Crosslinking of neurofilaments.* 2,5-Hexanedione was proposed to directly bind and subsequently crosslink neurofilaments (Graham, 1980; Graham and Abou-Donia, 1980). This crosslinking was also proposed as the mechanism by which neurofilaments accumulate above nodes of Ranvier and form giant axonal swellings that are characteristic of gamma ketone-induced neurotoxicity. Neurofilament crosslinking was demonstrated *in vitro* and *in vivo* with 2,5-HD and 3,4-dimethyl-2,5-hexanedione (Anthony et al., 1983). Recent studies have suggested that autooxidation of pyrroles with subsequent crosslinking of neurofilaments may be involved in the pathogenesis of GDIN. It is unclear, however, whether neurofilament crosslinking is the primary or secondary mechanism by which aliphatic hydrocarbons produce neurotoxicity.

d. *Decreased phosphorylation of neurofilaments.* The mechanism of GDIN may be related to the interference of gamma diketones with protein kinase-mediated phosphorylation of neurofilaments (Lapadula et al., 1988). It has been proposed that binding of 2,5-HD with lysine residues may block the phosphorylation of adjacent serine groups. In normal conditions, phosphorylated neurofilaments predominantly occur in the axon while dephosphorylated neurofilament proteins are primarily present in cell bodies. Recent studies have demonstrated that 2,5-HD diminished protein kinase phosphorylation of neurofilament *in vivo*.

e. *Diminished proteolysis of neurofilaments.* 2,5-Hexanedione may prevent normal proteolysis of neurofilament proteins either by its binding to the neurofilaments or by inhibition of proteases. Neurofilament proteins are degraded primarily in the terminal portions of the axon by Ca^{2+}-activated proteases. Neurofilament from 2,5-hexanedione-treated rats exhibited a diminished amount of neurofilament breakdown products (Cavanagh, 1982; Lapadula et al., 1988).

Although more than one mechanism for GDIN may be operating, the following cascade of events seem to be supported by present knowledge. *n*-Hexane is metabolized to 2,5-HD, which causes decreased kinase-mediated phosphorylation of neurofilament proteins. This leads to breakdown of the cytoskeletal matrix and dissociation of neurofilament proteins. This is consistent with increased fast axonal transport of neurofilament proteins in 2,5-HD treated animals, which may contribute to their accumulation in the distal parts of the axon. Accumulated neurofilaments would then react with 2,5-HD forming crosslinked neurofilaments and resulting in giant axonal swellings and distal axonopathy.

The neurotoxicity induced by *n*-hexane, a weak neurotoxic agent, is synergized by other nonneurotoxic aliphatic hydrocarbons. Thus, 2-butanone (methyl ethyl ketone, MEK) and 4-methyl-4-pentanone (methyl isobutyl ketone, MiBK), neither of which is a neurotoxicant, have increased the neurotoxic potency of *n*-hexane. The mechanism of this synergistic action is the induction of cytochrome P-450 mixed function oxidases that are responsible for the oxidative metabolism of *n*-hexane to the active neurotoxic metabolite 2,5-HD (Abou-Donia et al., 1985). These results indicate that neither MEK nor MiBK should be used concurrently with *n*-hexane, since it may lead to increased neurotoxic action of *n*-hexane resulting in neurologic dysfunctions. MiBK, which occurs naturally in oranges and grapes, has become an environmental contaminant because of improper disposal of industrial waste.

Alicyclic Compounds

These colorless liquid chemicals are saturated or unsaturated hydrocarbons with three or more carbon atom rings. They include cycloalkines, cycloparaffins, and naphthenes (Perbellini et al., 1980). They are present in petroleum solvents and are used in the manufacture of organic chemicals.

Action. Alicyclic hydrocarbons possess effects similar to those of aliphatic hydrocarbons but have greater depressive action on the central nervous system. Cyclohexane does not produce polyneuropathy. It is metabolized in man to cyclohexanol. Cyclopropane releases endogenous catecholamines; it is incompatible with epinephrine.

Solid Aliphatic Hydrocarbons

Long-chain aliphatic hydrocarbons of 16 carbon atoms or more are solid at room temperature. Paraffin is in a mixture of C_{16} hydrocarbons; higher hydrocarbons are used for candles, wax paper, cosmetics, polish and water proofing.

Action
   Solid aliphatic hydrocarbon do not produce acute toxicity.

HALOGENATED HYDROCARBONS

   Halogenated hydrocarbons are excellent solvents. They are not explosive with low flammability. Their acute effect is central nervous system depression. Their actions are summarized in Table 1.

Carbon Tetrachloride

Chemistry
   The chemical structure of carbon tetrachloride is CCl_4. It is clear, colorless, highly volatile, and nonflammable liquid. Pyrolysis of carbon tetrachloride produces phosgene and hydrogen chloride.

$$CCl_4 \xrightarrow{\Delta} COCl_2 + HCl$$
$$\text{Phosgene}$$

Synonyms are tetrachloromethane and perchloromethane.

Uses
   Carbon tetrachloride's major use is in the manufacturing of fluorocarbon refrigerants, solvents, and aerosol propellants. Safer chlorinated hydrocarbons, e.g., methylchloroform (1,1,1-trichloroethane) and perchloroethylene (tetrachloroethylene) have replaced CCl_4.

Disposition
   Carbon tetrachloride is absorbed through the lungs, gastrointestinal tract, and skin. It is concentrated in fat, and its concentration in the liver and kidney is higher than that of the blood (Stewart and Dodd, 1964). A total of 50–80% of the dose is exhaled by the lung as unchanged carbon tetrachloride, while only 4% is expired as carbon dioxide. It is metabolized via cytochrome P-450 to phosgene and trichloromethyl free radical, which is subsequently metabolized to chloroform and carbon dioxide as well as hexachloroethane as follows:

$$Cl \longleftarrow CCl_4 \longrightarrow COCl_2$$
$$\text{Phosgene}$$

$$CO_2 \leftarrow HCOOH \leftarrow CHCl_3 \longleftarrow {}^{\bullet}CCl_3 \longrightarrow C_2Cl_6$$

| Chloroform | Trichloromethyl free radical | Hexachloroethane |

TABLE 1.
Effects of Halogenated Hydrocarbons

	Use	CNS Depression	Heart Sensitization	Liver Injury	Kidney Injury	Active Metabolite
Carbon tetrachloride	Solvent	+	+	+++	++	·CCl₃
Chloromethane (ClCl₃)	Paint stripper	+	+	+++	+++	Phosgene
Dichloromethane (Cl₂CH₂)	Dry cleaning	+	–	±	–	CO
Trichloroethylene (Cl₂C=CHCl)	Spot remover	+	+	±	–	
Tetrachloroethylene (Perchloroethylene) (Cl₂C=CCl₂)	Spot remover	+	–	±	–	
1,1,1-Trichloroethane (Cl₃C–CH₃)	Spot remover	+	+	±	–	
1,1,2-Trichloroethane (Cl₂CH–CH₂Cl)	Spot remover	+	?	+	+	

Trichloromethyl free radical attacks proteins resulting in their crosslinking. It destroys membranes by attacking membrane lipids and producing lipid peroxidation (Gordis, 1969). An alternative hypothesis to the formation of the trichloromethyl free radical is through the formation of trichloromethylperoxy free radical ($Cl_3COO^{\cdot}$), which results from the reaction between oxygen and trichloromethyl free radical. A third hypothesis is that under anaerobic conditions carbon tetrachloride may be converted to carbene, $Cl_3C{:}$.

Action

Acute exposure to carbon tetrachloride causes central nervous system depression followed by hepatic and renal dysfunction. Symptoms are headache, weakness, lightheadedness, blurred vision, ataxia, and unconsciousness (Stevens and Forster, 1953). Cerebellum is particularly sensitive to CCl_4 actions. Effects on the optic system are characterized by bilateral visual field constriction, optic atrophy and amblyopia. Alcoholics are more sensitive to CNS effects. Death results from respiratory depression or from dysrhythmias produced by catecholamine sensitization.

Hepatotoxicity caused by carbon tetrachloride is characterized by acute fatty degeneration of the liver and hepatic necrosis in the midzonal and centrilobular regions (Moon, 1950). Carbon tetrachloride also produces renal damage of acute tubular necrosis, primarily the proximal tubules and the loop of Henle (Ehrenreich, 1977). Pulmonary hemorrhage and edema are produced by carbon tetrachloride. Chronic exposure to carbon tetrachloride causes hepatocellular carcinoma in hamsters and hepatomas in mice and rats.

Treatment

Clinical tests for CCl_4 toxicity are elevation of hepatic aminotransferases, hematuria, proteinuria, and oliguria. Attention should be given to respiratory depression and cardiac dysrhythmias. Gastrointestinal decontamination is carried out using syrup of ipecac/lavage and activated charcoal. Skin should be washed.

Chloroform

Metabolism

The chemical structure of chloroform is $CHCl_3$. It is a colorless, nonflammable, and highly volatile liquid. Thermal degradation and metabolism with cytochrome P-450 produces phosgene, carbon dioxide, and hydrochloric acid.

$$CHCl_3 \xrightarrow[\text{or cyt. P-450}]{\Delta/[O]} \underset{\text{Trichloromethanol}}{CCl_3OH} \xrightarrow{-HCl}$$

$$O = \underset{\text{Phosgene}}{CCl_2} \xrightarrow{H_2O} CO_2 + HCl$$

It has the synonyms trichloromethane and methenyl trichloride.

Uses

Chloroform is now restricted to industrial uses as a solvent or a chemical intermediate. In 1847 it was introduced as a general anesthetic, because it was less volatile and not as flammable as ether. In 1912 its use as an anesthetic was discontinued as it was found to cause death. In 1976 the FDA banned it as an additive in mouth wash, toothpaste, and cough syrup because it produced liver, kidney, and thyroid tumors in rats.

Action

Chloroform depresses the central nervous system resulting in nausea, lightheadedness, headache, malaise, and coma. It may cause transient change in gait and slight tremor (Schroder, 1965). It produces hepatotoxicity characterized by fatty infiltration and necrosis that is less severe than that produced by carbon tetrachloride. Also, unlike CCl_4, chloroform does not produce lipid peroxidation, and hepatotoxicity may result from covalent binding with phosgene. Hepatotoxicity is increased by cytochrome P-450 inducers such as phenobarbital. Chloroform depletes hepatic glutathione. Chloroform may cause renal toxicity 24–48 h after exposure, and it is characterized by proteinuria, hematuria, and cellular casts. Chloroform may also cause chemical pneumonitis (Timmis and Moser, 1975). Chloroform, like most chlorinated hydrocarbons, causes sensitization of the myocardium to endogenous catecholamines. Chronic exposure to chloroform produced hepatocellular carcinoma in mice and kidney tumors in rats.

Treatment

Decontamination of the gastrointestinal tract (using syrup of ipecac) and the skin is indicated. Supportive care is prescribed to treat cardiac dysrhythmias and respiratory depression. Renal failure may require dialysis.

HYDROXY COMPOUNDS

Methanol

Chemistry

Methanol, also known as methyl alcohol, has the structure CH_3OH. It is miscible with water and many organic solvents. It is produced during destructive distillation of wood.

Uses

Methanol is used as a solvent, antifreeze (95% CH_3OH), and windshield washer fluid (35–95%).

Disposition

Methanol is well absorbed from the gastrointestinal tract and to a lesser extent from skin and the lung (Oalase and Tephly, 1975). It is distributed similarly to ethanol and has a volume of distribution of 0.6–0.7 L/kg. The highest concentrations occur in the kidney, liver, and the gastrointestinal tract. Smaller concentrations are found in the brain, muscle, and adipose tissue.

Methanol is excreted mostly as metabolites following its metabolism with liver enzymes. Only 2–5% is excreted unchanged in the urine. Methanol is rapidly oxidized by alcohol dehydrogenase to formaldehyde, which is rapidly oxidized further by aldehyde dehydrogenase to formic acid. Formic acid is oxidized via a folate-dependent pathway to carbon dioxide. This latter step is faster in rats, which accounts for the low methanol toxicity in this species. Ethanol has 10–20 times greater affinity for alcohol dehydrogenase than methanol, which results in ethanol competing more effectively for metabolism by this enzyme. Thus, methanol is oxidized ten times more slowly than ethanol.

The serum half-life of methanol is 10–20 h which increases to 30–35 h following concurrent administration with ethanol.

Action

Methanol causes CNS depression similar to ethanol (Lacouture and Lovejoy, 1981). Its toxicity results from its oxidation to formaldehyde, which causes retinal damage that may lead to blindness, and formic acid, which causes acidosis. Methanol oxidation may decrease the intracellular NAD/NADH$^+$ ratio, which results in stimulation of anaerobic glycolysis and lactate production (Shahangian and Ashko, 1986).

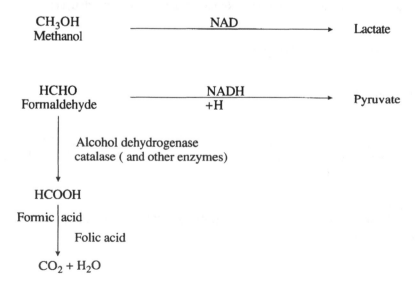

Symptoms

Methanol-induced CNS symptoms include headache, vertigo, lethargy, and confusion. These symptoms may progress to coma and convulsions in severe cases (Guggenheim et al., 1971). Severe metabolic acidosis may lead to dyspnea. Serious signs include bradycardia, shock, and anuria. It may cause putamen necrosis and a permanent Parkinsonian-like syndrome. It produces much less euphoria than ethanol. Death results from respiratory depression.

The methanol metabolite formaldehyde produced ocular toxicity characterized by blurred vision, decreased visual acuity, photophobia, dilated pupils, retinal edema, and hyperemia of the optic nerve.

Methanol causes nausea, vomiting, and abdominal pain. Hemorrhagic pancreatitis is produced. Patients may develop renal failure.

Treatment

Ethanol is the antidote for methanol toxicity. It acts by blocking its oxidation to formaldehyde and formic acid. Blood ethanol concentration should be measured between 100 and 150 mg/100 ml. Ethanol may be administered orally or via i.v. injection. Folic acid, 50 mg i.v. every 4 h for several days, has been recommended for treatment of methanol toxicity by increasing the oxidation of formic acid to CO_2. Gastrointestinal decontamination is carried out using ipecac. The patient should be monitored and stabilized.

Isopropanol

Chemistry

Isopropanol, also known as isopropyl alcohol and 2-propanol has the chemical structure CH_3–CHOH–CH_3. It is a clear, volatile liquid.

Uses

Isopropanol is used in industry as a solvent and disinfectant and in the home, as rubbing alcohol (70% isopropanol) and in skin and hair products.

Disposition

Isopropanol is absorbed rapidly from the gastrointestinal tract; 80% is absorbed in 30 min, and absorption is completed in 2 h. It is less absorbed through the skin (Lacouture, 1980). It distributes in body water with an apparent volume of distribution of 0.6–0.7 L/kg. About 20–50% of absorbed isopropanol is excreted via the kidney. Isopropanol is metabolized to acetone via alcohol dehydrogenase, which is eliminated through the kidney and the lung (Daniel et al., 1981). Acetone is oxidized further to acetic acid, formic acid, and CO_2. Isopropanol has a half-life of 2.5–3.2 h. Some isopropanol is excreted into the stomach and saliva.

Action

Isopropanol produces CNS and cardiovascular depression (Lacouture, 1980). It is the second most ingested alcohol behind ethanol. It was reported as the fifth most common cause of drug overdose.

Symptoms

CNS symptoms include dizziness, poor incoordination, headache, and confusion that may progress to coma (Lacouture et al., 1981). Isopropanol causes gastric irritation, which results in abdominal pain and vomiting. It produces hypotension resulting from peripheral vasodilation. Tachycardia may also result. Other symptoms may include acute tubular necrosis, hepatic dysfunction, hemolytic anemia, and myoglobinuria. Mild hypothermia may result from CNS peripheral vasodilation. The minimum toxic dose is 0.5 mg/kg, and the lethal dose is 2–4 ml/kg.

Treatment

There is no antidote. Treatment includes gut decontamination using ipecac within 2 h after ingestion. Supportive care is indicated to treat hypotension, first with fluids, then with vasopressors.

Ethylene Glycol

Chemistry

Ethylene glycol has the chemical name 1,2-ethanediol, and its chemical structure is $HO–CH_2–CH_2–OH$. It is a colorless, odorless, sweet-tasting, nonvolatile (bp 197°C) liquid that is highly water soluble.

Uses

Ethylene glycol is used as antifreeze, coolant, preservative, and glycerine substitute.

Disposition

Ethylene glycol is rapidly absorbed from the gastrointestinal tract but not from the lung or through the skin. It distributes in body water. About 20% of the dose is excreted unchanged in the urine; less than 1% of ethylene glycol is metabolized to oxalic acid. Thiamine and pyridoxine are cofactors for ethylene glycol metabolism. Liver enzymes oxidize ethylene glycol primarily to glycoaldehyde, glycolate, and then glyoxylate. These reactions require the reduction of NAD to NADH and change in the NAD/NADH ratio which shifts pyruvate to lactate. The acidic metabolites are more toxic than the ethylene glycol. The order of toxicity in descending order is glyoxalate > glycoaldehyde > ethylene glycol. The affinity of ethylene glycol for alcohol dehydrogenase is 100 times less than that of ethanol.

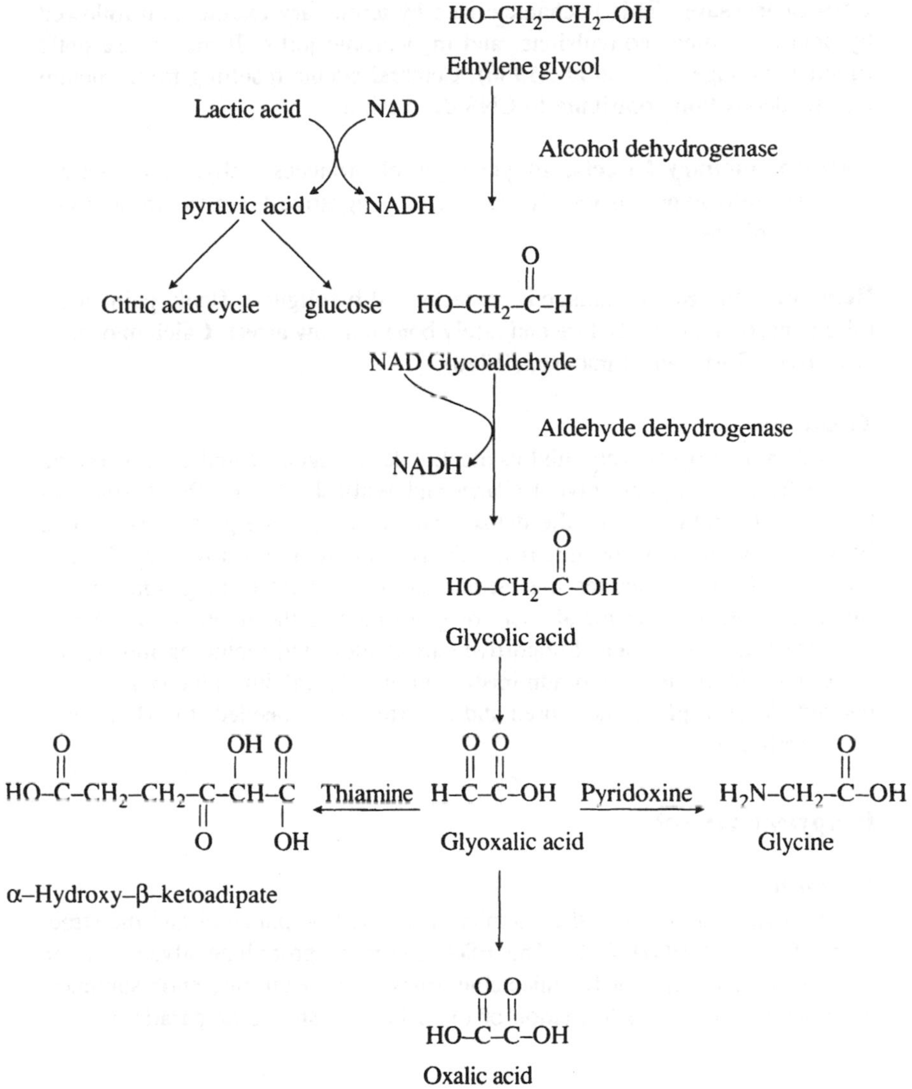

HO–CH₂–CH₂–OH

Ethylene glycol

Lactic acid   NAD

Alcohol dehydrogenase

pyruvic acid   NADH

Citric acid cycle   glucose   HO–CH₂–C̈–H

NAD Glycoaldehyde

Aldehyde dehydrogenase

NADH

HO–CH₂–C̈–OH

Glycolic acid

$$\underset{\substack{\\ \text{α–Hydroxy–β–ketoadipate}}}{\text{HO–C–CH}_2\text{–CH}_2\text{–C–CH–C}} \quad \xleftarrow{\text{Thiamine}} \quad \underset{\text{Glyoxalic acid}}{\text{H–C–C–OH}} \quad \xrightarrow{\text{Pyridoxine}} \quad \underset{\text{Glycine}}{\text{H}_2\text{N–CH}_2\text{–C–OH}}$$

HO–C–C–OH

Oxalic acid

Action

Ethylene glycol is a CNS depressant similar to that of ethanol. In addition ethylene glycol produces toxic metabolites (Clay and Murphy, 1977). The metabolic acidosis results from glycolic acid and lactic acid formation. Oxalic acid produces myocardial depression and acute tubular necrosis. Oxalic acid may chelate calcium ion resulting in hypocalcemia.

Symptoms

Symptoms are characterized by three phases.

CNS Depression. This is characterized by temporary exhilaration followed by acidosis, coma, convulsions, and myochronic jerks. It may cause optic function damage (Turk et al., 1986). Cerebral edema resulting from calcium oxalate deposition contribute to CNS depression.

Cardiopulmonary Effects. Ethylene glycol produces tachycardia, tachypnea, and mild hypertension. It may cause congestive heart failure and circulatory collapse.

Renal Effects. Renal damage is characterized by oliguria, flank, pain, acute tubular necrosis, renal failure and rarely bone marrow arrest. Calcium oxalate is formed (Burns and Finalyson, 1980).

Treatment

Ethanol competitively inhibits alcohol dehydrogenase and diminishes the metabolism of ethylene glycol (Parry and Wallach, 1974). Pyridoxine and thiamine are cofactors for the detoxification of ethylene glycol and should be given, 50 ml and 100 ml, respectively, i.m. four times per day for two days. Pyridoxine promotes the metabolism of glyoxalate to glycine, while thiamine promotes the metabolism of glyoxalate to the nontoxic metabolite α-hydroxy-β-ketoadipate. Supportive care is indicated including monitoring of serum calcium level and administration of 10% calcium gluconate i.v., if needed. Arterial pH is monitored and is corrected, if needed, to pH 7.2 with i.v. bicarbonate.

Propylene Glycol

Chemistry

Propylene glycol has the chemical name 1,2-propanediol and the structure $HO-CH_2-CHOH-CH_3$. The FDA considers propylene glycol safe in small doses as a diluent for injectable drugs, e.g., vitamins, antihistamines, barbiturates and in the formation of topical and cosmetic preparations.

Disposition

Dermal absorption is minimal. Approximately 45% is excreted unchanged by the kidney. The balance is metabolized by alcohol dehydrogenase to lactate, acetate, and pyruvate (Ruddick, 1972).

Action

Propylene glycol is a CNS depressant with as much as 1/3 the action of ethanol. It has low oral toxicity. Its metabolism in the liver produces lactic acid, which enters the glycolysis pathway and causes metabolic acidosis.

Symptoms

CNS depression may result in hypoglycemia, lactic acidosis, and seizures (Atulanantham and Genel, 1978).

Treatment

Decontamination with ipecac and supportive care is indicated.

Diethylene Glycol

Diethylene glycol has the chemical structure

$$HO-CH_2-CH_2-OCH_2-CH_2-OH$$

It is less toxic than ethylene glycol, since it is not metabolized to oxalate or formate. However, it is more nephrotoxic than ethylene glycol (Hardin and Lyon, 1984).

Treatment

Decontamination with ipecac and supportive measures is indicated.

Phenol

Chemistry

Phenol is an aromatic compound, monohydroxybenzene, and has the chemical formula C_6H_5OH. It is a white crystalline powder that is soluble in alcohol and oil but less soluble in water. Creosote is an oily mixture of phenolic compounds.

Uses

It is used in the manufacture of phenol-formaldehyde resins and plastics. It is also used in wood preservatives, disinfectants, and as an intermediate in chemical industry.

Disposition

Phenol is rapidly absorbed via the lung, skin, and gastrointestinal tract. Most absorbed phenol undergoes metabolic biotransformation in the liver to glucuronide and ethereal sulfate. The kidney excretes conjugated phenol metabolites and some unchanged phenol rapidly in the urine (Piotnowski, 1971).

Action

Phenol causes depression of the central nervous system. Initial phenol action may be seen as excitatory phase followed by coma and respiratory arrest (Baker et al., 1978). It also causes convulsions, hypotension, and ventricular tachycardia. Phenol also causes metabolic acidoses. Chronic exposure to phenol results in diarrhea, sore throat, pharyngeal sores, and dark urine resulting from oxidation products of phenol. Phenol denatures proteins and results in chemical burns to skin.

Treatment

Initial treatment should be the stabilization of breathing and circulation using breathing, i.v. fluids and ventilation. Gastrointestinal decontamination is carried out using activated charcoal.

PETROLEUM DISTILLATES

Petroleum distillates are crude oil by-products that are classified in Table 2 (Goldfrank et al., 1979). It should be noted that turpentine is not a petroleum distillate, but rather is an oil resin solvent extracted from pine resin by steam distillation.

Action

The target organs for petroleum hydrocarbons are the lung and the central nervous system (Eade et al., 1974). Aspirated hydrocarbons produce pulmonary insufficiency leading to death. Their effects are rarely fatal although they cause central nervous system depression leading to drowsiness, tremors, and convulsions. These symptoms result from hypoxia.

The narcotizing effect of aspirated hydrocarbons on the lung may result from inhibition of surfactant that results in alveolar instability, early distal airway closure, and eventual hypoxemia. Histopathological alterations of the lung are characterized by interstitial inflammation, atelectasis, hyperemia, vascular thrombiosis, bronchial and bronchiolar necrosis, intraalveolar hemorrhage, edema, and polymorphonuclear exudate.

Symptoms

Pulmonary symptoms resulting from aspiration of hydrocarbons include gasping, coughing, and choking. Central nervous system symptoms are lethargy, irritability, dizziness, and coma. Gastrointestinal tract symptoms are usually minor but include burning of the mouth, sore throat, nausea, and vomiting (Banner and Walson, 1983).

Although myocardial effects are rare, they may present as congestive failure secondary to a cardiomyopathy. Abuse of hydrocarbons results in sudden death from dysrhythmias due to sensitization of the myocardium to endogenous catecholamines.

Chronic exposure to hydrocarbons produces both glomerular damage (Goodpasture's syndrome) and renal tubular acidosis. Although these chem-

TABLE 2.
Classes of Petroleum Distillates

Class	Composition	Boiling Range °C
Petroleum ether (benzin or benzine)	*n*-Pentane and *n*-hexane	35–80
Petroleum naphtha (ligroin)	C_5 to C_{13} aliphatic hydrocarbons	30–238
Rubber solvent	C_5 to C_9 aliphatic hydrocarbons	38–149
Stoddard solvent	C_9 to C_{12} straight- and branched-chain hydrocarbons, naphthalene and higher aromatic hydrocarbons	152–210
VM (varnish makers') and PL (painters' naphtha)	C_7 to C_{10} aliphatic hydrocarbons	94–175
Gasoline	C_5 to C_{12} aliphatic hydrocarbons and xylene. Tetraethyl lead and alcohol may be added	40–225
Kerosene	C_{10} to C_{16} aliphatic hydrocarbons, xylene, and naphthalene	175–325
Mineral seal oil (furniture polish)	Saturated, higher-molecular-weight aliphatic hydrocarbons other than gasoline and kerosene	200–370
Diesel oil, fuel oil	C_9 and higher aliphatic hydrocarbons	less volatile than kerosene
Shoe polish	Chlorinated hydrocarbons, toluene	
Lighter fluid	Petroleum naphtha and kerosene	
Solvents and thinners	Petroleum ether, stoddard solvent (10–20% aromatic hydrocarbons) and VM and P naphthas	

icals did not produce death or permanent damage related to hepatic or renal function, temporary damage has been reported for liver, spleen, and kidney.

Treatment

Since the primary cause of death due to petroleum distillate ingestion is respiratory failure, particular attention should be given to maintain adequate breathing (Ng et al., 1974). Positive airway pressure may be initiated and intubation may be necessary. Epinephrine is not recommended to treat bronchospasm because of the possibility of myocardial sensitization to catecholamines. β-2 selective bronchodilator inhalants such as Alupent or salbutamol are recommended. Induction of emesis is not recommended because of the possibility that some of these compounds may be aspirated, resulting in chemical pneumonitis.

AROMATIC HYDROCARBONS

Benzene

Chemistry

Benzene is a colorless, highly flammable, and highly volatile liquid. Most of its use (85%) in the chemical industry is to produce styrene, phenol, and cyclohexane. It is also used in the manufacture of detergents, explosives, and pharmaceuticals. Benzene is a natural constituent of gasoline and averages 0.8% in American gasoline and up to 5% in other countries. Over two million workers are exposed to benzene annually. In 1979, approximately 12 billion pounds were produced. In addition, 1.6 billion pounds were imported.

Exposure

Humans are exposed to benzene principally by vapor inhalation or contact with skin. In 1976 it was estimated that 1.3 billion pounds of benzene were released into the atmosphere. Consequently, over 75% of the U.S. population may have been exposed to this chemical. Benzene has been detected in drinking water at concentrations of up to 10 ppm. It also is present in fruits, fish, vegetables, nuts, dairy products, beverages, and eggs. The National Cancer Institute estimated that a daily intake of benzene may reach 250 μg/day.

The TLV limit was set at 10 ppm by OSHA in 1971. However, in 1977, by an emergency temporary standard, TLV was lowered to 1 ppm with a maximum of 5 ppm for 15 min. These limits were subsequently overturned by the U.S. Supreme Court. The current TLV-TWA limit is 10 ppm with a ceiling value of 25 ppm.

Disposition

Benzene is absorbed rapidly via gastrointestinal and pulmonary routes and to a lesser extent through the skin (Bergman, 1979). Significant portion (10–50%) of absorbed benzene is exhaled through the lungs. Only small amounts are excreted in the urine and bile as benzene. It is metabolized in the liver by cytochrome P-450 mixed function oxidase system to phenols. It conjugates with glutathione to form phenylmercapturic acid. Benzene oxide is its most toxic metabolite and may be responsible for benzene-induced hematological abnormalities.

Action

Acute exposure to benzene results in central nervous system depression characterized by euphoria, headache, nausea, and ataxia. These symptoms may progress to a change in gait, coma, and convulsions. Acute effect of benzene may result in arrhythmias due to myocardial sensitization to endogenous catecholamines.

Chronic exposure by humans to low levels of benzene is associated with blood disorders such as aplastic anemia and leukemia (Braier, 1983). Aplastic anemia is the failure of blood-forming elements to mature, which is characterized by hypocellular bone marrow and pancytopenia. Bone marrow toxicity to benzene is characterized by a progressive decrease in each of the circulating formed elements of blood, i.e., erythrocytes (red blood cells), leukocytes (white blood cells), and thrombocytes (platelets). When all three cell types have been sufficiently absent, the condition is called pancytopenia. Aplastic anemia is the pancytopenia condition in the absence of functional marrow as the result of fatty replacement of bone marrow. Among leukocytes, granulocyte levels are usually more depressed by benzene than lymphocyte levels. Hydroquinone is the benzene metabolite responsible for aplastic anemia. This condition results in weakness, fatigue, increased infections, and poor blood coagulation, e.g., bleeding gums and gastrointestinal hemorrhage.

Epidemiological studies showed that workers exposed to benzene for more than five years have a risk of death from myelocytic and monocytic leukemia that was 21 times greater than control group. Leukemia involves the abnormal proliferation of hematopoietic cells. The leukemia most commonly associated with benzene exposure is acute myelogenous leukemia characterized by an increased number of cells morphologically similar to myeloblast (nonlymphocytic leukemia). Benzene epoxide is the metabolite suspected to cause leukemia. The carcinogenicity of benzene is supported by the chromosomal aberrations in workers, human cell cultures, and animals.

Experimental studies have shown that there are no animal models for benzene-induced leukemia. However, oral administration of benzene increased the incidence of zymbal gland carcinoma. Inhalation of benzene vapor by male mice increased the incidence of lymphoid tumors.

Treatment
Attention should be given to maintaining ventilation and detecting dys-rhythmias. Oxygen administration may be indicated. Gastrointestinal tract decontamination by emesis may be necessary. Chronic toxicity is treated symptomatically, i.e., chemotherapy and bone marrow transplant for leuke-mia and aplastic anemia, respectively.

Styrene

Chemistry
Styrene ($C_6H_5CH=CH_2$) is a colorless to yellowish, opaque oily liquid. It is moderately volatile, flammable, and soluble in lipid solvents. Synonyms are vinyl benzene, ethylene, styrol, cinnamene, and cinnemenol. More than 50% of the world's production of benzene is used in the manufacture of styrene. Styrene materials account for 20% of all plastic production, e.g., polystyrene, resins, and polyesters.

Disposition
The major route of styrene absorption is via the lung. Absorption also takes place from the gastrointestinal tract. Absorbed chemical is distributed throughout the body (Bergman, 1979).
Styrene is metabolized to mandelic acid (70%) and phenylglyoxylic acid (30%), which are excreted in the urine. It is also metabolized to styrene oxide which is more toxic than the parent compound. Hippuric acid is also excreted in the urine following exposure to styrene.

Action
The major action of styrene is on the central nervous system and results in neurobehavioral abnormalities (Hanninen, 1985). In addition to CNS depression styrene also causes mucosal irritation. Symptoms include con-junctivitis, nasal irritation, nausea, drowsiness, lightheadedness, ataxia, in-coordination, and loss of memory. Chronic exposure has also resulted in peripheral neuropathy characterized by hypoesthesias and decreased peroneal nerve conduction velocities.
Styrene causes mutagenicity and chromosomal abnormalities. Some stud-ies have shown that the metabolite styrene oxide is carcinogenic.

Treatment
Following ingestion, syrup of ipecac is administered to induce emesis. Care should be taken to prevent aspiration, which may result in hydrocarbon pneumonitis. Aminophylline is then used for bronchospasm treatment. Cate-cholamines, e.g., epinephrine, are contraindicated.

Toluene

Chemistry

Toluene is a clear, colorless, flammable liquid that has a boiling point of 110.6°C. Synonyms are methylbenzene, toluol, phenylmethane, and methyl benzol.

Toluene is used as a solvent for paints, lacquers, thinners, coatings, and glue. It is also used in the production of other chemicals such as toluene diisocyanate, phenol, benzene, benzoic acid, nitrotoluene, saccharin and fuel (Benignus, 1981).

Disposition

Most exposure is via inhalation. Approximately 50% of the dose is retained in the lungs. It is completely absorbed from the gastrointestinal tract. It is slowly absorbed through skin (Toftgard and Gustafsson, 1980).

Toluene is highly distributed through fatty tissues. Its biological half-life ranges from 1 hour for fatty tissues to minutes in highly perfused organs. No studies were carried out on placental transfer of toluene.

Approximately 18% of absorbed toluene is expired via the lungs. Only 0.06% is eliminated unchanged in the urine. It is metabolized via cytochrome P-450 mixed function oxidase system to benzoyl alcohol, which is further oxidized via alcohol and aldehyde dehydrogenese to benzaldehyde and benzoic acid, respectively (Wallen et al., 1985). Benzoic acid conjugates with glycine to form hippuric acid that is excreted in the urine. Some chemicals inhibit toluene metabolic biotransformation, i.e., n-hexane, methylene chloride, benzene, styrene, and trichloroethylene.

Action

Acute exposure to toluene, i.e., 800 ppm, causes central nervous system depression with the following symptoms: drowsiness, tiredness, headache, dizziness, and nausea. These symptoms may progress to ataxia and confusion. Exposure for a short time to concentrations of 10,000–30,000 ppm produces unconsciousness (Boor and Hurtig, 1977).

Long term inhalation of toluene by abusers produced cerebellar ataxia, and chronic encephalopathy characterized by ataxia, tremors, and emotional lability (Knox and Nelson, 1966). Swedish studies have implicated exposure to toluene as causative for increased incidence of neuroasthenia, which is characterized by insomnia, fatigue, anxiety, depression and isolation; cognitive dysfunction manifested as short term memory impairment, inability to concentrate, and emotional instability; and psychomotor impairment exhibited as decreased reaction, manual dexterity, and perceptual speed (Iketa and Miyaka, 1978). There are doubts that these effects were related only to toluene because other chemicals were present. On the other hand, exposure of rats to 4,000 ppm for 2 h/day for 60 days resulted in cortical dysfunction.

Peripheral neuropathy reported for toluene in humans is more likely to have resulted from exposure to *n*-hexane that was present as a contaminant. Animal studies showed that toluene protected animals from *n*-hexane neurotoxicity (Abou-Donia, unpublished data). Kidney damage has been reported in sniffers of glue that may have contained toluene and other solvents. The damage included distal renal tubular acidosis, high anion gap metabolic acidosis, hypokalemia, hematuria, proteinuria, and pyuria. Effects on liver and lung seem to be minimal.

In contrast, death due to high exposure to toluene may result from its sensitization of the myocardium to the arrhythmogenic action of endogenous catecholamine. Toluene has neither carcinogenic nor mutagenic action.

Treatment
Treatment of toluene exposure is symptomatic.

Xylene

Chemistry
Xylene is produced by distillation of coal tar and petroleum products. The three isomers *ortho, meta,* and *para* constitute the commercial product with *meta* being the major isomer. Commercial products of toluene also contain ethyl benzene, toluene, phenol, and trimethylbenzene. Xylene is a clear, colorless flammable liquid with a chemical formula of $C_6H_4(CH_3)_2$. Synonyms are xylol and dimethylbenzene. It is used as a solvent, cleanser, degreaser, and plane fuel. It is used in chemical manufacturing of phthalic acid, synthetic fibers, plastics, and enamel.

Disposition
Xylene is rapidly absorbed through the lung, with approximately 65% retained in the lung (Bergman, 1979). It is also absorbed through the gastrointestinal tract and the skin. Following absorption, the highest concentrations were present in the adrenal gland, bone marrow, spleen, brain, and blood. Only 3–6% of absorbed xylene is exhaled via the lung. Most of the remaining xylene is oxidized to methylbenzoic acid, which is conjugated with glycine to form methylhippuric acid. Only 2% of absorbed xylene is excreted in urine as xylenols. The plasma half-life of xylene is 4 hours (Riihimaki et al., 1979).

Action
The major action of xylene is the depression of the central nervous system resulting in lightheadedness, nausea, headache, and ataxia at low doses (Lawwerys, 1983). High exposure to xylene produces confusion, respiratory depression, and coma. It causes conjunctivitis, nasal irritation, sore

throat, and respiratory irritation. Concentrations as high as 10,000 ppm produce reversible but mild increase in hepatic aminotransferase activity and reversible renal failure.

Treatment

Initial treatment is support of ventilation, including oxygen if needed, to treat central nervous system depression. If ingested, emesis should be induced by the use of syrup of ipecac. If aspired, xylene causes chemical pneumonitis.

REFERENCES

Abdel-Rahman, M.S., L.B. Hetland, and D. Couri (1976). Toxicity and metabolism of methyl *n*-butyl ketone. *Amer. Ind. Hyg. Assoc.* 37: 95–102.

Abdo, K.M., D.G. Graham, P.R. Timmons, and M.B. Abou-Donia (1982). Neurotoxicity of continuous (90 days) inhalation of technical grade methyl butyl ketone in hens. *J. Toxicol. Environ. Health* 9: 199–215.

Abou-Donia, M.B. and D.M. Lapadula (1988). Cytoskeletal proteins as targets for organophosphorus compound and aliphatic hexacarbon-induced neurotoxicity. *Toxicology* 49: 469–477.

Abou-Donia, M.B. and D.M. Lapadula (1989). Cytoskeletal proteins and axonal neuropathies. *Comme. Toxicol.* 3: 427–444.

Abou-Donia, M.B., D.M. Lapadula, G. Campbell, and P.R. Timmons (1985). The synergism of *n*-hexane-induced neurotoxicity by methyl isobutyl ketone following subchronic (90 days) inhalation in hens: induction of hepatic microsomal cytochrome P-450. *Toxicol. Appl. Pharmacol.* 81: 1–16.

Abou-Donia, M.B., H.M. Makkawy, and D.G. Graham (1982). The relative neurotoxicities of *n*-hexane, methyl *n*-butyl ketone, 2,5-hexanedione following oral or intraperitoneal administration in hens. *Toxicol. Appl. Pharmacol.* 62: 369–389.

Anthony, D.C., K. Boekelheide, C.W. Anderson, and D.G. Graham (1983). The effect of 3,4-dimethyl substitution on the neurotoxicity of 2,5-hexanedione. II. Dimethyl substitution accelerates pyrrole formation and protein crosslinking. *Toxicol. Appl. Pharmacol.* 71: 372–382.

Atulanantham, K. and G. Genel (1978). Central nervous system toxicity associated with ingestion of propylene glycol. *J. Pediatr.* 93: 515–516.

Aviado, D.M. and M.A. Belej (1974). Toxicity of aerosol propellants in the respiratory and circulatory systems. I. Cardiac arrhythmia in the mouse. *Toxicology* 2: 31–42.

Baker, E.L., P.J. Landrigan, and P.E. Bertozzi (1978). Phenol poisoning due to contaminated drinking water. *Arch. Environ. Health* 33: 89–94.

Banner, W. and P.D. Walson (1983). Systemic toxicity following gasoline aspiration. *Am. J. Emerg. Med.* 3: 292–294.

Benignus, V.A. (1981). Neurobehavior effects of toluene: a review. *Neurobehav. Toxicol. Teratol.* 3: 407–415.

Bergman, K. (1979). Whole body autoradiography and allied tracer techniques in distribution and elimination studies of some organic solvents. *Scand. J. Work Environ. Health, Suppl. 1:* 29–92.

Boor, J.W. and H.I. Hurtig (1977). Persistent cerebellar ataxia after exposure to toluene. *Ann. Neurol.* 2: 440–442.

Braier, L. (1983). An hypothesis for the induction of leukemia by benzene. *Arch. Toxicol., Suppl. 6:* 42–46.

Burns, J.R. and B. Finalyson (1980). Changes in calcium oxalate crystal morphology as a function of concentration. *Invest. Urol.,* 18: 174–177.

Cavanagh, J.B. (1982). The pattern of recovery of axons in the nervous system of rats following 2,5-hexanediol intoxication: a question of rheology. *Neuropath. Appl. Neurobiol.* 8: 19–34.

Clay, K.K. and R.C. Murphy (1977). On the metabolic acidosis of ethylene glycol intoxication. *Toxicol. Appl. Pharmacol.* 39: 39–49.

Daniel, D.R., B.H. McAnally, and J.C. Garriott (1981). Isopropyl alcohol metabolism after acute intoxication in humans. *J. Anal. Toxicol.* 5: 110–112.

DeCaprio, A.P., E.J. Olajos, and P. Weber (1982). Covalent binding of a neurotoxic *n*-hexane metabolite: conversion of primary amines to substituted pyrrole adducts by 2,5-hexanedione. *Toxicol. Appl. Pharmacol.* 65: 440–450.

Eade, N.R., L.M. Taussig, and M.I. Marks (1974). Hydrocarbon pneumonitis. *Pediatrics* 54: 351–357.

Ehrenreich, T. (1977). Renal disease from exposure to solvents. *Ann. Clin. Lab. Sci.* 7: 6–16.

Goldfrank, L., R. Kirstein, and E. Bresnitz (1979). Gasoline and other hydrocarbons. *Hosp. Physician* 14: 32–39.

Gordis, E. (1969). Lipid metabolites of carbon tetrachloride. *J. Clin. Invest.* 48: 203–209.

Graham, D.G. (1980). Hexane neuropathy: a proposal for pathogenesis of a hazard of occupational exposure and inhalant abuse. *Chem.-Biol. Interact.* 32: 339–345.

Graham, D.G. and M.B. Abou-Donia (1980). Studies on the molecular pathogenesis of hexane neuropathy. I. Evaluation of the inhibition of glyceraldehyde-3-phosphate dehydrogenase by 2,5-hexanedione. *J. Toxicol. Environ. Health* 6: 623–631.

Guggenheim, M.A., J.R. Couch, and W.W. Weinberg (1971). Motor dysfunction as a permanent complication of methanol ingestion. *Arch. Neurol.* 24: 550–554.

Hanninen, H. (1985). Twenty-five years of behavioral toxicology within occupational medicine: a personal account. *Am. J. Ind. Med.* 7: 19–30.

Hardin, B.D. and J.P. Lyon (1984). Summary and overview: NIOSH symposium on toxic effects of glycol ethers. *Environ. Health Perspect.* 57: 273–275.

Iketa, T. and H. Miyaka (1978). Decreased learning in rats following repeated exposures to toluene: preliminary report. *Toxicology Lett.* 1: 235–239.

Knox, J.W. and J.R. Nelson (1966). Permanent encephalopathy from toluene inhalation. *N. Engl. J. Med.* 275: 1494–1496.

Lacouture, P.G. (1980). Isopropyl alcohol. *Clin. Toxicol. Rev.* 3:3.

Lacouture, P.G. and F.H. Lovejoy (1981). Methanol. *Clin. Toxicol. Rev.* 3(12): 1–3.

Lapadula, D.M., E. Suwita, and M.B. Abou-Donia (1988). Evidence for multiple mechanisms responsible for 2,5-hexanedione neuropathy. *Brain Res.* 458: 123–131.

Lawwerys, R.R. (1983). *Industrial Chemical Exposure: Guidelines for Biological Monitoring.* Biomedical Publications, Davis, CA, p. 67.

Moon, H.D. (1950). The pathology of fatal carbon tetrachloride poisoning with special reference to the histogenesis of the hepatic and renal lesions. *Am. J. Pathol.* 26: 1041–1057.

Ng, R.C., H. Darwish, and P.A. Stewart (1974). Emergency treatment of petroleum distillate and turpentine ingestion. *Can. Med. Assoc. J.* 111: 537–538.

Oalase, M. and T.R. Tephly (1975). Metabolism of formate in the rat. *J. Toxicol. Environ. Health* 1: 13–24.

Parry, M.F. and R. Wallach (1974). Ethylene glycol poisoning. *Am. J. Med.* 57: 143–150.

Partanen, T. (1970). Coronary heart disease among workers exposed to carbon disulfide. *Br. J. Ind. Med.* 27: 313–325.

Perbellini, L., F. Brugnone, and I. Pavan (1980). Identification of the metabolites of *n*-hexane, cyclohexane, and their isomers in man's urine. *Toxicol. Appl. Pharmacol.* 53: 220–229.

Piotnowski, J.K. (1971). Evaluation of exposure to phenol: absorption of phenol vapor in the lungs and through the skin and excretion of phenol in the urine. *Br. J. Ind. Med.* 28: 172–178.

Riihimaki, V., P. Pfaffli, and K. Savolainen (1979). Kinetics of *m*-xylene in man. Influence of intermittent physical exercise and changing environmental concentrations on kinetics. *Scand. J. Work Environ. Health* 5: 232–248.

Ruddick, J.A. (1972). Toxicology, metabolism and biochemistry of 1,2-propanediol. *Toxicol. Appl. Pharmacol.* 21: 102–111.

Sabri, M.I., C.L. Moore, and P.S. Spencer (1979). Studies on the biochemical basis of distal axonopathies. I. Inhibition of glycolysis of neurotoxic hexacarbon compound. *J. Neurochem.* 32: 683–689.

Schroder, H.G. (1965). Acute and delayed chloroform poisoning. *Br. J. Anaesth.* 37: 972–975.

Shahangian, S. and K.O. Ashko (1986). Formic and lactic acidosis in a fatal case of methanol intoxication. *Clin. Chem.* 32: 395–397.

Spencer, P.S. and H.H. Schaumberg (1975). Ultrastructural studies of the dying back process. *J. Neuropathol. Exp. Neurol.* 36: 276–299.

Stevens, H. and P.M. Forster (1953). Effect of carbon tetrachloride on the nervous system. *AMA Arch. Neurol. Psychiatry* 70: 635–649.

Stewart, R.D. and H.C. Dodd (1964). Absorption of carbon tetrachloride, trichloroethylene, tetrachloroethylene, methylene chloride and 1,1,1-trichloroethane through the human skin. *Am. Ind. Hyg. Assoc.* 25: 439–446.

Timmis, R.M. and K.M. Moser (1975). Toxicity secondary to intravenously administered chloroform in humans. *Arch. Intern. Med.* 135: 1601–1603.

Toftgard, R. and J.A. Gustafsson (1980). Biotransformation of organic solvents. A review. *Scand. J. Work Environ. Health* 6: 1–18.

Turk, J., L. Morrell, and L.V. Avioli (1986). Ethylene glycol intoxication. *Arch. Intern. Med.* 146: 1601–1603.

Wallen, M., S. Holm, and M.B. Nordqvist (1985). Coexposure to toluene and *p*-xylene in man: uptake and elimination. *Br. J. Ind. Med.* 42: 111–116.

Chapter

3

Gases and Vapors

Mohamed B. Abou-Donia
Departments of Pharmacology and Neurobiology
Duke University Medical Center
Durham, North Carolina

CARBON MONOXIDE

Chemistry

Carbon monoxide is a colorless, odorless gas. It has a specific gravity of 0.97 and is present in the atmosphere at a concentration of 0.001% (10 ppm). It does not cause irritation.

Carbon monoxide is produced *in vivo* by catabolism of the α-methane carbon atom in the protoporphyrin ring of hemoglobin. Endogenous CO forms carboxyhemoglobin (COHb) of 0.4–0.7%. Smoking one pack of cigarettes per day produces 5–6% COHb (Stewart, 1975).

Cars without catalytic converters produce up to 9% CO: catalytic converters reduce this level to below 1%. Lethal concentrations of CO can be produced in a closed garage in 10 min. While natural gas contains no CO, gas supplied from coal contains 18–30% CO.

The use of methylene chloride, a paint remover, for 3 h results in COHb levels of 5–10% *in vivo*. Annual fatalities from CO poisoning range from 3500–4000 cases, which makes it the leading cause of poisoning deaths in the United States. Occupational exposure is limited to a TLV-TWA value of 35 ppm and a short term exposure limit (STEL) of 200 ppm.

Action

Carbon monoxide exerts toxicity by binding to hemoglobin to form carboxyhemoglobin (Stewart, 1975). This impairs the oxygen carrying capacity of hemoglobin and interferes with the delivery and utilization of oxygen which leads to cellular hypoxia. Carbon monoxide has 200 times more affinity for hemoglobin than O_2 (Coburn, 1979).

$$Hb(Fe^{+2}) \xrightarrow{\ O_2\ } O_2(Fe^{+2})Hb \text{ oxyhemoglobin}$$

$$Hb(Fe^{+2}) \xrightarrow{\ CO\ } CO(Fe^{+2})Hb \text{ carboxyhemoglobin}$$

COHb increases oxygen binding to hemoglobin, which impairs oxygen release. This is evident by the shift of the oxyhemoglobin disassociation curve to the left, which diminishes the release of oxygen from oxyhemoglobin to the tissues.

Although carbon monoxide binds to cytochrome oxidase *in vitro*, its *in vivo* binding may be small, because oxygen has a much greater affinity for cytochrome oxidase than CO.

Symptoms

Carbon monoxide toxicity depends on its concentration in the atmosphere, which is related to its COHb level as summarized in Table 1 (Stewart, 1975).

CNS damage produced by CO is related to its hypoxic effect. Thus, histologically, pathologic lesions following CO poisoning are indistinguishable from those produced by other causes of hypoxias, e.g., cardiorespiratory arrest, hypoglycemia, and cyanide poisoning. CO produces lesions characterized by cerebral edema, hemorrhage, focal necrosis, and perivascular infarction. A characteristic lesion of CO is bilateral necrosis of globus pallidus. Pathologic alterations in the cerebral gray matter are also present in the substantia nigra, hippocampus, cerebral cortex, and cerebellum. In rare cases, after initial recovery, a postanoxic demyelination, i.e., anoxic leukocephalopathy, occurs and results in irritability, confusion, coma, and death.

CO severely affects the heart. It causes pathomyocardial necrosis ischemia-related lesions. Ischemic symptoms are characterized by chest pain, dyspnea, diaphoresis and nausea. Myocardial infarction may occur leading to hypotension. CO may cause pulmonary edema due to its direct effect on capillary membranes, or a secondary effect of hypoxia on the heart or alveolar membrane. CO may also cause hemolytic anemia. A characteristic finding of CO poisoning at autopsy is the cherry red skin of lips and mucous membranes. COHb level in blood is used to document exposure to CO.

TABLE 1. Effect of CO Atmospheric Concentration on Blood COHb and CO Toxicity Symptoms

CO in atmosphere		COHb in	
%	ppm	blood (%)	Symptoms
0.007	70	10	No effort except shortness of breath in vigorous exercise
0.012	120	20	Shortness of breath in moderate exercise; occasional headache
0.022	220	30	Headache, irritability, fatigue, dizziness, dimness of vision
0.035–0.052	350–520	40–50	Headache, confusion, collapse, fainting in exercise
0.080–0.122	800–1220	60–70	Unconsciousness, convulsions, respiratory failure, death upon prolonged exposure
0.195	1950	80	Rapid death

Treatment

Administration of 100% oxygen and hyperbaric oxygen (HBO) is an indication (Neubauer, 1979). This treatment reduces COHb half-life from 5–6 h to 30–60 min or 20–30 min for 100 O_2 or hyperbaric O_2, respectively. Sometimes a mixture of 95% O_2 and 5% CO_2 is administered. In severe poisoning, exchange perfusion is indicated. Supportive care is indicated for neurologic and cardiovascular symptoms.

CYANIDE

Chemistry

Cyanide occurs as a gas or as a liquid hydrogen cyanide (HCN) that is also known as prussic acid.

Cyanide normally exists in the blood resulting from the metabolism of vitamin B_{12}, food, and smoking. It is present in many chemicals. It is produced *in vivo* from Laetrile. Long term use of the tropical plant *cassava* results in chronic cyanide exposure and produces tropical ataxia neuropathy (Osintokun et al., 1970).

Disposition

Cyanide gas is rapidly absorbed by inhalation and produces symptoms in seconds and death in minutes. Cyanide is also rapidly absorbed following ingestion; symptoms and death follow in minutes to hours. The lethal adult dose is 50 mg.

Although Laetrile is not approved by the FDA for cancer treatment, it is available in 22 states, including California. It is made from amygdalin plants. One gram of Laetrile contains the equivalent of 60 mg of cyanide. Laetrile overdose produces cyanide poisoning. Cyanide binds to blood protein. Its RBC/plasma ratio is 100:1. Its volume of distribution is 1.5 L/kg.

Metabolism

Cyanide is metabolized by the enzyme rhodanese in the presence of thiosulfate to thiocyanate, that is excreted by the kidney. The supply of thiosulfate is limited, and its amount is rate limiting.

$$CN^- \;+\; Na_2S_2O_3 \;\xrightarrow{\text{rhodanese}}\; SCN^- \;+\; Na_2SO_3$$

| cyanide | sodium thiosulfate | thiocyanate | sodium sulfite |

Rhodanese is present mostly in the mitochondria in high concentrations in the liver and kidney.

Alternately, cyanide may react with hydroxycobalamin (vitamin B_{12a}) to form the nontoxic cyanocobalamin (vitamin B_{12}).

$$CN^- \;+\; \text{hydroxycobalamin} \;\longrightarrow\; \text{cyanocobalamin}$$

Vitamin B_{12a} Vitamin B_{12}

Only small amounts of cyanide are excreted via the lung and sweat. Ingested cyanide passes through the portal system, where it is detoxified by the liver by the first-pass effect.

Action

Cyanide binds to ferric iron in the cytochrome a-a^3 complex. This results in the inhibition of the final step of oxidative phosphorylation and stops aerobic metabolism. Thus, asphyxia is the cause of death from cyanide poisoning. The result of cyanide poisoning is the accumulation of lactic acid due to the conversion of carbohydrate metabolism from pyruvate to lactate. Cyanide does not produce cyanohemoglobin.

Acute lethal doses of cyanide primarily affect the central nervous system. Initially, cyanide increases respiration by stimulating the peripheral chemoreceptors. It also slows the heart by stimulating the carotid body receptors. Cyanide produces necrosis more often than demyelination, and the changes occur primarily in the white matter.

Related Conditions

Leber's Disease

Persons with Leber's disease are unable to metabolize and detoxify cyanide to thiocyanate. Thus, this hereditary disease results in visual abnormalities, i.e., optic atrophy, that are associated with heavy smoking and vitamin B_{12} deficiency. This congenital condition may render the optic nerve more vulnerable to cyanide; it may be treated effectively with hydroxycobalamin.

Toxicity of Nitroprusside

Nitroprusside, $Na_2Fe(CN)_5NO$, produces its toxicity by oxidizing a heme group to ferric iron and by reacting with a cyanide group to form cyanomethemoglobin. CN in cyanomethemoglobin is inactive. The other four cyanide groups bind to cytochrome oxidase.

Cassava Toxicity

The tropical plant, cassava, produces ataxic neuropathy that may result from cyanide toxicity (Osintokun, 1970). This condition is not associated with vitamin B_{12} deficiency. Hydroxycobalamin is not effective in treating this condition.

Clinical Condition

Cyanide poisoning results in cellular hypoxia. Within 30 seconds of inhalation, cyanide produces flushing, headache, tachypnea, and dizziness. These symptoms progress to difficult breathing, coma, convulsion, seizures, and death within 10 min. Despite respiratory depression, there is absence of cyanosis. Cyanide also depresses the cardiovascular system resulting in an initial tachycardia followed by bradycardia. Rapid onset of coma accompanied by metabolic acidosis are diagnostic for cyanide poisoning.

Treatments

Support of respiration and circulation is indicated by a diminishing 100% oxygen. Hyperbaric oxygen is not more effective than normobaric oxygen in the treatment of cyanide poisoning. Gut decontamination is carried out using gastric lavage and charcoal (Vogel and Sultan, 1981). Antidotal treatment with nitrites, e.g., amyl and sodium nitrites, depends on the oxidation of blood iron to form methemoglobin, which attracts cyanide from the heme group of cytochrome oxidase and allows thiosulfate to detoxify cyanide. The antidotal action of thiosulfate is potentiated by chlorpromazine and phenoxybenzamine.

$$CN^- + \text{cytochrome oxidase} - Fe^{3+} \rightleftharpoons \text{cytochrome oxidase} - FeCN$$

$$NaNO_2 + (O) + HbFe^{2+} \longrightarrow HbFe^{3+} + NaNO_2$$

$$HbFe^{3+} + \text{cytochrome oxidase} - FeCN \longrightarrow HbFeCN$$

$$+ \text{cytochrome oxidase} - Fe^{3+}$$

Thiosulfate is administered with nitrites. It is used to convert cyanide to the nontoxic thiocyanate, a reversible reaction that is catalyzed by rhodanase, a sulfur transferase enzyme present in the mitochondria.

$$HbFeCN + \underset{\substack{\text{sodium} \\ \text{thiosulfate}}}{Na_2S_2O_3} \xrightarrow[\text{(mitochondria)}]{\text{rhodanese}} HbFe^{3+} + \underset{\substack{\text{sodium} \\ \text{sulfite}}}{Na_2SO_3} + \underset{\text{thiocyanate}}{SCN^-}$$

Hydroxycobalamin (vitamin B_{12a}) is effective in treating cyanide poisoning in animals. Cyanide reacts with vitamin B_{12a} to form vitamin B_{12}, which is excreted by the kidney.

Dicobalt edetate (Co_2EDTA) is used in Britain and France as a second-line antidotal treatment for cyanide. It chelates cyanide without producing methemoglobinemia.

HYDROGEN SULFIDE

Chemistry

Hydrogen sulfide has the chemical formula H_2S. It is a colorless gas, with a strong rotten egg odor detectable at 0.2–0.3 ppm. It has a specific gravity of 1.19 g/cm³. It reacts with water and produces sulfur dioxide and elemental sulfur. It dissociates to hydrosulfide (HS^-, about one third at physiological pH) and traces of sulfide S^{2-}.

Hydrogen sulfide occurs naturally in caves, volcanoes, and bacterial decomposition of sulfur in gastrointestinal tract, soil and sewers (Milby, 1962). It is a by-product of the petroleum industry, tanning, rubber vulcanizing, viscose rayon, paper mills, mines, silk, hot sulfur springs, burning of wool, hair, meats, and hides, and production of heavy water for nuclear reactors.

Disposition

The lung is the major route of entry of hydrogen sulfide; skin absorption is minimal. A negligible portion is excreted via the lung. H_2S is mainly excreted in the urine after its oxidation to sulfate, methylation, and reaction with metalloprotein or disulfide-containing proteins. The primary sites for H_2S detoxification is red blood cell and liver mitochondria (Thoman, 1969).

$$H_2S \xrightarrow[\text{Thiol-}S\text{-methyl transferase}]{\text{Metalloprotein (Fe,Cu)}} \text{Methemoglobin}$$

$$CH_3SH \xrightarrow[S\text{-adenosylmethionine}]{\text{Thiol-}S\text{-methyl transferase}} CH_3S\text{--}CH_3$$

In vitro hydrogen sulfide is primarily oxidized via sulfide oxidase to thiosulfate ($S_2O_3^{2-}$), which is excreted in the urine. Thiosulfate may be further oxidized to sulfite (SO_3^{2-}) and sulfate (SO_4^{2-}). H_2S does not produce sulfhemoglobin.

Action

Hydrogen sulfide exerts its toxicity by inhibiting cytochrome oxidase resulting is disruption of cytochrome oxidase and leading to cellular hypoxia (Nicholls, 1975). This also results in anaerobic metabolism, which causes lactate accumulation and metabolic acidosis. H_2S causes respiratory depression and produces death by respiratory arrest and hypoxia.

Clinical Condition

Acute exposure of H_2S causes depression of the central nervous system characterized by headache, lethargy, vertigo, and coma (Thoman, 1969). Other symptoms may appear, e.g., loss of consciousness, dizziness, nausea, vomiting, sore throat, conjunctivitis, leg and hand weakness, dyspnea, convulsions, pulmonary edema, and cyanosis. Local irritation may produce keratoconjunctivitis, rhinitis, pharyngitis, bronchitis, and pneumonia. It may also affect the cardiac system resulting in dysrhythmias and myocardial depression.

Chronic exposure results in headache, weakness, nausea, vomiting, and weight loss. Spasticity, cerebellar ataxia, tremor, and cyanosis may also be produced.

Treatment

The patient should be removed by exposure area. The patient's condition is stabilized by maintaining adequate ventilation and circulation (Marcus, 1983). Oxygen administration is indicated. A muscle relaxant, i.e., succinylcholine, may be administered to control violent convulsion movements. Antidotal treatment is carried out by nitrites, which produce methemoglobin. H_2S moves away from cytochrome oxidase and reacts with methemoglobin to form sulfmethemoglobin which undergoes rapid spontaneous detoxification in the body.

OTHER METHEMOGLOBINEMIA-PRODUCING CHEMICALS

Many classes of chemicals produce methemoglobin by oxidizing Fe^{2+} in hemoglobin to Fe^{3+} (Lovejoy, 1984). These classes include nitrites, nitrates, anilin compounds, anilids, nitrosamines, nitro compounds, phenols, quiniones, methanol, aromatic amines, chlorates, antipyrine, arsine, and lidocaine.

Sources

Inorganic nitrates, i.e., KNO_3 and $NaNO_3$, also known as saltpeter, and inorganic nitrites are present in contaminated water (Lovejoy, 1984). Organic nitrates, e.g., glycerol trinitrate, and organic nitrites, e.g., amyl nitrite, are used medically. Nitrosamines are present in the gastrointestinal tract of persons who consume sodium nitrite; they react with secondary amines *in vivo* to form nitrosamines. This nitrosation reaction is inhibited by ascorbic acid, caffeine, the amino acid cysteine and histidine, glutathione, and butylated hydroxytoluene (BHT). Aniline, $C_6H_5NH_2$, is an industrial chemical that is used as an intermediate in the manufacture of pharmaceuticals, dyestuff, perfumes, shoe polish, and organic chemicals.

Action

While nitrates are weaker inducers of methemoglobin, nitrites are strong oxidizers of hemoglobin ($HbFe^{2+}$) to methemoglobin ($HbFe^{3+}$) (Kearney et al., 1984). The mechanisms of toxicity resulting from the formation of methemoglobin include the following:

1. It reduces the oxygen-carrying capacity of the blood, since methemoglobin cannot carry oxygen.
2. The presence of Fe^{3+} reduces the release of oxygen in tissues by shifting the oxyhemoglobin dissociation curve to the left.

Because RBC lacks the tricarboxylic acid cycle, it depends for energy on glycolysis and the hexos monophosphate shunt (Lovejoy, 1984). A normal methemoglobin level is 1–2%. Endogenous methemoglobin is reduced to hemoglobin via two mechanisms:

1. NADH-dependent methemoglobin reductase (diaphorese I). This enzymatic system catalyzes the reduction of 95% endogenous methemoglobin. NADH is produced by glycolysis.

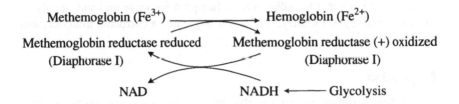

2. NADPH-dependent methcmoglobin reductasc (diaphorase II). This enzymatic system reduces less than 5% of endogenous methemoglobin. NADPH is produced by the hexose monophosphate shunt. The pathway is limited by the lack of endogenous cofactor. Methylene blue may be used as an exogenous factor.

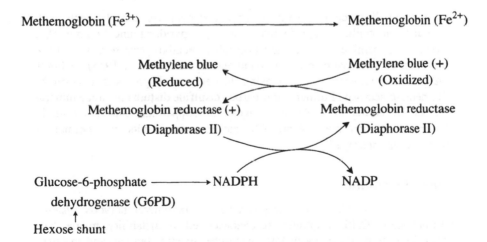

Since diaphorase II reduces only a minor amount of methemoglobin, a reduction in G6PD does not result in increased methemoglobumia. Methemoglobin turns the blood a chocolate brown color.

CARBON DISULFIDE

Chemistry

Carbon disulfide, CS_2, is a clear, colorless, volatile liquid that has a strong sulfur odor similar to that of decaying cabbage. It is highly flammable and is oxidized to sulfur dioxide on combustion.

The primary source of carbon disulfide is manufacturing. Most exposure takes place in the viscose rayon manufacturing industry. It is used as an intermediate chemical for adhesives, a fumigant for grain and soil, a solvent in rubber and rayon industry, a remover of grease, a corrosion inhibitor, and an insecticide.

Disposition

Carbon disulfide is rapidly absorbed via the lung and gastrointestinal tract. Ingestion of 15 ml may be fatal to an adult. Most of the dose (80–90%) is metabolized in the body and excreted by the lung. Only a small portion is excreted in the urine. CS_2 metabolizes in the urine include thiourea, 2-mercapto-2-thiazolin-zone, and 2-thiazolidine-4-carboxylic acid. Disulfiram is metabolized in the body to CS_2.

Action

Carbon disulfide is a central nervous system toxicant. It reacts with several nucleophilic groups in the body, e.g., pyridoxamine (vitamin B_6), cerebral monoamine oxidase, and dopamine decarboxylase resulting in the inhibition of enzymatic reaction (Kleinfeld and Tabershaw, 1955). It binds with microsomal enzymes, decreasing their activity. It produces a centrilobular hepatic necrosis. Furthermore, carbon disulfide disturbs the trace mineral balance by acting as a chelating agent for copper and zinc resulting in disruption of their metabolism. Chronic exposure results in a permanent distal axonal neuropathy.

Clinical Condition

Carbon disulfide is a strong central and peripheral nervous systems neurotoxicant. CNS symptoms are characterized by irritability, mania, hallucinations, tremors, memory loss, headache, nausea, fatigue, and malaise. These effects may be irreversible. Other symptoms include peripheral distal axonopathy and cranial nerve dysfunction.

Chronic exposure results in increased blood cholesterol, retinopathy, i.e., microaneurysm, optic neuropathy, decreased glucose tolerance, reduced serum thyroxine levels and Parkinsonism (Partanen, 1970).

Treatment

No antidotal treatment is indicated. Patients should be removed from exposure. Attention should be given to breathing and circulation. Treatment is symptomatic.

AIR POLLUTANTS

Air pollutants are produced by vehicles (51%), other sources of fuel combustion (16%), industrial processes (15%), miscellaneous sources including forest fires (14%), and solid waste disposal processes (4%) (De-Nevers, 1983).

Classes of Pollutants

Direct Pollutants

Direct pollutants are active chemicals that produce their biological actions directly without changes in the air. These chemicals include sulfur oxides, carbon monoxide, nitrogen oxides, and organic, and inorganic particulate matter, i.e., dusts, smoke, fumes, and mists.

Sources. Carbon monoxide, which accounts for more pollutants than any other chemical, is produced mostly by the internal combustion engine. Sulfur dioxide's main source is the combustion of fossil fuels. Nitrogen oxides are formed by motor vehicles and industrial and internal heating. Catalytic converters diminish the amounts of nitrogen oxide, hydrocarbons, and carbon monoxide. However, they result in the oxidation of sulfur dioxide to sulfur trioxide (SO_3), which reacts with water to form sulfuric acid (H_2SO_4).

The air over London is characterized with high levels of carbon monoxide and sulfur oxides.

Photochemical Pollutants

Photochemical pollutants, or smog, exert their action after the conversion to their active form in the atmosphere. These chemicals, i.e., hydrocarbons derived from car exhaust, react with SO_3 to form harmful free radicals and harmful chemicals including ozone, peroxyacetyl nitrate, and aldehydes, i.e., formaldehyde and acrolein.

Sources. Oxidants such as ozone and peroxyacetyl nitrate result from the interaction of nitrogen oxides and sulfur oxides with ultraviolet.

$$NO_2 \xrightarrow{\text{UV}} NO + O$$

$$O + O_2 \longrightarrow O_3$$

$$O_3 + NO \longrightarrow NO_2 + O_2$$

Peroxacetyl nitrate (PAN) results from the reaction of peroxyacetyl radical and NO_2.

$$\underset{\displaystyle CH_3-C-\dot{O}}{\overset{\displaystyle \overset{O}{\|}}{}} + NO_2 \longrightarrow \underset{\displaystyle CH_3C-ONO_2}{\overset{\displaystyle \overset{O}{\|}}{}}$$

Air over areas such as Los Angeles and Houston are characterized by photochemical pollutants. The brown color of this type is due to NO_2.

Health Effects. The adverse effects of air pollutants on health are dependent upon concentration, duration of exposure, and other ill health conditions, i.e., cardiopulmonary disease. Premature infants, the newborn, the elderly, and persons with chronic cardiac or pulmonary diseases are most sensitive to the adverse health effects of air pollutants.

Oxides of Sulfur

Sulfur in the atmosphere exists essentially in three forms: sulfur dioxide, SO_2, produced mainly from the combustion of coal, which accounts for approximately 25% of sulfur in the atmosphere; hydrogen sulfide, H_2S, which comes primarily from natural sources; and various sulfates that are produced from the oxidation of SO_2. SO_2 alone is not so harmful unless it is in the presence of other particulates, which result in the formation of sulfuric acid (H_2SO_4):

$$SO_2 + O_3 \longrightarrow SO_3 + O_2$$

$$SO_3 + H_2O \longrightarrow H_2SO_4$$

Sulfur Dioxide

Sulfur dioxide and sulfates aggravate the condition of persons with chronic respiratory diseases. This may result from the deposition of hydrogen ion on bronchial lining.

Oxides of Nitrogen

The oxides of nitrogen exist in the following form:

Nitric oxide, NO
Nitrous oxide, N_2O
Nitrogen dioxide, NO_2
Nitrogen trioxide, NO_3
Nitric anhydride, N_2O_5
Nitrous anhydride

Only NO and NO_2 are the significantly made oxides; they are referred to as NO_x, pronounced "nox."

Nitric Oxide (NO)

Although nitric oxide is formed in the combustion of all fossil fuels, its primary sources are cars.

Nitrogen Dioxide (NO_2)

Nitric oxide is rapidly oxidized in the atmosphere to nitrogen dioxide (NO_2). This oxidation process is greatly accelerated by sunlight and the presence of organic materials in the air.

Health Effects of Nitrogen Oxides

Nitrogen oxides produce interstitial edema and epithelial proliferation. High levels may cause fibrosis and emphysema. NO_2 has a more adverse health effect on humans than NO. It has affinity for hemoglobin, thus reducing its carrying capacity. It forms acid in the lung, which increases its toxicity.

Ozone

Adverse effects of ozone result from oxidizing thiol-containing compounds and unsaturated fatty acids. It results in respiratory irritation and decreased forced expiratory volume (FEV) following exposure to above 0.3 ppm. It causes eye irritation and bronchitis. It increases the risk of mutogenesis and fetal toxicity.

Peroxide Nitrate

Peroxide nitrate causes lacrimation and pulmonary irritation.

Aldehyde

Exposure to formaldehyde or acrolein causes pulmonary irritation and bronchospasm.

Carbon Monoxide

High levels of carbon monoxide reduce the oxygen-carrying capacity of hemoglobin and impair its transport. This action may effect patients with cardiac diseases but has little effect on healthy individuals. CO is generated from car exhausts and cigarette smoking. Heavy smokers have a 7% level saturation of CO in the blood.

REFERENCES

Coburn, R.F. (1979). Mechanisms of carbon monoxide toxicity. *Prev. Med.* 8: 310–322.

DeNevers, N. (1983). Community air pollution. In *Environmental and Occupational Medicine*, W.N. Rom, Ed., Little, Brown, Boston, pp. 797–810.

Kearney, T.E., A.S. Manoguerra, and J.V. Dunford (1984). Chemically induced methemoglobinemia from aniline poisoning. *West. J. Med.* 140: 282–286.

Kleinfeld, M. and I.R. Tabershaw (1955). Carbon disulfide poisoning. Report of two cases. *JAMA* 159: 677–679.

Lovejoy, F.H. (1984). Methemoglobinemia. *Clin. Toxicol. Rev.* 6(5): 1–2.

Marcus, S.M. (1983). Hydrogen sulfide. *Clin. Toxicol. Rev.* 5(4): 1–2.

Milby, T.H. (1962). Hydrogen sulfide intoxication: review of the literature and report of unusual accident resulting in two cases of nonfatal poisoning. *J. Occup. Med.* 4: 431–437.

Neubauer, R.A. (1979). Carbon monoxide and hyperbaric oxygen. *Arch. Intern. Med.* 139: 829.

Nicholls, P. (1975). The effect of sulphide on cytochrome a_3, isosteric and allosteric shifts of the reduced alpha peak. *Biochim. Biophys. Acta*, 396: 24–35.

Osintokun, B.D., A.O.G. Adenja, and A. Aladehoyinbo (1970). Free cyanide levels in tropical ataxic neuropathy. *Lancet* 2: 372–373.

Partanen, T. (1970). Coronary heart disease among workers exposed to carbon disulfide. *Br. J. Ind. Med.* 27: 313–325.

Stewart, R.D. (1975). The effect of carbon monoxide on humans. *Annu. Rev. Pharmacol.* 15: 409–422.

Thoman, M. (1969). Sewer gas: hydrogen sulfide intoxication. *Clin. Toxicol.* 2: 383–386.

Vogel, S.N. and T.R. Sultan (1981). Cyanide poisoning. *Clin. Toxicol.* 18: 367–383.

Pesticides

Mohamed B. Abou-Donia
Departments of Pharmacology and Neurobiology
Duke University Medical Center
Durham, North Carolina

INTRODUCTION

Pesticides are natural or synthetic chemicals used to kill or interfere with some form of life. Many of these chemicals are applied intentionally in the environment. Most of the pesticides are harmful to nontarget species including humans and endangered species (Matsumura, 1985).

Pesticides are defined in the Federal Insecticide, Fungicide, and Rodenticide Act (FIFRA, 1947; amended 1959) as "any substance or mixture of substances intended for preventing, destroying, repelling, or mitigating any insects, rodents, nematodes, fungi, or weeds, or any other form of life declared as pests; any substance or mixture of substances intended for use as plant regulator, defoliant, or dessicant."

CLASSIFICATION OF PESTICIDES

Pesticide chemicals can be classified according to the living system they control.

Insecticides
1. Inorganic
   a. Arsenic compounds
   b. Fluorides

      c. Thallium
      d. Selenium
      e. Metaldehyde
      f. Mercury
      g. Phosphorus
      h. Sodium borate
      i. Hydrocyanic acid (cyanide)
      j. Antimony
2. Botanicals
      a. Nicotine
      b. Pyrethrin
      c. Ryania
      d. Rotenone
      e. Sabadilla
3. Biologicals
    Bacillus thuringiensis
4. Petroleum
5. Synthetic organic insecticides
      a. Chlorinated hydrocarbons
          (1) DDT-type compounds
          (2) Lindane (cyclohexane hexachloride, BHC)
          (3) Cyclodiene compounds
          (4) Chlorinated camphene
      b. Organophosphorus insecticides
          (1) Aliphatic phosphates
          (2) Vinyl phosphates
          (3) Aromatic phosphates
          (4) Pyrophosphates
          (5) Phosphonates
      c. Carbamates
          (1) Naphthylcarbamates
          (2) Phenylcarbamates
          (3) Heterocyclic dimethylcarbamates
          (4) Heterocyclic methylcarbamates
          (5) Oximes
      d. Thiocyanates
      e. Pyrethroids
          (1) Type I
              (a) Tetramethrin
              (b) Resmethrin
              (c) Permethrin
          (2) Type II
              (a) Deltamethrin
              (b) Cypermethrin
              (c) Fenvalerate

Miticides
1. Sulfites, sulfones, sulfides, sulfonates
2. Dinitrophenols
3. Dicofol (Kelthane)
4. Others

Rodenticides
1. Inorganic compounds
    a. Arsenic
    b. Thallium
    c. Phosphorus
    d. Barium carbonate
    e. Zinc phosphide
    f. Vacor (*N*-3-pyridylmethyl-*N'*-*P*-nitrophenyl urea)
2. Organic compounds
    a. Sodium fluoroacetate
    b. 1-Naphthyl thiourea
    c. Warfarin (Pival, Valone)
    d. Red squill
    e. Strychnine sulfate
    f. Dicarboximide

Fungicides
1. Inorganic compounds containing
    a. Copper
    b. Mercury
    c. Chromium
    d. Zinc
    e. Other metallic compounds
    f. Sulfur
2. Organic compounds
    a. Dithiocarbamates
    b. Phthalimides
    c. Karathane
    d. Dodine
    e. Quinones
    f. Pentachlorophenol
    g. Hexachlorobenzene (HCB)
    h. Tetramethylthiuram disulfide (Thiram)

Herbicides
1. Inorganic compounds
    a. Sodium chlorate
    b. Potassium cyanate
    c. Sodium arsenite
    d. Caustic acids and alkalis

2. Organic compounds
   a. Petroleum distillates (diesel oil, crude oil, etc.)
   b. Phenoxy type
      (1) 2,4-Dichlorophenoxy acetic acid (2,4-D)
      (2) 2,4,5-Trichlorophenoxy acetic acid (2,4,5-T)
      (3) Other phenoxy
   c. Phenylureas
   d. Carbamate (thiol, *N*-phenyl)
   e. Dinitrophenols
   f. Triazines
   g. Benzoic acid derivatives
   h. Phosphorus (aliphatic phosphites, phosphates)
   i. Amides
   j. Trifluralin (Treflan)
   k. Quaternary amines:
      (1) Diquat
      (2) Paraquat

Fumigants
1. Cyanides
2. Carbon tetrachloride
3. Naphthalene
4. Paradichlorobenzene
5. Methyl bromide, chloride, and iodide
6. Dimethylphthalate
7. Indolane
8. Diethyltoluamide

Defoliants and dessicants
1. Phosphites
2. Sulfuric acid (H_2SO_4)

Repellents (insects, birds)
1. Alkyl isothiocyanate

Plant growth regulators
1. Naphthylacetic acid

INSECTICIDES

Introduction

Insecticides are chemicals used to control insects. These chemicals may be further classified according to their mode of action: 1) physical toxicants, 2) cytoplasmic toxicants, 3) metabolic inhibitors, and 4) nerve toxicants.

Another classification of insecticides is based on the route of entry into the body: 1) stomach toxicants, which enter via ingestion; 2) contact toxicants, which come into contact with skin; and 3) fumigants, which enter lungs through respiration.

Most modern synthetic insecticides interrupt the nervous system via one of the following mechanisms: 1) interference with synaptic transmission by inhibition of AChE such as organophosphorus compounds and carbamates, or interference with nerve receptors such as nicotine and its analogues; 2) interference with axonal transmission by affecting ion permeability with chlorinated hydrocarbon insecticides and pyrethroids; and 3) degeneration of axon and myelin induced by delayed neurotoxic organophosphorus compounds (OPIDN). This effect is not desired in pesticides, and chemicals showing this action are not registered for use.

The Nervous System in Mammals and Insects

The nervous system, in both mammals and insects, consists of two subdivisions: central and peripheral. In mammals, the central nervous system consists of the brain and spinal cord, whereas it is composed of the brain and central nerve cord in insects. In both mammals and insects, the somatic and autonomic nervous systems form the peripheral nervous system.

Mammalian Peripheral Nervous System

The Somatic Nervous System. In mammals, the somatic system consists of incoming or sensory (afferent) and outgoing or motor (efferent) pathways. Motor nerves transmit at the neuromuscular junction with ACh as the neurotransmitter.

The Autonomic Nervous System. This system controls the movements of the smooth muscles of internal organs and the stimulation of various glands. It consists of two subdivisions, the sympathetic and parasympathetic systems. These systems are involuntary and act in opposite directions. The sympathetic system is characterized by a large ganglia chained together just outside the central nervous system. These ganglia contain nicotinic receptors and are mediated with ACh. The neurotransmitter between the postganglionic sympathetic axon and the organ is norepinephrine. On the other hand, the parasympathetic system has smaller ganglia present near the organs. The neurotransmitter in the parasympathetic system is ACh with the nicotinic receptor in the ganglion and muscarinic receptor in the post ganglionic axon and the organ.

Insect Nervous System

The nervous system in insects is much simpler than that of mammals, and there are also some basic differences.

1. Cholinergic system is not present in the insect's peripheral nervous system, i.e., no ACh in the neuromuscular junction.
2. Autonomic nervous system is controlled by hormones.
3. Ganglia are not present in the peripheral nervous system.
4. The ganglia in insects correspond to the mammalian central nervous system.
5. Acetylcholine is the neurotransmitter in central nervous system.
6. There is no myelin in the insect nervous system.
7. A tough lipoid sheath known as "nerve sheath" surrounds the insect nerve cord (CNS). It functions as the brain-blood barrier in mammals by limiting the penetration of polar or ionized molecules.
8. The insect nervous system is supplied with oxygen by direct diffusion through tracheal systems.

For a chemical to be an effective insecticide, it should be lipid soluble and have no charge, so that it can penetrate the lipoid sheath surrounding the nervous system.

Chlorinated Hydrocarbon Insecticides

Insecticide Formulations
Insecticide formulations consist of technical grade insecticide in dry forms or in different organic solvents, such as kerosene, toluene, or petroleum solvents. The organic solvents have intrinsic toxicity, and they may also potentiate or synergize the toxicity of the insecticide in formulation.

Classification of Chlorinated Hydrocarbon Insecticides
Chlorinated hydrocarbon insecticides may be divided into the following sugroups:

DDT-Type Compounds. This group of chemicals includes DDT, DDD, acaralate, acarol, bulan, prolan, chlorobenzilate, dimite, dicofol, methoxychlor, and perthane.

Hexachlorocyclohexanes. These chemicals were mistakenly called hexachlorobenzenes. They include seven isomers, depending on the presence of the chlorine atoms in relation to the plane of the cyclohexane structure; axial (a) above and below or equatorial (e) in the plane of the cychlohexane plane. These isomers are: α, aaeee; β, eeeeee; γ, aaaeee, δ, aeeeee; χ, aeeaee; η, aeaaee; and θ, aeaeee. The only isomer with insecticidal activity is the γ isomer, which is called lindane and has an acute oral LD_{50} of 125 mg/kg.

Chlorocyclodines. This subgroup includes aldrin, dieldrin, endrin, heptachlor, and chlordane.

Mirex and Kepone

Terpene Polychlorinated Insecticides. Toxaphene and strobane.

Action
Chlorinated hydrocarbons may be grouped into two groups according to their acute oral toxicity.

Highly Toxic Compounds. Aldrin, dieldrin, and endrin.

Moderately Toxic. DDT, endosulfan, lindane, heptachlor, kepone, chlordane, strobane, dicofol (kelthane), chlorobenzylate, methoxychor, and mirex, all of which are insecticides or acaricides; hexachlorobenzene (HCB), which is a fungicide.

Mode of Action of Chlorinated Hydrocarbon Insecticides. These insecticides interfere with axonal transmission of nerve impulses, which results in the disruption of the central nervous system functions. This leads to the following consequences: behavioral changes, sensory and equilibrium disturbances, involuntary muscle activity, and depression of the central nervous system, particularly respiratory centers.

Action of DDT

Action of DDT on the Nervous System. Small doses of DDT cause hyperactivity, while large doses produce tremor and convulsions (Narahashi and Yamasaki, 1960). DDT acts directly on the axon and slow sodium channel closure after they have opened during passage of nerve impulse (Narahashi, 1979; Farley et al., 1979). This leads to prolonged action potential followed by repetitive discharges. This is manifested as ataxia, hyperexcitability, convulsions, respiratory failure and death. DDT-induced tremor may result from an action on the brain or a direct effect on the spinal cord and peripheral nerves.

DDT increases turnover of serotonin. Chlorinated hydrocarbon insecticides, including DDT, inhibit (Na^+-K^+)-adenosine triphosphatase, which plays an important role in the active transport of ions across the nerve membrane more than Ca-Mg^{2+}-ATPase. The role of this inhibition in the mode of action of chlorinated hydrocarbons is not clear since DDE, a DDT metabolite with no insecticidal activity, also inhibits this enzyme. DDT has no effect on cholinesterases, various oxidases, or dehydrogenases.

Endocrine Effects of DDT. DDT-induced estrogenic and androgenic activity has been attributed to the action of its isomer o,p'-DDT, a contaminant in the technical product. *o,p'*-DDT exerts its endocrine action by a direct effect on the target organs, i.e., ovary and testis, by interfering with the receptor system. Furthermore, *o,p'*-DDT may act on the hypothalamo-pituitary axis. This interference results in disruption of the normal feedback mechanisms that regulate systemic hormone levels.

Thermoregulatory Effects of DDT. DDT produces breakdown of the thermoregulatory mechanisms in treated rats. This results in acute hyperthermia that contributes to its toxicity and lethal action.

Induction of Microsomal Enzymes. DDT, DDD, DDE as well as many chlorinated hydrocarbon insecticides are able to induce microsomal cytochrome P-450 enzymes in many species (Abou-Donia and Menzel, 1968b, 1976).

Action of Hexachlorocyclohexane (BHC)
Lindane acts as a stimulant to the mammalian central nervous system. Although lindane may cause a neurotoxic action similar to that of DDT, its mode of action is not identical to DDT. The neuroexcitatory action of lindane was postulated to result from its antagonistic action of γ-aminobutyric acid (GABA) (Matsumura, 1985).

Action of Cyclodiene Insecticides
This group of chlorinated hydrocarbon insecticides includes aldrin, dieldrin, chlordane, and heptachlor. These compounds affect the rate of respiration more effectively than DDT, but less than lindane. For all compounds, there is a latent period before the increase in rate of oxygen consumption, which reaches a peak at a time that correlates well with a sudden increase in respiratory activity.

Some individuals exposed to low levels of chlordane developed megaloblastic anemia. Endrin produces more liver toxicity than other chlorinated hydrocarbon insecticides.

Action of Chlordecone (Kepone)
Chlordecone contains one carbonyl group that makes it less lipophilic than DDT or mirex. It increases the rate of serotonin turnover as indicated by the elevation of 5-hydroxyindolacetic acid. It induces poor thermoregulatory effects in treated animals. It damages hypothalamic homeostatic and adaptive mechanisms and interferes with the neurotransmitters GABA and catecholamines. It inhibits ATPase. Workers exposed to chlordecone developed nervousness, tremor, incoordination, weakness, and infertility.

Chlordecone has estrogenic action that results in the blocking of the estrus cycle of female rats. It interferes with the estradiol receptors in the brain. Chlordecone-induced estrogenic activity may be related to the decrease in the hypothalamo-pituitary axis of pituitary metenkephalin and elevation of prolactin and luteinizing hormones.

Signs and Symptoms of Poisoning of Chlorinated Hydrocarbon Insecticides in Humans

Acute Intoxication

1. *Ingestion.* Soon after ingestion of toxic doses, salivation, nausea, vomiting and abdominal pain may occur.
2. *Skin absorption.* Dermatitis, apprehension, twitching, tremors, confusion, and convulsion often accompany toxicity.
3. *Inhalation.* Irritation of the eyes, nose, and throat; blurring of vision; cough; and pulmonary edema occur. Respiratory depression results from the action of the insecticide and the petroleum solvents in which these insecticides are usually dissolved. Interference of convulsion with respiration may lead to cyanosis.

Chronic Exposure. It is characterized by anorexia, loss of weight, skin irritation, damage of liver and kidney, emaciation, and disturbance of the central nervous system.

Insecticide formations contain organic solvents such as kerosene or benzene. These solvents increase the action of the central nervous system, causing stimulation and resulting in restlessness, hyperirritability, incoordination, muscle spasms, tremors, clonic and tonic convulsions, depression, collapse, cyanosis, labored respiration and death due to respiratory failure.

Treament

1. Tissue oxygenation and airway clearance by aspiration of secretions.
2. Control of convulsions using diazepam (Valium), pentobarbitol, phenytoin, thiopental, or succinylcholine.
3. Skin decontamination by bathing and shampooing with soap and water.
4. In case of ingestion, gastric lavage with 2–4 l tap water is indicated (catharsis with 30 g (1 oz.) sodium sulfate in one cup of water). Activated charcoal (30–40 g in 3–4 oz. of water) is administered through stomach tube to ensure the removal of the remaining insecticide.
5. If the ingested insecticide is principally in a hydrocarbon solvent (e.g., kerosene) emesis should not be induced. Emesis may result in the aspiration of the solvent, which causes chemical pneumonitis.

6. Cholestyramine resin has been used to enhance the biliary-gastrointestinal excretion of chlorinated hydrocarbon insecticides using 3–8 g per dose 4 times daily before meals and at bedtime.
7. Calcium gluconate (10% in 10 ml ampules) is given intravenously every 4 hours.
8. A balanced high carbohydrate, protein, and vitamin diet is indicated during convalescence to minimize liver injury.
9. Milk, oil, laxatives or cream should not be used, since they enhance absorption of chlorinated hydrocarbons.
10. Epinephrine should not be used.

Current Status of Chlorinated Hydrocarbon Insecticides

Although Müller's discovery in 1939 that DDT had a wide spectrum of insecticidal action, by the middle of the 1960s DDT was banned from most of the developed world largely because of its persistence in the environment. Likewise chlordane, heptachlor, and toxaphane have been banned because they were shown to be carcinogenic in experimental animals. Kepone use was also banned after it caused brain and liver damage in workers.

Organophosphorus Compounds

A large number of organophosphorus compounds are manufactured for numerous uses. In 1981, more than 396.5 million pounds of phosphorus-containing compounds were produced in the United States. Originally, organophosphorus compounds as inhibitors of acetylcholinesterase were developed for use as insecticides, which include over 50,000 compounds. Later some of these chemicals were used as potential nerve agents in warfare (Abou-Donia, 1985).

The Uses of Organophosphorus Compounds

Nerve Agents. Nerve agents were first synthesized by Schrader in Germany during World War II. Their extreme toxicity results from their irreversible inhibition of acetylcholinesterase (AChE). These chemicals are divided into two classes, the G- and V-agents as follows:

1. G-agents include Tabun (GA, *O*-ethyl *N,N*-dimethylphosphoramidocyanidate); Sarin (GB, *O*-isopropyl methylphosphonofluoridate), Soman (GD, *O*-1,2,2-trimethylpropyl methylphosphonofluoridate); and GF, (*O*-cyclohexyl methylphosphonofluoridate).
2. V-agents include VX, (*O*-ethyl *S*-(2-diisopropylaminoethyl) methylphosphonothioate) and VM, (*O*-ethyl *S*-2-diethylaminoethyl) methylphosphonothioate.

Pesticides. This group of chemicals include thousands of pesticides used as insecticides, acaricides, fungicides, or cotton defoliants.

1. *Insecticides and acaricides.* During World War II, tetraethylpyrophosphate (TEPP) was developed in Germany to replace nicotine, which was not immediately available. Due to its extreme toxicity, however, other insecticides such as parathion were prepared. Later, less toxic compounds were synthesized including chlorthion and malathion. Malathion exhibits low toxicity to mammals and high insecticidal activity. This selective toxicity results from (a) its oxidation (activation) to malaoxon, a more toxic metabolite in insects and (b) its rapid detoxification in mammals via hydrolysis to the less toxic product malathion acid.

   a. *Contact insecticides.* Generally, organophosphorus insecticides are contact poisons. Some of them may also act as stomach and fumigant insecticides.

   b. *Systemic insecticides.* These chemicals are absorbed by plants through the foliage or the roots. Following their absorption and distribution through the plant, they protect the plant from insects. This class of insecticides include parathion, malathion, and diazinon.

2. *Veterinary pesticides.* Veterinary pesticides are used to control parasites of domestic animals such as insects, mites, and helminthes. These pesticides may be administered orally or dermally. Examples of this group are haloxon, coroxon, coumophos, and fenchlorphos (Ronnel).

3. *Nematocides.* This group of pesticides control nematodes in the soil by applying them through the water. Chemicals included in this group are prophos (MoCap) and zinophos.

4. *Insect chemosterilants.* Organophosphorus chemosterilants are alkylating agents. Chemicals belonging to this group have no practical use for insect control because they may also have mutagenic, teratogenic, and carcinogenic effects. Most of these chemicals are aziridine derivatives of phosphoric acid such as TEPA, thio-TEPA, metepa, and apholate. Cyclophosphamide, an alkylating agent also has anticancer activity.

   Aziridines act by inhibiting DNA polymerase which results in diminishing the biosynthesis of DNA. The primary site of alkylation of nucleic acids is the N-7 of the guanine moiety. Diminished DNA biosynthesis may also result from the action of thio-TEPA, which inhibits the conversion of purine ribonucleotides into deoxy compounds for incorporation into DNA.

5. *Fungicides.* Most organophosphorus fungicides are phosphorothiolates, phosphonothiolates, or phosphoroamidic acid derivatives. They include kitazin, inezin, edifenphos, and phosbutyl. Kitazin inhibits the incorporation of glucosamine into the wall chitin, which blocks the biosynthesis of fungal wall chiten.

6. *Herbicides.* This group of organophosphorus pesticides include bensulide, zytron, cremart, and amiprophos. The mechanisms of the herbicidal action of these compounds are multiple and not well characterized. The herbicide ethephon acts by generating ethylene in plants, which induces fruit ripening, abscission, and flowering. The related compound *O*-ethylpropylphosphonic acid reduces the internode elongation that retards plant growth. Glyphosate, an herbicide used for control of both annual and perennial weeds, may interfere with the biosynthesis of phenylalanine in plants, probably by inhibiting chorismate mutase and/or prephenate dehydratase. Two aliphatic organophosphorus compounds used as cotton defoliants are DEF (butifus) and merphos (Folex).

7. *Rodenticides.* Because of the high cholinergic toxicity of organophosphorus compounds, which prevents the consumption of a lethal dose by an animal, only a few compounds are used as rodenticides. An example of this class of chemicals is gophacide.

8. *Insecticide synergists.* Propyl-2-propynylphenylphosphonate inhibits microsomal mixed-function oxidases, a property that makes it useful as a synergist for pyrethroids. Also, DEF, which is a potent nonspecific esterase inhibitor, is capable of potentiating the insecticidal potency of malathion.

9. *Insect repellents.* Some phosphoramidates such as *O*-*n*-butyl-*O*-cyclohexenyl-*N*-*N*-diethylphosphoramidate function as insect repellents.

Therapeutic Use. DFP is used for glaucoma treatment because of its anticholinesterase activity, which leads to a decrease in intraocular pressure in primary glaucoma. This compound is extremely potent and has a long duration. The organophosphorus compound echothiophate has replaced DFP in the treatment of glaucoma.

Flame Retardants. Since flammability standards for children's sleepware were established in 1972 in the United States, organophosphorus compounds have been used to make fabrics flame retardant. These chemicals include *tris*-PB and fyrol FR-2. *Tris*-PB is carcinogenic in mice, while fyrol FR-2 and some of its metabolites are mutagenic.

Other Industrial Uses. Many organophosphorus compounds are used as plasticizers, stabilizers/antioxidants, antiwear and antifriction additives in numerous synthetic lubricants, and chemical intermediates for synthesis of

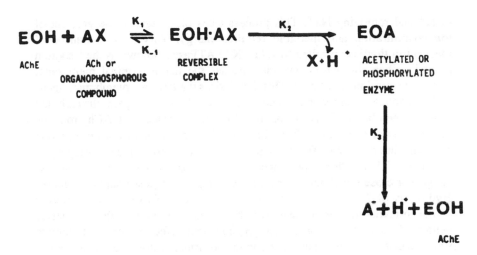

Figure 1. Acetylcholinesterase interaction with acetylcholine and organophosphorus esters.

pharmaceuticals and pesticides. Most of these chemicals have low acute toxicity. Some such as tri-*o*-cresyl phosphate (TOCP) and triphenyl phosphite (TPP), however, produce delayed neurotoxicity.

Mechanisms of Action of Organophosphorus Compound Acute Toxicity

Organophosphorus esters cause acute toxicity by inhibiting AChE (Figure 1). Acetylcholine is the neurotransmitter used in the cholinergic system localized in the following areas of the peripheral nervous system: (1) all skeletal neuromuscular junctions; (2) all synapses between parasympathetic preganglionic and postganglionic nerves; (3) all synapses made by sympathetic preganglionic and postganglionic nerves; (4) synapses on sweat glands and on a very few blood vessels (sympathetic vasodilator fibers) made by sympathetic postganglionic nerves; (5) autonomic effect sites, innervated by postganglionic parasympathetic fibers; (6) some synapses in the central nervous system. Thus, it is evident that the cholinergic system controls all peripheral nerve functions except for sensory systems.

Mechanisms of Cholinergic Neurotransmission. The axonal plasma membrane or axolemma, has different permeability to Na^+ and K^+. In normal state, the interval K^+ concentration far exceeds that outside, and Na^+ concentration is much less than outside which results in a resting potential of -75 mV inside the axon. In the presence of an axonal stimulus, Na^+ ions inflow and K^+ ions outflow across the axonal membrane. This results in the depolarization of axonal membrane and the propagation of the action potential along the axon through the movement of these ions across the membrane in a chain reaction. The excited membrane returns to its normal state by taking in K^+ and removing Na^+. This process takes place against the gradient of ion concentration and therefore requires energy. This "active transport" mechanism that is controlled by (Na^+-K^+)-ATPase is known as the "sodium pump." When the action potential reaches the nerve terminal membrane, neurotransmission is initiated. This depolarizes the terminal membrane, opens Ca^{2+} channels (ionophores), and allows Ca^{2+} ions to pass through the terminal membrane. This mechanism triggers the release of ACh from the nerve terminal (Iversen, 1975). Released ACh diffuses within the synaptic cleft and interacts specifically with the ACh receptor on the postsynaptic membrane. This ACh-receptor interaction may activate a biochemical process or open or close ion channel, or both, resulting in the postsynaptic response. Hydrolysis of ACh by AChE terminates its action. This enzyme is present on the postsynaptic membrane and on the plasma membrane of the cholinergic neuron. ACh may have a greater physiological effect at some cholinergic synapses than at others. Thus, AChE inhibitors usually cause selective actions.

Cholinergic Action

1. *Acetylcholine (ACh).* Choline acetyltransferase catalyzes the synthesis of ACh from choline and acetyl-CoA in the cholinergic nerve terminal (Rossier, 1977).

$$\text{Choline} + \text{Acetyl-CoA} \xrightarrow{\text{Choline acetyltransferase}} \text{ACh}$$

Choline is not synthesized in the neuron. Instead, choline is produced in the liver from phosphatidyl choline, which is a metabolic break down product of ingested phospholipids. Cholinergic terminals take up choline by a high affinity process. Acetyl-CoA required for ACh synthesis is very limited. Usually metabolic inhibitors and hypoxia diminish ACh synthesis in the brain long before energy reserves are exhausted.

2. *Acetylcholine receptors (ACh-R).* There are two classes of ACh-R, muscarinic and nicotinic (Mayer, 1980).
    a. *Muscarinic receptors.* Muscarinic receptors can be subdivided into two subclasses, M_1 and M_2, according to their selectivity for certain agonists and antagonists. Atropine acts as a muscarinic antagonist by blocking all muscarinic responses to injected ACh and cholinomimetic drugs, at excitatory sites such as the intestine and inhibitory sites such as the heart.
        i. *M_1 receptors*
            1. Present in corpus striatum, cerebral cortex, hippocampus, and autonomic ganglia
            2. Have a high affinity for binding to pirenzepine, which acts as M_1 specific agonist

      3. Can be preferentially stimulated by the muscarinic agonist McN-A-343

      4. May take part in the regulation of Ca^{2+} flux and the production of phosphorylated derivatives of inositol.

  ii. *M_2 receptors*

      1. Are predominant in heart, cerebellum, ileum, and smooth muscle (automonic effector cells)

      2. Have low affinity for pirenzepine

      3. Can be preferentially stimulated by bethanechol

      4. May be activated by the inhibition of adenylate cyclase. Such activation may inhibit the release of the neurotransmitter at the cholinergic and adrenergic nerve terminals

      5. ACh increases K^+ permeability (inhibition) in the heart and often decreases K^+ permeability (excitation) in the brain. Activation of muscarinic receptors in glands, smooth muscle, and perhaps other locations is associated with breakdown of membrane phosphoinositide. Breakdown products, i.e., inositol triphosphate, increase intracellular Ca^{2+}, and diacylglycerol activates protein phosphorylation.

  b. *Nicotinic receptors.* Nicotinic receptors are present in skeletal muscle, all ganglia including the adrenal medulla and some synapses in CNS (spinal cord and optic tectum). Nicotinic receptors may be subdivided into two subclasses (N_G receptors and N_M receptors):

  i. *N_G receptors (ganglionic receptors)*

      1. Present in ganglia

      2. React selectivity with hexamethonium

  ii. *N_M receptors (neuromuscular receptors)*

      1. Present at the neuromuscular of striated muscles

      2. React selectively with decamethonium

3. *Action of acetylcholinesterase (AChE).* The action of ACh is terminated by its hydrolysis with AChE. This enzyme contains two active sites: an anionic site that interacts with the quaternary nitrogen atom and the three methyl groups of the choline moiety of ACh, and an esteratic site, which binds the carbonyl carbon atom of the acetyl moiety of ACh. The hydroxyl group of a serine amino acid residue in the esteratic site of AChE forms a covalent bond with the carbonyl groups of ACh. This acetylated enzyme is then dehydrolyzed to yield acetic acid and the free enzyme. This hydrolysis takes place in less than 0.1 msec.

Organophosphorus Esters Action. Organophosphorus esters inhibit AChE by phosphorylating the eserine hydroxyl group at the esteratic site. The phosphoric or phosphonic acid ester formed with the enzyme is hydrolyzed very slowly, and sometimes the esterification is virtually irreversible (Holmstedt, 1963). In this case, the duration of inhibition is too long, because AChE recovery is determined by the time required to synthesize new AChE. The reaction of AChE with organophosphorus esters is analogous to that with the substrate, ACh, which may be represented by the scheme in Figure 1. This reaction involves three steps: (1) complex formation that is controlled by the affinity constant $K_a = k_{-1}/k_1$; (2) acetylation or phosphorylation, k_2; (3) deacetylation or dephosphorylation, k_3.

In both reactions k_3 is the rate-limiting step. With the natural substrate ACh, k_2 and k_3 are very fast, and the total reaction takes place in milliseconds, which results in the generation of free enzyme. The overall reaction is measured by the turnover number, that is the number of substrate molecules hydrolyzed per minute by one molecule of enzyme. This number has been determined to be 300,000 and 0.008 for ACh and dimethyl phosphate, respectively. When organophosphorus esters react with AChE, k_2 is moderately fast, while k_3 is extremely slow ($k_2 > k_3$); thus the phosphorylated enzyme EOA accumulates, and the amount of EOH.AX is minimal at any time.

"Aging" of Phosphorylated AChE. Phosphorylation of AChE with some organophosphorus esters virtually results in an irreversible inhibition of the enzymes (Aldridge, 1976). The rate of regeneration of plasma or red blood cell AChE that has been phosphorylated with DFP coincides with the rate of resynthesis of new enzymes. In this case, phosphorylated AChE undergoes "aging," a process that involves the loss of one of the alkyl groups on the phosphorylated enzyme, resulting in the negatively charged monoalkyl enzyme.

Reactivation of Phosphorylated AChE. Oxime derivatives such as 2-pyridine aldoxime methiodide (2-PAM or pralidoxime) are very effective in removing organophosphorus esters from AChE. These compounds contain a quaternary nitrogen atom that interacts with the anionic site of AChE and a hydroxylamine group that binds the phosphate to pralidoxime thus activating AChE at the neuromuscular junction, although it does not cross the blood-brain barrier. The aged enzyme cannot be activated, thus 2-PAM should be given soon after exposure.

Tolerance of Acute Sublethal Effects of Organophosphorus Compounds. Repeated administration of sublethal doses of some organophosphorus compounds to experimental animals initially produces acute toxicity (DuBois, 1965). In time, however, the animals no longer developed signs of acute cholinergic effects despite continued administration of these chemicals. When examined at termination, these seemingly normal animals have

TABLE 1.
Signs and Symptoms of Organophosphorus Insecticide Poisoning

Action Site	Signs and Symptoms
Central nervous system	Giddiness (a whirling, dizzy sensation), anxiety, CNS stimulation at low to moderate doses due to sparing of ACh from hydrolysis, depression at high doses, apathy, confusion, restlessness, headache, dizziness, anoxia, insomnia, ataxia, absence of reflexes, Cheyne-Stokes respiration, depression of respiratory and circulatory centers, electroencephalographic (EEG) changes, convulsion, and coma
Muscarinic receptor	
Sweat glands	Sweating
Salivation glands	Excessive salivation
Lacrimation glands	Lacrimation (tearing)
Pupils	Constricted pupils (pinpoint, miosis)
Ciliary body	Blurred vision
Bronchi	Wheezing and increased bronchial secretion, cough, pulmonary edema
Cardiovascular system	Bradycardia (slow heart beat), fall in blood pressure
Urinary bladder	Urinary incontinence
Gastrointestine	Abdominal pain, vomiting, diarrhea, fecal incontinence
Nicotinic receptors	
Neuromuscular junction	Fasciculations, cramps, weakness, muscle twitching, respiratory difficulty, tightness in chest tremor, paralysis, cyanosis, arrest
Sympathetic ganglia	Tachycardia, elevated blood pressure (since only few cholinergic synapses are present in the vasculature, they have no control over blood pressure and the predominant action is stimulation of the sympathetic ganglia, i.e., increased blood pressure)

markedly inhibited AChE activity in blood and nervous tissue and elevated levels of ACh in their nervous system. This adaptation phenomenon has been explained by the development of tolerance of ACh receptor or ACh or a decrease in the total number of ACh receptors.

Signs and Symptoms of Organophosphorus Insecticide Poisoning

Organophosphorus insecticides inhibit AChE in the central and peripheral nervous systems resulting in the accumulation of endogenous ACh at synapses and neuromuscular junctions (Mayer, 1980). Excess ACh initially causes excitation and then paralysis of the cholinergic transmission and consists of: (1) the central nervous system, (2) the parasympathetic nerve endings and a few sympathetic nerve endings such as the sweat glands (muscarinic effects), and (3) the somatic nerves and the ganglionic synapses of autonomic ganglia (autonomic effects). The resulting signs and symptoms are those produced by excessive and continued stimulation of the muscarinic, nicotinic, and a central nervous system ACh receptors outlined in Table 1.

The severity of the clinical manifestations of poisoning depends on the compound and extent and route of exposure, as follows:

Mild. The first complaints are usually fatigue, giddiness (a whirling, dizzy sensation), and sweating. This may be accompanied by anorexia, headache, weakness, anxiety, tremors of tongue and eyelids, miosis (constriction of the pupils), impairment of visual acuity, and tightness in the chest. Serum butyrylcholinesterase (BuChE) is inhibited to 20–50% of normal. Prognosis is good.

Moderate. In moderate poisoning, earlier signs and symptoms may be followed by nausea, salivation, lacrimation, abdominal cramps, vomiting, sweating, slow pulse, bradycardia (muscarinic effect), drop in blood pressure, and muscular tremors. Serum BuChE inhibited to 10–20% of normal. Recovery is expected with treatment.

Severe. Severe poisoning results in diarrhea, pinpoint and nonreactive pupils, muscular twitching, increase in bronchial secretion, respiratory difficulty, pulmonary edema, cyanosis, loss of sphincter and urinary bladder control, tachycardia (nicotinic effect), elevated blood pressure, convulsions, coma, heart block, and possibly death. Serum BuChE is inhibited to less than 10% of normal. Severe exposure is fatal if not treated.

Other Considerations. Body temperature is not elevated with organophosphorus poisoning. The pulse rate may be initially increased, but it slows in later stages. The blood pressure may be mildly elevated. Tendon reflexes are usually within normal limits, and nuchal rigidity and abnormal reflexes do not occur. Brain hypoxia from AChE inhibition, even for a short time, may result in prolonged CNS dysfunction.

Death
1. Death may occur 5 min to 24 h after a single exposure, depending upon the chemical, dose, and route of exposure.
2. The primary cause of death is asphyxia, which follows respiratory failure with a secondary cardiovascular effect. Respiratory failure results from one or more of the following mechanisms:
   a. Muscarinic bronchoconstriction and laryngospasm
   b. Muscarinic excessive tracheobronchial and salivary secretions
   c. Nicotinic paralysis of diaphragm and respiratory muscles
   d. Central nervous system depression and paralysis of respiratory centers.

Treatment of Organophosphorus Poisoning
The following treatment measures should be performed for organophosphorus poisoning.

Clearing airway by removal of secretions

Administration of oxygen and initiation of artificial respiration

Atropine sulfate administration via intravenous or intramuscular injection, if i.v. injection is not possible. Atropine acts as an antidote for acetylcholine in muscarinic receptors. It is ineffective against nicotic action or central nervous system effects. For adults, including children over 12 years: 0.4–2.0 mg repeated every 15 min until atropinization, characterized by tachycardia (pulse of 140 per min), flushing, dry mouth, dilated pupils, is reached. A mild degree of atropinization is maintained for at least 48 h. Miosis, nausea, bradycardia, and other cholinergic signs are indicative of inadequate atropinization. For children under 12 years, 0.05 mg/kg atropine sulfate is repeated every 15 min to achieve atropinization, which is maintained with repeated dosages of 0.02–0.05 mg/kg. If atropine toxicity signs, e.g., fever, muscle fibrillations, and delirium develop, it should be discontinued temporarily. Atropine should not be given to cyanotic patient. Artificial respiration should precede atropine treatment.

Pralidoxime (2-PAM) should only supplement atropine therapy and should never replace it. It is an oxime, a class of chemicals that hydrolyzes phosphorylated AChE resulting in a free enzyme. Reactivation of AChE occurs rapidly (within 10–15 min), while plasma ChE requires 12–15 days to return to a normal level. Pralidoxim is administered early, usually less than 36 h after poisoning. It relieves nicotinic and central nervous system signs. An adult dose, including children 12 years is 1.0 g intravenously at a rate not to exceed 0.5 g per min. This treatment is repeated in 1–2 hours, then at 10–12 h intervals if needed. If i.v. injection is not employed, pralidoxime may be administered by deep intramuscular injection.

Neither atropine sulfate nor pralidoxime should be administered prophylactically to persons exposed to organophosphorus pesticides

For ingestion, stomach is lavaged with 5% sodium bicarbonate if not vomiting

Decontamination of the skin should be accomplished by washing with alkaline soap and water to enhance the hydrolysis of the organophosphorus ester. In the event of convulsions, sodium thiopental (2.5% solution) should be given intravenously. Also diazepam (Valium®), 5–10 mg for adults, 0.1 mg/kg for children 6 years or 23 kg, should be injected intravenously.
The following drugs should be avoided in organophosphorus poisoning: morphine, theophylline, aminophylline, succinylcholine, and tranquilizers of the reserpine or phenothiazine drug groups.

Organophosphorus Compound-Induced Delayed Neurotoxicity

Although the immediate hazard associated with organophosphorus compounds is related to their ability to inhibit AChE, several of these compounds are also capable of producing delayed neurotoxicity (Smith et al., 1930; Abou-Donia, 1981). Organophosphorus compound-induced delayed neurotoxicity (OPIDN) was first produced in man following exposure to tri-*o*-cresyl phosphate (TOCP).

Characteristics of OPIDN

1. Most delayed neurotoxic organophosphorus compounds are AChE inhibitors, but not all AChE inhibitors cause OPIDN.
2. Organophosphorus compounds capable of producing OPIDN are, generally, potent inhibitors of nonspecific esterases *in vivo*.
3. There is a delay period of 6–14 days before onset of neurologic deficit characterized by ataxia and paralysis.
4. Neuropathological changes are seen in the sciatic, peroneal, and tibial nerves; spinal cord; and medulla; but not in higher brain.
5. Onset of lesions begins at the distal part of the long fibers and large diameter spinal cord and large diameter peripheral nerves.
6. Histopathologic lesions are seen as Wallerian-type degeneration of the axon with subsequent degeneration of myelin.
7. Some animal species, such as cats, dogs, cows, water buffaloes, sheep, and chickens are susceptible to OPIDN, while others, such as rodents and some primates are less sensitive. The adult chicken has become the test animal to study this effect.
8. There is an age-related sensitivity to OPIDN; young chicks are insensitive to a single dose of delayed neurotoxic organophosphorus compounds, and old chickens are more sensitive than are young adults.

Human Cases. TOCP, a plasticizer used in lacquers and varnishes, first caused OPIDN in humans and later in sensitive species. This was observed at the end of the nineteenth century, and since then an estimated 40,000 human cases of delayed organophosphorus pesticides have caused OPIDN in workers (Abou-Donia and Lapadula, 1990). Organophosphorus pesticides that have been implicated in human cases of OPIDN are mipafox, leptophos, methamidophos, trichlorphon, trichlornat, EPN, and chlorpyrifos.

Animal Studies. Most studies on OPIDN have been carried out in the adult chicken (Abou-Donia, 1981). Because of the health hazard of delayed neurotoxic organophosphorus compounds, the U.S. Environmental Protection Agency requires the testing of these compounds for the potential to produce OPIDN in the chicken. The cat, another sensitive species, has been used in some studies on OPIDN. Other organophosphorus compounds that have been shown to produce OPIDN in animals are: DEF, merphos, S-seven, haloxon, coumaphos, cyanofenphos, disbromoleptophos, leptophos oxon, EPN oxon.

The development, severity, and progression of OPIDN are dependent upon the following factors: the chemical structure, dosage, route of administration, and frequency and duration of exposure. Pharmacokinetics and metabolism of organophosphorus compounds play an important role in their ability to produce OPIDN. Thus, insensitive species are capable of metabolizing and excreting (mainly in the urine) neurotoxic organophosphorus, e.g., TOCP, leptophos, and EPN as degradation products, faster than sensitive species. Some of these chemicals may also undergo metabolic activation to more active neurotoxic agents; TOCP is metabolized to the more potent neurotoxicant saligenin cyclic-*o*-tolyl phosphate.

Mechanisms of OPIDN. Although OPIDN has been known since the end of the nineteenth century, studies on its molecular pathogenesis have started only recently. The common characteristics of delayed neurotoxic organophosphorus compounds are (1) they contain a phosphorus atom and (2) they are direct or indirect inhibitors of esterases. Numerous studies have been carried out to characterize the neurologic dysfunctions, electrophysiologic deficits, neuropathologic lesions, and biochemical changes that occur in OPIDN.

Early attempts to study the mechanisms of action of OPIDN hypothesized that neurotoxic organophosphorus esters induce delayed neurotoxicity by inhibiting an esterase in the nervous system. Both brain acetylcholinesterase (Bloch and Hottinger, 1943) and pseudocholinesterase (Earl and Thompson, 1952) were hypothesized as the target for OPIDN but subsequently were ruled out. *In vivo* and *in vitro* studies have demonstrated that a small portion (about 6%) of the total phenylvalerate-hydrolyzing activity in hen brain is sensitive to inhibition by delayed neurotoxic organophosphorus compounds such as mipafox but not by nondelayed neurotoxic esters such as paraoxon. This enzymatic activity was called neurotoxic esterase (NTE) (Johnson, 1969). To produce delayed neurotoxicity, an organophosphorus compound should cause at least 70% inhibition of hen brain NTE 24 h after administration of an LD_{50} dose. The NTE theory also hypothesized the necessity for its "aging" by organophosphorus compounds that produce OPIDN. Although NTE offers a good test to screen organophosphorus compounds for the delayed neurotoxicity potential, it does not seem to be involved in the mechanisms of OPIDN. There is no hypothesis as to how the inhibition and aging of NTE would lead to the pathological lesions and neurologic dysfunctions. Furthermore, NTE has not been isolated, and its physiological and biochemical functions are unknown. Also, NTE is present and becomes inhibited with a single dose of delayed neurotoxic compounds in young chicks and in rats without subsequent development of OPIDN.

To elucidate the molecular mechanisms for OPIDN, a hypothesis that delayed neurotoxic organophosphorus compounds produce OPIDN by phosphorylating a protein kinase that catalyzes endogenous phosphorylation of a neuronal enzyme or cytoskeletal proteins, thereby adversely affecting the regulation of normal neuronal processes and resulting in axonal degeneration has been investigated.

Recent studies have demonstrated that in OPIDN there is an increased CaM kinase II-catalyzed phosphorylation of cytoskeletal proteins: α- and β-tubulin, microtubule-associated proteins-2 (MAP-2), and the neurofilament triplet proteins (Abou-Donia et al., 1984; Patton et al., 1983, 1985, 1986; Suwita et al., 1986a,b).

Other investigators have shown that the consequences of increased cytoskeletal protein phosphorylation are (1) increased phosphorylation of MAP-2 depolymerizes cold labile microtubules, (2) increased phosphorylation of tubulin results in aggregated polymers, and (3) increased phosphorylation of α- and β-tubulin, MAP-2, and neurofilament triplet proteins interferes with their assembly (normal interactions with themselves and with each other). These studies indicate that increased Cam kinase II-dependent phosphorylation of cytoskeletal proteins lead to their aggregation into solid masses instead of formation of polymers.

The morphologic similarity between the early histopathologic changes in OPIDN which were seen as aggregation and accumulation of microtubules and neurofilaments accompanied by proliferation of smooth endoplasmic reticulum (Bischoff, 1970), and those seen by increased Cam kinase-II-mediated phosphorylation of cytoskeletal proteins add support to the involvement of cytoskeletal proteins in the mechanism of OPIDN (Abou-Donia et al., 1988; Abou-Donia and Lapadula, 1990).

Carbamates

Carbamates are widely used as insecticides and miticides on crops, fruits, vegetables, and forests. Like other organosynthetic pesticides, they are absorbed through oral, dermal, and inhalation exposure. They are absorbed less by the skin than by other routes. They exert their insecticidal action by inhibiting acetylcholinesterase (Murphy, 1986). Unlike organophosphorus insecticides, carbamates inhibit AChE reversibly as the result of the *in vivo* spontaneous hydrolysis of the carbamylated AChE enzyme. This leads to less severe and shorter-duration action. The signs of carbamate intoxication in animals and symptoms in humans are similar to those of organophosphorus insecticides. Some carbamates, e.g., aldicarb, are absorbed by plant roots and accumulate in the foliage, flowers, and fruits. This may result in systemic toxicity to humans.

Action

Carbamates with high acute oral toxicity, i.e., $LD_{50} < 50$ mg/kg, in order of decreasing toxicity are aldicarb (Temik) > carbofuran (Furadan) > methomyl (Lannate) > aminocarb (Matacil) > dimetilan. Moderately toxic carbamates with acute oral $LD_{50} > 50$ mg/kg are, in decreasing order, methiocarb (Mesurol) > propoxur (Baygon) > carbaryl (Sevin).

Metabolism

Aromatic carbamates undergo hydrolysis and hydroxylation of the aromatic ring followed by conjugation mostly in the liver. Metabolic products are excreted by the kidneys.

Symptoms

Symptoms are similar to those produced by organophosphorus insecticides resulting from inhibition of acetylcholinesterase and the accumulation of acetylcholine at the nicotinic and muscarinic acetylcholine receptors in the peripheral and central nervous systems.

Treatment

Like organophosphorus poisoning, the speed at which treatment is carried out is imperative.

1. Atropine is intravenously injected at 1 to 4 mg doses. A 2-mg dose injection is repeated when symptoms begin to occur at 15–60 min intervals. Excessive salivation is an indication that more atropine is needed.
2. Airways should be kept open. Aspiration is carried out and oxygen is used. Endotracheal tube is inserted. Tracheotomy and artificial respiration are done if needed.
3. For ingested carbamates, stomach lavage with 5% sodium bicarbonate is carried out if there is no vomiting. Contaminated skin is washed with soap and water, and eyes are washed with isotonic saline. Care should be taken to prevent the spread of the carbamate by wearing rubber gloves while washing contact area.
4. 2-PAM and other oximes are contraindicated, since they may increase AChE inhibition. Theophylline, aminophylline, barbiturates, and epinephrine are contraindicated. Epinephrine may produce ventricular fibrillation (cardiac arrhythmias) that results from myocardium sensitization.

Botanical Insecticides

Nicotine

Nicotine was first used as an insecticide in 1763. It is obtained from the leaves of *Nicotiana tabacum* (2–5%) and *Nicotiana rustica* (5–14%). Nicotine alkaloid and nicotine sulfate have been used as contact, stomach, and fumigant insecticides. Nicotine acts to stimulate receptors in autonomic ganglia, at the neuromuscular junction, and in some pathways of the central nervous system.

The oral LD_{50} of nicotine in rats is 10–60 mg/kg. Human poisoning is followed by salivation; vomiting (from ganglionic stimulation); muscle weakness; fibrillation (by stimulation at the neuromuscular junction); and ultimately, clonic convulsions and cessation of respiratory function (effects in the central nervous system).

Treatment of nicotine poisoning is by use of anticonvulsants.

Nicotine is oxidized and hydroxylated by microsomal oxidases that yield less toxic metabolites.

Rotenone

Rotenone is extracted from the roots of the plant *Derris elliptica* and *Derris malaccensis,* which has been used in Africa as a fish bait. It acts by inhibiting oxidative phosphorylation by inhibition of $NADH_2$ dehydrogenase, which results in blockage of the oxidation of glutamate, α-ketoglutarate, and pyruvate by the NAD system. Its oral LD_{50} in rats is 100–300 mg/kg. Its oral toxicity, however, in man is low; it takes 10–100 g to kill a 70 kg man. Orally ingested rotenone produces gastrointestinal irritation, nausea, and vomiting. Rotenone is more toxic by inhalation and causes respiratory stimulation followed by depression and convulsions.

Pyrethroids

Chemistry. Pyrethroids are classified into two groups: naturally-occurring compounds present in the flowers of the genus *Chrysanthemum* and synthetic compounds (Casida et al., 1983). Both groups have been used as insecticides; natural pyrethroids since 1800 and synthetic pyrethroids developed during the past half century.

Pyrethrums. Extraction of chrysanthemum flowers (0.6–3.0% pyrethrum extract) consists of four esters resulting from the combinations of two alcohols, pyrethrolone and cinerolone, and two acids, chrysanthemum monocarboxylic acid (acid I) and chrysanthemum dicarboxylic acid monomethyl ester (acid II). The four esters are

pyrethrin I = acid I + pyrethrolone
pyrethrin II = acid II + pyrethrolone
cinerin I = acid I + cinerolone
cinerin II = acid II + cinerolone

While effective in controlling insects, pyrethrums have low mammalian toxicity; acute oral LD_{50} in rats is 1,500 mg/kg. These chemicals have no persistent residues and lack the stability to allow their use in controlling crop insects.

Synthetic Pyrethroids. Over the past half century, the chemical structure of synthetic pyrethroids has been optimized to produce insecticides that are both effective and have long-lasting action. These compounds include permethrin, resmethrin, cypermethrin, decamethrin, tetramethrin, phenothrin, cypherothrin, kadethrin, and fenvalerate. Due to their long residual activity, premethrin, cypermethrin, deltamethrin, and fenvalerate are used to control insects for several weeks after a single application. Also, some of the synthetic pyrethroids undergo less metabolic deactivation than natural pyrethroids.

Selectivity Toxicity. Pyrethroids are more selectively toxic to insects than mammals and are generally considered among the safest of the highly potent insecticides. The average ratios for oral rat LD_{50} to insect topical LD_{50} (mg/kg) for various classes of insecticides are methyl carbamates, 16; organophosphorus insecticides, 33; chlorinated hydrocarbons, 91; and pyrethroids, 4500 (Casida et al., 1983).

Although pyrethroids are less toxic to mammals than other types of insecticides, they are as toxic to shrimp, lobster, and fish as is DDT and some other chlorinated hydrocarbons.

Mode of Action of Pyrethroids. Pyrethroids are neurotoxic agents. The major target site of these chemicals are sodium channels in nerve membranes (Narahashi, 1971). Nerve preparations containing synapses are very sensitive to pyrethroids' actions. Applying a stimulus to the presynaptic nerve at skeletal neuromuscular junctions following treatment with pyrethroids evoked repetitive end plate potentials. The action resulted from repetitive discharges in the presynaptic nerve fibers that are shown by focal (extracellular) recording of both pre- and postsynaptic responses. These repetitive afterdischarges are induced by a single stimulus, as demonstrated in studies using intracellular microelectrode with isolated giant fiber preparations exposed to pyrethroids. Voltage clamp experiments revealed that the depolarizing afterpotential caused by allethrin resulted from the slowing of the sodium channel inactivation mechanism. Similar results were obtained with tetramethrin, which apparently modifies sodium channels at their resting or closed states.

Classification of Pyrethroids. Based on their chemical structure and toxic response, pyrethroids have been classified into two groups, type I and II (Gammon et al., 1981). Alternately, pyrethroids may be classified on the basis of signs of toxicity. Thus, type I produces tremor and is classified as (T), while type II produces choreoathetosis with salivation and is called (CS).

1. *Type I pyrethroids.* Compounds belonging to this group are characterized as follows:
   a. There is no α-cyano group. Many conventional pyrethroids belong to this group, i.e., pyrethrins, allethrin, tetramethrin, kadethrin, resmethrin, phenothrin, and permethrin.
   b. It modifies sodium channels, causing repetitive discharges to be produced in a nerve fiber that results from prolongation of sodium current. This action is accompanied by a large membrane depolarization, thus no impulse conduction block takes place.
   c. Although repetitive discharges may take place in any part of the nervous system, the synaptic transmission is most affected when presynaptic nerve terminals are affected. Consequently, the central nervous system and peripheral ganglia are very sensitive to pyrethroid actions since they contain a large number of synapses. Thus pyrethroid effects on the nervous system result in hyperexcitation, ataxia, tremors, and convulsions of animals.
   d. Pyrethroids belonging to this group are intrinsically 5–10 times more potent than those of the type II group.
   e. Ca^{2+} may play an important role in the mechanism of action of pyrethroids, since they are involved in controlling Na^+ inactivation in nerves.
2. *Type II pyrethroids.* Compounds in this group are characterized by the following:
   a. An α-cyano group is present, i.e., α-cyanophenoxybenzyl alcohol moiety. This group of pyrethroids includes fenvalerate, deltamethrin, cypermethrin, cyphenothrin, and fenopropathrin.
   b. They cause an extremely prolonged sodium current resulting in depolarization of the nerve membrane. They cause only nerve blockage in sensory axons with no repetitive firing.
   c. Since sensory neurons tend to discharge when depolarized, the depolarizing action of type II pyrethroids have a profound action on the sensory nervous system. This has been demonstrated as a tickling sensation on the face of humans exposed to chemicals belonging to type II pyrethroids.
   d. In addition to the sensory nervous system, the depolarization produced by these compounds causes an increased release of a

neurotransmitter (i.e., acetylcholine), resulting in cholinergic signs such as salivation, choreoathetosis, and tremors.

e. Type II pyrethroids may affect GABA receptors, since the benzodiazepine diazepam delays the onset of type II but not type I action.

Symptoms in Humans. Pyrethroids are contact insecticides. They have no stomach poison effect because they are rapidly hydrolyzed in the gastrointestinal tract to nontoxic products.

Symptoms of pyrethroid toxicity depend on the route of exposure.

1. *Skin.* Skin exposure causes contact dermatitis.
2. *Inhalation.* Symptoms include asthmatic wheezing; a stuffy, runny nose and scratchy throat; sudden bronchospasm; swelling of oral and laryngeal mucous membranes; and shock (anaphylaxis). Hypersensitivity pneumonitis is also common; it is characterized by delayed dyspnea, cough, and fever. Massive inhalation of pyrethroids may cause nervous irritability, tremors, or ataxia (Flannigan and Tucker, 1985).

Treatment
1. Skin decontamination is accomplished with a soap and water wash.
2. Asthma or anaphylaxis (allergic reaction) is treated with adrenalin or histamines and decongestants for less severe allergic reactions.
3. Hypersensitivity pneumonitis may be treated with oxygen, steroids, antibiotics, and bed rest.
4. Ingested pyrethroids are treated with large doses of activated charcoal followed by cathartic doses of sodium or magnesium sulfate.
5. Diazepam is given to control nervousness and tremors.

HERBICIDES

Chlorophenoxy Compounds

This group of herbicides includes dichlorophenoxyacetic acid (2,4-D); 2,4,5-trichlorophenoxyacetic acid (2,4,5-T); and 2-(2,4,5-trichlorophenoxy)propionic acid (2,4,5-TP, silvex). Agent orange is a 50% mixture of each of 2,4-D and 2,4,5-T compounds.

Use

Chlorophenoxy herbicides are used in agriculture for control of broadleaf weeds and in the control of weeds along highways. They exert their herbicidal action by acting as growth regulators in plants.

Action

Chlorophenoxy herbicides have no hormonal action in animals. The mean adult lethal dose in humans was estimated to be 28 g. An intravenous injection of 2000 mg of sodium 2,4-D caused no clinical effects, whereas 3000 mg produced serious toxic signs (Seabury, 1963). Poisoning symptoms in humans include chemical hepatitis; metabolic acidosis; stiffness of the legs and arms; ataxia; paralysis; coma; ventricular fibrillation; cardiac arrest; and eventually, death. Peripheral neuropathy was reported in humans exposed to 2,4-D. Local effects of this compound include burning skin sensation, cough, emesis, chest pain, abdominal pain, and diarrhea (O'Reilly, 1984).

Treatment

For ingestion, stomach lavage with tap water. For skin contact, exposed area is washed. Supportive treatment. Quinidine sulfate or quinine to relieve myotonia or suppress abnormal ventricular cardiac rhythm.

TCDD (Dioxin)

TCDD (2,3,7,8-tetrachlorodibenzo-*p*-dioxin) is a by-product in the synthesis of 2,4,5-T from the contaminant 2,4,5-trichlorophenol, which is also a contaminant of the antibacterial agent hexachlorophene. TCDD was also identified as contaminant of the herbicide Agent orange. Veterans of the Vietnam War who were exposed to Agent orange are concerned of possible residual health effects related to TCDD.

TCDD and related chlorinated dioxins are extremely toxic with LD_{50} values on the order of 1–100 μg/kg. The acute toxicity of TCDD varies greatly with species.

Species	LD_{50} (μg/kg)
Female guinea pig	0.1
Female rat	45
Hamster	3,000

Occupational exposures and industrial accidents have shown that TCCD effects include hepatic enlargement, chloracne, neuromuscular symptoms, porphyria, tissue wasting, loss of body fat, microsomal enzyme induction, immune depression, teratogenic and fetotoxic effects and increased tumor incidence, i.e., soft tissue sarcomas and lymphomas (Safe, 1986).

Nitrophenolic and Chlorophenolic Compounds

This group of herbicides is also used as fungicides. Examples of these compounds are 4,6-dinitro-*o*-cresol and pentachlorophenol.

Action

These chemicals act by uncoupling oxidative phosphorylation. This results in a relatively uncontrolled production of body heat, with depletion of carbohydrate and fat stores, dehydration, and tachycardia. Hot environments exacerbate the symptoms, which include weakness, malaise, profuse sweating, warm flushed skin, headache, and diaphoresis that progress to fever, tachycardia, apprehension, and dyspnea and eventually worsens to delirium, coma, and convulsions; apprehension, anxiety, manic behavior, or unconsciousness reflect severe cerebral injury. Yellow staining of the skin is indicative of nitrophenolic compound poisoning (Bidstrup and Payne, 1951).

Pentachlorophenol may contain TCDD. It is used extensively as a wood preservative and has caused problems in buildings containing improperly treated wood.

Treatment

1. Treatment is nonspecific except cold baths or sponges.
2. In case of ingestion, stomach lavage is recommended as well as the use of activated charcoal within the first several hours.
3. Contaminated skin should be thoroughly washed.
4. Aspirin should not be used because it increases the uncoupling of oxidative phosphorylation.

Dipyridy Herbicides

Paraquat

Paraquat (1,1'-dimethyl-4,4'-dipyridium) is the most toxic herbicide. It exerts its herbicidal action by disrupting photosynthesis via inhibiting the reduction of NADP to NADPH and by interfering with the electron transfer system.

Toxicity. The lethal adult oral doses in humans ranges from 3 to 6 g of paraquat despite its poor gastrointestinal tract absorption, i.e., 5–10% of dose. Although dermal absorption is minimal, fatal dermal exposures have been reported after a single or chronic contamination with commercial concentrations (Crome, 1986). Inhalation of paraquat spray is unlikely since droplets have diameters exceeding 5 μm. The diameter range is too large to reach alveolar membrane and cause direct or systemic toxicity.

Most of ingested paraquat is excreted without metabolism. Rats excrete 96% of the dose within 3 days in the urine. Biliary excretion accounts for a small fraction of the dose.

Action. Several studies have shown that paraquat converts molecular oxygen to superoxide radicals (O_2^-), hydroperoxy radicals (HO_2^-), and hydrogen peroxide, which disrupt cell structure and functions (Bus et al., 1976).

Although superoxide dismutase catabolizes the superoxide radical and glutathione reductase reduces lipid hydroperoxides to less toxic lipid alcohols, the use of these enzymes as therapeutics has not been successful.

Symptoms of Paraquat Toxicity. Paraquat injures the epithelial tissues, skin, nails, cornea, liver, kidney, the gastrointestinal tract, respiratory tract, and the brain. Mild toxicity results in dermatitis, conjunctivitis, cough, and headache. Moderate toxicity results in the development of progressive pulmonary fibrosis leading to hypoxia and death. Ingestion of large amounts results in immediate severe gastroenteritis followed by failure of several systems over a few days, i.e., acute renal tubular necrosis, centrilobular hepatic necrosis, cerebral and adrenal hemorrhage, myocardial necrosis, pulmonary congestion and focal hemorrhage (Onyeama and Oehme, 1984).

Paraquat is actively concentrated in the pneumocytes of the lung tissue. Initially, there is swelling and fragmentation of membranous pneumocytes (type I), followed by degeneration of granual pneumocytes (type II). Several days after poisoning, these cells die, followed by a rapid proliferation of connective tissue cells which fill the alveolar spaces. This leads to death from asphyxia.

Treatment. There is no antidotal treatment. Absorption of ingested paraquat may be minimized by gastric lavage, active charcoal, saline chartics, and forced diuresis (Tabei et al., 1982). Dialysis may be required. Since oxygen accelerates the pathological process caused by paraquat, it should not be administered unless arterial PO_2 drops below 60–70 mm Hg.

Diquat

Diquat is less toxic than paraquat as indicated by their oral LD_{50} of 400 mg/kg and 25–50 mg/kg, respectively. Diquat is less likely than paraquat to produce death. Its main target organs are gastrointestinal tract, kidney, liver, and the central nervous system. Diquat is not concentrated in lung tissue and does not cause pulmonary fibrosis, however, it may cause pulmonary hemorrhages (Clark and Hurst, 1970).

Chlorate Salts

Sodium chlorate is also a nonselective herbicide. Although it has relatively low toxicity, it may be fatal if ingested. The oral lethal dose in adults is 20–35 g. Potassium chlorate is more toxic (lethal dose is 12 g in adults; 5 g in children; 1 g in infants) than sodium chlorate (Helliwell and Nunn, 1979).

Action

Sodium chlorate strongly oxidizes hemoglobin to methemoglobin and produces hemolytic anemia and direct nephrotoxic effects. These effects result in tissue hypoxia due to methemoglobinemia, hyperkalemia from severe hemolysis, and acute renal failure leading to coma and death.

Symptoms

Mild symptoms of chlorate poisoning include nausea, vomiting, diarrhea, and abdominal pain. Severe toxicity produces cyanosis, dyspnea, and coma as well as renal failure.

Treatment

Decontamination is accomplished using gastric lavage and absorbants. Methemoglobinemia may be treated by intravenous injection of methylene blue. Oxygen may be administered. Sodium thiosulfate orally or intravenously inactivates chlorate ion.

Atrazine

Atrazine (2-chloro-4-ethylamino-6-isopropylamino-5-triazine) is a widely used herbicide. Oral LD_{50} in rats is 1.2 g/kg. In cattle, an oral 400 mg/kg dose caused diffuse subcutaneous hemorrhage, liver necrosis, ataxia, and death. This compound was implicated in producing sensorimotor polyneuropathy in a farmer exposed dermally to atrazine (Castano et al., 1982).

FUNGICIDES

Commonly used fungicides generally have low acute toxicity.

Thiocarbamates and Dithiocarbamates

These chemicals include the following classes:

Bis dithiocarbamates, e.g., thiram.
Metallo bis dithiocarbamates, e.g., Ziram (Zn), Nabam (Na), and Ferbam (Fe).
Ethylen bis dithiocarbamates, e.g., Maneb (Mn) and Zineb (Zn).
Monothiocarbamates, e.g., butylate, cycloate (Ro-neet), EPTC (EPTAM), diallate (Avadex), and triallate (Far-Go, Avadex-BW).

Action

None of these chemicals have cholinergic or delayed neurotoxic effects. The metallo bis dithiocarbamates and ethylene bis dithiocarbamates are moderately irritating to skin and respiratory mucous membranes. Thiram and antabuse inhibit aldehyde dehydrogenase. However, the other classes of thio- and dithiocarbamates do not inhibit this enzyme. High chronic dosages of iron and zinc bis dimethyl dithiocarbamates produce anatomic central nervous system damage. The nervous system action may be related to CS_2 liberated from these chemicals *in vivo*. Ethylene bis dithiocarbamates do not produce neurotoxicity. On the other hand, large doses of monothiocarbamate herbicides may produce paralysis (Fishbein, 1976).

Treatment

Treatment is affected by decontamination of the skin and stomach by intravenous injection of ascorbic acid, which acts as a hydrogen donor and may antagonize the action of unreacted dithiocarbamate compounds. Alcoholic beverages should be avoided.

Phthalimides

The *N*-sulfinyl phthalimide fungicides have low acute toxicity in humans. They include captan, captafol, and folpet.

These chemicals sensitize the skin and the respiratory tract resulting in dermatitis similar to that produced by poison oak and allergic conjunctivitis. They are teratogenic and mutagenic. These chemicals are structurally related to thalidomide, a known human teratogen.

Hexachlorbenzene

Hexachlorbenzene is used as a fungicide to control smut diseases on cereal grains, primarly wheat seeds.

Action

Hexachlorbenzene produces porphyria cutanea tardia with symptoms of neurotoxic, visceral, arthritic, cutaneous, and hepatic diseases. It is excreted in breast milk and may be consumed by infants who are the most severely affected by its action, which may progress to weakness, convulsions, and death (Courtney, 1979).

Treatment

Treatment is mostly symptomatic and supportive.

RODENTICIDES

Anticoagulants

Anticoagulant rodenticides belong to two types: coumarin and indanediones (Goldfrank et al., 1981).

Coumarin type: warfarin, coumafuryl, and warficide.
Indanedione type: diphacinone, chlorphacinone, and pindane.

Action

These compounds cause death by gastrointestinal hemorrhage. They depress synthesis in the liver of vitamin K_1-dependent clotting factors (II, VII, IX, and X) by inhibiting the vitamin K_1 2,3-reductase. Direct damage to capillary occurs concurrently.

Unlike coumarin anticoagulants, the indanediones cause symptoms and signs of neurologic and cardiopulmonary injury in rats, which may result in death before hemorrhage occurs (Fristedt and Sterner, 1965).

Symptoms

Ingestion of small doses remains asymptomatic. Large doses produce hematuria, nose bleed, hematoma, bleeding gums, abdominal pain and back pain.

Treatment

Specific antidote is vitamin K_1 (Mephyton), 25 mg i.m. Vitamin K_3 (Menadione) and K_4 (Menadiol) are not effective. Activated charcoal and cathartics are used for decontamination. Fresh-frozen plasma and fresh blood may be needed.

Sodium Monofluoroacetate (1080)

This potent rodenticide that was developed during World War II is naturally present in the poisonous South African plant gifblaar (*Dichaptelaum cymosum*), the South American genus *Palicourea*, and the Australian genera *Gastrolobium, Oxylobium,* and *Acacia*.

Action

The mean human oral lethal dose ranges from 2–10 mg/kg. It is absorbed from the gastrointestinal tract, lungs, and mucous membranes (Egekeze and Oehme, 1979).

The monofluoroacetate metabolite, the fluorocitrate, inhibits two of the Krebs cycle enzymes, i.e., aconitase, which catalyzes citrate metabolism, and succinate dehydrogenase, which catalyzes succinate metabolism. This inhibition blocks the Krebs cycle and reduces glucose metabolism, energy stores, and cellular respiration. Heart, kidney, and brain tissues with high metabolic rate are very susceptible to the toxicity of monofluoroacetate. Neither fluoroacetate nor fluorine is responsible for monofluoroacetate toxicity.

Symptoms

Early symptoms include nausea, vomiting, and abdominal pain followed by anxiety, agitation, muscle spasm, seizures, and coma. Tachycardia and hypotension may lead to fibrillation. Suicidal cases with monofluoroacetate revealed cerebellar degeneration and cerebral atrophy (Chung, 1984).

Treatment

Decontamination is effected by using gastric lavage, activated charcoal, or cathartics. Adequate respiratory status should be maintained. Glycerol monoacetate has been used successfully as an antidote in monkeys, but less successfully in humans.

Zinc Phosphide

The oral LD_{50} of zinc phosphide is 40 mg/kg. Following its ingestion, phosphine gas is generated in the stomach.

Symptoms

Toxicity of zinc phosphide is characterized by nausea, vomiting, headache, lightheadedness, and dyspnea. These symptoms progress to pulmonary edema, hypertension, and convulsions. Myocardial necrosis, vascular damage, and pulmonary edema were evident on postmortem examination.

Treatment

No antidotal treatment is indicated. Decontamination is carried out using gastric lavage, activated charcoal, and cathartics within the first several hours after exposure.

Strychnine

Strychnine is an alkaloid that naturally occurs in the seed of the Indian tree *Strychnos nux vomica*, which contains about 1–1.5% strychnine. The lethal dose in humans is 5–8 mg/kg. The previous use of this chemical as appetite suppressant and aphrodisiac has been discontinued.

It is rapidly absorbed from the gastrointestinal tract and vasal mucosa but not from the skin. Strychnine is detoxified by hepatic microsomal enzymes.

Action

Strychnine exert its toxic action by competitively antagonizing the central inhibitory neurotransmitter glycine by blocking its postsynaptic uptake at the spinal cord and brain stem receptors (McGuigan, 1983). This results in hyperexcitation of muscle groups from lack of normal inhibition.

High concentratins of strychnine act on ion permeability of neurons similar to that of local anesthetics. It excites all levels of the central nervous system, but its action is more in the spinal cord than the stimulation of higher brain centers.

Symptoms

The onset of symptoms is rapid—within 15–20 min. Mild toxicity produces apprehension, nausea, muscle, and twitching. Severe toxicity causes extensor spasm, interference with breathing, convulsions, symmetrical seizures, and coma (Edmunds et al., 1986).

Treatment

Decontamination is carried out using activated charcoal (50 g) immediately along with a cathartic. Gastric lavage may produce muscle spasm. Vomiting is not recommended because of the aspiration potential. Adequate airway ventilation should be maintained. Oxygen may be administered. To prevent endotracheal intubation-induced muscle spasms, succinylcholine is administered, which produces neuromuscular paralysis of the airway. Diazepam and phenobarbital are used to control convulsions. No specific antidote is available.

Vacor

Vacor Rat Killer (PNU, *N*-3-pyridylmethyl-*N*-*P*-nitrophenylurea) use was suspended in 1979 because of its induction of insulinopenic diabetes mellitus in exposed humans. The lethal dose in humans is 5 mg/kg.

Action

Vacor exerts its toxic action by reducing intracellular synthesis of NAD and NADH. It destroys pancreatic beta cells resulting in pronounced insulinopenia in the presence of normal glucagon levels (Miller et al., 1978).

Symptoms

Vacor-induced symptoms resemble those of diabetic ketoacidosis characterized by triphasic response of serum glucose levels: transient hyperglycemia, hypoglycemia lasting up to 48 h, and permanent glucose intolerance and ketoacidosis (Johnson et al., 1980). Symptoms are seen as nausea, vomiting, abdominal pain, diffuse myalgias, polyuria, polydypsia, malaise, weakness, and dyspnea. A peanut odor is characteristic of vacor poisoning. Nervous system damage is demonstrated as autonomic neuropathy characterized by dysphagia, impotence, urinary retention, and constipation or diarrhea. Also, distal motor and sensory peripheral neuropathy as well as metabolic encephalopathy may ensue (Lewitt, 1980).

Treatment

The treatment of vacor poisoning is similar to that of diabetic ketoacidosis by administering intravenous fluids, low-dose continuous insulin therapy, potassium replacement, acid-base monitoring. Nicotinamide (Vitamin B$_3$, not niacin or nicotinic acid) may be used as antidote. Gastrointestinal decontamination is carried out using ipecac/lavage, activated charcoal, and cathartics within the first few hours after exposure.

RELATED CHEMICALS

Hexachlorophene

Properties and Uses

The chemical nomenclature of hexachlorophene is 2,2'-methylene bis[3,4,6-trichlorophenol]. It is a crystalline compound. Since 1934 it has been used as a bacteriostatic agent. Many cosmetic products contained hexachlorophene in the 1950s and 1960s, such as shaving creams, soaps, shampoos, deodorants, and feminine hygiene products. During the 1960s and early 1970s, the dermal toxicity of hexachlorophene was recognized, particularly for newly-born babies (Shuman et al., 1974).

Disposition

Hexachlorophene is absorbed from the gastrointestinal tract and the skin. More absorption takes place through the skin of small or premature infants. It is metabolized by the liver to glucuronide, which is excreted in the urine with some of the unchanged compound.

Action

The exact mechanism of neurotoxicity is unknown. Hexachlorophene causes demyelination of the central nervous system resulting in lesions that are spongioform changes and are confined to the white matter (Kimbrough, 1973). The changes are characterized by the presence of empty vacuoles produced by the splitting and separation of myelin lamellae (Gowdy and Ulsamer, 1976). Vaculation of diffuse areas of the cerebellum and cerebrum occurs in infants. Similar lesions are produced by isoniazid, triethyltin, cuprizone and phenelizinc.

Clinical Condition

Ingestion of hexachlorophene results in nausea, vomiting, and diarrhea, which may result in dehydration and hypotension (Shuman et al., 1974). Neurologic deficits develop later as lethargy, facial twitching, fever, blurred vision, blindness, hyperflexia, convulsion, and coma, which appear 12–18 h after ingestion. These symptoms are followed by cardiac dysrhythmias, apnea, and cardiac arrest 48–60 h after ingestion. Recovery begins by the third day.

Dermal absorption of hexachlorophene produces nausea, vomiting, irritability, diplopia, hypertonicity, anorexia, weakness, and fever. Severe poisoning results in papilloedema, retinal hemorrhage, convulsions, coma, and death.

In mice, hexachlorophene crosses the placenta and accumulates in neural tissues, suggesting potential teratogenic action.

Treatment

Diazepam is used to treat seizures and respiratory depression, the immediate life-threatening problems. Hypotension resulting from dehydration may be treated by electrolyte replacement. Gastrointestinal decontamination is carried by ipecac.

Acrylamide

Chemistry

Acrylamide has the chemical structure $CH_2=CHCONH_2$. It is a colorless solid with the following synonyms: propenamide, acrylamide monomer, acrylic amide.

Acrylamide is used to waterproof soil. It is also used in the paper and permanent press fabric industries.

Action

Acrylamide is a nervous system toxicant (Auld and Bedwell, 1967). It causes central-peripheral distal axonopathy or "dying back." Although it inhibits glycolytic enzymes, e.g., enolase, recent studies cast doubts on the hypothesis that acrylamide neurotoxicity results from the reduction in metabolic energy. Acrylamide binds specifically to cytoskeletal proteins and to MAP-2 in particular. Polymer acrylamide is nontoxic (Lapadula et al., 1989; Carrington et al., 1990).

Clinical Condition

Acute exposure to large doses affects the midbrain and cerebrum resulting in dysarthria, tremor, positive Romberg signs, and change in gait. It may produce visual disturbances, such as reduction of red and green discrimination, and a hypertensive retinopathy. Peripheral neuropathy may also result from acute exposure to acrylamide.

Chronic exposure to acrylamide produces a motor and sensory distal axonopathy. It is characterized by ataxia, dysarthria, and tremor. Other symptoms include weakness, paresthesia, fatigue, lethargy, decreased pin prick sensation, decreased reflexes, and positive Romberg sign. Acrylamide neuropathy is characterized by profuse sweating of the palms. Improvement and recovery depend on the level, duration, and frequency of exposure as well as the severity of the condition (Garland and Patterson, 1976).

Treatment

Patients should be removed from exposure. Treatment is symptomatic.

Polychlorinated Biphenyls

Chemistry

Polychlorinated biphenyls are known as PCBs; chlorinated biphenyls; chlorobiphenyls, Aroclor, Clophen, Fenclor, Kanechlor, Phenoclor Pyrolene, Santotherm (Schneider, 1979). Aroclors are characterized by a four digit number. The first two digits indicate that the mixture contains biphenyls (12), the last two digits give the weight percent of chlorine in the mixture, e.g., Aroclor 1245 contains biphenyls with approximate 45% chlorine.

BCPs are heavy oils, chemically inert, heat resistant, nonflammable, and electricity conducting.

Action

In February 1968, in Yusho, Japan, 1200 persons consumed rice oil contaminated with "heat exchanger fluid" in which PCBs were the major ingredients (Schneider, 1979). The following symptoms resulted: skin rash, swelling and discharge from the eyes, weakness, and disturbance in the liver. Other signs and symptoms include black pigmentation of the face, eyelids, lips, and gums; discoloration and deformation of nails and toes. General symptoms include loss of appetite, nausea, vomiting, weakness, and numbness of the extremities. Ten years after the incident the following symptoms persisted: dullness, headache, stomach ache, numbness and pain in the extremities, swelling and pain in the joints, coughing and bronchitis symptoms.

Offspring of mice fed tetrachlorobiphenyl during pregnancy (at a dose that did not cause toxicity in the mother) developed a "spinning syndrome" by chasing their tails at a rate of 40–150 turns per min. PCBs produced liver tremor when fed to rodents. These chemicals are very potent inducers of cytochrome P-450.

Polybrominated Biphenyls

A representative of this group, Fliemaster BP-6 (2,2',4,4',5,5'-hexabromobiphenyl), is a solid that softens at 72°C and decomposes above 300°C. It is used as a flame retardant (Schneider, 1979).

In 1973, through a mix-up at the factory, Fliemaster BP-6 rather than Nutrimaster (magnesium oxide) was shipped to a farm cooperative that mixed dairy feed and distributed it to farmers around the state. As a result of feeding this contaminated feed, whole herds of dairy cattle developed loss of appetite, weakness, open sores that would not heal, decreased milk production, sterility, and giving birth to dead calves.

These cattle produced PBB-contaminated milk and meat. By 1976, the following animals died or were destroyed: 29,000 dairy cattle, 1,470 sheep, and almost 1.5 million chickens. Farm families, who had consumed the largest quantities of the chemicals, complained of headaches, fatigue, joint pain, and numbness in fingers and toes (Schneider, 1979). In laboratory animals, PBBs produced liver tumors when fed to rodents.

REFERENCES

Abou-Donia, M.B. (1981). Organophosphorus ester-induced delayed neurotoxicity. *Annu. Rev. Pharmacol. Toxicol.* 21: 511–548.

Abou-Donia, M.B. (1985). Biochemical toxicology of organophosphorus compounds. In *Neurotoxicology*. K. Blum and L. Manzo, Eds., Marcel Dekker, New York, pp. 423–444.

Abou-Donia, M.B. and D.M. Lapadula (1990). Mechanisms of organophosphorus ester-induced delayed neurotoxicity: type I and type II. *Annu. Rev. Pharmacol. Toxicol.* 30: 405–440.

Abou-Donia, M.B., D.M. Lapadula, and E. Suwita (1988). Cytoskeletal proteins as targets for organophosphorus compound and aliphatic hexacarbon-induced neurotoxicity. *Toxicology* 49: 469–477.

Abou-Donia, M.B. and D.B. Menzel (1968b). Chick microsomal oxidases. Isolation, properties, and stimulation by embryonic exposure to 1,1,1-trichloro-2,2-bis(*p*-chlorophenyl)ethane. *Biochemistry* 7: 3788–3794.

Abou-Donia, M.B. and D.B. Menzel (1976). DDT: The degradation of ring-labeled ^{14}C-DDT to ^{14}Co$_2$ in the rat. *Experientia* 32: 500–501.

Abou-Donia, M.B., S.E. Patton, and D.M. Lapadula (1984). Possible role of endogenous protein phosphorylation in organophosphorus compound-induced delayed neurotoxicity. *Proc. Symp. Cellular and Molecular Neurotoxicity*. T. Narahashi, Ed., Raven Press, N.Y., pp. 265–283.

Aldridge, W.N. (1976). Survey of major points of interest about reactions of cholinesterases. *Croat. Chem. Acta* 47: 225–233.

Auld, R.B. and S.F. Bedwell (1967). Peripheral neuropathy with sympathetic overactivity from industrial contact with acrylamide. *Can. Med. Assoc. J.* 96: 652–654.

Bidstrup, P.L. and D.J.H. Payne (1951). Poisoning by dimitro-ortho-cresol. Report of eight fatal cases occurring in Great Britain. *Br. Med. J.* 2: 16–19.

Bischoff, A. (1970). Ultrastructure of tri-ortho-cresyl phosphate poisoning in the chicken. *Acta Neuropathol. (Berlin)* 15: 142–155.

Bloch, H. and A. Hottinger (1943). Uber die spezifitat der cholinesterase-hemmung durch tri-o-kresyl phosphat. *Z. Vitaminforsch.* 13: 90.

Bus, J.S., S.D. Aust, and J.E. Gibson (1976). Paraquat toxicity: proposed mechanism of action involving lipid peroxidation. *Environ. Health Perspec.* 16: 139–146.

Carrington, C.D., D.M. Lapadula, L. Dulak, M. Friedman, and M.B. Abou-Donia (1990). *In vivo* binding of [^{14}C]acrylamide to proteins in the mouse nervous system. *Neurochem. Int.* 18: 191–197.

Casida, J.E., D.W. Gawmar, A.H. Glickman, and L.J. Lawrence (1983). Mechanisms of selective action of pyrethroid insecticides. *Annu. Rev. Pharmacol. Toxicol.* 23: 413–438.

Castano, P., V.F. Ferrario, and L. Vizzotto (1982). Sciatic fibers in albino rats after atrazine treatment: a morpho-quantitative study. *Ital. J. Tissue React.* 4: 269–275.

Chung, H.M. (1984). Acute renal failure caused by acute monofluoroacetate poisoning. *Vet. Hum. Toxicol. (Suppl. 2)* 26: 29–32.

Clark, D.G. and E.W. Hurst (1970). The toxicity of diquat. *Br. J. Ind. Med.* 27: 51–55.

Courtney, K.D. (1979). Hexachlorobenzene (HCB): a review. *Environ. Res.* 20: 225–266.

Crome, P. (1986). Paraquat poisoning. *Lancet* 1: 333–334.

DuBois, K.P. (1965). Low level organophosphorus residues in the diet. *Arch. Environ. Health* 10: 837–841.

Earl, C.J. and R.H.S. Thompson (1952). The inhibitory action of tri-ortho-cresyl phosphate and cholinesterases. *Br. J. Pharmacol.* 7: 261–269.

Edmunds, M., T.M.T. Sheehan, and W. Van Hoff (1986). Strychnine poisoning: clinical and toxicological observations on a non-fatal case. *Clin. Toxicol.* 24: 245–255.

Egekeze, J.O. and F.W. Oehme (1979). Sodium monofluoroacetate (SMFA, compound 1080): a literature review. *Vet. Hum. Toxicol.* 21: 411–416.

Farley, J.M., T. Narahashi, and G. Holan (1979). The mechanism of action of a DDT analog on the crayfish neuromuscular junction. *Neurotoxicology* 1: 191–207.

Fishbein, L. (1976). Environmental health aspects of fungicides. Dithiocarbamates. *J. Environ. Health* 1: 713–735.

Flannigan, S.A. and S.B. Tucker (1985). Variation in cutaneous sensation between synthetic pyrethroid insecticides. *Contact Dermatitis* 13: 140–147.

Fristedt, B. and N. Sterner (1965). Warfarin intoxication from percutaneous absorption. *Arch. Environ. Health* 11: 205–208.

Gammon, D.W., M.A. Brown, and J.E. Casida (1981). Two classes of pyrethroid action in the cockroach. *Pestic. Biochem. Physiol.* 15: 181–191.

Garland, T.O. and M.W.H. Patterson (1976). Six cases of acrylamide poisoning. *Br. Med. J.* 4: 134–138.

Goldfrank, L., N. Flomenbaum, and R.S. Weisman (1981). Rodenticides. *Hosp. Physician* 17: 81–99.

Gowdy, J.M. and A.G. Ulsamer (1976). Hexachlorophene lesions in newborn infants. *Am. J. Dis. Child.* 130: 247–250.

Helliwell, M. and J. Nunn (1979). Mortality in sodium chlorate poisoning. *Br. Med. J.* 1: 1119.

Holmstedt, B. (1963). Structure-activity relationships of the organophosphorus anticholinesterase agents. In *Cholinesterases and Anticholinesterase Agents*. G.B. Koelle, Ed., *Handbuch der Experimentellen Pharmakologie*, Springer-Verlag, Berlin, pp. 428–485.

Iversen, L.L. (1975). Uptake processes for biogenic amines. In *Handbook of Psychopharmacology, Vol. 3*. L.L. Iversen, S.D. Iversen, and S. Snyder, Eds., Plenum Press, New York, pp. 381–442.

Johnson, D., P. Kubie, and C. Levitt (1980). Accidental ingestion of vacor rodenticide. *Am. J. Dis. Child.* 134: 161–164.

Johnson, M.K. (1969). A phosphorylation site in brain and the delayed neurotoxic effect of some organophosphorus compounds. *Biochem. J.* 111: 487–495.

Kimbrough, R.D. (1973). Review of recent evidence of toxic effects of hexachlorophene. *Pediatrics* 51: 391–394.

Lapadula, D.M., M. Bowe, C.D., Carrington, L. Dulak, M. Friedman, and M.B. Abou-Donia (1989). *In vitro* binding of [^{14}C]acrylamide to neurofilament and microtubule proteins of rats. *Brain Res.* 481: 157–161.

Lewitt, P.A. (1980). The neurotoxicity of rat poison vacor. *N. Engl. J. Med.* 302: 73–77.

Matsumura, F. (1985). Toxicology of Insecticides, 2nd ed. Plenum Press, New York, pp. 111–196.

Mayer, S.E. (1980). Neurohumoral transmission and the autonomic nervous system. In *Goodman and Gilman's The Pharmacological Basis of Therapeutics*. A.G. Gilman, L.S. Goodman, and A. Gilman, Eds., MacMillan, New York, pp. 56–90.

McGuigan, M.A. (1983). Strychnine. *Clin. Toxicol. Rev.* 6: 1–2.

Miller, J.U., J.D. Stokes, and C. Silpipat (1978). Diabetes mellitus and autonomic dysfunction after vacor rodenticide ingestion. *Diabetes Care* 1: 73–76.

Murphy, S.D. (1986). Toxic effects of pesticides. In *Casarett and Doull's Toxicology: The Basic Science of Poisons*, 3rd ed. C.D. Claassen, M.O. Amdur, and J. Doull, Eds., MacMillan, New York, pp. 519–581.

Narahashi, T. (1971). Mode of action of pyrethroids. *Bull. World Health Org.* 44: 337–345.

Narahashi, T. (1979). Nerve membrane ionic channels as the target site of insecticides. In *Neurotoxicity of Insecticides and Phermenes*. T. Narahashi, Ed., Plenum New York, pp. 211–243.

Narahashi, T. and T. Yamasaki (1960). Mechanism of increase in negative afterpotential by dicophanum (DDT) in the giant axons of the cockroach. *J. Physiol. (London)* 152: 122–140.

Onyeama, H.P. and F.W. Oehme (1984). A literature review of paraquat toxicity. *Vet. Hum. Toxicol.* 26: 494–502.

O'Reilly, J.R. (1984). Prolonged coma and delayed peripheral neuropathy after ingestion of phenoxyacetic acid weed killers. *Postgrad. Med. J.* 60: 76–81.

Patton, S.E., J.P. O'Callaghan, D.B. Miller, and M.B. Abou-Donia (1983). The effect of oral administration of tri-*o*cresyl phosphate on *in vitro* phosphorylation of membrane and cytosolic proteins from chicken brain. *J. Neurochem.* 41: 897–901.

Patton, S.E., D.M. Lapadula, J.P. O'Callaghan, D.B. Miller, and M.B. Abou-Donia (1985). Changes in *in vitro* brain and spinal cord protein phosphorylation after a single oral administration of tri-*o*-cresyl phosphate in hens. *J. Neurochem.* 45: 1567–1577.

Patton, S.E., D.M. Lapadula, and M.B. Abou-Donia (1986). The relationship of tri-*n*-cresyl phosphate-induced delayed neurotoxicity to enhancement of *in vitro* phosphorylation of hen brain and spinal cord proteins. *J. Pharmacol. Exp. Ther.* 239: 597–605.

Rossier, J. (1977). Choline acetyltransferase: a review with special reference to its cellular and subcellular localization. *Int. Rev. Neurobiol.* 20: 283–337.

Safe, S.H. (1986). Comparative toxicology and mechanism of action of polychlorinated dibenzo-*p*-dioxins and dibenzofurans. *Annu. Rev. Pharmacol. Toxicol.* 26: 371–399.

Schneider, M.J. (1979). Persistent poisons. In *Chemical Pollutants in the Environment*. The New York Academy of Sciences, New York.

Seabury, J.H. (1963). Toxicity of 2,4-dichlorophenoxyacetic acid for man and dog. *Arch. Environ. Health* 7: 202–209.

Shuman, R.M., R.W. Leech, and E.D. Alvord (1974). Neurotoxicity of hexachlorophene in the human. I. A clinicopathologic study of 248 children. *Pediatrics* 54: 689–695.

Smith, M.I., E. Elvove, P.J. Valer, W.H. Frazier, and G.E. Mallory (1930). Pharmacological and chemical studies of the cause of so-called ginger paralysis. *U.S. Public Health Rep.* 45: 1703–1716.

Suwita, E., D.M. Lapadula, and M.B. Abou-Donia (1986a). Calcium and calmodulin stimulated *in vitro* phosphorylation of rooster brain tubulin and MAP-2 following a single oral dose of tri-*o*-cresyl phosphate. *Brain Res.* 374: 199–203.

Suwita, E., D.M. Lapadula, and M.B. Abou-Donia (1986b). Calcium and calmodulin-enhanced *in vitro* phosphorylation of hen brain cold-stable microtubules and spinal cord neurofilament triplet proteins following a single oral dose of tri-*o*-cresyl phosphate. *Proc. Natl. Acad. Sci.* 83: 6174–6178.

Tabei, K., Y. Asano, and S. Hosoda (1982). Efficacy of charcoal hemoperfusion in paraquat poisoning. *Artif. Organs* 6: 37–42.

Chapter 5

Drugs of Abuse

Mohamed B. Abou-Donia
Departments of Pharmacology and Neurobiology
Duke University Medical Center
Durham, North Carolina

CONTROLLED SUBSTANCE ACT

In the United States drugs are grouped by the Controlled Substance Act into the following five schedules:

Schedule I

Substances in this group have no accepted medical use and have a high abuse potential. This group includes heroin, marihuana, LSD, morphine methylsulfonate, and acetylmethadol.

Schedule II

Substances in this group have a high abuse potential with the ability to induce severe psychic or physical dependence but with a limited medical application. Examples are amphetamines, opium, morphine, codeine, methadone, cocaine, amobarbital, pentobarbital, secobarbital, phenylacetone, and phenylcyclidine.

Schedule III

Substances in this group have an abuse potential less than those in Schedule II. Examples are derivatives of barbituric acid that do not belong to Schedule II, glutethimide, nalorphine, and benzphetamine.

Schedule IV

Substances in this group have an abuse potential less than those in Schedule III. Examples are barbital, phenobarbital, methylphenobarbital, chloral hydrate, meprobamate, paraldehyde, diazepam, and its related compounds.

Schedule V

Substances in this group have an abuse potential less than those listed in Schedule IV. This group includes antitussive and antidiarrheal agents that may not require prescriptions.

OPIATES

Definition

Opiate

Naturally occurring compounds present in the extract of opium poppy.

Opioid

Directly acting drugs whose effects are antagonized by naloxone.

Morphine is a constituent of opium. When its phenolic acid hydroxyl groups are acetylated, it yields heroin. Methylation of its phenolic group gives rise to codeine. Methadone is a synthetic diphenylpropylamine with analgesic potency similar to that of morphine. Methadone is used in detoxification programs for physical dependence on narcotics.

Naloxone is a specific antagonist at all three major opiate receptor sites, i.e., mu, kappa, and sigma. It is used to antagonize opiate symptoms.

Heroin is the most commonly used opiate. It is rapidly converted into morphine in the body. Rapid intravenous injection of opiates produces a warm flushing of the skin and sensation in the lower abdomen described by addicts as similar in intensity and quality to sexual orgasm. This feeling continues for about 45 s and is known as a "rush," "kick," or "thrill." Later a dreamy-state high is obtained.

Opioid Peptides

These natural opioids (Martin, 1984) are classified into three groups: proenkephalin A that produces enkephalines, proenkephalin B (pro-dynorphin) that produces dynorphins, and proopiomelanocortin that produces endorphins (corticotropin and β-lipotropin). These natural analgesics are normally inactive, and act only in stress conditions, e.g., anesthesia, surgery, pain, hypoxia, shock, and neonatal states.

Symptoms

Opiates have little toxicity; they reduce pain, aggression, and sexual drive. Thus, it is unlikely that they would induce crime. However, addicts may commit crimes to obtain money to buy these drugs.

Signs and symptoms associated with overdose of opiates include pinpoint pupils, hypothermia, hypotension, bradycardia, pulmonary edema, respiratory depression, seizures, and coma.

Tolerance to Opiates

Tolerance to opiates is characterized by a shortened duration of action and a decreased intensity of CNS depressant effects such as analgesia, euphoria, and sedation, and there is an increase in the size of the lethal dose.

There are two mechanisms of tolerance to opiates:

1. Rapid metabolism, deactivation, and excretion.
2. Adaptive nerve cells to drug action.

The result of tolerance is that an increased dose of drug is required to produce a given effect; this frequently leads to overdosing and toxicity.

Dependence

It is unlikely that long term opiate dependency causes mental deterioration or psychiatric disorder. The major problems associated with opiate dependence include use of contaminated needles, impulsive behavior, unemployment, and family disruption.

Withdrawal Syndrome

Onset of withdrawal symptoms occurs 6–8 h after the last dose and is accompanied by lacrimation, rhinorrhea, yawning, and perspiration (Council Reports, 1972). At 12–14 h symptoms include restless sleep followed by an awake period characterized by dilated pupils, anorexia, gooseflesh, restlessness, irritability, and tremor. These symptoms peak at 36–48 h and continue to 72 h, and then gradually diminish over the next 5–10 days.

At the peak, symptoms include increased irritability, insomnia, anorexia, anxiety, violet yawning, sneezing, weakness, depression, nausea, vomiting, intestinal spasm, diarrhea, tachycardia, increased blood pressure, and alternating chilliness and flushing. The skin may look like plucked turkey flesh.

Symptoms in Pregnant Opiate-Dependent Women

Opiate-dependent women frequently become sterile. If they get pregnant, they may abort or have low birth-weight infants, i.e., less than 2.5 kg. These patients may experience poor self image, depression, and hostility. Other problems frequently encountered are anemia, infection, hepatitis, subcutaneous abscesses, and venereal diseases.

Neonatal Opiate Withdrawal

Opiate-addicted women give birth to infants who have the following withdrawal symptoms: irritability, tremors, hyperactivity, sleeplessness, sweating, frequent sneezing, frequent yawning, vomiting that may result in dehydration, electrolyte imbalance and low-grade fever, diarrhea, and convulsions. They may also have hypoglycemia that can produce jitteriness, irritability, tremor, and tachypnea.

Treatment of Opiate-Dependence Withdrawal

1. Withdrawal treatment should take place in specialized facilities, drug free, and in the absence of family and friends
2. Administration of methadone
3. Tranquilizers (benzodiazepines) at bedtime
4. Antispasmodics for treatment of nausea, vomiting, abdominal cramps, or diarrhea.

Total withdrawal usually takes 1–2 weeks.

Phencyclidine

Phencyclidine (PCP) was introduced as a veterinary anesthetic. Since the late 1970s it has become very popular as a psychotomimetic drug. Other analogs include TCP, the thiophene analog of PCP, and its action appears to be similar in humans. PCE or cyclohexamine shares psychopharmacological actions with PCP. Ketamine is a derivative of phencyclidine that is legally used as anesthetic under the name Ketalar. PHP or phenylcyclohexylpyrrolidine produces pharmacologic actions similar to those of PCP when smoked. PCP has the following slang equivalents: angel dust, DOA, peace pill, and hog.

Use

PCP is used by ingestion, inhalation (sprinkled on smoking preparations or inhaled directly), and injection. It is inexpensive to produce and may bring great monetary profit.

Action

PCP acts as an anesthetic, stimulant, depressant, hallucinogenic, and analgesic. The specific pharmacologic actions of PCP are listed below.

Dopaminergic Action. PCP-induced schizophrenia may result from its inhibiting presynaptic dopamine uptake, release of dopamine, and stimulation of adenylate cyclase and tyrosine hydroxylase (Gravey, 1979).

Anticholinergic Action. PCP has both cholinergic (e.g., hypersalivation) and anticholinergic (e.g., urinary retention) actions resulting from its selective and reversible binding to the active sites of acetylcholinesterase and the central acetylcholine receptor, respectively (Snyder, 1979).

Adrenergic Action. PCP is an adrenergic agonist with sympathomimetic activity. It produces blood pressure elevation. At high doses it causes myocardial depression and reduces contractibility and velocity of contraction.

Renal Failure. PCP does not have a direct toxic action on the kidney. However, it may cause renal dysfunction resulting from hypotension or myoglobinuria.

Teratogenic Action. Retrospective epidemiologic studies have suggested that PCP use in humans results in an increased incidence of limb defects.

Immunotoxicity. PCP binds to B cells and T-helper lymphocytes, which results in the reduction of both humoral and cellular immune responses.

Tolerance/Withdrawal Effects

PCP produces tolerance in humans that results in the need to use much larger doses to reach the same effects. Also, abrupt discontinuation results in withdrawal symptoms such as memory loss, chronic fatigue, anxiety, irritability, and depression.

Symptoms

PCP use results in the following neurologic symptoms: blurred vision, variable pupil size, reduced pain and temperature sensation, ataxia, tremors, muscle weakness, slurred speech, drowsiness, convulsions, and coma (Liden et al., 1975). Psychological symptoms include amnesia, anxiety, agitation, euphoria, hallucinations, disordered thought process, and activation of latent psychopathology. PCP produces the following cardiovascular symptoms: increased pulse rate and increased blood pressure. Gastrointestinal effects include nausea and vomiting. Renal toxicity of PCP is manifested as increased urine output. Respiratory effects are bronchospasm and increased respiratory rate. PCP also results in the following autonomic symptoms: urinary retention, hyperthermia, and diaphoresis.

PCP is well absorbed when administered orally, or by inhalation or nasal routes. It has a large volume of distribution, i.e., 6.2 L/kg indicating extensive binding in tissues resulting from its high lipid solubility. It has a mean half-life of 24 or 21 h following smoking or ingestion. PCP undergoes metabolism yielding hydroxylated metabolites that are excreted as conjugates (mostly in urine). PCP is transferred across the placenta.

Treatment

Attention should be given to airway and respiration. Comatose or catatonic patients should be given intravenous naloxone and glucose. Intravenous diazepam is the treatment for apnea, while haloperidol and chlorpromazine are effective neuroleptic drugs for PCP-induced psychosis.

MARIHUANA

Marihuana (also marijuana) is the dried leaves and flowers of the Indian hemp plant, *Cannabis sativa*. The principal psychoactive substance is Δ^9-tetrahydrocannabinol or THC. This active ingredient is found in all parts of the plant with decreasing order in the flowers, leaves, stems, seeds, and roots. Usually cigarettes are formed from cut dried plants. Marihuana plants in the U.S. contain low THC content ranging from 1–3%. A 500 mg marihuana cigarette contains between 5–25 mg of THC. Hashish is the dried resin prepared from the flower tops of the plant and contains varying concentrations of THC up to 10%. Sinsemilla is unpollenated female marihuana that contains 5% THC. This plant, which grows widely, especially in California, accounts for 85% of domestic production in the U.S.

Use

Marihuana ranks as the second or third most commonly used recreational drug in the U.S. In 1979, an estimated 20–25% of Americans had used marihuana at least once. In 1981, 46% of high school seniors had sampled the drug, and 20% of the U.S. active military personnel had used it at least once.

A THC preparation, Dronabinol, is used for the treatment of nausea and vomiting in chemotherapy patients. It is classified as a Schedule II drug in the U.S.

Disposition

A range of 20–50% of a single inhaled THC dose in absorbed. Much less is absorbed after an oral administration; and only 3–6% of THC is eliminated in the feces. Low oral bioavailability of THC results from large first-pass effect through liver metabolism. Peak plasma concentration of THC is reached 7–8 min after the start of smoking, then declines; euphoria appears about 20 min after peak plasma occurs. On the other hand, plasma THC level peaks about 45 min following ingestion and remains relatively constant for 4–6 h. Euphoria begins 30–60 min after ingestion and peaks within 2–3 h. Due to its high lipid solubility, THC concentration in plasma rapidly declines and is redistributed to various tissues. The volume of distribution of an intravenous dose in 10 L/kg. THC almost quantitatively binds to proteins. Its mean terminal half-life is 20 h in users and 25–57 h in nonusers (Hunt and Jones, 1980).

THC is metabolized by liver enzymes to an active metabolite, 11-OH-Δ^9-THC. Only a small fraction of THC is excreted via urine: from 13–16% of the parent compound in three days. The major excretion route of THC is the feces, thus after three days 30% and 50% of the dose is eliminated in the feces following intravenous and oral administration, respectively.

Although THC crossed the placenta in experimental animals, its concentration in the fetal blood was 5–6 times lower than that in maternal blood (Blackard and Tennes, 1984). Also, maternal milk from chronic heavy marihuana users contained THC eightfold that of plasma.

Action

Marihuana has a multisystem action as follows.

Central Nervous System

The mechanism of THC on the CNS is not known. In experimental animals serotonin, norepinephrine, and dopamine metabolism was affected but not the cholinergic system. It is postulated that THC may carry out its CNS action by changing the balance between serotonin and acetylcholine (Moss et al., 1978).

Within a few minutes of beginning smoking or marihuana, the following psychotomimetic effects occur: change in mood, sensory perception, cognition, sensorium, motor coordination, and self perception. These effects peak after 20 min and last for 2–3 h (Hollister, 1971). The intensity and duration of these effects depend on the dose, route of administration, and individual's expectations, previous experience, and premorbid personality. THC also causes impairment of short term, but not long term memory, attention, and systemic thinking. It decreases balance, steadiness, muscle strength, cognition, psychomotor coordination, and reaction speed. Initial responses are euphoria and relaxation at low doses, which progress to depersonalization, pressured speech, paranoia, anxiety, and manic psychosis, and visual hallucination at high doses. Other autonomic and CNS effects include drowsiness, dry mouth, dizziness, disorientation, inability to concentrate, schizophrenia, paranoia, and suicidal ideas.

Following chronic use of marihuana, adolescents develop a condition known as "amotivational syndrome." This condition is characterized by apathy, inability to carry out tasks, easy frustration, poor concentration, absence of goals, low expectations of academic achievement, less compatibility with parents and friends, and less involvement in conventional institutions. Other characteristics include more independence, more problem role models, increased tolerance of deviant behavior, and greater involvement problem acts such as public drunkenness.

Cardiovascular System
THC causes tachycardia and augmented left ventricular performance. It also causes elevated plasma norepinephrine.

Pulmonary System
Inhalation of marihuana smoke results in acute bronchodilation. Paraquat sprayed on marihuana plants did not cause damage because it was pyrolyzed by high combustion level in marihuana cigarettes.

Eye
Marihuana smoking results in a slight pupillary constriction; however, reddened conjuctiva are the most reliable and sensitive indicator of marihuana effect (Shapiro, 1974).

Gastrointestinal Tract
Marihuana use stimulates appetite, especially for sweet food, causes dry mouth and may lead to nausea, especially in first-time users. THC has antiemetic properties.

Tolerance

Chronic use of marihuana produces tolerance to both tachycardia and psychotropic effects.

Withdrawal

A mild withdrawal reaction has been reported following discontinued use of marihuana among heavy users. Withdrawal symptoms are characterized by restlessness, sleeplessness, decreased appetite, nausea, irritability, sweating, and increased dreaming.

Addiction

Addiction that is characterized by psychological dependence, tolerance, and withdrawal may develop, usually in patients that are addicted to opiates and alcohol. Although most marihuana users are casual users, some individuals feel compelled to use it daily to cope with life. These individuals are psychologically dependent on marihuana.

Urinary Analysis

Immunoassay analytical methods for THC analysis, EMIT and SYVA, cross react with dihydroxy-THC, polar acids, and some other marihuana related compounds (O'Connor and Rejent, 1981). The sensitivity of the EMIT method is 20 ng/ml urine. Both methods have a 4% incidence of false-positive results. High-performance liquid chromatography with electrochemical detector and gas-liquid chromatography-mass spectrometry have detection limits for THC of 5 ng/ml.

Positive urine test neither correlates with the amount taken nor with psychomotor impairment. Following a single marihuana cigarette, THC and related compounds are detected for as long as seven days. The detection period may be extended for 1–2 days following the smoking of an additional 1–2 cigarettes. Since THC is lipid soluble, its metabolites may be detectable in the urine for several weeks after smoking. In chronic users, cannabinoids were detected in urine for as many as 77 days. Prolonged heavy passive inhalation of marihuana may result in a positive cannabinoid urine test (Cone and Johnson, 1986).

Treatment

The physical condition of the patient should be stabilized, using usual measures such as glucose, naloxone, and thiamine. Marihuana-induced psychoses usually resolve within 6 h after a single administration. However, chronic users may experience this condition for as long as a week. Patients should be kept in a well-lit room with some familiar people. Diazepam may be used to sedate the patient. Haloperidol is used to treat psychosis. Physostigmine must be avoided, becuase it produces severe depression after intake of marihuana characterized by uselessness, hopelessness, apathy, sadness, psychomotor retardation, dysphoria, and thoughts of suicide. This condition is treated with 1 mg of atropine.

COCAINE

Cocaine is a naturally occurring, potent central nervous system stimulant. It is present in leaves of the tropical plant *Erythroxylum coca* (0.5–2.0%) that is found in Peru, Ecuador, Bolivia, and to a lesser extent in Mexico, the West Indies, and Java; and the plant *Erythroxylum novogranatense* found in Colombia, the Caribbean coast of South America and the north coast of Peru. In 1886, a cocaine-caffeine mixture was marketed under the name of Coca-Cola. In 1906, the coca leaves were decocainized before making the drink. The coca plant is not the cocoa plant which contains caffeine rather than cocaine. Cocaine was characterized in 1959 as one of at least fourteen alkaloids extracted from the leaves of coca plants. Cocaine may be synthesized chemically and can be recognized from the natural product (laboratory isomer only) by the presence of diastereoisomers or its dextraenantiomer.

Medical Use

Cocaine has dual use for vasoconstriction and local anesthetic actions for ear, nose and throat surgery. It has been used as a topical anesthetic of the cornea. It also has been added to analgesic mixtures for terminally ill patients, e.g., Brompton's cocktail (Twycross, 1977).

Abuse

Almost ten million persons over eleven years of age reported using cocaine in 1980 in the U.S. Among high school seniors, its use increased from 6% in 1976 to 20% in1982. In 1983, the price of 1 g of cocaine was as low as $50 in New York City, and was comparable to the price of 28 g of marihuana.

Only two million of the twenty million people (10%) who reported trying cocaine progressed to habitual use. The routes of cocaine administration are nasal, 40%; freebase smoking, 30%; injection, 20%; and combined routes, 10%. Cocaine-related deaths resulted from all routes of administration including intravenous, nasal, vaginal, and oral.

Sniffing, Snorting, or Insufflation

This route is favored by occasional users. The euphoric effects occur within minutes and last 20–45 min. After the euphoria subsides, depression, restlessness, and irritability appear, which lead to more cocaine use. The maximum dose of intranasal cocaine is 80–200 mg.

Freebasing

The cocaine freebase is more volatile and may be smoked in a cigarette or water pipe. "Crack" is the popular name for the crystalline freebase. The smoked dose is approximately 300 mg, is rapidly absorbed by the lungs and produces immediate pleasure. This route of use has great potential for dependency and overdose.

Intravenous

This route of use is expensive and produces euphoria and adverse effects similar to those produced by freebasing.

Disposition

Almost all of the intranasal dose is absorbed within 4 h. A 100 mg intranasal dose gives rise to euphoric effects similar to those of a 25 mg intravenous dose. Oral administration produces a peak subjective high that is more intense than by insufflation but is delayed to 45–90 min instead of 15–60 min via intranasal. Both routes of use have 60% bioavailability at doses of 2 mg/kg. Freebase cocaine is rapidly and completely absorbed following inhalation and produces euphoria within 6–11 min. Intravenous injection has similar kinetics to freebasing, with euphoric effects appearing several minutes earlier after intravenous administration (Spiehler and Reed, 1985).

The volume of distribution of cocaine is 1.2–1.9 L/kg. Highest concentrations are found in urine and kidney, followed by brain, blood, liver, and bile. It corsses the blood-brain barrier and has a brain/blood ratio of 20 within 1–2 h. Cocaine is hydrolyzed by plasma butyrylcholinesterase to benzoylecgonine. The polar metabolites are extensively excreted in the urine, with less than 10% eliminated unchanged in urine. The elimination half-life values for cocaine are 75 min, 48 min, and 54 min for intranasal, oral, and intravenous routes, respectively.

Pregnancy

The spontaneous abortion rate was increased in a group of pregnant cocaine addicts. This may have resulted from placental constriction and increased uterine contractibility by cocaine. These mothers had infants with significant depression and poor response to external stimuli (Fantel and MacPhail, 1982).

Tolerance

Chronic use of cocaine leads to tolerance of the euphoric and physiological effects.

Withdrawal

Discontinued use of cocaine leads to physical withdrawal symptoms such as depression, insomnia, headache, fatigue, irritability, gastrointestinal distress, i.e., nausea and vomiting, agitation, and paranoia.

Action

Cocaine is a potent CNS stimulant and sympathomimetic agent. It produces its action by either potentiating norepinephrine and epinephrine or by depressing central inhibitory pathways (Kelsner and Nickerson, 1969). Acute cocaine action results in increased dopaminergic neurotransmission. However, chronic use may deplete brain dopamine leading users to crave cocaine.

Symptoms

Central Nervous System

Initial symptoms are euphoria, hyperactivity, and restlessness. These symptoms progress to tremor, hypreflexia, and convulsions. In high doses, following the excitatory phase, depression ensues which is characterized by coma, hyporeflexia, and respiratory and cardiovascular depression. Death results from either respiratory depression or cardiac arrest (Post and Kopandra, 1975).

Cardiovascular System

Cocaine causes increased blood pressure, pulse rate, tachycardia, and peripheral vascular constriction resulting in hypertension. These sympathomimetic actions may result from the accumulation of norepinepherine at adrenergic receptors due to cocaine blockade of norepinephrine and epinephrine uptake (Simpson and Edwards, 1986).

Respiratory System

The initial stimulatory effect of cocaine tends to increase ventillation resulting in tachypnea and increased tidal volumes. High doses cause respiratory depression, cyanosis, and hypoxemia. Death results from respiratory depression in addition to ventricular fibrillation and cardiovascular collapse.

Eye

Cocaine causes dilated pupils by both local and central α-adrenergic effects on the radial muscles of the iris.

Temperature Regulation

Large doses of cocaine elevate body temperature as high as 114°F in overdose cases.

Gastrointestinal Tract

Stimulation of the central vomiting centers by cocaine results in nausea and vomiting. Its effect on the intestinal wall produces diarrhea and abdominal cramps.

Hepatic necrosis in experimental animals by cocaine results from depletion of glutathione stores and subsequent formation of reactive intermediate metabolites (Evans and Harbison, 1978).

Treatment

Death resulting from cocaine overdose takes place within minutes of exposure. Patients who arrive at hospitals have a good likelihood to survive. Life threatening symptoms are seizures, respiratory arrest, hyperthermia, and dysrhythmias. It is vital to secure adequate airway. Seizure is treated with diazepam (0.1–0.3 mg/kg). Hyperthermia is decreased by an ice bath. Hypertension and tachycardia are treated with propranolol in moderate overdoses showing signs of stimulation. Severe overdosing results in depression with no cardiovascular stimulation.

AMPHETAMINES

Amphetamine (Benzedrine) is a racemic β-phenylisopropylamine that was synthesized in 1887. Other amphetamine compunds include: methylenedioxyamphetamine (MDA), *p*-methoxyamphetamine hydrochloride (PMA), and 3,4-methylenedioxymethamphetamine (MDMA). Amphetamines produce strong central nervous system stimulation that lasts longer than those of cocaine. Among abused drugs, amphetamines rank about tenth in the U.S. An estimated thirteen million persons abuse these drugs. They are commonly used among college students, athletes, and truck drivers. Amphetamines are classified as DEA Schedule II.

The following amphetamine analogs are classified as Schedule III: Benzphetamine HCl, Chlorphentemine HCl, Clortermine, and Phendimetrazine tartrate. Schedule IV compounds include Diethylpropion HCl, Fenfluramine HCl, phentermine, and methylphenidate.

Medical Use

Amphetamines and methylphenidate are used for adult narcolepsy and attention deficit disorder (ADD) in adults. Amphetamines are also used for a short term (8–12 weeks) suppression of appetite. Schedule III and IV analogs of amphetamines are anorectic and are approved for short term weight loss. The legal use of amphetamines for obesity accounts for 80% of their use.

Abuse

Amphetamines are taken at 5–20 mg oral doses to combat fatigue, elevate mood, or prolong awakening. In contrast to persons who use these drugs to enhance performance, others who have difficulty interacting socially use these drugs to increase their internal arousal mechanism, thus reducing the need for external stimuli. On rare occasions, hallucinations and sudden death may occur during heavy exercising in warm weather (Bailey and Manoguerra, 1980).

Long term use of daily doses of 20–40 mg amphetamine results in tolerance and leads to increased daily doses of 50–150 mg. Reduction of the dose usually results in depression and lethargy (Cohen, 1975).

Intravenous injection of amphetamines result in hyperactivity, euphoria, and increased sexual pleasure compared to orgasm, a condition known as "flash." Repeated injections of 1000 mg amphetamine up to 5 times within 24 h is known as a "speed run." During this phase, some individuals may have difficulty achieving orgasm or ejaculation. High doses of amphetamine result in paranoid psychosis. These persons ignore their appearance and become unreliable, irritable, unstable, and prone to committing suicide. Acute toxicity of amphetamines varies with individuals. Thus, while an oral dose of 1.5 mg/kg methamphetamine produced death in some individuals, others have survived the ingestion of 28 mg/kg (Smith and Fisher, 1970).

Action

Central Nervous System

Amphetamines stimulate the central nervous system and lead to euphoria, alertness, and increased confidence and self esteem. Large doses cause anxiety, dysphoria, confusion, depression, nausea, vomiting, headache, sweating, and confusion. Amphetamines result in increased awakening and reduced food consumption and fatigue, without changing the metabolic rate (Nicholson and Stone, 1980).

Cardiovascular System

Small doses of amphetamines cause slow heart rate, while large doses produce tachycardia, palpitations, and dysrhythmias. They increase both systolic and diastolic blood pressure.

Mechanism of Action

Amphetamine action is related to increasing the synaptic concentration of the neurotransmitters dopamine and norepinephrine. Amphetamine-induced schizophrenic effects result from the stimulation of central dopaminergic pathways (Trulson and Jacobs, 1979). On the other hand, increased physical and sexual activities result from amphetamine-induced elevation of synaptic norepinephrine.

Tolerance

Both oral and intravenous use of amphetamines lead to development of tolerance. When tolerance develops as much as 50–150 mg may be used daily. Also, it may result in injections of as much as 1 g every 2–3 h. Daily doses of 5–8 g may be used.

Withdrawal

Withdrawal symptoms of amphetamines are characterized by apathy, lethargy, anxiety, sleep disorder, depression, and suicidal tendency (Hodding et al., 1980).

Treatment

There is no antidotal treatment for amphetamine intoxication. Ingested amphetamines are treated by decontaminating the gut using syrup of ipecac or lavage, activated charcoal, or cathartics (Oderda and Schwartz, 1979). Urinary acidification renders amphetamine a weak base to an ionized form and results in increased excretion of unmetabolized drug.

LYSERGIC ACID DIETHYLAMIDE (LSD)

Lysergic acid diethylamide (LSD) is a psychedelic drug, and comprises a group of mind-altering compounds that produces visual illusions and hallucinations. The grain fungus *Claviceps purpurea* is the natural source for the precursor lysergic acid. LSD is classified as a Schedule I drug (high abuse potential, no medical use) in the United States. Lysergic acid, however, is classified as a Schedule III drug. LSD is colorless, tasteless, odorless, and water soluble. It is a very potent drug. Doses as small as 25 µg are effective. It is formulated as tablets, capsules, powder, solution, or is impregnated in a variety of substances, e.g., gelatin, sugar cubes, or blotting paper.

Tolerance

Tolerance to LSD develops after 3–4 daily doses and disappears 4–5 days after its withdrawal (Trulson and Crisp, 1981). LSD tolerance results from alterations in the central nervous system rather than an increase in its metabolism.

Mode of Action

LSD is a strong peripheral serotonin antagonist; however, it is a mixed serotonin agonist/antagonist in the central nervous system (Bennett and Snyder, 1976). LSD is a dopamine agonist for striatum receptors. LSD and serotonin bind to the same areas in the brain: the hippocampus, corpus striatum, cerebral cortex, and, to a much lesser extent, and cerebellum (Kuhn et al., 1978).

Clinical Symptoms

LSD consistently produces pupil dilation. There is slight increase in heart rate, body temperature, and respiration (Cohen, 1967). Blood pressure does not change. Sympathomimetic effects include piloerection, uterine concentration, and bronchiole smooth muscle constriction. Parasympathetic effects present are salivation, lacrimation, nausea, and vomiting. Neuromuscular effects are evident as weakness, tremor, hyperreflexia, and ataxia.

LSD alters body image and concrete thinking and causes depersonalization. It impairs sequential thinking and intelligence testing. Users become quiet, passive, self centered, and withdrawn. They may develop hostility. Users with psychotic history are prone to schizophrenia and hallucination. Following a period of abstinence, visual alterations, time distortions, and body image changes may take place. Impaired color perception may be developed that lasts as long as two years after chronic LSD use (Friedman and Hirsch, 1971).

Treatment

There is no specific antidotal treatment for LSD intoxication. Since ingested LSD is rapidly absorbed, gut decontamination is usually not useful. Because of its short half-life, elimination enhancement is not used. The patient should be stabilized, and should be placed in a safe, quiet environment. Sedation may be induced by administering 5–10 mg i.v. diazepam (Solursh and Clement, 1968).

BARBITURATES

Barbiturates were introduced in 1903 as sedative-hypnotic drugs. Because of the risk for overdose and abuse, they have been replaced by the safer benzodiazopines.

Classification

Barbiturates are derivatives of barbituric acid, which has no central nervous system effects. They are classified into four groups depending on their elimination half-lives in animals and duration of action as follows:

1. *Ultrashort-acting.* Thiopental, methohexital, and thiamylal.
2. *Short-acting.* Pentobarbital, secobarbital, and hexobarbital.
3. *Intermediate-acting.* Amobarbital, aprobarbital, and butabarbital.
4. *Long-acting.* Barbital, mephobarbital, phenobarbital, and metharbital.

Medical Uses

Barbiturates are used to treat anxiety, gastrointestinal upset, pain, and sleep disorders. Long-acting drugs are used as anticonvulsants. Ultrashort-acting barbiturates are used as anesthetics.

Action

The depressant effect of barbiturates of the central nervous system is probably mediated via the inhibitory GABA synapses in the brain (Gaudreault, 1981). Also, they may selectively depress noradrenergic activity. Death results from respiratory depression. Barbiturates depress myocardial contractility and cause vasodilation and hypotension. They are powerful inducers of hepatic microsomal cytochrome P-450 enzymes.

Barbiturate Abuse

Intoxication with barbiturates is characterized by slurred speech, ataxia, lethargy, nystagmus, headache, parathesia, vertigo, and confusion (Preskom and Schwin, 1980). When coma ensues, pupils become constricted, but later dilate. In early coma, respiratory depression occurs. Other symptoms are medullary depression, weak and rapid pulse, cyanosis, cold and clammy skin, decreased urine output, confusion, and shock. Later stages are characterized by hallucination, delirium, hyperpyrexia, and finally death.

Physical Dependence and Tolerance

A marked degree of both physical dependence and tolerance develops to all barbiturates. Withdrawal symptoms are weakness, restlessness, insomnia, vomiting, hyperthermia, hypotension, confusion, disorientation, agitation, delusions, hallucinations, tremor, ataxia, hyperreflexia, and eventually convulsive seizure (Smith and Wesson, 1970).

Treatment

No antidotal treatment is available. Patient should be stabilized and provided with supportive care (Goldberg and Berlinger, 1982). Since the major cuase of death is respiratory arrest, adequate respiration should be maintained including administration of oxygen. Mentally depressed patients should be given glucose, naloxone, and thiamine. Gut decontamination is indicated within 6–8 h after ingestion using repeated doses of activated charcoal. Also urinary alkalinization increases phenobarbital excretion 5–10 times.

BENZODIAZEPINES

Chemistry

Benzodiazepines are classified according to their elimination half-life into three groups.

1. *Long-acting (>24 h)*. Chlordiazepoxide (Librium), chlorazepate (Tranxene, clonazepam (Clonopin), diazepam (Valium), flurazepam (Dalmane), and prazepam (Centrex).
2. *Intermediate-acting (10–24 h)*. Alprazolam (Zanax), lorazepam (Ativan), and oxazepam (Serax).
3. *Short-acting (<10 h)*. Midazolam (Versed), temazepam (Restoril), and triazolam (Halcion).

Action

Benzodiazepines potentiate the action of GABA by binding to the polysynaptic terminals where GABA is released and results in hyperpolarization (Tall et al., 1980).

Medical Use

Due to their efficacy, safety, and low cost, benzodiazepines are the most prescribed drugs in the world (Fagan and Illsley, 1985). They are used to treat anxiety. They produce less suppression of the cardiorespiratory and central nervous systems. Ingested benzodiazepines are rapidly absorbed from the gastrointestinal tract with peak plasma level occurring 1–3 h after intake.

Toxic Effects

Although large oral doses of benzodiazepines cause sedation and ataxia, they seldom result in death. Their action is potentiated by other sedative-hypnotic drugs, ethanol and antipsychotic drugs, resulting in increased sedation and respiratory depression. Ethanol also increases diazepam absorption from the gastrointestinal tract.

Benzodiazepines cross the placenta and should not be used in the first trimester. They are also excreted in the mother's milk and may produce lethargy and poor feeding in neonates.

Withdrawal

Long term use of large therapeutic doses (45 mg/day) results in withdrawal (Murphy et al., 1984). Symptoms develop 1–11 days following discontinuation and last 3–4 days. Severity of withdrawal symptoms is less than that of ethanol and barbiturates. Symptoms are anxiety, insomnia, headache, muscle spasm, anorexia, vomiting, nausea, tremor, postural hypotension, and weakness. Withdrawals to higher doses (60–300 mg/day) include psychosis, agitation, confusion, hallucinations, delirium, motor dysfunction, and seizures.

Treatment

Gut contamination takes place by induction of emesis or lavage and the administration of activated charcoal. Antidotal treatment is limited by side effects of antidotes (Allen and Gough, 1983). Sedative action of diazepam is treated with Doxapram (100 mg i.v.). Benzodiazepine-induced coma is treated with naloxone (Bell, 1975). Central nervous system depression is controlled by physostigmine. A specific antagonist (fumazenil) is now available in the U.S. for large benzodiazepine overdoses.

REFERENCES

Allen, C.J. and K.R. Gough (1983). The effect of doxapram on heavy sedation produced by intravenous diazepam. *Br. Med. J.* 286: 1181–1182.

Bailey, D.N. and A.S. Manoguerra (1980). Survey of drug abuse patterns and toxicology analysis in an emergency room population. *J. Anal. Toxicol.* 4: 199–203.

Bell, E.F. (1975). The use of naloxone in the treatment of diazepam poisoning. *J. Pediatr.* 87: 803–804.

Bennett, J.P. and S.H. Snyder (1976). Serotonin and lysergic acid diethylamide binding in rat brain membranes: relationship to postsynaptic serotonin receptors. *Mol. Pharmacol.* 12: 373–389.

Blackard, C. and K. Tennes (1984). Human placental transfer of cannabinoids. *N. Engl. J. Med.* 311: 797.

Cohen, S. (1975). Amphetamine abuse. *JAMA* 231: 414–415.

Cone, E.J. and R.E. Johnson (1986). Contact highs and urinary cannabinoid excretion after passive exposure to marijuana smoke. *Clin. Pharmacol. Ther.* 40: 245–256.

Council Reports (1972). Treatment of morphine-type dependence by withdrawal methods. *JAMA* 219: 1611–1615.

Evans, M.A. and R.D. Harbison (1978). Cocaine induced hepatotoxicity in mice. *Toxicol. Appl. Pharmacol.* 45: 739–754.

Fagan, D.R. and S.S. Illsley (1985). Benzodiazepine hypnotics. A comparative review of recently approved agents. *Hosp. Formul. Manage.* 20: 491–499.

Fantel, A.G. and G.J. MacPhail (1982). The teratogenicity of cocaine. *Teratology* 26: 17–19.

Friedman, S.A. and S.E. Hirsch (1971). Extreme hyperthermia after LSD ingestion. *JAMA* 217: 1549–1550.

Gaudreault, P. (1981). Barbiturates. *Clin. Toxicol. Rev.* 3: 1.

Goldberg, M.J. and W.G. Berlinger (1982). Treatment of phenobarbital overdose with activated charcoal. *JAMA* 247: 2400–2401.

Gravey, R.E. (1979). PCP (phencydidine): an update. *J. Psychedelic Drugs* 11: 265–275.

Hodding, G.C., M. Jann, and I.P. Ackerman (1980). Drug withdrawal syndromes. A literature review. *West. J. Med.* 133: 383–391.

Hollister, L.E. (1971). Marihuana in man: three years later. *Science* 172: 21–29.

Hunt, C.A. and R.T. Jones (1980). Tolerance and deposition of tetrahydrocannabinol in man. *J. Pharmacol. Exp. Ther.* 215: 35–44.

Kelsner, S. and M. Nickerson (1969). Mechanism of cocaine potentiation of responses to amines. *Br. J. Pharmacol.* 35: 428–439.

Kuhn, D.M., F.J. White, and J.B. Appel (1978). The discriminative stimulus properties of LSD: mechanisms of action. *Neuropharmacology* 17: 257–263.

Liden, C.B., F.H. Lovejoy, and C.E. Costello (1975). Phencyclidine: nine cases of poisoning. *JAMA* 234: 513–516.

Martin, W.R. (1984). Pharmacology of opioids. *Pharmacol. Rev.* 35: 285–323.

Moss, D.E., P.L. Peck, and R. Salome (1978). Tetrahydrocannabinol and acetylcholinesterase. *Pharmacol. Biochem. Behav.* 8: 763–765.

Murphy, S.M., R.T. Owen, and P.J. Tyrer (1984). Withdrawal symptoms after six weeks treatment with diazepam. *Lancet* 2: 1389.

Nicholson, A.N. and B.M. Stone (1980). Heterocyclic amphetamine derivatives and caffeine on sleep in man. *Br. J. Clin. Pharmacol.* 9: 195–203.

O'Connor, J.E. and T.A. Rejent (1981). EMIT cannabinoid assay: confirmation by RIA and GC/MS. *J. Anal. Toxicol.* 5: 168–173.

Oderda, G.M. and W.K. Schwartz (1979). Management of poisonings with central nervous system stimulants. *Clin. Toxicol. Consult.* 1: 73–88.

Post, R.M. and R.T. Kopandra (1975). Cocaine, kindling and reverse tolerance. *Lancet* 1: 409–410.

Preskom, S.H. and R.L. Schwin (1980). Analgesic abuse and the barbiturate abstinence syndrome. *JAMA* 244: 369–370.

Shaprio, D. (1974). The ocular manifestations of cannabinols. *Ophthalmologica* 168: 366–369.

Simpson, R.W. and W.D. Edwards (1986). Pathogenesis of cocaine-induced ischemic heart disease. Autopsy findings in a 21-year-old man. *Arch. Pathol. Lab. Med.* 110: 479–484.

Smith, D.E., and C.M. Fisher (1970). An analysis of 310 cases of acute high dose methamphetamine toxicity in Haight-Ashbury. *Clin. Toxicol.* 3: 117–124.

Smith, D.E. and D.R. Wesson (1970). A new method for treatment of barbiturate dependence. *JAMA* 213: 294–295.

Snyder, S.H. (1979). Phencyclidine. *Nature* 285: 355–356.

Solursh, L.P. and W.R. Clement (1968). Hallucinogenic drug abuse. Manifestations and management. *Can. Med. Assoc. J.* 98: 407–413.

Spiehler, V.R. and D. Reed (1985). Brain concentrations of cocaine and benzoylecgonine in fatal cases. *J. Forensic Sci.* 30: 1003–1011.

Tall, J.F., S.M. Paul, and P. Skolnick (1980). Receptors for the age of anxiety: pharmacology of the benzodiazepines. *Science* 207: 274–281.

Trulson, M.E. and T. Crisp (1981). Tolerance develops to LSD while the drug is exerting its maximal behavioral effects. Implications for the neural basis of tolerance. *Eur. J. Pharmacol.* 96: 317–320.

Trulson, M.E. and B.L. Jacobs (1979). Long term amphetamine treatment decreases brain serotonin metabolism: implications for theories of schizophrenia. *Science* 205: 1295–1297.

Twycross, R. (1977). Value of cocaine in opiate containing elixirs. *Br. Med. J.* 2: 1348.

Naturally Occurring Toxins

Mohamed B. Abou-Donia
Departments of Pharmacology and Neurobiology
Duke University Medical Center
Durham, North Carolina

INTRODUCTION

There are thousands of species of venomous or poisonous animals. A venomous animal has a secretory gland where poison is secreted and delivers it during a biting or stinging. A poisonous animal contains poison in its body, and it is unable to deliver it. Naturally occuring toxins may have several toxicity targets, e.g., the nervous system, cardiovascular system, respiratory system, etc. (Kingsbury, 1964; Hardin and Arena, 1974; Lampe, 1986; Russell, 1986).

NERVOUS SYSTEM TOXINS

Acetylcholine Receptors

Muscarinic Receptor Agonists

Muscarine mimics the action of ACh at the muscarinic receptor sites resulting in stimulation of the parasympathetic nervous system. Muscarine occurs naturally in the mushroom *Amanita muscaria* and *A. phalloides*. It results in increased sweating, salivation, bronchial secretion, gastric secretion and motility, a slower heart rate, and bronchoconstriction. It also causes nausea, emesis, painful colic, and watery (sometimes bloody) diarrhea. Muscarine is generally not lethal, but may exacerbate asthma, coronary insufficiency, or peptic ulcer.

Some mushrooms contain a disulfiram-like substance that inhibits aldehyde dehydrogenase, a liver enzyme responsible for oxidizing aldehydes to carboxylic acids. Subsequent ethanol ingestion results in the accumulation of acetaldehyde and leads to respiratory difficulties, nausea, and hypotension. The selective hepatic toxicity of *A. phalloides* appears to result from damaging liver cells, by inhibiting RNA polymerase II.

Muscarinic Receptor Antagonists

Atropine antagonizes the action of ACh by blocking its binding at the muscarinic receptors. Atropine is found in the belladonna alkaloids present in the seeds and flowers of Jimson weed (*Datura stramonium*). Early symptoms are mydriasis with blurring of vision and dryness of the mouth followed by speech difficulty. The skin becomes hot, dry, and flushed with occasional rash in the face, ears, and neck. Severe poisoning is characterized by hyperpyrexia, delirium, and hallucinations. Convulsions may appear in young children followed by coma. Atropine may cause death by inducing convulsions and circulatory collapse by acting at central muscarinic receptors.

Nicotinic Receptor Agonist

Nicotine mimics the action of ACh at the nicotinic receptor sites at the neuromuscular junction and preganglionic cholinergic receptors. It is present in tobacco plants. It causes a spontaneous emesis with 15–60 min of ingestion. This is accompanied by profuse salivation, abdominal cramps, and diarrhea. Other symptoms are headache, confusion, incoordination, mydriasis, and tachycardia. Death results from respiratory failure.

Nicotinic Receptor Antagonists

d-Tubocurarine antagonizes the action of ACh at the nicotinic receptors by blocking the binding of ACh to these receptors. *d*-Tubocurarine is the active neurotoxicant of curare that occurs in the South American plants whose extracts are used as arrowhead poison. It competes with ACh in the skeletal muscle end plate. The major action of this compound is a prolonged flaccid paralysis of striated muscles. Death results from respiratory paralysis. The antidotal treatment is the administration of reversible AChE inhibitors such as neostigmine and physostigmine. α-Bungarotoxin, a component of some South American snake venoms, binds irreversibly to neuromuscular nicotinic receptors.

Sodium Channels

Some marine toxins act by binding to the major voltage-sensitive sodium channels which are responsible for the propagation of electrical impulses in both neurons and cardiac muscle cells. These toxins include tetrodotoxin (pufferfish), saxitoxin (shellfish), and maculotoxin (cephalopods). They interfere with increased sodium ion permeability of the axon during excitation by blocking the passage of sodium ions through the channel, thereby decreasing sodium conductance. This results in blocking of sodium ion inward movement during cell depolarization. They also block the skeletal muscle membrane but have no direct effect on the neuromuscular junction except for that on nerve ending and the muscle membrane. This action results in a massive release of acetylcholine with subsequent development of cholinergic symptoms. In humans, symptoms of tetrodotoxin poisoning are weakness, dizziness, and parathesia of the lips, tongue, and throat. Onset of symptoms is usually very rapid, e.g., within 5–30 min of exposure. Other symptoms include salivation, vomiting, bradycardia, dyspnea, cyanosis, shock and flaccid paralysis. These symptoms are treated with administration of oxygen, intravenous fluids, atropine, and activated charcoal.

Batrachotoxin holds sodium channels in an open position which increases sodium ion conductance. This neurotoxin has neurotoxic and cardiotoxic actions leading to cardiac arrest with a lethal dose in humans of less than 200 μg. Its toxicity is antagonized by tetrodotoxin and saxitoxin. Also, ciguatoxin, which is present in many fish species, increases sodium ion conductance, but probably acts by a different mechanism.

Tetrodotoxin occurs in ocean sunfish, some puffers, porcupinefish, as well as certain amphibian species of the family Salamandridae and the blue-ringed octopus. It is also present in the skin and eggs of frogs. Other toxins that occur in frog skin and eggs include chiriquitoxin, batrachotoxin, homobatrachotoxin, and dihydrobatrachotoxin. Saxitoxin or paralytic shellfish poison (PSP) is present in certain molluscs, arthropods, mussels, and other marine animals that have ingested toxic protistan, e.g., protozoans, algae, diatoms, bacteria, yeasts, and fungi. Red tide or red water results from the presence of protistan in high concentration, e.g., 20,000 organisms per ml of water. This results in red color that is attributed to the presence of xanthophyll. Other toxins found in red tide are gonyautoxin II (GTX$_2$) and gonyautoxin III (GTX$_3$). These toxins cause paralysis in animals consuming ingested shellfish as well as mortality of marine animals.

GABA Receptors

Muscimol, a neurotoxin that is present in mushrooms, acts as a specific agonist for GABA-A receptors, resulting in drowsiness and dizziness, which may be accompanied by sleep. This may progress to increased motor activity, tremors, and delusions.

Microtubules

Colchicine binds to tubulin, and prevents its polymerization to form microtubules. This results in the interruption of axonal transport and of miosis. Colchicine occurs in crocus or meadow saffron plants (*Colchicum* spp.). Poisoning results in vascular damage, thrombocytopenia, bone marrow depression, hypothermia, and muscle weakness.

CARDIAC SYSTEM TOXINS

Toxins that affect the nervous system, in particular those that act on ACh receptors and those that bind to sodium channels, also affect the heart, either directly or indirectly. Both tetrodotoxin and batrachotoxin cause death mostly by a direct action on the heart.

Cardioactive glycoside-like digitalis, which occurs in the plants oleander (*Nerium oleander*), foxglove (*Digitalis purpurea*), and lily-of-the-valley (*Convallara majalis*), acts more exclusively on the heart. Digitalis contains two cardiac glycosides, digoxin and digitoxin, which inhibit Na^+/K^+ ATPase. This action results in an increase in the force of the contraction of the cardiac muscle. At lower concentrations, digitalis can have a therapeutic effect on patients with congestive heart failure. Slightly higher doses, however, can cause nausea, dizziness, bloody diarrhea, circulatory irregularities, slowed heart rate, convulsions, unconsciousness, and respiratory arrest which may lead to death.

VASCULAR SYSTEM TOXINS

Snake venoms contain toxins that affect blood flow. They include hydrolytic enzymes which break down vascular walls and red blood cells. Thrombin-like enzymes that often occur in snake venoms prevent blood clotting. This may lead to fatal hypotension and shock.

Snake venoms contain the following enzymes: proteolytic enzymes, arginine ester hydrolase, thrombin-like enzyme, collagenase, hyaluronidase, phospholipase A_2, phospholipase B, phospholipase C, lactate dehydrogenase, phosphomonoesterase, phosphodiesterase, acetylcholinesterase, RNase, DNase, 5'-nucleotidase, NAD-nucleotidase, and L-amino acid oxidase.

Naturally occurring amines in some venoms and plants such as histamine and catecholamines may affect local blood flow around the wound by causing either vasodilation or vasoconstriction.

IMMUNE SYSTEM TOXINS

In the United States, bee stings cause more death than from toxins of all other animals combined. It is noteworthy that the hydrolytic enzymes and peptides present in bee venom are not toxic. Sensitization to repeated stings, however, may result in anaphylactic reactions involving cardiovascular, respiratory, or nervous systems. Allergic response may kill in a few minutes to hours. Most other insects produce local inflammation.

SKIN TOXINS

Venoms contain hydrolytic enzymes which break down and kill the cells surrounding the wound resulting in dermatonecrosis that may also affect underlying muscle. Dermal irritation (urticaria) may result from amines present in venoms or plants. Delayed contact sensitivity is a syndrome that results from poison ivy and poison oak.

Some plants, e.g., carrots, parsnips, and citrus, contain furocoumarins which are able to penetrate moist skin. This results in the sensitization of the skin to ultraviolet light, and the selective burning of the sensitized area following exposure to sunlight or fluorescent light. The ultraviolet may act by inducing binding of the furocoumarins to DNA.

REFERENCES

Hardin, J.W. and J.M. Arena (1974). *Human Poisoning From Native and Cultivated Plants,* 2nd ed. Duke University Press, Durham, NC.

Kingsbury, J.M. (1964). *Poisoning Plants of the United States and Canada.* Prentice-Hall, New York.

Lampe, K.F. (1986). Toxic effects of plant toxins. In *Casarett and Doull's Toxicology The Basic Science of Poisons,* 3rd ed., C.D. Klaassen, M.D. Amdur, and J. Doull, Eds., MacMillan, New York, pp. 757–767.

Russell, F.E. (1986). Toxic effects of animal toxins. In *Casarett and Doull's Toxicology The Basic Science of Poisons,* 3rd ed., C.D. Klaassen, M.D. Amdur, and J. Doull, Eds., MacMillan, New York, pp. 706–756.

Principles and Methods for Evaluating Neurotoxicity

Mohamed B. Abou-Donia
Departments of Pharmacoloty and Neurobiology
Duke University Medical Center
Durham, North Carolina

INTRODUCTION

Neurotoxicity procedures are designed to determine the possible adverse effects that may occur on the nervous system when exposed to a test chemical. These studies may involve acute testing resulting from a single exposure or may involve subchronic testing related to multiple dosages. Testing procedures usually include the route of exposure expected for humans. Usually one or more animal species are used as experimental animals. The selection of animal species for neurotoxicity studies should take into consideration, whether possible, the sensitity of the species to the test compound. The choice of neurotoxicity test depends on whether the purpose of the study is to evaluate the neurotoxicity potential of a new compound, i.e., prospective study, or to confirm in laboratory animal a neurologic dysfunction incidence in humans, i.e., retrospective study.

EXPERIMENTAL DESIGN FOR NEUROTOXICITY STUDIES

Test Compound

Chemical Structure

The nature and extent of the neurotoxic action induced by a chemical depends largely upon the chemical structure of the compound, particularly the presence of a reactive functional group or groups in the molecule. The known reaction of these functional groups with groups on critical proteins such as enzymes or receptors should provide some understanding of the mechanisms of the neurotoxic action of test compounds. The chemical structure of a test compound must be known before designing neurotoxicity testing. It gives the opportunity to predict the neurotoxicity potential of the test chemical based on prior information about chemically related compounds. It is also important in selection of the test animal, e.g., the adult chicken is used to study organophosphorus compounds for delayed neurotoxicity (OPIDN; Abou-Donia, 1981). Knowledge of the chemical structure of test-compound should allow the prediction of the potential known metabolic pathways and the ability of the parent compound or its metabolites to penetrate blood-brain barrier. Very minor changes in the chemical structure of a compound may result in a profound change on its neurotoxic potential. For example, while tri-*o*-cresyl phosphate (TOCP) can produce delayed neurotoxicity, its *meta* and *para* isomers are void of this property (Smith et al., 1930). Also, while 2,5-hexanedione induces neurotoxicity, 2,4-hexanedione does not (Spencer and Schaumburg, 1980). Evaluation of neurotoxicity by analogy with chemically related compounds requires a great deal of information on structure-neurotoxicity relationship (Abou-Donia et al., 1981).

Physiochemical Properties

The chemical and physical properties listed in Table 1 provide useful information for neurotoxicity testing (Zbinden, 1972). Properties such as chemical stability at various pH values and stability to light and heat are important considerations for storing of the chemical prior to administration to test animals. Photochemical reactions (Crosby, 1972) and ester hydrolysis (Eto, 1974) may change the neurotoxic action of test compound.

Physical properties such as organic solvent/water partition coefficient and pK are important for determining the absorption, disposition, and distribution of a compound in test animal. Neurotoxic compounds that are highly soluble in lipid solvents such as DDT (Abou-Donia and Menzel, 1968) and leptophos (Abou-Donia 1979, 1980) are readily absorbed from the gastrointestinal tract and skin.

The presence of a formal charge or the extent of ionization of an organic acid or base influence their passage across biological membranes (LaDu et

TABLE 1.
Physiochemical Properties of
Test Compounds

Molecular weight
Molecular formula
Structural formula
Electrophile/neutrophile
Acidity/alkalinity
Particle characteristics
Density
Corrosivity
Solubility in water
Solubility in lipid solvents
Melting point
Boiling point
Vapor pressure
Dissociation constant (pK)
Stability at various pH values
Stability to heat and light

al., 1971). The unionized lipid soluble form of the organic compound is readily transferable across lipoprotein membranes. The extent of ionization of a weak organic acid or base, which determines the ionized/unionized ratio of the compound, depends on the pH of the environment.

In neurotoxicity studies *via* inhalation, particle size, shape, and density are of particular importance in determining the site of deposition and the rates of clearance from the respiratory tract (Hatch and Gross, 1964). Vapor pressure of a test compound is also significant in the design of inhalation studies, since it determines the extent of vaporization. Also, the particle size of orally administered suspensions greatly affects their neurotoxicity.

Impurities

In the design of neurotoxicity studies, it is critical to know the purity of the test compound. Often when a technical grade chemical is tested, the neurotoxicity observed may result from or be influenced by trace contaminants. For example, technical grade leptophos (*O*-4-bromo-2,5-dichlorophenyl *O*-methyl phenylphosphonothioate) was more potent than pure compound in producing delayed neurotoxicity (Abou-Donia et al., 1980).

Dose

Definitions

Dose is the amount of test chemical administered. It is expressed as weight of test substance (g or mg) per unit weight of test animal (e.g., mg/kg). Dosage is a general term comprising the dose, its frequency, and the duration of administration. Dose-neurotoxic effect is the relationship between the dose and the extent of neurotoxic action on test animal. Dose-response is the relationship between the dose and the proportion of a population sample showing neurotoxicity.

Selection of Doses

To study the extent of the neurotoxic action, the highest dose should produce significant neurologic deficit without a high incidence of mortality. On the other hand, the lowest dose should produce no detectable neurotoxic effects (no-observed adverse-effect level, NOEL). Usually at least one intermediate dose is used to obtain a dose-neurotoxic effect relationship.

Selection of the doses for neurotoxicity studies may be obtained from the results of prior acute studies that determined LD_{50} values. Usually, a fraction of the LD_{50} dose is used that may be nonlethal but still may produce neurotoxicity during the experimental period.

Preparation of Chemical for Administration

If the test compound is water soluble, it should be dissolved in water or saline (0.9% NaCl in water). When the chemical is not water soluble, it should be dissolved in a vegetable oil such as corn oil or an organic solvent such as polyethylene glycol 400. Emulsions of water insoluble chemicals may be prepared using polyethylene glycol and Cremophor-EL for intravenous injection (Abou-Donia et al., 1989). In large animals, gelatin capsules may be used.

Controls

Control animals that have not been treated with test compound are usually included in each test to determine the effects in experimental animals that have resulted from exposure to test compound. Control animals should be selected randomly from the same animal population that was used for test animals to guarantee the proper interpretation of results.

Test Animals

Species Selection

The animal species whose response to the neurotoxic effects of test compound is most similar to humans should be used. Because such information may not be available, it is suggested that two species be tested initially: the rat and the adult chicken. While the rat is sensitive to neurotoxicity produced by most classes of chemicals, it does not exhibit signs of neurologic deficit following treatment with organophosphorus compounds (Abou-Donia, 1981). On the other hand, both sexes of the adult chicken are sensitive and readily exhibit signs of delayed neurotoxicity after a single or multiple doses of organophosphorus compounds.

There are numerous studies that provide evidence of wide variations among species in their rates of metabolic biotransformation of xenobiotics (Williams, 1967). It is important to characterize the metabolic biotransformation of neurotoxicants in several animal species during early toxicity studies. This will allow the use of the species that closely resembles humans. Other factors are also involved such as variation in the sensitivity of the neurotoxic target among different species. The acute toxicity of some organophosphorus insecticides may be related to the sensitivity of the target enzyme, i.e., acetylcholinesterase, than to differences in hepatic metabolic rates (Murphy et al., 1968). When the neurotoxic insecticide leptophos was orally administered to nonsusceptible species (e.g., mice [Holmstead et al., 1973] and rats [Whitacre et al., 1976; Hassan et al., 1977]), it was rapidly metabolized and excreted as degradation products mainly in the urine. In contrast, this compound persisted in hen tissue and had a half-life 38 times that in the mouse (Abou-Donia, 1980).

Age of Animals

When testing for the neurotoxicity of a chemical, it is useful to use all age groups. Use of a neonate or fetus allows the detection of neurotoxic effects of chemicals on early stages of life. Adult animals and aged animals should also be used to assess the neurotoxic effects for better extrapolation to all age range in humans. Many potential variables are age dependent. These variables include the maturity of the blood-brain barrier. Young animals have immature blood-brain barrier that allows many chemicals to cross into the brain. Generally, the nervous system is vulnerable to attack with neurotoxic agents during early age as well as later in life, e.g., geriatric patients. Furthermore, drug metabolizing enzymes that play a critical role in the activation and inactivation of neurotoxic agents change with age resulting in increased or decreased neurotoxic action.

Development of some neurotoxic effects are age dependent (Abou-Donia, 1981). Thus, organophosphorus compounds that produce type I organophosphorus compound-induced delayed neurotoxicity (OPIDN) such as TOCP do not induce OPIDN in young chicks, with 70 days being the earliest age to develop OPIDN (Abou-Donia et al., 1982). Thus, the adult hen is required for testing these chemicals for OPIDN. On the other hand, chemicals that produce type II OPIDN, such as triphenyl phosphite, produce OPIDN in young chicks and in adult chickens (Abou-Donia and Lapadula, 1989).

Number of Animals

The number of animals to use varies depending on the study, but should allow for appropriate statistical analysis of the results. In neurotoxicity studies, where both neuropathological and biochemical evaluations are to be carried out, the number of test animals should be large enough to allow for half of the animals to be used for each test. For example, ten animals will be used per sex per dose level, five of which should be used for histopathological assessments and the other five for biochemical evaluation.

Cyclic Variation in Function or Response

Many physiological functions or responses undergo cyclic peaks and regressions of activity (Altman and Dittmer, 1966). The duration of these cycles may be diurnal (24 h) or longer. The cycles may be controlled by intrinsic factors or may be affected by environmental variables such as light and temperature. Most diurnal variations relate to eating and sleeping habits (Boyd, 1972). Since rats are nocturnal feeders, their stomachs would have more food early in the morning compared with the afternoon, which may influence acute effects following oral administration. Some but not all of the effects are related to chemical absorption from the stomach. Mice given pentobarbital showed a diurnal cycle with respect to the anesthesia with longest duration of effects at 14:00 h and the shortest (40–60% of that at 14:00 h) duration at 02:00 h (Davis, 1962).

The induction of drug metabolism by phenobarbital and drug metabolism in rats showed a seasonal or circannual variation (Beuthin and Bousquet, 1970). Also, locomotor activity is greatest at night in rodents (Boyd, 1972).

Temperature

Wide variations in environmental temperature and relative humidity may impair the general health and increase the infection of test animals (Doull, 1972). Also, it may change their response to the compound. Some physiological parameters, such as ventilation, circulation, body water, and intermediary metabolism, which determine the absorption, disposition, and action of test compounds may be slowed by varying environmental temperature. Also, fluctuations in environmental temperature may result in functional changes that might be mistakenly attributed to the action of the test compound (Murphy, 1969).

Diet

Diet plays an important role in the toxicity of test compounds. Low protein diet resulted in increased toxicity of several pesticides in rats (Boyd, 1969). On the other hand, protein-deficient diets protected rats against acute hepatic toxicity of carbon tetrachloride (McLean and McLean, 1969). Low-protein diets usually result in the reduction of the activity of hepatic mixed-function oxidases. The effect of low-protein diet on increasing or decreasing the toxicity of a test chemical will depend on whether microsomal biotransformation results in more or less toxic metabolites. Natural diets having alfalfa meal which contains flavonoid compounds, may account for the induction of aryl hydrocarbon hydroxylase in rats fed these diets. Also,

TABLE 2.
Parameters Recorded in Acute Neurotoxicity Studies

Time to onset of clinical signs of neurotoxicity
Progress of clinical signs with time
Weight changes
Observation of skin, eyes, mucous membranes, etc.
Changes in behavior
Signs of autonomic nervous system effect, such as tearing, salivation, diarrhea
Changes in respiratory rate and depth
Cardiovascular changes such as flushing
Central nervous system changes such as tremors, convulsion, and coma
Time of death
Necropsy results
Histopathological findings of the brain, spinal cord, and sciatic nerve

---

contaminants such as pesticide residues in animal diets may markedly influence the effect of test chemicals.

Duration of Exposure

Neurotoxicity studies may be conducted following a single or multiple exposures. Duration of exposure differs according to the scope of study and whether it is an acute, subacute, subchronic, or chronic study.

Acute Neurotoxicity

Acute neurotoxicity studies are carried out by administration of a single dose. The maximum dose level is accepted to be 5000 mg/kg for oral administration. Acute neurotoxicity studies are carried out to determine the neurotoxic hazard of exposure to large doses. Signs of neurotoxicity may develop very shortly after dosing or after a delay. Animals treated with a large dose of an organophosphate ester usually exhibit signs of cholinergic effects soon after administration (Abou-Donia, 1985). In contrast, delayed neurotoxic organophosphates such as TOCP produce neurologic deficits after a latent period of 6–14 days in hens. Usually animals are observed at 1, 2, and 4 h and daily for 21 days inside cages and when moving freely outside. Animals are weighed weekly. The parameters that are obtained and recorded in acute neurotoxicity studies are listed in Table 2.

Subacute Neurotoxicity

The aim of subacute studies is to determine the dose levels for subchronic studies. Several dose levels, usually a fraction of the LD_{50} dose, are employed. The dose range falls between 10–25% of the LD_{50}. The duration of exposure is usually 2–4 weeks. The same information recorded in the acute studies (Table 2) are gathered in subacute studies. Maximum dose levels are

5% of the diet, or 2.5 g/kg of body weight per day for oral administration, 2 g/kg of body weight for dermal exposures, and 20 mg/m³ (6 h/day) for inhalation of particulates, gases, and vapors.

Subchronic Neurotoxicity

Subchronic neurotoxicity studies are carried out for 90 days (13 weeks). These studies are carried out to determine the short term risk to exposure to low levels of the neurotoxic chemical. At least three doses are used, including a dose that produces severe neurotoxicity, a no-observed adverse-effect dose, and at least one intermediate dose which is determined in subacute dose, and at least one intermediate dose which is deterined in subacute studies. Animals are observed daily and weighed weekly. All parameters listed in Table 2 are recorded.

Chronic Neurotoxicity

The purpose of this test is to find any neurotoxic effects of a long period of time to assess the potential for long term, low dose exposure to neurotoxic chemicals similar to what might be present in food or the workplace. The doses chosen are based on the results of subchronic toxicity studies. They may last for 6 months or the lifetime of the animal species being tested (18–24 months for rats, 6–18 months for mice, 6–12 months for chickens, usually). The number of animals used is larger than in subchronic studies to assure enough survive the test period to provide adequate data for statistical analysis.

Route of Exposure

Generally, the test chemical should be administered by the route by which humans would be exposed.

Oral Administration

To test chemicals for neurotoxicity potential via the oral route, it is preferred that they are administered by gavage rather than mixed with the diet. Often, animals are starved for 24 h before administration. When rats or mice are used the volume administered by the oral route should not greatly exceed 0.005 ml/g of body weight, i.e., 1.0 ml per 200 g rat or 0.1 ml per 20 g mouse (Matsumura, 1985). Usually a constant concentration is administered for various doses rather than a constant volume. In some chronic and subchronic studies, the test compound is fed to the animals by adding it to the diet. Feeding has the advantage of resembling the normal route of ingestion of toxic substances such as pesticides in the food. This method has the following disadvantages: (1) the estimation of the exact ingested dose cannot be done accurately, (2) it is difficult to prepare homogenously chemical-impregnated feed, (3) the chemical may be unstable or react with some component in the feed, (4) the test chemical may affect the acceptability of the food and the animals' appetite, and (5) as time passes, the animal may become ill and eat less, thus consuming less test chemical.

Dermal Administration

The skin is an important port of entry for chemicals in cases of occupational exposure. Lipid soluble organic neurotoxicants are capable of penetrating the skin, resulting in systemic exposure. Chemical can be applied with or without occlusive dressing. Extrapolation of the results from dermal studies in animals to humans are limited due to the following reasons.

1. Rabbit skin is more permeable than human skin for most chemicals.
2. There are large differences in the permeability of various anatomic sites of skin to chemical, e.g., the palm allowed approximately the same penetration of ^{14}C-labeled parathion, malathion, and carbaryl as the forearm (Maibach et al., 1971). A fourfold greater penetration was observed in follicle-rich sites, such as the scalp, postauricular area and forehead, whereas the scrotum allowed almost total absorption.
3. Neither rodent nor rabbit skin mimics human skin since they do not perspire.
4. Use of occlusive dressings on test compound exaggerates human exposure.

Percutaneous penetration is generally considered to be by diffusion. The extent of absorption of chemicals depends primarily on their physical properties, and, to a lesser degree, their chemical properties. The condition of the skin, whether normal or damaged, also contributes to the level of absorption (Feldman and Maibach, 1969). The rate at which the test compound is lost to the environment is an important factor. The surface loss rate for most unprotected chemicals exceeds absorption rates. Usually less than 50% of a chemical is absorbed through the skin (Feldman and Maibach, 1970). The loss of applied leptophos from the comb of hens ranged from 53.3–64.5%, 12 and 20 days, respectively, after dermal application (Abou-Donia, 1979). The fraction of the dermally applied dose of leptophos absorbed, however, was twice that of the same dose when administered orally (Abou-Donia, 1980). Similarly, the delayed neurotoxicant insecticide *O*-ethyl *O*-4-nitrophenyl phenylphosphonothioate (EPN) was more efficiently absorbed through skin (Abou-Donia et al., 1983) than through gastrointestinal tract (Reichert et al., 1978).

Inhalation

Inhalation of chemical vapors is a very important route of exposure in the workplace. Estimation of the dose is very complicated in inhalation studies. If the test compound is a fluid that has an appreciable vapor pressure, it may be administered by passing air through the solution under controlled-temperature conditions. Unless the particle size is less than 2 μm, it will not reach the terminal alveolar sacs in the lungs. The concentration of test compound vapor in PPM is determined as follows:

$$ppm = \frac{\text{solvents volume (ml)} \times \text{density (mg/ml)} \times 24.45 \times 10^6}{\text{exposure chamber volume (liters)} \times MW}$$

$$\text{also } ppm = \frac{\text{mg/m}^3 \times 24.45 \text{ 1}}{MW}$$

where ppm = parts per million; 24.45 = gas constant at 25°C and 760 mm of mercury barometric pressure; MW = molecular weight of the substance; m = meter.

Injections

Injection administration may be carried out with one of the following procedures: intravenous, i.v.; intraperitoneal, i.p.; intramuscular, i.m.; or subcutaneous injection, s.c. When employing injection, care must be taken to choose the right vehicle and to use proper speed of injection. Intravenous injection is the most direct method of application. Common vehicles used are emulsions containing 15–20% of either vegetable oil or polyethylene glycol 300, or both, and an emulsifying agent, such as Cremophor-EL, and water, preferably saline. Intravenous injection should be limited to a dose of 0.1–0.5 ml for rodents, and the speed of injection must be slow (Matsumura, 1985). Intravenous injection of large volume of innocuous solvents such as water or saline could be fatal. The i.v. LD_{50} of distilled water in the mouse is 0.044 ml/g body weight and that of isotonic saline is 0.068 ml/g body weight.

JOINT NEUROTOXIC ACTION OF A COMBINATION OF CHEMICALS

The joint action of neurotoxic chemicals may result from alteration of a chemical or physical properties of the molecules, resulting in changes in their neurotoxic effects, interaction at the neurotoxicity target, indirect interactions via modification of xenobiotic metabolizing enzymes that render the compounds to more or less active neurotoxicants, or interaction at the absorption site leading to changes in the body burden of the neurotoxic chemical.

As early as 1910, the joint action of combining drugs was studied by Burgi, who stated, "In combining drugs with the same end-effect, the resulting activity is additive when the sites of action of the components are identical and superadditive if these are different.'' He also defined the term ''potentiation'' as the augmentative effect of one compound by the other.

When an organism is exposed to two or more neurotoxicants, their joint neurotoxic action may be

independent—the neurotoxicants produce different effects or have different mechanisms of neurotoxic action;

additive—the magnitude of the combined effect produced by two or more chemicals is numerically equal to the sum of the effects of each chemical when used alone;

more than additive—when the combined effect of two or more chemicals is greater than that the chemicals would produce individually (Rentz, 1932);

synergism—one chemical has little or no intrinsic neurotoxicity when used alone;

potentiation—when both chemicals have intrinsic neurotoxicity;

less than additive (often called antagonism or inhibition)—the neurotoxic action produced by applying two or more chemicals is numerically greater than would be expected from simple summation of the effects induced by each chemical.

Joint neurotoxicity action is affected by many factors such as the time intervals between exposure, frequency, and duration of exposure. Furthermore, the dose size is very important since the joint neurotoxic action at lethal dose levels may be very different from that produced at low dose level.

Most statistical models for joint action have been proposed for cases in which two or more chemicals are administered simultaneously. A model was developed to predict the joint action of a mixture of chemicals that act at the same site, produce the same type of acute toxicity, and have parallel regression lines of probits against log doses (Finney 1952, 1971). The equation for the median effective dose (ED_{50}) for a mixture of three chemicals is

$$\frac{1}{ED_{50(A,B,C)}} = \frac{f_A}{ED_{50(A)}} + \frac{f_B}{ED_{50(B)}} + \frac{f_C}{ED_{50(C)}}$$

where f_A, f_B, and f_C are the fractions of substances A, B, and C in the mixture. A predicted ED_{50} for the mixture is calculated from the individual ED_{50s} and is compared with the actual ED_{50} of the mixture determined experimentally. An equal value indicates additive joint action. A smaller experimental ED_{50} than predicted indicates a more than additive response (synergism or potentiation); a greater experimental ED_{50} demonstrates a less than active additive response (antagonism). This equation satisfactorily predicted the joint action of some industrial chemicals (Smyth et al., 1969).

Availability of information concerning the metabolic bioactivation or inactivation of various compounds in a mixture enhances the possibility of predicting the type of neurotoxic action. Abou-Donia et al. (1985a,b,c) studied the joint neurotoxic action of binary mixtures of chemicals. In these studies they quantified the neurotoxicity of each treatment by calculating the neurotoxicity index (NTI). The type of joint neurotoxicity was determined using the coneurotoxicity coefficient (CNC) for mixtures of chemicals.

Neurotoxicity Index (NTI)

Neurotoxic potency of test treatment was quantified by ranking the animals, starting with minimal changes, in the following three parameters: (1) the time of onset of neurologic dysfunction, (2) the severity of neurologic deficit, and (3) the severity and frequency of histopathologic changes (Jonckheere, 1954). First, animals were sorted and assigned ranks within each of these categories. In the case of ranking ties, the mean rank of the animals involved was assigned to each of these animals. NTI for each treatment was calculated as the mean of the three ranks of animals in each of the three parameters (Abou-Donia et al., 1985a).

Coneurotoxicity Coefficient (CNC)

Joint neurotoxic action for two or more chemicals was determined using the following equation:

$$CNC = \frac{\text{experimentally determined NTI for chemicals } 1, 2, \ldots, n}{NTI_1 + NTI_2 + \cdots + NTI_n}$$

where NTI_1, NTI_2, and NTI_n are neurotoxicity indices for chemicals 1, 2, and n, respectively (Abou-Donia et al., 1985a).

When the CNC value is larger than 1, a more than additive response (synergism or potentiation) occurs. When the CNC is less than one, a less than additive response (antagonism) is indicated. A CNC value of one demonstrates additive effect.

Abou-Donia et al. (1985b) reported that simultaneous subchronic (90 days) exposure to vapors of the weak neurotoxicant *n*-hexane and the non-neurotoxic solvent methyl isobutyl ketone (MiBk) markedly increased the neurotoxic action of *n*-hexane in hens. While continuous inhalation of 1000 ppm *n*-hexane produced mild neurotoxicity, the same concentration of MiBk failed to cause any neurotoxic effects. In contrast concurrent exposure to 1000 ppm of *n*-hexane and 250, 500, or 1000 ppm MiBk led to increased neurotoxicity in a dose-dependent manner.

The increased neurotoxicity of *n*-hexane with MiBK depended upon MiBK concentration, duration of exposure, and length of the period between

TABLE 3.
Neurotoxicity Index of
n-Hexane and/or
MiBK in Hens

Concentration		
n-hexane	MiBk	NTI
0	1000	4.2
1000	0	6.2
1000	100	24.6
1000	500	28.1
1000	250	24.3
1000	1000	19.2

onset of neurotoxicity signs and termination of experiment. The extent of neurotoxicity was reflected in the time of onset of neurologic deficit, severity of the clinical condition, and histopathological alterations, all of which are encompassed in the neurotoxicity index. The NTI for various treatment with *n*-hexane and MiBk is listed in Table 3 (Abou-Donia et al., 1985a).

The coneurotoxicity index for simultaneous exposure to 1000 ppm each of *n*-hexane and MiBk was 2.3, indicating a more than additive response. Since MiBk is not a neurotoxic chemical, the joint neurotoxic action of MiBk and *n*-hexane is classified as synergistic. Since MiBk induces chicken hepatic microsomal P-450, it may synergize the neurotoxic action of *n*-hexane by enhancing its metabolic activation to the neurotoxicant 2,5-hexanedione. Studies have confirmed this conclusion (Habig et al., 1989).

The type of joint neurotoxic action of inhaled methyl butyl ketone (MBK; methyl *n*-butylketone: methyl isobutyl ketone, 7:3) and dermally applied *O*-ethyl *O*-4-nitrophenyl phenylphosphonothioate (EPN) was assessed in adult hens (Abou-Donia et al., 1985a). Concurrent exposure to EPN and MBK resulted in increased neurotoxicity than either chemical would produce individually. This joint neurotoxic action is considered potentiation, since both chemicals have considerable intrinsic neurotoxicity. Although both chemicals produce central-peripheral axonopathy resulting from Wallerian-type degeneration of the axon and myelin, their mode of action is different. MBK-induced neurotoxicity is characterized by the accumulation of 10-nm neurofilaments resulting in giant axonal swellings and axonal degeneration. On the other hand, EPN results in increased activity of CaM kinase II and enhanced phosphorylation of cytoskeletal proteins leading to dissociation of cytoskeletal elements and their precipitation into condensed proteins (Abou-Donia and Lapadula, 1989). Neuropathologic lesions resulting from these compounds have different distribution and are morphologically distinct. Since the molecular mechanisms are different, their potentiating effect could not be considered neurotoxicity-site interaction per se. It was postulated that the potentiation of EPN and MBK resulted from the metabolic activation of each neurotoxicant by the other chemical via induction of cytochrome P-450. EPN has been shown to produce about a 200% increase in hepatic microsomal P-450 (Lasker et al., 1982). This could lead to an increased metabolic activation of MBK to 2,5-hexanedione, which is a more potent neurotoxic chemical (Habig et al., 1989). Also, MBK may induce hepatic P-450 leading to more bioactivation of EPN to EPN oxone, a compound that is a more potent neurotoxicant than EPN. Another factor that may have contributed to the potentiating effect of concurrent exposure to EPN and MBK is the local trauma produced by these chemicals in nerve tissue that might increase vascular permeability. This in turn would enhance the entry from circulation of EPN and MBK or their active metabolites and thereby enhance their neurotoxic effects locally.

Statistics

In designing a protocol for neurotoxicity studies, it is essential to use a sufficient number of animals to allow statistically valid conclusions in comparing the response of test and control animals and to allow statistical extrapolation to larger populations. Statistical procedures are of two types: (1) parametric, such as t test and F-test, are restricted to data that have specific frequency distributions and (2) nonparametric, such as signed rank test and rank run test, are not based on any assumptions about distribution (Jonckheere, 1954).

The degree of confidence required and the extent of experimental variations will determine the number of animals that must be used to make statistically valid conclusions. Also, the uniformity of the test animals plays a role in determining the number of animals. Furthermore, experimental variations are very important in deciding the number of animals.

REFERENCES

Abou-Donia, M.B. (1978). Role of acid phosphatase in delayed neurotoxicity induced by leptophos in hens. *Biochem. Pharmacol.* 27: 2055–2058.

Abou-Donia, M.B. (1979). Pharmacokinetcis and metabolism of a topically-applied dose of O-4-bromo-2,5-dichlorophenyl O-methyl phenylphosphonothioate in hens. *Toxicol. Appl. Pharmacol.* 51: 311–328.

Abou-Donia, M.B. (1980). Metabolism and pharmacokinetics of a single oral dose of O-4-bromo-2,5-dichlorophenyl O-methyl phenylphosphonothioate (leptophos) in hens. *Toxicol. Appl. Pharmacol.* 55: 131–145.

Abou-Donia, M.B. (1981). Organophosphorus ester-induced delayed neurotoxicity. *Annu. Rev. Pharmacol. Toxicol.* 21: 511–548.

Abou-Donia, M.B. (1985). Biochemical toxicology of organophosphorus compounds. In *Neurotoxicology*. K. Blum and L. Manzo, Eds., Marcel Dekker, New York, pp. 423–444.

Abou-Donia, M.B., D.G. Graham, M.A. Ashry, and P.R. Timmons (1980). Delayed neurotoxicity of leptophos and related compounds: differential effects of subchronic oral administration of pure, technical grade, and degradation products on the hen. *Toxicol. Appl. Pharmacol.* 53: 150–163.

Abou-Donia, M.B., D.G. Graham, H.M. Makkawy, and K.M. Abdo (1983). Effect of subchronic dermal application of *O*-ethyl *O*-4-nitrophenyl phenylphosphono-thioate on producing delayed neurotoxicity in hens. *Neurotoxicology* 4: 274–260.

Abou-Donia, M.B. and D.M. Lapadula (1989). Cytoskeletal proteins and axonal transport. *Comm. Toxicology* 3: 427–444.

Abou-Donia, M.B., D.M. Lapadula, G.M. Campbell, and K.M. Abdo (1985a). The joint neurotoxic action of inhaled methyl butyl ketone vapor and dermally applied o-ethyl o-4-nitrophenyl phenylphosphonothioate in hens: potentiating effect. *Toxicol. Appl. Pharmacol.* 79: 69–82.

Abou-Donia, M.B., D.M. Lapadula, G.M. Campbell, and P.R. Timmons (1985b). The synergism of *n*-hexane-induced neurotoxicity by methyl *iso*-butyl ketone following subchronic (90 days) inhalation in hens: induction of hepatic microsomal cytochrome P-450. *Toxicol. Appl. Pharmacol.* 81: 1–16.

Abou-Donia, M.B., M.H. Makkawy, and G.M. Campbell (1985c). The pattern of neurotoxicity produced by dermal application of *n*-hexane, methyl *n*-butyl ketone, 2,5-hexanedione alone and in combination with *O*-ethyl O-4-nitrophenyl phenyl-phosphonothioate in hens. *J. Toxicol. Environ. Health* 16: 85–100.

Abou-Donia, M.B., H.M. Makkawy, A.E. Salama, and D.G. Graham (1982). The effect of ages of hen and their sensitivity to delayed neurotoxicity induced by a single oral dose of tri-*o*-tolyl phosphate. *The Toxicologist* 2: 178.

Abou-Donia, M.B. and D.B. Menzel (1968). The *in vivo* metabolism of DDT, DDD and DDE in the chick by embryonic injection and dietary ingestion. *Biochem. Pharmacol.* 17: 2143–2161.

Abou-Donia, M.B., M.A. Othman, and P. Obih (1989). Interspecies comparison of pharmacokinetic profile and bioavailability of (±)-Gossypol in male Fischer-344 rats and male B6C3F mice. *Toxicology* 55: 37–51.

Abou-Donia, M.B. and S.H. Pressig (1976). Delayed neurotoxicity of leptophos: toxic effects on nervous system of hens. *Toxicol. Appl. Pharmacol.* 35: 269–282.

Abou-Donia, M.B., B.L. Reichert, and M.A. Ashry (1983). The absorption, distribution, excretion and metabolism of a single oral dose of *O*-ethyl *O*-4-nitrophenyl phenylphosphonothioate in hens. *Toxicol. Appl. Pharmacol.* 70: 18–28.

Altman, P.L. and D.S. Dittmer, Eds., (1966). *Environmental Biology*. Federation of American Societies for Experiment Biology, Bethesda, MD, pp. 565–608.

Beuthin, P.K. and W.F. Bousquet (1970). Long-term variation in basal and pheno-barbital-stimulated oxidative drug metabolism in the rat. *Biochem. Pharmacol.* 19: 620–625.

Boyd, E.M. (1969). Dietary protein and pesticide toxicity in male weanling rats. *Bull. World Health Org.* 40: 801–805.

Boyd, E.M. (1972). *Predictive Toxicometrics*. Williams & Wilkins, Baltimore, p. 408.

Burgi, E. (1910). Die wirkung von narcotica-kombinationen. *Dtsch. Med. Wochenschr.* 36: 20–23.

Crosby, D.G. (1972). Environmental photooxidation of pesticides. In *Proceedings of a Conference on Degradation of Synthetic Organic Molecules in the Biosphere,* San Francisco, June 1971. National Academy of Science, Washington, D.C., pp. 260–278.

Davis, W.M. (1962). Day-night periodicity in pentobarbital response of mice and the influence of socio-physiological conditions. *Experientia (Basel)* 18: 235–237.

Doull, J. (1972). The effect of physical environmental factors on drug response. In *Toxicology—the Basic Science of Poisons.* MacMillan, New York, pp. 133–147.

Eto, M. (1974). Organophosphorus Pesticides: Organic and Biological Chemistry. CRC Press, Boca Raton, FL, p. 387.

Feldman, R.J. and H.I. Maibach (1970). Absorption of some organic compounds through the skin of man. *J. Invest. Dermatol.* 54: 399–404.

Finney, D.G. (1952). *Probit Analysis,* 2nd ed. Cambridge University Press, London.

Finney, D.G. (1971). *Probit Analysis,* 3rd ed. Cambridge University Press, London.

Habig, C., M.B. Abou-Donia, and D.M. Lapadula (1989). Cytochrome P-450 induction in chickens exposed simultaneously to *n*-hexane and methyl *iso*-butyl ketone. *The Toxicologist* 9: 194.

Hassan, A., F.M. Abdel-Hamid, and S.F. Mohammed (1977). Metabolism of ^{14}C-leptophos in the rat. *Arch. Environm. Contam. Toxicol.* 6: 447–454.

Hatch, T.F. and P. Gross (1964). *Pulmonary Disposition and Retention of Inhaled Aerosols.* Academic Press, New York, p. 184.

Holmstead, R.L., T.R. Fukuto, and R.B. March (1973). The metabolism of *O*-(4-bromo-2,5-dichlorophenyl) *O*-methyl phenylphosphonothioate (Leptophos) in white mice and on cotton plants. *Arch. Environ. Contam. Toxicol.* 1: 133–147.

Johnson, M.K. (1977). Improved assay of NTE for screening organophosphates for delayed neurotoxicity potential. *Arch. Toxicol.* 37: 113–115.

Jonckheere, A.R. (1954). A distribution-free *k*-sample test against ordered alternatives. *Biometrika* 41: 133–145.

Lowry, O.H., N.J. Rosebrouger, A.L. Farr, and R.J. Randall (1951). Protein measurement with the Folin phenol reagent. *J. Biol. Chem.* 193: 265–275.

LaDu, B.N., G. Mandel, and E.L. Way, Eds. (1971). *Fundamentals of Drug Metabolism and Drug Disposition.* Williams & Wilkins, Baltimore, p. 615.

Lasker, J.M., D.G. Graham, and M.B. Abou-Donia (1982). Differential metabolism of *O*-ethyl *O*-4-nitrophenyl phenylphosphonothioate by rat and chicken hepatic microsomes. *Biochem. Pharmacol.* 31: 1961–1967.

Maibach, H.I., R.J. Feldman, T.H. Milby, and W.F. Serat (1971). Regional variation in percutaneous penetration in man. *Arch. Environ. Health* 23: 208–211.

Matsumura, F. (1985). *Toxicology of Insecticides,* 2nd ed. Plenum Press, New York, pp. 19–30.

Murphy, S.D., R.R. Louwerys, and K.L. Cheeven (1968). Comparative anticholinesterase action of organophosphorus insecticides in vertebrates. *Toxicol. Appl. Pharmacol.* 12: 22–35.

Murphy, S.D. (1969). Some relationships between effects of insecticides and other stress conditions. *Ann. N.Y. Acad. Sci.* 160: 366–377.

McLean, A.E. and E.K. McLean (1969). Diet and toxicity. *Br. Med. Bull.* 25: 278–281.

Rentz, E. (1932). Zur systematic und nomenklatur der komination swirkungen. *Arch. Int. Pharmacodyn. Ther.* 43: 337–361.

Smith, M.I., E. Elvoe, P.J. Valer, W.H. Frazier, and G.E. Mallory (1930). Pharmacological and chemical studies of the cause of the so-called ginger paralysis. U.S. Public Health Rep. 45, 1703–1716.

Smyth, H.F. (1959). The toxicological basis of threshold limit values. I. Experience with threshold limit values based on animal data. *Am. Ind. Hyg. Assoc. J.* 20: 341–345.

Spencer, P.S. and H.H. Schaumburg (1980). *Experimental and Clinical Neurotoxicology.* Williams & Wilkins, Baltimore, pp. 456–475.

United States Envrionmental Protection Agency (1985). Registration of pesticides in the United States. Proposed guidelines. Subpart G—Neurotoxicity. *Fed. Reg.* 50: 39458–39470.

Whitacre, D.M., M. Badie, B.A. Schwemmer, and L.I. Diaz (1976). Metabolism of ^{14}C-leptophos and ^{14}C-4-bromo-2,5-dichlorophenol in rats: a multiple dosing study. *Bull. Environ. Contam. Toxicol.* 16: 689–696.

Williams, R.T. (1967). Comparative patterns of drug metabolism. In *Proceedings of an International Symposium on Comparative Pharmacology. Fed. Proc.* 26(4): 1029–1046.

Zbinden, B. (1963). Experimental and clinical aspects of drug toxicity. *Adv. Pharmacol.* 2: 1–112.

Zbinden, G. (1973). *Progress in Toxicology: Special Topics.* Springer-Verlag, New York, 88 pp.

Neurobehavioral Toxicology

Hugh A. Tilson
Neurotoxicology Division
U.S. Environmental Protection Agency
Research Triangle Park, North Carolina

G. Jean Harry
Systems Toxicity Branch
National Institute of Environmental Health Sciences
Research Triangle Park, North Carolina

INTRODUCTION

Importance of the Central Nervous System

The central nervous system (CNS) receives and integrates input and then responds to maintain bodily functions. The complex interaction of the nervous system with other organ systems suggests that it should be highly vulnerable to the deleterious effects of chemical and physical agents. The actual measurement of CNS dysfunction is made difficult since there are so many functions that might be assessed and there are few commonly accepted guidelines as to how to correlate many neurobehavioral changes with specific histopathological or neurochemical alterations.

There are numerous examples underscoring the importance of functional alterations following exposure to agents with known neurotoxicity. Accidental exposure to agents in the environment or workplace to a wide variety of agents, including heavy metals, pesticides, and solvents is known to have deleterious effects on sensory, motor and/or cognitive processes of humans.

At the present time, the number of chemicals causing behavioral or neurological alterations is not known. Anger (1984) reported that of 588 chemicals listed by the American Conference of Governmental Industrial Hygienists in their 1982 publication on threshold limit values for chemical substances and physical agents in the workplace, 29% or 167 have threshold limit values based in some manner on direct neurological or behavioral effects. It has also been estimated by the National Academy of Sciences that inadequate toxicological information exists for hazard assessments on 78% of the nearly 13,000 commercial chemicals with production volumes in excess of 1 million pounds per year. If it is assumed that one fourth of these chemicals may produce neurotoxicity, then there are an estimated 2500 chemicals being produced at approximately 1 million pounds per year with undefined neurotoxic potential. Moreover, the actual incidence of problems associated with exposure to low levels of neurotoxicants is not known since many of the signs and symptoms are vague, ambiguous, or subjective. As concluded in a recent report on neurotoxicology by the Office of Technology Assessment (1990) there is a reasonable possibility for subtle and uncharacterized neurotoxic effects in the population.

Rationale for Use of Behavioral Procedures in Toxicology

Behavioral procedures derived largely from experimental and behavioral psychology have been used in pharmacology and toxicology for several years (Evans and Weiss, 1978; Norton, 1982; Alder and Zbinden, 1983; Tilson and Mitchell, 1984; Tilson, 1987, 1990a; Cory-Slechta, 1989; Moser, 1989). Behavior has been defined as the net sensorimotor and integrative processes occurring in the nervous system and it follows that an alteration in behavior might be a relatively sensitive indicator of exposure. Thus, under some circumstances, the dose effect curve for some behavioral alterations might lie to the left of some endpoints of toxicity.

Behavioral endpoints are also important in toxicological studies since they are generally noninvasive and can be used to assess subjects repeatedly during the course of chronic exposure. This may be crucial if exposure to a chemical produces a subtle loss of brain capacity, such as that seen normally with aging. It has been suggested that if the usual age-related loss of neurons was accelerated by a rate as low as 0.1% per year due to repeated exposure to a chemical, then significant deterioriation might occur, particularly if such exposure were to begin at a relatively young age and continued for several years (Evans and Weiss, 1978).

Obviously, functional indicators might be useful in detecting and characterizing effects of agents that act directly on the nervous system. However, the use of behavioral procedures need not be limited to the study of specific neurotoxicants; changes in behavior might also be useful endpoints in the study of toxicants that affect organs other than the nervous system.

The purpose of this Chapter is to define the general concepts of behavioral analysis and discuss behavior within the context of its usage in neurotoxicology. Several procedures will be described and examples of how these methods have been used to study neurotoxicity will be presented. Furthermore, several critical issues related to the use of behavioral techniques in toxicology will be discussed.

CLASSIFICATION OF BEHAVIORAL APPROACHES

Definition of Behavior

Behavior has been defined as the movement of an organism or its parts within a temporal and spatial context (see Tilson and Harry, 1982). Thus, behavior is thought to be comprised of units called responses, which covary with effective controlling variables called stimuli. The functional analysis of behavior is concerned with the relationship between stimuli, behavior and the consequences of the behavior in the environment. Behavioral responses may be divided into two categories, including respondent (elicited) and operant (emitted) as summarized in Table 1. In addition, responses may also be unlearned (unconditioned) or learned (conditioned). Responses show specific physical properties, such as topography, rate, force or latency, all of which are the dependent variables measured in behavioral experiments.

Respondents are elicited by a known environmental stimulus, usually one with a specific temporal relationship to the occurrence of the response. The frequency of the response depends primarily upon that of the eliciting stimulus. Examples of unconditioned respondent behaviors include kineses, taxes, reflexes and species-specific behaviors. Unlearned operant or emitted responses are not elicited by a single, identifiable, temporally-cued stimulus in the environment. These responses occur within the context of many environmental stimuli, but there is no single eliciting stimulus, as in the case of respondent behaviors. Horizontally directed exploratory motor activity is an example of an unconditioned operant response.

Most behavior is modified by learning. Behaviorists have identified two types of learning, respondent (classical) and operant (instrumental). Respondent or classical conditioning refers to a set of operational procedures in which there is the approximate simultaneous presentation of two stimuli; one of the stimuli belongs to a genetically determined stimulus-response relationship such as a reflex. With repeated pairing of the two stimuli, there is an increase in the strength of another reflex, the conditioned reflex, composed of a response resembling the one in the original reflex and elicited by the originally neutral stimulus. An example of a classicaly conditioned response is the conditioned withdrawal reflex. If a brief shock is applied through elec-

TABLE 1. Classification of Behavior

A. Unconditioned (unlearned) behavior
   1. Respondent
      a. Elicited by known observable stimulus
      b. Responses typically include those of smooth muscles, glandular secretions, autonomic responses, environment-elicited effector responses
      c. Data are measures of response magnitude, probability, latency, or related to intensity of eliciting stimulus
      d. Taxonomy of respondents include the following:
         (1) Kinesis, environment-directed, and movement is random
         (2) Taxis, stimulus-directed, and movement is specific response of whole organism
         (3) Reflex, object-directed, and movement involves specific effect or system
         (4) Species-specific, stimulus-specific, and movements are sequences of behaviors (fixed-action patterns)
   2. Operant
      a. Emitted, with no known observable eliciting stimulus
      b. Responses typically include those mediated by CNS, such as skeletal muscular movements that operate on and change the environment
      c. Data are measures of response probability or frequency
B. Conditioned behavior
   1. Classically conditioned (respondent or type S learning): response (CR) is elicited by a new stimulus (CS) as the result of close temporal pairing of that stimulus (CS) with another stimulus (US), which originally elicited the response (UR)
   2. Instrumentally conditioned (operant or type R learning): response (R) changes in frequency of occurence as a function of the response consequence (SR)

trodes to one limb of a restrained animal, a reflexive withdrawal response is elicited. If the onset of a light repeatedly precedes the presentation of the shock, the light eventually comes to elicit a conditioned escape response (limb withdrawal).

Operant or instrumental conditioning involves the pairing of a response with a stimulus. When the occurrence of an operant response is followed by the presentation of a reinforcing stimulus, the probability of recurrence of the response increases. A reinforcer is any stimulus that increases the probability of a response. An example of an instrumentally conditioned response is a lever press by a food-deprived rat for a reinforcement of a drop of milk. By the process of operant conditioning, a relatively low probability response such as pressing a lever increases in frequency following the presentation of the milk reinforcer.

Primary and Secondary Behavior Tests

One method of classifying behavioral procedures is according to their desired use in the experiment (Tilson and Mitchell, 1984). Techniques in-

TABLE 2.
Some Behavioral Endpoints Quantifiable in Animals

Motor	Activity changes
	Incoordination
	Weakness and paralysis
	Abnormal movement and posture
	Tremor
	On-going performance
Sensory	Primary sensory deficits (auditory, gustatory, olfactory, visual, somatosensory)
	Pain
	Equilibrium disorders
Arousal or reactivity	Increased irritability or reactivity; change in CNS excitability
Cognitive	Associative and nonassociative learning; spatial learning
General	Performance changes
	Reproductive behavior
	Consummatory

tended to measure the presence or absence of an effect are usually different from those used to assess the degree of toxicity or the lowest level required to produce an effect. Screening procedures designed to allow large numbers of animals to be tested may not require extensive training of the animals and are relatively simple to perform. However, these techniques are frequently labor intensive, often require subjective (unautomated) measures, generally yield quantal data, and may not be as sensitive to subtle effects as other tests.

Some tests may be designed to determine more definite or precise endpoints and are usually employed in studies concerning mechanism of action or the estimation of the least effective dose. These procedures are sometimes called secondary tests and may require special behavioral equipment, pretraining of experimental animals, or the use of motivational factors such as negative reinforcement or food deprivation. Secondary procedures are usually automated and generate graded or continuous data making them amenable to repeated measures, experimental designs and parametric data analyses.

Tests Based Upon Function

Another way of classifying behavioral procedures is based upon the neurobehavioral functions that might be affected by exposure to a neurotoxicant, i.e., chemicals can affect a wide range of behavioral and neurological functions, including motor, sensory, cognitive, CNS excitability and general changes (Damstra, 1978; Anger, 1984; Weiss, 1985; Kulig, 1989; Mattsson et al., 1989). Table 2 contains a summary of behavioral and neurological signs observed in humans that can be measured in animals. Behavioral toxicologists have tended to select procedures capable of detecting and quantifying some aspect of these behavioral and neurological functions.

Apical Tests

Frequently, behavioral toxicologists employ several tests in an attempt to generate a profile of effects permitting a more accurate characterization of a chemical's activity. In some cases, a single test requiring the successful integration of intact subsystems, sometimes referred to as an apical test, is appropriate (Butcher, 1976). An example of an apical test is performance on an operant schedule of reinforcement. Such a procedure typically uses intermittent reinforcement of a defined response and establishes a dependency between the occurrence of a specific response such as a lever press and the presentation of a specific stimulus such as food. Deficits in operant responding produced by exposure to a chemical may be due to alterations in any one or more neurobehavioral functions (i.e., sensory, motor, motivational, associative). Cabe and Eckerman (1982) have characterized the determination of chemical effects on learning as "apical," i.e., an effect on learning might account for observed alterations on behavior mediated by nonassociative processes.

EXAMPLES OF NEUROBEHAVIORAL TESTS

Batteries of Tests

Frequently, behavioral procedures are used in a battery of tests to assess chemical-induced alteration in neurological functioning. The observation and recording of neurotoxic signs have been commonly used in pharmacology. In the description of frogs and mice exposed to various substances, Fuehner (1932) listed many neurobehavioral effects including insecure gait, vocalization, myoclonic bursts, and numbness. Neurobehavioral checklists used in acute toxicity testing have been used for many years (Boyd, 1959; Zbinden, 1963; Balazs, 1970; Marshall and Teitelbaum, 1974; Moser, 1989; Haggerty, 1989). For the assessment of psychoactive, neurological and autonomic signs, Irwin (1968) proposed a battery of tests designed to detect chemical-induced alterations in sensorimotor function. Marshall and colleagues (1974) described a battery of tests to assess toxicant-induced alteration in vision, audition, pain perception and olfaction.

Evans and Weiss (1978) described a three-tiered approach initially consisting of observational assessments of toxicity such as ratings of locomotor impairment, the presence or absence of tremor, ptosis, and convulsions, alterations in various reflexes and autonomic dysfunction. Preliminary tests

TABLE 3.
Summary of Measures in a Function Observational Battery,
and the Type of Data Produced by Each

Home Cage and Open Field	Interactive
Posture (D)	Ease of removal (R)
Convulsions, tremors (D)	Handling reactivity (R)
Palpebral closure (R)	Palpebral closure (R)
Lacrimation (R)	Approach response (R)
Piloerection (Q)	Click response (R)
Salivation (R)	Touch response (R)
Vocalizations (Q)	Tail pinch response (R)
Rearing (C)	Righting reflex (R)
Urination (C)	Landing foot splay (I)
Defecation (C)	Forelimb grip strength (I)
Gait (D,R)	Hindlimb grip strength (I)
Arousal (R)	Pupil response (Q)
Mobility (R)	
Stereotypy (D)	
Bizarre behavior (D)	

Note: D, descriptive data; R, rank order data; Q, quantal data; I, interval data; C, count data.

were followed by secondary tests to assess specific sensory and motor functions. Gad (1982) has also proposed a series of tests utilizing simple measurements of sensorimotor function. Alder and Zbinden (1983) have recently proposed a neurobehavioral checklist for use in toxicological studies in rats. Their battery of tests is similar to those suggested by Irwin (1968) for the mouse. Furthermore, the tests are selected for use in acute and repeated-dose toxicity experiments and designed for cage side observations during the course of exposure. Several laboratories utilize a battery of neurobehavioral tests in screening for neurotoxicity (Haggerty, 1989; Kulig, 1989; O'Donoghue, 1989; Moser, 1989). Table 3 summarizes measures characteristic of those used in a functional observational battery.

Pavlenko (1975) proposed three phases in a battery of testing. First, simple methods are used to assess changes in orienting, defensive and other reflexes. The second phase is intended to determine the threshold and subthreshold amounts of chemical necessary to affect higher order nervous system function such as conditioned reflexes. In the last phase of assessment, functional stress tests are recommended to determine minor or latent compensatory alterations and to study mechanisms of toxicity.

Reiter and colleagues (1981) proposed the development of a behavioral toxicity index based upon the acute LD_{50} and experimentally derived ED_{50} values for a variety of behavioral tests, including motor activity in a figure-eight maze, schedule-controlled operant behavior, conditioned taste aversion and activity in a radial arm maze. Mitchell and Tilson (1982) proposed a battery of tests chosen to evaluate a wide range of neurobehavioral functions, from simple reflexes to more complex processes such as sensory and motor function, changes in reactivity and associative processes. Such a battery has been used by the National Toxicology Program to assess several chemicals (Tilson, 1990b).

In summary, there is general agreement that behavioral endpoints, unconditioned or conditioned respondents or operants, may be used to measure neurotoxicity in a sequential testing. At the screening level, single or apical tests may be very useful as early indicators of neural dysfunction, while more specialized tests used to determine the nature of any observed effects or the level of exposure at which they occur.

Tests of Motor Function

Motor function can be affected in several ways following exposure to neurotoxicants. Table 4 summarizes representative neuromuscular defects known to be produced by some chemical agents.

Spontaneous Motor Activity

As reviewed by Reiter and MacPhail (1979) and Rafales (1986), this behavior has been used extensively in behavioral toxicology both as a measure of motor dysfunction and as an apical test. Movement within the living space or environment occurs at a relatively high frequency and appears to be sensitive to the effects of chemicals. Motor activity is a complex behavior consisting of a variety of motor acts, such as sniffing, grooming, rearing, and ambulation. Since activity is not a singular measure, changes in the frequency of this behavior could reflect toxicant-induced changes in any one or more sensorimotor functions, arousal, or motivational states.

Many types of devices have been designed to measure activity (Reiter and MacPhail, 1979), but the figure-eight maze has been used extensively and successfully to measure toxicant-induced changes in behavior. In general, this apparatus is typically used to evaluate the effects of a chemical as measured during a relatively brief period of time (minutes or hours), although it can be used to measure diurnal activity changes (Reiter, 1983). MacPhail et al. (1989) have recently reviewed the use of automated motor activity measurements with regard to a wide range of chemical exposure and issues such as sensitivity, reliability, efficiency, and, to some extent, specificity.

Other investigators have used computer-assisted techniques in order to continuously monitor spontaneous locomotor patterns of activity continuously. For example, Elsner et al. (1979) reported that methyl mercury decreased activity during the night portion of the diurnal cycle of rats. The importance of naturally occurring cycles to the expression and measurement

TABLE 4. Examples of Tests For Neurotoxicity

Function	Procedure	Representative Agents
Neuromuscular		
Weakness	Grip strength; swimming endurance; suspension from rod; discriminative motor function, hindlimb splay	*n*-Hexane, methyl butylketone, carbaryl
Incoordination	Rotorod, gait measurements	3-Acetylpyridine, ethanol
Tremor	Rating scale, spectral analysis	Chlordecone, type I pyrethroids, DDT
Myoclonia, spasms	Rating scale, spectral analysis	DDT, type II pyrethroids
Sensory		
Auditory	Discriminated conditioning; reflex modification	Toluene, trimethyltin
Visual toxicity	Discriminated conditioning	Methyl mercury
Somatosensory toxicity	Discriminated conditioning	Acrylamide
Pain sensitivity	Discriminated conditioning (titration); functional observation battery	Parathion
Olfactory toxicity	Discriminated conditioning	3-Methylindole, methylbromide
Learning/Memory		
Habituation	Startle reflex	Diisopropylflurophosphate (DFP)
Classical conditioning	Nictitating membrane	Aluminum
	Conditioned flavor aversion	Carbaryl
	Passive avoidance	Trimethyltin, IDPN
	Olfactory conditioning	Neonatal trimethyltin
Operant or instrumental conditioning	One-way avoidance	Chlordecone
	Two-way avoidance	Neonatal lead
	Y-maze avoidance	Hypervitaminoisis A
	Biel water maze	Styrene
	Morris water maze	DFP
	Radial arm maze	Trimethyltin
	Delayed matching-to-sample	DFP
	Repeated acquisition	Carbaryl
	Visual discrimination learning	Lead

of motor activity has been observed by many investigators. For example, Ruppert et al. (1982) exposed adult rats to various doses of trimethyltin and measured activity in the figure-eight maze at hourly intervals for 23 h. On days 49–51 after an acute exposure to trimethyltin, animals were hyperactive during all phases of the diurnal cycle (Figure 1); however, this was not pronounced during the night portion of the cycle. In another study indicating

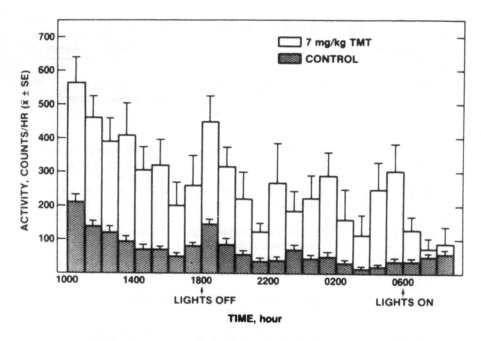

Figure 1. Figure-eight maze activity for hourly intervals during a 23-h test for control (N = 10) and 7 mg/kg trimethyltin-dosed rats (N = 9). Values are mean ± S.E. for photocell counts. This figure illustrates diurnal cyclicity normally present in rats over a 24-h period and that trimethyltin augmented motor activity normally occurring during both portions of the cycle. (Reprinted with permission, Ruppert et al., 1982.)

that the diurnal cycle may either mask or possibly reveal the presence of toxicity, offspring of pregnant mice exposed to 3,4,3',4'-tetrachlorobiphenyl during gestation were observed to be hyperactive and exhibit a circling or spinning syndrome (Tilson et al., 1979). Analysis of the motor activity exhibited in their home cage revealed that the affected mice were hyperactive during the nocturnal portion of the light cycle, but exhibited little or no hyperactivity during the normally equiescent day portion of the cycle.

Tests of Motor Coordination

Many procedures have been used to assess chemical-induced alterations in motor coordination, including negative geotaxis (Pryor et al., 1983), rope climbing (Carlini et al., 1967) and performance on an inclined plane (Graham et al., 1957). One of the more frequently used tests to quantify motor dysfunction in rodents is performance on a rotating rod. As reviewed by Bogo et al. (1981), numerous variations of this task have been used to assess agents for toxicological effects. For example, Bogo et al. (1981) used the rotorod to measure the cumulative toxicity of acrylamide, an agent known to produce peripheral neuropathy associated with muscle weakness. These

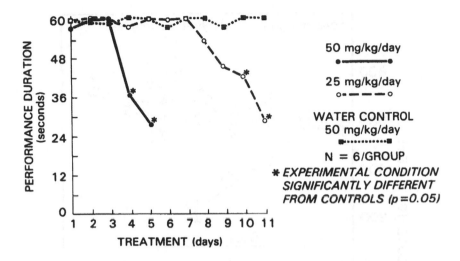

Figure 2. Cumulative acrylamide effects on rotorod performance in rats (N = 6/group). Subjects were dosed daily after testing until an effect was seen. This figure illustrates the use of the rotorod procedure to measure motor dysfunction in rats. (Reprinted with permission, Bogo et al., 1981.)

investigators dosed rats with 25 or 50 mg/kg of acrylamide, measured performance on the rotorod daily, and showed that rats receiving larger doses of acrylamide lost capability to perform on the rotating rod faster than those receiving the lower dose (Figure 2). The rotating rod test is used widely in pharmacology and is regarded by many as convenient technique in toxicological experiments (Alder and Zbinden, 1973).

Tests for Weakness and Paralysis

One of the earlier indicators of exposure to many neurotoxicants and psychopharmacological agents is muscle weakness or decrements in grip strength and various tests have been devised to assess this effect. These tests include valuation by performance and/or endurance (Kniazuk and Molitor, 1944; Klaus and Erdmann, 1978). Bhagat and Wheeler (1973) and suspension from a horizontal rod (Molinergo and Orsetti, 1976). Meyer et al. (1979) described a procedure that uses mechanical strain gauges to measure strength in the fore- and hindlimbs of both rats and mice. The fore- and hindlimb grip test appears to be sensitive to agents that produce axonopathopies or act acutely as muscle relaxants. For example, Pryor et al. (1983) found that repeated dosing with acrylamide, a chemical known to produce muscle weakness and decreased grip strength in humans, produced dose-related decreases in grip strength (Figure 3). Hindlimb grip strength was affected to a greater extent than forelimb grip strength, an observation consistent with the known pathological progression of this chemical.

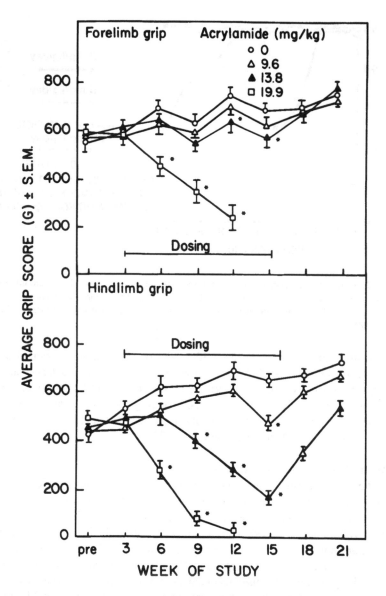

Figure 3. Effects of repeated exposure to acrylamide on fore- and hindlimb grip strength. Testing occurred every 3 weeks during dosing and 3 and 6 weeks after cessation of exposure. The date are means ± grip scores (grams) for 9 rats/group. This figure represents the use of a relatively simple device to measure neuromuscular strength in rats exposed to neurotoxicants. (Reprinted with permission, Pryor et al., 1983.)

Abnormal Movement and Posture

Neurotoxicants frequently result in abnormal posture and gait. Several techniques have been devised to quantify these neurological signs and they are summarized by Jolicoeur et al. (1979). These investigators described a battery of tests, including locomotor activity, assessment of catalepsy and rigidity, hindlimb splay, analysis of gait and reflex analysis. Of the tests described, two appeared to be relatively sensitive indicators of the effects of agents known to produce peripheral neuropathy or ataxia such as acrylamide and 3-acetylpyridine.

Another frequently used measure of neuromuscular dysfunction, particularly that produced by agents that produce "dying-back" axonopathies, is based on the observation that exposure to such agents usually results in a splaying of the hindlimbs. This neurological response can be quantified in rats by inking the hindpaws and either dropping them from a constant height onto absorbent paper or quantifying gait characteristics (stride width, length, and angle between steps) as the animals walk (Lee and Peters, 1976). Edwards and Parker (1977) used the hindlimb splay procedure to determine the onset of acrylamide-induced neurological dysfunction. Schallert et al. (1978) used gait analysis to measure motor disturbances produced by 6-hydroxydopamine, a cytotoxicant that, when administered directly into certain brain sites, produces degeneration of catecholaminergic pathways that can control certain components of movement.

Measures of Tremor

Many neurodegenerative diseases, psychopharmacological agents and neurotoxicants are characterized by signs of tremor (Stein and Lee, 1981). Several procedures have been developed to quantify tremor following exposure to environmental agents (Gerhart et al., 1985). In animals, tremor and stereotypic behaviors are frequently assessed using qualitative or semi-quantitative rating scales. In some cases, attempts have been made to automate tremor produced by chemicals using transducers to generate a signal that can be analyzed using routine statistical procedures. Gerhart et al. (1985) described a procedure in which a tremulous rat is placed on a platform attached to load cell and the analog output generated by whole body tremor is quantified by a spectral analyzer performing a fast Fourier analysis of the data. Using such a procedure, Gerhart et al. differentiated the power spectra generated by pharmacological agents such as oxotremorine and harmine and tremorigenic insecticides such as chlordecone. Recently, this procedure was used to determine the dose and time-related effects of various neurotoxicants (Gerhart et al., 1985; Hudson et al., 1985; Tilson et al., 1985; Newland, 1988). The data in Figure 4 show that the intensity of tremor in rats produced by the insecticide, p,p'-DDT, increases as a function of dose and time after administration of the chemical.

On-Going Performance

Techniques available from behavioral psychology have been used to assess the effects of chemicals on motor function or rats. For example, Falk (1970) gradually trained rats to press a lever, which was attached to a force

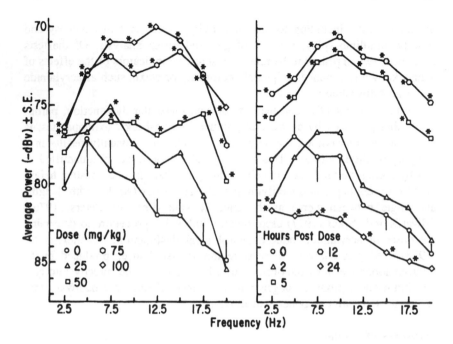

Figure 4. Dose- and time-related effects of p,p'-DDT on spectral analyzed movement of rats. Animals were given various amounts of p,p'-DDT and tremor measured 12 hours after dosing or given 75 mg/kg and tremor measured at various times after dosing. Data are means ± S.E. of power units derived from fast Fourier transformation of motor movements. This figure illustrates the use of a relatively simple procedure to detect and quantify tremor, which is a commonly observed neurological manifestation. (Redrawn from Hudson et al., 1985.)

transducer, within a specific range of forces (15–20 g) for a given duration (1.5 s) to obtain food reinforcement. Such a procedure has been used to study fine motor control in animals dependent upon ethanol (Samson and Falk, 1974). The use of schedule-controlled behavior in toxicology is discussed in greater detail in another section.

Tests For Sensory Effects

Alterations in sensory processes, such as paresthesias or visual or auditory impairments, are frequently among the first signs of toxicity reported by humans exposed to toxicants. Several attempts have been made to develop objective tests for sensory dysfunction in laboratory animals and include screening tests, reflex modification, and procedures based instrumental conditioning. Table 4 provides examples of chemicals known to affect sensory functions.

Screening Procedures

In addition to the tests described previously in the section on functional observational batteries, there have been several tests devised to screen for sensory deficits. For example, the "visual cliff" procedure, which measures whether or not an animal will choose to step onto a nearby platform or floor ("shallow" floor) as compared to one perceived to be farther away ("deep" floor), has been used to assess depth perception (Langman et al., 1975). Another simple test of visual function is the optokinetic drum, which relies on the optokinetic nystagmus or optomotor response. Although this response is believed to be a measure of visual acuity (Wallman, 1975), its validity in neurotoxicological studies has not been adequately demonstrated.

Other screening tests rely on orientation or other responses to the presentation of a stimulus. The acoustic startle response, for example, has been used to study the ototoxic effects of antibiotics (Harpur, 1974). The startle reflex consists of at least two components, sensory and motor, and additional experiments are required in order to ascertain specificity of effects. Several screening procedures adopted from psychopharmacology, such as the flinch-jump technique (Evans, 1961) and hot-plate procedure (Pryor et al., 1983) have been used to measure toxicant-induced changes in reactivity to noxious stimulation. For example, Walsh et al. (1984a) used the hot-plate procedure to study the relative toxicity of four organometal compounds. In these experiments, rats were given a single dose of trialkyltin or lead and changes in responsiveness to a 55°C hot-plate were measured. The results of these experiments showed that these chemicals produced a dose- and time-related decrease in sensitivity (increased latencies to respond) to the thermal stimulus.

Reflex Modulation

Recently, reflex modulation procedures have been used to assess sensory dysfunction. Presentation of an irrelevant stimulus prior to another stimulus that elicits a reflex can, under some circumstances, inhibit that reflex (Hoffman and Ison, 1980). It is important to note that stimuli capable of inhibiting the reflex are near the threshold and changes in intensity have marked effects upon reflex modulation. Likewise, the reflex modulation observed by the presentation of the irrelevant stimulus does not depend upon prior association with the eliciting stimulus and appears to be mediated at some level within the brainstem. Recently, this procedure, termed the prepulse inhibition paradigm, has been used to evaluate auditory processing (Marsh et al., 1978) and in toxicology and pharmacology to assess sensory alterations. Young and Fechter (1983) demonstrated that neomycin, an aminoglycoside antibiotic, resulted in a shift in the auditory threshold as measured by the prepulse inhibition procedure. Fechter and Young (1983) also studied the effects of triethyltin in the prepulse paradigm. By determining the intensity of the pure tone necessary to produce 15% inhibition of the reflex, that there was no

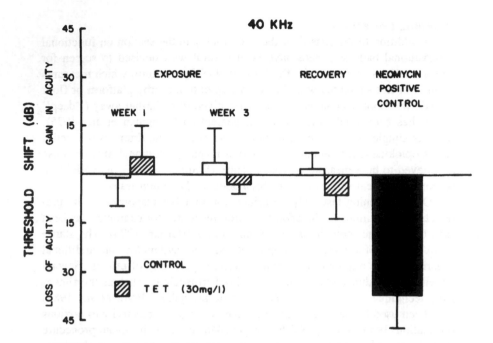

Figure 5. Shift in detection threshold for 40-kHz tone for control-exposed (clear bars) and triethyltin-exposed (striped bars) rats. Neomycin, an ototoxic antibiotic, was used as a positive control and its effect is seen on the righthand side of the graph. This figure illustrates the use of the prepulse procedure to detect hearing loss of an agent that produces loss of hair cells in the cochlea, while an agent such as triethyltin that produces primarily motor deficits has no apparent sensory effects. (Reprinted with permission, Fechter and Young, 1983.)

shift in detection thresholds for a 40 kHz tone during or after exposure to triethyltin, while there was a highly significant loss of acuity following neomycin (Figure 5). In spite of the lack of any effect of triethyltin or auditory acuity, tin-induced neuromuscular deficits such as hindlimb splay and weakness were observed. Wecker and Ison (1984) used the prepulse inhibition paradigm to show that alcohol does not affect loudness perception, but may disrupt temporal relationships with these primary auditory pathways. The use of prepulse inhibition techniques in behavioral toxicology has been reviewed recently by Crofton and Sheets (1989).

Instrumental Conditioning

Mazes and similar types of apparatus have been used to test for alteration in the performance of tasks based upon discrimination of sensory cues (Zenick et al., 1978; Winneke et al., 1977). Zenick et al. (1978) trained rats exposed to lead during development to escape on the basis of visual discrimination from a T-maze containing water. Lead-exposed rats showed an increased number of errors when brightness (black vs. white arms of the maze) or shapes (circles vs. triangles) were used as cues to the position of the arm containing an escape platform. Lead-induced alterations in the motor capability of the rats to perform the task were not evident.

Pryor et al. (1983) utilized a multisensory conditioned avoidance response to measure three sensory modalities concurrently in the same animal. Rats were trained to climb or pull a rope to escape and then to avoid a noxious footshock applied to the grids of the floor of the test chamber. By a process of training, the conditioned response is brought under the control of a tone (4 kHz), a low-intensity nonaversive current on the floor (0.125 mA) and a change in the intensity of the house light in the chamber. A quasipsychophysical curve was established for each modality before exposure to various toxic agents.

The most commonly used instrumental procedures are those derived from operant behavioral psychology. In such studies, animals are motivated by food or some other reinforcer to make a response only in the presence of specific stimuli. A graded stimulus-intensity response function curve can be generated by varying some parameter of the stimulus, such as intensity. Toxicant-induced effects on visual (Merigan, 1979), auditory (Chiba and Ando, 1976; Stebbins and Moody, 1979) and reactivity to electric shock (Weiss and Laties, 1961; Tilson and Burne, 1981) have been studied using operant procedures.

A good example of the use of operant procedures to study toxicant-induced changes in sensory thresholds is the study by Maurissen et al. (1983), who investigated the effects of acrylamide on somatosensory thresholds in monkeys. These investigators trained monkeys to make a response with one hand whenever a vibratory or small electrical stimulus was applied to the fingertip of the other hand. Marked changes in vibration sensitivity were noted during the course of repeated exposure to acrylamide, but sensitivity to the electric shock was not affected. These studies are important in that they demonstrate in an animal species the characteristic sensory deficits (loss of vibration sense) seen in humans exposed to neurotoxicants such as acrylamide.

Another example of the use of operant procedures to measure toxicant-induced changes in sensory processes is that of Stebbins and Moody (1979), who used trained animals to respond in the presence of a stimulus such as a tone. By changing the intensity of the tone at various frequencies, these investigators were able to determine psychophysical or auditory threshold curves for an individual animal. Figure 6 shows the effects of repeated dosing with an ototoxic aminoglycoside antibiotic. Kanamycin given daily caused progressive hearing loss beginning with higher frequencies and spreading to lower frequencies as time of exposure increased. Eventually, complete hearing loss accompanied by loss of hair cells in the cochlea was observed. Similar findings have been observed with other species using this procedure.

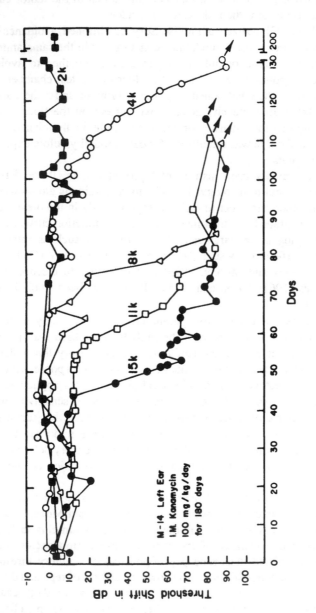

Figure 6. Progressive changes in threshold for a macaque monkey for different acoustic frequencies during and following daily kanamycin treatment. The zero line represents normal hearing at all frequencies prior to drug treatment. This figure illustrates the use of operant psychophysical procedures to measure sensory deficits in animals. (Reprinted with permission, Stebbins and Moody, 1979.)

One feature of many chemicals is that they are sometimes irritating. Such effects usually involve the skin, eyes, respiratory pathways and gastrointestinal tract. It has been proposed that behavioral procedures be used to estimate the potential for chemicals to produce irritation (Wood, 1979, 1981). In such procedures, experimental animals are trained to make a response, such as a nose poke to interrupt a photobeam, to terminate the presence of irritating agents from the test environment. This bioassay has been used to determine the relative irritability of several chemicals, including ozone, chlorine, toluene, acetic acid and ammonia.

Tests for Arousal or Reactivity

Frequently reported indicators of neurotoxicity are nervousness, irritability, "emotionality" and altered reactivity to environmental stimulation.

Startle Reflex

Changes in responsiveness to external stimulation such as noise or movement are often reported following exposure to toxicants. One way to quantify this effect is to measure the acoustic startle reflex. For example, Crofton and Reiter (1984) studied the effects of deltamethrin and cismethrin, type II and type I pyrethrins, respectively, on the acoustic startle response in rats. These investigators found that the effects of these two agents on startle responsiveness were different, i.e., cismethrin increased, while deltamethrin decreased acoustic startle responsiveness. The effect of these two agents on the augmentation of the startle reflex as a function of background noise (i.e., sensitization) was also studied. If sensitization of the startle response is calculated as the difference in response amplitude between 80 and 50 dB background noise, the two pyrethrins had significantly different effects on this behavioral process.

The ability of the startle reflex to measure changes in CNS function is further illustrated in Figure 7. Rats received either corn oil vehicle or 25 or 50 mg/kg of chlordecone, an organochlorine insecticide that produces behavioral hyperreactivity and tremor. Five hours after dosing, the rats were tested for the responsiveness to a 120 dB, 8 kHz tone over a 36 trial test period. Some rats were exposed to 70 dB white noise for 15 min prior to testing, while others were exposed to 90 dB prior to testing. Chlordecone produced a dose-related augmentation of the acoustic startle response under the 70 dB condition. Prior exposure to the 90 dB tone resulted in an apparent exacerbation of chlordecone's effect, suggesting that chlordecone interacted with the process of sensitization resulting from preexposure to the more intense background noise.

Change in Seizure Susceptibility

Exposure to toxicants frequently can alter the excitability of the nervous system and this can be quantified by measuring changes in seizure thresholds.

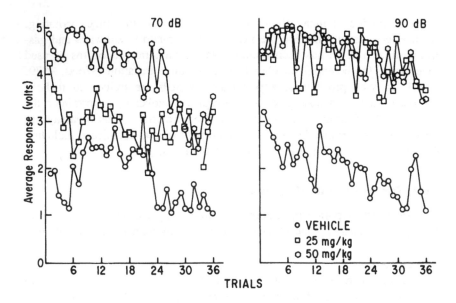

Figure 7. The effects of various doses of chlordecone on startle responsiveness of rats exposed to two different levels of background white noise prior to the experiment. Data are means of 10 rats for startle magnitude (volts) elicited by a 120 dB, 8 kHz tone given for 36 consecutive trials. This figure illustrates the use of the acoustic startle procedure to measure CNS excitability and the interaction between neurotoxicants and processes, such as habituation and sensitization.

For example, changes in threshold to produce maximal electroshock seizure (MES) have been demonstrated in rats exposed during development to lead (Fox et al., 1979). Dyer et al. (1982) have reported that rats exposed to trimethyltin were more sensitive to the effects of metrazol, suggesting a general increase in seizure susceptibility.

Another procedure used to study changes in nervous system excitability is "kindling," which is a procedure using repeated presentations of sub-threshold electrical stimuli applied to specific regions of the brain. "Kindling" is said to have occurred when seizures are elicited by presentation of a previously ineffective stimulus. Some toxicants, such as lindane and dieldrin, have been shown to be proconvulsants; that is, repeated exposure to low, nonconvulsive doses of lindane decreases the number of amygdaloid stimulations necessary to produce "kindling" (Joy et al., 1982).

"Emotionality"

Procedures that ostensibly measure "emotionality" of animals exposed to toxicants are many and varied (Archer, 1973; Alder and Zbinden, 1983). However, as pointed out by Alder and Zbinden, none of these techniques have been validated properly for use in toxicology and there is some question as to the actual "emotion" measured by such procedures (Barnett and Cowan, 1976).

Tests for Learning and Memory

Behavioral toxicologists have employed a wide variety of tests to assess chemical effects on associative and cognitive function in laboratory animals (Heise, 1984; Miller and Eckerman, 1986; Peele, 1989). The effects of chemicals on learning and memory must be inferred from the change in behavior following exposure relative to that prior to exposure. Acquisition may be defined as an enduring change in behavior, while memory can be defined as the preservation of the learned behavior over time. Measurement of changes in learning and memory must be separated from other kinds of changes in performance involving nonassociative processes. Thus, before learning and memory can be said to be affected, other controlling variables such as motivation, sensory or motor changes must be considered. Furthermore, apparent toxicant-induced changes in learning and memory should be demonstrated over a range of stimulus and response conditions. Table 4 contains examples of several chemicals with known effects on learning and memory.

Nonassociative Learning

The most simple form of learning is habituation which consists of a gradual decrease in the magnitude or frequency of a response following repeated presentations of a stimulus (Thompson and Spencer, 1966). A good example of a study on chemical-induced alterations in habituation is that of Overstreet (1977), who exposed rats to diisopropyl fluorophosphate (DFP), an irreversible inhibitor of cholinesterase. Rats were placed in a small cage to which was attached a small strain gauge. A 1 kHz, 100 dB tone was presented for 1 s for 42 trials per session, each trial separated by 90 s. Responding was determined prior to, as well as after drug administration, and in some experiments, multiple sessions were run with the same animals. Exposure to DFP had no apparent effect on the magnitude of the initial startle response, but DFP-exposed rats responded more than controls on subsequent trials, i.e., it decreased habituation. These investigators also showed that physostigmine, a reversible inhibitor of cholinesterase, had a similar effect, suggesting that the cholinergic system may be involved in the modulation of habituation processes.

Associative Learning

Classical Conditioning. As described in a previous section, classical conditioning involves the pairing of two stimuli to form a conditioned response. One classically conditioned response used in behavioral pharmacology and toxicology is the conditioned nictitating membrane response in the rabbit. In this procedure, a restrained rabbit is presented with a mild electric shock (unconditioned stimulus) to the skin of the paraorbital region after presentation of a tone or light conditional stimulus. The unconditioned response, movement of the nictitating membrane, eventually occurs following presentation of the conditional stimulus. Yokel (1983) dosed rabbits repeatedly with aluminum and found that treated rabbits learned the conditioned response less well than controls. One important feature of this experiment was the fact that baseline rates of the nictitating membrane extension (unconditioned stimulus-response reflex) was not affected prior to conditioning nor was sensitivity to the electric shock; these control manipulations tended to rule out aluminum-induced changes in sensory or motor function.

Another classically conditioned response used in toxicology is the conditioned taste aversion. If a rat ingests a novel substance and becomes ill, it will tend to ingest less of that substance in the future. This form of learning consists of pairing a novel stimulus, such as a sweet taste, with a toxic effect, which is the unconditioned stimulus. A single pairing of the conditioned and unconditioned stimuli frequently is sufficient to produce the conditioned response, long-lasting aversion has been suggested as a screen for toxicity (MacPhail, 1982; Riley and Tuck, 1985). In their review on the use of flavor aversions, Riley and Tuck (1985) conclude that most known toxins produce conditioned taste aversions, although there are still a number of criticisms of this procedure. For instance, there may be a possibility of a high incidence of false-positives as it pertains to toxicology; it is known that many psychoactive drugs not generally considered to be toxicants can produce flavor aversions.

The conditioned suppression method involves the presentation of a stimulus such as a mild electric shock which tends to disrupt or suppress ongoing behavior. Disruption of responding might be regarded as an unconditioned response and pairing a previously neutral stimulus with the shock eventually to elicit pausing by a process of classical conditioning. Although the procedure seems to be potentially useful in assessing the effects of chemicals on acquisition and/or retention of classical conditioning, it has not been used much in this context. In behavioral pharmacology, it has been used to evaluate the psychoactive properties of anxiolytics (Cook and Davidson, 1973) and in toxicology to assess chemicals on sensory processing (Chiba and Ando, 1976).

Instrumental Conditioning. As discussed in a previous section, instrumental or operant learning involves pairing of a response with a reinforcing event; learning involves making a response to obtain positive reinforcement such as food or removal or termination of aversive stimuli such as electric footshock.

  1. Procedures using negative reinforcement.
      a. *Passive avoidance.* A frequently used test of learning is the passive avoidance procedure. In this test, the experimental animal receives one or more training trials in which it is placed

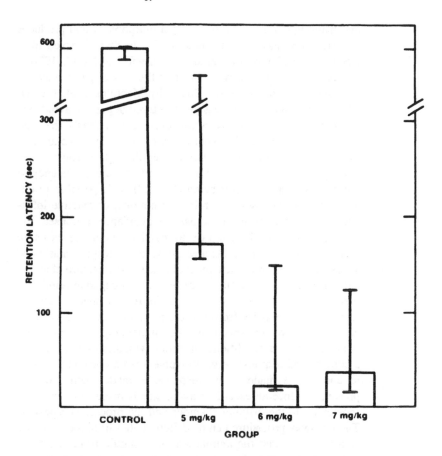

Figure 8. Effects of various doses of trimethyltin on 24-h retention latencies of a step-through passive avoidance task of rats. Bars represent the median retention latencies with interquartile ranges. This figure illustrates the use of the passive avoidance procedure to measure short term memory deficits in animals. (Reprinted with permission, Walsh et al., 1982.)

on a raised platform or in a lighted chamber. Within a period of time, the animal will step down from the platform or leave the lighted chamber to enter another chamber at which time a noxious stimulus is presented. At some later time, the animal is replaced upon the platform or in the lighted chamber; a longer step-down or step-through latency on the test trial compared to the initial trial is taken as an indication of learning. Walsh et al. (1982a) dosed rats with trimethyltin and 21 days later trained them in a passive avoidance step-through procedure. The tin-exposed rats showed significant impairment of retention when tested 24 h later (Figure 8). This observation is significant in that trimethyltin affects hippocampal morphology and produces an effect on passive avoidance similar to that observed after electrolytic lesioning of that area (O'Keefe and Nadel, 1978).

A significant improvement upon the typical passive avoidance paradigm has been developed by Mactutus et al. (1982), who reported a multiple measure step-through passive avoidance procedure to determine retention deficits in animals that had been exposed neonatally to chlordecone. By using several measures of responding (i.e., initial step-through latencies, frequency of vacillatory behaviors, as well as the usual latency to reenter the chamber after being shocked), it is possible to determine more precisely the role of emotional or reactive influences in the mediation of the passive avoidance response.

b. *Active avoidance*. In contrast to passive avoidance tasks in which an animal withholds a response to avoid presentation of a negative reinforcer, active avoidance tasks require that the animal perform a specific response to avoid negative reinforcement. One-way shock avoidance tasks require that the animal move unidirectionally from one chamber to another to avoid or escape negative reinforcement. The impending onset of shock is signalled by a conditional stimulus which is extinguished if a conditioned response (e.g., movement from one compartment to another) is made; if no response is made, then shock is presented. Once an escape or avoidance is made, the animal is replaced into the original chamber and the process is repeated. The one-way procedure can be differentiated from the two-way shuttle box in that the animal learns to shuttle from one compartment to another in order to escape or avoid negative reinforcement. Unlike one-way avoidance, the animals must learn to return to a compartment where they have been negatively reinforced. Sobotka et al. (1975) reported that rats exposed neonatally to lead acetate performed as well as controls on a one-way shock avoidance task, but displayed significant deficits when required to learn a two-way task.

Acquisition of avoidance responding is dependent upon many variables present in the test environment such as shock intensity, characteristic of the conditional stimuli and the configuration of the test chamber. For example, Tilson et al. (1982) varied the sizes of the two compartments used in the conditioning of a two-way shuttle box response in rats exposed previously to triethyllead. In this experiment, the sizes of the two compartments were different. Rats having received triethyllead performed significantly better over a 60 trial acquisition session and subsequent analysis found that the lead-exposed rats adapted

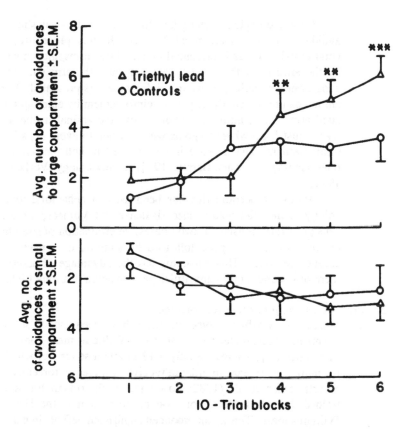

Figure 9. Effects of triethyllead on the direction of avoidance responding in a two-way shuttle box. Data are mean number of avoidance responses to either a large or small compartment per 10-trial blocks ± S.E. This figure illustrates the use of the shuttle box to measure effects of neurotoxicants on acquisition and performance in laboratory animals and the significance of the testing environment in the detection of such effects. (Reprinted with permission, Tilson et al., 1982.)

a strategy of retracing, i.e, starting from the smaller box, avoiding the larger box and then retracing back to the smaller box before the start of the next trial (Figure 9). This observation suggests that nonassociative factors may have contributed to the observed enhanced rate of acquisition in the lead-exposed rats. However, a subsequent experiment indicated that triethyllead did not change sensitivity to the electric footshock as measured by a flinch-jump procedure. As in the case of an impaired passive avoidance response, facilitation of a two-way shuttle response is sometimes taken as an indication of brain damage. The facilitatory effects of septal and hippocampal lesions on active avoidance are well known (King, 1958).

A more complex learning task than one- or two-way shuttle avoidance is the symmetrical Y-maze. In this procedure, a conditional stimulus is presented in one of two arms unoccupied by the animal. If the animal does not enter the arm with the conditioned stimulus, negative reinforcement is presented. The animal can terminate the negative reinforcement by entering the cued arm. Thus, unlike the shuttle box, the animal in the Y-maze must learn where to go as well as when to avoid negative reinforcement. The Y-maze has been used in both behavioral pharmacology (Ray and Barrett, 1975) and toxicology (Vorhees, 1974).

Another procedure that has been used in behavioral toxicology is the Biel water maze (Butcher and Vorhees, 1979). This procedure consists of two phases, determination of straight channel swimming speed followed by solution of a maze to escape the water. This apparatus has the advantage of assessment of general motor capabilities prior to actual learning trials.

2. Procedures using positive reinforcement.
    a. *Mazes.* Many different types of mazes have been used to study chemical effects on learning. The Hebb-Williams maze consists of a series of problems, usually a 12-problem sequence, which the animal must learn in order to receive positive reinforcement. Swartzwelder et al. (1982) dosed rats with trimethyltin and trained them to perform for food reinforcement in the Hebb-Williams maze. This agent produced significant deficits in maze performance (increases in number of total and perseverative errors); these effects were associated with marked thinning of the pyramidal cell fields in the hippocampus.

    Another maze used in behavioral toxicology is the radial arm maze (RAM). The RAM is a spatial learning task in which animals are required to recall a series of previously entered and nonentered feeding sites during a free-choice test session (Olton et al., 1979). In this test, the most effective response strategy is not to enter arms of the maze that have been previously entered and the food removed or arms in which food has never been present. The maze usually consists of a circular starting arena from which arms radiate outward. Performance in this maze is thought to demonstrate the existence of two memory components, working and reference memory. Walsh et al. (1982b) reported that trimethyltin-exposed rats were impaired in the performance of this task and that this effect was associated with damage to the hippocampus (Figure 10). Recently, Walsh et al. (1984b) reported that bilateral administration of AF64A,

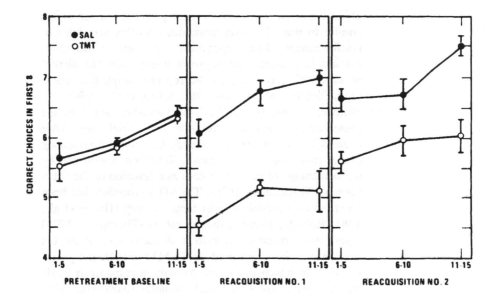

Figure 10. The effect of trimethyltin on performance of a radial arm maze. Data are mean number of nonrepeated feeder entries in the first eight choices. Animals were trained for 15 sessions, then received trimethyltin or vehicle and retested under different conditions. One interpretation of this data is that trimethyltin disrupted working, but not reference spatial memory. (Reprinted with permission, Walsh et al., 1982.)

a cholinergic neurotoxin, into the lateral cerebroventricles impaired RAM performance and that this effect was associated with cholinotoxicity in the hippocampus.

b. *Operant discrimination procedures.*

(i) *Repeated acquisition.* The procedure known as repeated acquisition of response chains (Thompson, 1973) requires that a series of problems be solved in which responding must occur in a particular order, which may vary from session to session. The procedure generates a pattern of within sessions acquisition which can remain stable for relatively long periods of time. For example, Schrot et al. (1984) trained rats on a repeated acquisition of behavioral chains procedure in which food reinforcement depended on the correct completion of a four member response sequence on three separate response levers. The animals were exposed to carbon monoxide periodically during the tests. These investigators found that carbon monoxide exposure resulted in pausing following completion of a sequence, but did not affect accuracy. Thus, exposure to carbon monoxide appeared to disrupt baseline performance, an effect similar to that of carbon monoxide on other schedules of reinforcement. Other experiments have shown that pharmacological agents and microwave exposure can disrupt accuracy of responding on the repeated acquisition paradigm (Schrot and Thomas, 1983; Schrot et al., 1980).

(ii). *Matched-to-Sample (MTS)*. This procedure involves the presentation of a sample stimulus for a brief time. After some delay the subject must identify the sample from among one or more comparison stimuli. Retention is indicated by the percentage of correct choices as a function of the delay interval (Wasserman, 1976). The MTS procedure has been regarded as a model of short term memory (Hogan et al., 1981; Roberts, 1976), although others (Thompson, 1978) regard the procedure to reflect stimulus control by the sample stimulus. Variations on the MTS include the go/no-go MTS in which a decision to make a response is made on the presentation of a sample stimulus (Konorski, 1957) and the continuous nonmatching-to-sample (Pontecorvo, 1983) in which a variable number of trials with one stimulus alternates with a variable number of trials of a second stimulus and the first response on a trial following a stimulus change is reinforced. The effects of various psychoactive drugs on MTS performance has been summarized by McMillan (1981).

In a recent series of studies on the effects of chemicals on MTS in pigeons, Idemudia and McMillan (1985a,b) exposed trained birds to trimethyltin and found a dose-dependent decrease in matching accuracy which was delay duration-dependent; the rate of responding was less affected. These data are consistent with the interpretation that the MTS to sample can assess short term memory since other investigators have also found that trimethyltin affects memory as assessed by other behavioral tests. Histopathological evidence also indicates that trimethyltin causes damage to the hippocampus which may be involved in processes related to learning the memory. Furthermore, Idemudia and McMillan (1985b) found that triethyltin affected performance, but not accuracy, of MTS performance; triethyltin has little specificity for the hippocampus and related regions.

Schedule-Controlled Operant Behavior

Responses are not usually reinforced on a one-to-one basis inside or outside the laboratory. Systematic arrangement of reinforcement contingencies following responses has evolved into descriptive classes called schedules of reinforcement. In general, there are four main types of schedules, including simple, compound, complex, and higher order, which define the relationship between response and reinforcement. A full description of the schedules can be found elsewhere (Ferster and Skinner, 1957).

Schedule-controlled operant responding has been used frequently in behavioral toxicology (Rice, 1988) because it is sensitive to a wide range of chemicals, including methyl mercury (Armstrong et al., 1963; Evans et al., 1975; Laties and Evans, 1980), solvents (Colotla et al., 1979; Glowa, 1985), pesticides (Anger and Wilson, 1980; Leander and MacPhail, 1980; Dietz and McMillan, 1979a,b; Bloom et al., 1983; MacPhail, 1985), acrylamide (Tilson et al., 1980; Rafales et al., 1982; Daniel and Evans, 1985), carbon monoxide (Geller et al., 1979; MacMillan and Miller, 1974), carbon disulfide (Levine, 1976) organic (Tilson et al., 1982) and inorganic lead (Barthalmus et al., 1977; Rice, 1985; Cory-Slechta and Weiss, 1985) and triethyl- and trimethyltin (Swartzwelder et al., 1981; DeHaven et al., 1982; Tilson and Burne, 1981; Wenger et al., 1984). Another reason for the use of schedule-controlled performance is that the experimental animal frequently serves as its own control. This aspect has several advantages including the potential to study a few animals extensively over relatively long periods of time. Such a strategy can be of value in attempts to determine the onset and recovery of an effect (Weiss et al., 1983). In addition, complex behavioral analysis may provide a more dependable basis for extrapolation to humans.

The multiple fixed-ratio, fixed-interval schedule of reinforcement has been widely used in pharmacology and toxicology because it generates different rates and patterns of responding within the same animal during the same testing session. For instance, responding under the fixed ratio component of the schedule is typified by a relatively continuous rate responding punctuated by short pauses following reinforcement. Responding under the fixed interval component is characterized by a low rate of responding after reward followed by an increase in response rate until reinforcement occurs. Pharmacological and toxicological agents have been shown to affect both rates and patterns of responding selectively. Swartzwelder et al. (1981) reported that a single dose of trimethyltin can have long lasting suppressant effects on fixed-ratio schedules of reinforcement, while triethyltin produces progressive decreases in the number of responses in rats trained on a fixed-ratio or fixed-interval schedule of reinforcement (DeHaven et al., 1982). In a recent report, Wenger et al. (1984) found that mice trained to respond under a multiple fixed-ratio 30, fixed-interval 600-s schedule of milk reward showed disrupted responding for up to 6–7 weeks after a single injection of trimethyltin. Figure 11 shows that the rate of responding on the FI component of the schedule was generally increased for several days after exposure to trimethyltin; FR response rate was initially decreased, but was near control rates after one week postinjection. This result shows that neurotoxicants can influence operant performance differentially depending upon the schedule of reinforcement.

Naturally Occurring Behaviors

One approach to the use of naturally occurring behaviors is to measure the repertoire of responses animals exhibit in their homecage environment. The frequency and patterning of motor activity and consummatory responses were quantified by Bushnell and Evans (1985) using a computer monitoring system. These investigators assessed diurnal patterns of feeding, drinking, locomotor activity and rearing in rats for up to two weeks after a single dose of trimethyltin. Food and water consumption decreased and increased, respectively, immediately after dosing and the diurnal patterns of drinking and rearing were also disrupted.

Silverman et al. (1981) recorded species-specific behaviors, such as exploration, sex-related activities, aggression and submission, of rats in an observation cage and measured toxicant-induced changes in the frequency of these behaviors. Silverman et al. reported that exposure to methyl mercury via the diet significantly affected these behaviors.

Another naturally occurring behavior is aggression, which is frequently mentioned in toxicology studies. Although aggression has operationally been defined as actual inflicting or threat to inflict damage (Sheard, 1977), actual quantification of the response can be somewhat subjective. Predatory aggression usually occurs without autonomic signs and is typically a spontaneous attack on some other species. Elicited aggression usually occurs with autonomic signs and is facilitated by environmental manipulations such as isolation or electric footshock. Dominance/submissive postures are frequently mentioned as indicators of aggression. Miczek (1974) has quantified this behavior in rats.

Special Application of Behavioral Procedures

Pharmacological Challenges

Pharmacological manipulations are used frequently in behavioral toxicology as a method of determining neurotoxicant-induced alteration in nervous system function. The basis for this approach is that exposure might alter the dynamic equilibrium of the nervous system, causing compensation

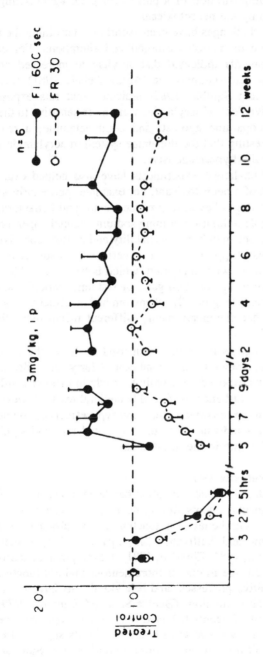

Figure 11. The rate of responding by mice on Fixed Interval and Fixed Ratio components of a multiple schedule following a single dose of trimethyltin. This figure illustrates that neurotoxicants may affect different components of operant-controlled responding differentially depending on the schedule of reinforcement and time after exposure. (Reprinted with permission, Wenger et al., 1984.)

by existing homeostatic mechanisms. These adaptive changes might be un-covered following administration of a pharmacological agent acting on the same system affected by the neurotoxicant.

Pharmacological challenges have been useful in determining the possible functional significance of neurotoxicant-induced alterations. For example, neurochemical experiments indicated that acrylamide increased dopamine receptor binding in the neostriatum of rats (Agrawal et al., 1981). Subsequent experiments (Tilson and Squibb, 1982) indicated that rats exposed to a behaviorally ineffective dose of acrylamide were more sensitive to the effects of apomorphine, a dopamine agonist, and d-amphetamine, a releaser of catecholamines, suggesting that the dopamine system in acrylamide-exposed animals was functionally hyperresponsive.

Pharmacological challenge experiments have also helped examine the suspected mechanism of a neurotoxicant. For example, permethrin and p,p'-DDT are two insecticides believed to produce tremor and behavioral hyper-excitability in the whole animal by holding sodium channels open once they are open, which increases membrane excitability and causes repetitive firing. This interpretation was supported by an experiment (Tilson et al., 1985) showing that pretreatment with phenytoin, which is believed to block repet-itive firing by binding to inactivation gates of sodium, markedly attenuated the effects of these agents (Figure 12). Phenytoin had no effect or exacerbated the effects of two other chemicals having different mechanisms than per-methrin or p,p'-DDT.

Pharmacological agents have also been used to detect the presence of toxicity not seen under other testing conditions (Harry and Tilson, 1982). For example, adult rats exposed neonatally to triethyltin showed little if any signs of neurological disturbance in several behavioral tests. However, these rats were found to be more sensitive to the stereotypic effects of apomorphine; subsequent experiments indicated that the tin-treated rats had significantly increased dopamine binding in the neostriatum.

Developmental Neurotoxicology

During the last decade, much attention has been drawn to the fact that exposure to environmental agents during development can result in long-lasting alterations in neurobehavioral function in the absence of physical malformations (Barlow and Sullivan, 1975; Spyker, 1975; Rodier, 1978; Adams and Buelke-Sam, 1981; Zbinden, 1981). Developmental neurotoxicity can be manifest as a change in one or more neurobehavioral functions, i.e., sensorimotor or cognitive processes, and behavioral tests have been used to detect alterations in these functions (Buelke-Sam and Kimmel, 1979). Table 5 contains a partial list of agents believed to have developmental neurotox-icity. There have been several tests or batteries of tests suggested to assess the consequences of developmental exposure (Barlow and Sullivan, 1975;

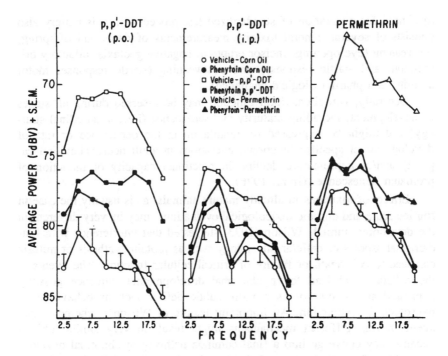

Figure 12. The effects of pretreatment with the anticonvulsant phenytoin on the tremorigenic activity produced by p,p'-DDT given orally or by i.p. injection and by permethrin, a pyrethroid believed to have a mechanism of action similar to that of p,p'-DDT. This figure illustrates the use of a pharmacological challenge to study the potential mechanism of a neurobehavioral effect produced by a neurotoxicant. (Reprinted with permission, Tilson et al., 1985.)

TABLE 5. Partial List of Environmental Agents Believed to Have Developmental Neurotoxicity

Alcohols	Methanol, ethanol
Antimitotics	X-radiation, azacyridine methyl-azomethanol
Insecticides	DDT, kepone, organophosphates
Metals	Lead, methyl mercury, cadmium
Polyhalogenated hydrocarbons	PCBs, PBBs
Solvents	Carbon disulfide, toluene

Spyker, 1975; Grant, 1976; Tesh, 1977; Nelson, 1978; Rodier, 1978; Buelke-Sam and Kimmel, 1979; Butcher and Vorhees, 1979; Dews and Wenger, 1979; Reiter et al., 1980; Adams and Buelke-Sam, 1981; Zbinden, 1981; Mactutus et al., 1982, 1984; Mactutus and Tilson, 1985). One good example is the Psychoteratogenicity Screening Test Battery for rats (Butcher and Vorhees, 1979; Vorhees, 1983), which includes measurement of body weights of the dams and offspring and several preweaning and postweaning tests. Recently, a collaborative behavioral teratology study was initiated in response to a need for evaluation of standardized test procedures. This battery also consists of several general toxicity measurements of dams and offspring, preweaning (eye opening, incisor eruption, negative geotaxis, olfactory orientation, and startle response) and postweaning (startle response, motor activity, and pharmacological challenge).

Generally, neurobehavioral function may be assessed during all stages of development, including maturity and senescence (i.e., longitudinal strategy) and might be expressed by retardation in the occurrence or rate of development of specific functions, alterations in adult neurobehavioral capacity and/or a premature decline in functional capacity or induction of premature senescence (Grant, 1976).

Effects of toxicants in already mature animals, as is usually the case in routine acute and chronic toxicological evaluations, may be very different in the developing animal. Wilson (1973) suggested that interference with fundamental aspects of cellular physiology such as protein synthesis or mitosis can lead to cell death or failure of critical cellular function. The events at the cellular level lead to an abnormal developmental sequence possibly expressed as lethality or as a more subtle deficit such as behavioral or neurological dysfunction. An important feature of Wilson's scheme is that agents having different mechanisms might result in very similar effects because they converge into a final common pathway of abnormal development. Therefore, the nature of the functional change might depend upon what portion of the brain is developing in the CNS at the time of exposure. Rodier (1979) has illustrated this principle well by showing that the type of behavioral effect produced by a specific agent such as X-radiation or exposure to an antimitotic drug can depend upon the embryonic stage development.

CONSIDERATIONS IN THE USE OF BEHAVIORAL PROCEDURES

Mechanism of Action

One problem concerning the use of behavioral tests in pharmacology and toxicology is that it is sometimes difficult to isolate the relative contributions of the various sensory, motor, arousal, or cognitive factors contributing to an observed change in behavior. For example, if a toxicant produces a decrease in the magnitude of the startle reflex to an acoustic stimulus, the conclusion that the chemical has affected hearing is only one of several possible interpretations of the data. The neurotoxicant may have altered the capacity of the animal to respond to the stimulus (motor weakness) or may have interfered with the acoustic startle reflex circuit lying in the brainstem. It is frequently necessary to conduct additional experiments or run appropriate control groups in order to define the behavioral mechanism of action.

Defining an Adverse Effect

In toxicology, there is considerable disagreement concerning the definition of an adverse effect whenever behavioral endpoints are employed. Many have advocated that any evidence of a behavioral change constitutes an adverse effect, while others require evidence of an irreversible decrease in performance or enhanced susceptibility to the deleterious effects of other environmental influences. The Office of Toxic Substances of the Environmental Protection Agency (1982) has defined neurotoxicity as any adverse effect on the structure or function of the central and/or peripheral nervous system related to exposure to a chemical substance. Furthermore, a toxic effect is defined as adverse change in the structure or function of an experimental animal as a result of exposure to a chemical substance.

Adverse effects are difficult to assess due to the inherent variability of biological data and an imprecise definition of "normal" limits of functioning. Discrimination between adverse and nonadverse effects requires an understanding of the importance of reversible changes and the ability to detect and quantify deviations from "normal."

Functional Reserve

Compensatory mechanisms inherent in a biological system may cloud the presence of toxicant-induced damage as assessed by functional tests. One approach to this problem might be to incorporate in test procedures one or more conditions in which the system or organism is placed under some type of stress. The combination of the test substance plus the stress might reveal the presence of latent toxicity or may result in a greater deficit in performance than might otherwise be seen without the stressor.

Statistical Considerations

Behavioral and other functional tests are confronted with the usual array of statistical considerations that confront any biological endpoint. For example, when it is desirable to test for the presence of a specific functional change, it is important to use tests that are valid in the sense that a test measures what it is intended to measure. It may also be important that tests have predictive validity if the results of the test are to be used to predict adverse health effects in humans. In addition, tests should be able to detect effects if they are present and be reproducible across subjects, samples of subjects and across laboratories.

SUMMARY

Because of the relative sensitivity to some chemicals, their generally noninvasive characteristics, and the ability to detect toxicity in general, behavioral measures are being used more frequently in toxicology as indicators of the functional state of the nervous system. Such functions include motor, sensory, cognitive, performance, and naturally occurring processes that are known to be disrupted in humans accidentally exposed to neurotoxicants. Currently, many procedures and techniques derived from behavioral pharmacology and experimental psychology are in use that can quantify and characterize neurotoxicant-induced neurobehavioral processes in laboratory animals. Several problems are evident when using behavioral endpoints in toxicology, including the identification of the behavioral mechanism of action, definition of an adverse effect, problem of functional reserve and the reliability, reproducibility, sensitivity, and validity of such tests. On the other hand, behavioral procedures show promise for those that study mechanism of action of toxic chemicals, as well as for the provisional assessment of chemicals for potential neurotoxicity.

ACKNOWLEDGMENTS

The authors acknowledge the assistance of Ms. Loretta Moore in the preparation of this manuscript. Some of the work described in this manuscript were presented at the Symposium on "Predicting Neurotoxicity from Preclinical Data," held in Washington, D.C. on October 1, 1985. The authors are also grateful for the thoughtful suggestions of Drs. L. Reiter and T. Sobotka who reviewed an earlier draft of this manuscript.

REFERENCES

Adams, J. and J. Buelke-Sam (1981). Behavioral assessment of the postnatal animal: testing and methods development. In *Developmental Toxicology*. C.A. Kimmel and J. Buelke-Sam, Eds., Raven Press, New York, pp. 233–258.

Agrawal, A.K., P.K. Seth, R.E. Squibb, H.A. Tilson, L.L. Uphouse, and S.C. Bondy (1981). Neurotransmitter receptors in brain regions of acrylamide-treated rats. I. Effects of a single exposure to acrylamide. *Pharmacol. Biochem. Behav.* 14: 527–531.

Alder, S. and G. Zbinden (1973). Use of pharmacological screening tests in subacute neurotoxic studies of isoniazid, pyridoxine HCl and hexachlorophene. *Agents Actions* 3/4: 233–243.

Alder, S. and G. Zbinden (1983). Neurobehavioral tests in single- and repeated-dose toxicity studies in small rodents. *Arch. Toxicol.* 54: 1–23.

Anger, W.K. (1984). Neurobehavioral testing of chemicals: impact on recommended standards. *Neurobehav. Toxicol. Teratol.* 6: 147–153.

Anger, W.K. and S.M. Wilson (1980). Effects of carbaryl on variable interval response rates in rats. *Neurobehav. Toxicol.* 2: 21–24.

Archer, J. (1973). Tests for emotionality in rats and mice. *Rev. Anim. Behav.* 21: 205–235.

Armstrong, R.D., L.J. Leach, P.R. Belluscio, E.A. Moore, H.C. Hodge, and J.K. Scott (1963). Behavioral changes in the pigeon following the inhalation of mercury vapor. *Am. Ind. Hyg. Assoc. J.* 24: 366–375.

Balazs, T. (1970). Measurement of acute toxicity. In *Methods in Toxicology.* G.E. Paget, Ed., Blackwell Scientific, Oxford, pp. 49–81.

Barlow, S.M. and F.M. Sullivan (1975). Behavioral toxicology. In *Teratology: Trends and Application.* C.L. Berry and D.E. Poswillo, Eds., Springer-Verlag, New York, pp. 103–120.

Barnett, S.A. and P.E. Cowan (1976). Activity, exploration, curiosity and fear: an ethological study. *Interdiscip. Sci. Rev.* 1: 43–62.

Barthalmas, G.T., J.D. Leander, D.E. McMillan, P. Mushak, and M.R. Krigman (1977). Chronic effects of lead on schedule-controlled pigeon behavior. *Toxicol. Appl. Pharmacol.* 42: 271–284.

Bhagat, B. and M. Wheeler (1973). Effect of nicotine on the swimming endurance of rats. *Neuropharmacology* 12: 1161–1165.

Bloom, A.S., C.G. Staatz, and T. Dieringer (1983). Pyrethroid effects on operant responding and feeding. *Neurobehav. Toxicol. Teratol.* 5: 321–324.

Bogo, V., T.A. Hill, and R.W. Young (1981). Comparison of accelerod and rotorod sensitivity in detecting ethanol- and acrylamide-induced performance decrement in rats: review of experimental considerations of rotating rod systems. *Neurotoxicology* 2: 765–787.

Boyd, E.M. (1959). The acute oral toxicity of acetylsalicylic acid. *Toxicol. Appl. Pharmacol.* 1: 229–239.

Buelke Sam, J. and C.A. Kimmel (1979). Development and standardization of screening methods for behavior teratology. *Teratology* 20: 17–30.

Bushnell, P.J. and H.L. Evans (1985). Effects of trimethyltin on homecage behavior of rats. *Toxicol. Appl. Pharmacol.* 79: 134–142.

Butcher, R.E. (1976). Behavioral testing as a method for assessing risk. *Environ. Health Perspect.* 18: 75–78.

Butcher, R.E. and C.V. Vorhees (1979). A preliminary test battery for the investigation of the behavioral teratology of selected psychotropic drugs. *Neurobehav. Toxicol.* 1: 207–212.

Cabe, P.A. and D.A. Eckerman (1982). Assessment of learning and memory dysfunction in agent-exposed animals. In *Nervous System Toxicology.* C.L. Mitchell, Ed., Raven, New York, pp. 637–649.

Carlini, E.A., M. Teresa, A. Silva, L.C. Cesare, and R.M. Endo (1967). Effects of chronic administration of β-(3,4-dimethoxyphenyl)-ethylamine and β-(3,4-trimethoxyphenyl)-ethylamine on the climbing rope performance of rats. *Med. Pharmacol. Exp.* 17: 534–542.

Chiba, S. and K. Ando (1976). Effects of chronic administration of kanamycin on conditioned suppression to auditory stimulus in rats. *Jpn. J. Pharmacol.* 26: 419–426.

Colotla, V.A., S. Bautista, M. Lorenzana-Jimenez, and R. Rodriguez (1979). Effects of solvents on schedule-controlled behavior. *Neurobehav. Toxicol.* 1: 113–118.

Cook, L. and A.B. Davidson (1973). Effects of behaviorally active drugs in a conflict-punishment procedure in rats. In *The Benzodiazepines*. S. Garattini, E. Mussini, and L.O. Randall, Eds., Raven Press, New York, pp. 327–345.

Cory-Slechta, D.A. and B. Weiss (1985). Alterations in schedule-controlled behavior of rodents correlated with prolonged lead exposure. In *Behavioral Pharmacology: The Current Status*. L.S. Seiden and R.L. Balster, Eds., Alan R. Liss, New York, pp. 487–501.

Cory-Slechta, D.A. (1989). Behavioral measures of neurotoxicity. *Neurotoxicology* 10: 271–295.

Crofton, K.M. and L.W. Reiter (1984). Effects of two pyrethroid insecticides on motor activity and the acoustic startle response in the rat. *Toxicol. Appl. Pharmacol.* 75: 318–328.

Crofton, K.M. and L.P. Sheets (1989). Evaluation of sensory system function using reflex malification of the startle response. *J. Am. Coll. Toxicol.* 8: 199–211.

Damstra, T. (1978). Environmental chemicals and nervous system dysfunction. *Yale J. Biol. Med.* 5: 457–468.

Daniel, S.A. and H.L. Evans (1985). Effects of acrylamide on multiple behavioral endpoints in the pigeon. *Neurobehav. Toxicol. Teratol.* 7: 267–273.

DeHaven, D.L., M.J. Wayner, F.C. Barone, and S.M. Evans (1982). Effects of trimethyltin on schedule dependent and schedule induced behaviors under different schedules of reinforcement. *Neurobehav. Toxicol. Teratol.* 4: 231–239.

Dews, P. and G. Wenger (1979). Testing for behavioral effects of agents. *Neurobehav. Toxicol.* 1: 119–127.

Dietz, D.D. and D.E. McMillan (1979a). Comparative effects of mirex and kepone on schedule-controlled behavior in the rat. I. Multiple fixed-ratio 12 fixed interval 2-min schedule. *Neurotoxicology* 1: 369–385.

Dietz, D.D. and D.E. McMillan (1979b). Comparative effects of mirex and kepone on schedule-controlled behavior in the rat. II. Spaced-responding, fixed-ratio, and unsignalled avoidance schedules. *Neurotoxicology* 1: 387–402.

Dyer, R.S., W.F. Wonderlin, and T.J. Walsh (1982). Increased seizure susceptibility following trimethyltin administration in rats. *Neurobehav. Toxicol. Teratol.* 4: 203–208.

Edwards, P.M. and V.H. Parker (1977). A simple sensitive and objective method for early assessment of acrylamide neuropathy in rats. *Toxicol. Appl. Pharmacol.* 40: 589–591.

Elsner, J., R. Looser, and G. Zbinden (1979). Quantitative analysis of rat behavior patterns in a residential maze. *Neurobehav. Toxicol. Teratol.* 1: 163–174.

Environmental Protection Agency (1982). Health Effects Tests Guidelines, EPA 560/6-82-001, National Technical Information Service, Springfield, VA.

Evans, H.L., V.G. Laties, and B. Weiss (1975). Behavioral effects of mercury and methylmercury. *Fed. Proc. Fed. Am. Soc. Exp. Biol.* 34: 1858–1867.

Evans, H.L. and B. Weiss (1978). Behavioral toxicology. In *Contemporary Research in Behavioral Pharmacology*. D.E. Blackman and D.J. Sanger, Eds., Plenum Press, New York, pp. 449–487.

Evans, W.O. (1961). A new technique for the investigation of some analgesic drugs on a reflexive behavior in the rat. *Psychopharmacologia* 2: 318–325.

Falk, J.L. (1970). The behavioral measurement of fine motor control: effects of pharmacological agents. In *Readings in Behavioral Pharmacology*. T. Thompson et al., Eds., Appleton-Century-Crofts, New York, pp. 223–236.

Fechter, L.D. and J.S. Young (1983). Discrimination of auditory from nonauditory toxicity by reflex modulation audiometry: effects of triethyltin. *Toxicol. Appl. Pharmacol.* 70: 216–227.

Ferster, C.B. and B.F. Skinner (1957). *Schedules of Reinforcement*. Appleton-Century-Crofts, New York.

Fox, D.A., S.R. Overmann, and D.E. Woolley (1979). Neurobehavioral ontogeny of neonatally lead-exposed rats. II. Maximal electroshock seizures in developing and adult rats. *Neurotoxicology* 1: 49–70.

Fuehner, H. (1932). Beitrage zur vergleichenden Pharmakologie. I. Die giftigen und todlichen gagen einiger substanzen fur frosche und mause. *Naunyn-Schmiedeberg's Arch. Pharmacol.* 166: 437–471.

Gad, S.C. (1982). A neuromuscular screen for use in industrial toxicology. *J. Toxicol. Environ. Health* 9: 691–704.

Geller, I., V. Mendez, M. Hamilton, R.J. Hartmann, and E. Gause (1979). Effects of carbon monoxide on operant behavior of laboratory rats and baboons. *Neurobehav. Toxicol.* 1: 179–184.

Gerhart, J.M., J.S. Hong, and H.A. Tilson (1985). Studies on the mechanism of chlordecone-induced tremor in rats. *Neurotoxicology* 6: 211–230.

Glowa, J.R. (1985). Behavioral effects of volatile organic solvents. In *Behavioral Pharmacology: The Current Status*. L.S. Seiden and R.L. Balster, Eds., Alan R. Liss, New York, pp. 537–552.

Graham, R.C.B., F.C. Lu, and M.G. Allmark (1957). Combined effect of tranquilizing drugs and alcohol on rats. *Fed. Proc. Fed. Am. Soc. Exp. Biol.* 16: 302.

Grant, L.O. (1976). Research strategies for behavioral teratology. *Environ. Health Perspect.* 18: 85–94.

Haggerty, G.C. (1989). Development of Tier I neurobehavioral testing capabilities for incorporation into pivotal rodent safety assessment studies. *J. Am. Coll. Toxicol.* 8: 53–70.

Harpur, E.S. (1974). The effect of ototoxic drugs on the acoustic startle reaction of the rat. *Br. J. Pharmacol.* 52: 137.

Harry, G.J. and H.A. Tilson (1982). Postpartum exposure to triethyltin produced long-term alterations in responsiveness to apomorphine. *Neurotoxicology* 3: 64–71.

Heise, G.A. (1984). Behavioral methods for measuring effects of drugs on learning and memory in animals. *Med. Res. Rev.* 4: 535–558.

Hoffman, H.S. and J.R. Ison (1980). Reflex modification in the domain of startle. I. Some empirical findings and their implications for how the nervous system system processes sensory input. *Psychol. Rev.* 87: 175–189.

Hogan, D.E., C.A. Edwards, and T.R. Zentall (1981). Delayed matching in the pigeon: interference produced by the prior delayed matching trial. *Anim. Behav.* 9: 398–400.

Hudson, P.M., P.H. Chen, H.A. Tilson, and J.S. Hong (1985). Effects of p,p'-DDT on the rat brain concentrations of biogenic amine and amino acid neurotransmitters and their association with p,p'-DDT-induced tremor and hyperthermia. *J. Neurochem.* 45: 1349–1355.

Idemudia, S.O. and D.E. McMillan (1985a). Effects of chemicals on delayed matching behavior in pigeons. III. Effects of triethyltin. *Neurotoxicology*, in press.

Idemudia, S.O. and D.E. McMillan (1985b). Effects of chemicals on delayed matching in pigeons. IV. Effects of trimethyltin. *Neurotoxicology*, in press.

Irwin, S. (1968). Comprehensive observational assessment. Ia. A systematic quantitative procedure for assessing the behavioral and physiologic state of the mouse. *Psychopharmacologia* 13: 222–257.

Jolicoeur, F.B., D.B. Rondeau, A. Barbeau, and M.J. Wayner (1979). Comparison of neurobehavioral effects induced by various experimental models of ataxia in the rat. *Neurobehav. Toxicol.* 1: 175–178.

Joy, R.M., L.G. Stark, and T.E. Albertson (1982). Proconvulsant effects of lindane: enhancement of amygdaloid kindling in the rat. *Neurobehav. Toxicol. Teratol.* 4: 347–354.

King, F.A. (1985). Effects of septal and amygdaloid lesions on emotional behavior and conditioned avoidance responses in the rat. *J. Nerv. Ment. Dis.* 126: 57–63.

Klaus, S. and A. Erdmann (1978). Zur durchfuhrung und auswertung des schwimmtests bei der maus. Wiss. Z. Ernst-Moritz-Arndt-Univ. Greifswald, *Med. Reihe* 27: 135–137.

Kniazuk, M. and H. Molitor (1944). The influence of thiamin-deficiency on work performance in rats. *J. Pharmacol. Exp. Ther.* 80: 362–372.

Konorski, J.A. (1957). A new method of physiological investigation of recent memory in animals. *Bull. Pol. Acad. Sci.* 7: 115–117.

Kulig, B.M. (1989). A neurofunctional test battery for evaluating the effects of long-term exposure to chemicals. *J. Am. Coll. Toxicol.* 8: 71–84.

Langman, J., W.S. Webster, and P.M. Rodier (1975). Morphological and behavioral abnormalities caused by insults to CNS in the perinatal period. In *Teratology: Trends and Applications*. C.L. Berry and D.E. Poswillo, Eds., Springer-Verlag, New York, pp. 182–200.

Laties, V.G. and H.L. Evans (1980). Methylmercury-induced changes in operant discrimination in the pigeon. *J. Pharmacol. Exp. Ther.* 214: 620–628.

Leander, J.D. and R.C. MacPhail (1980). Effect of chlordimeform (a formamidine pesticide) on schedule-controlled responding of pigeons. *Neurobehav. Toxicol.* 2: 315–321.

Lee, C.C. and P.J. Peters (1976). Neurotoxicity and behavioral effects of thiamine in rats. *Environ. Health Perspect.* 17: 35–43.

Levine, T.E. (1976). Effects of carbon disulfide and FLA-63 on operant behavior in pigeons. *J. Pharmacol. Exp. Ther.* 199: 669–678.

MacPhail, R.C. (1982). Studies on the flavor aversions induced by trialkyltin compounds. *Neurobehav. Toxicol. Teratol.* 4: 225–230.

MacPhail, R.C. (1985). Effects of pesticides on schedule-controlled behavior. In: *Behavioral Pharmacology: The Current Status*. L.S. Seiden and R.L. Balster, Eds., Alan R. Liss, New York, pp. 519–535.

MacPhail, R.C., D.B. Peele, and K.M. Crofton (1989). Motor activity and screening for neurotoxicity. *J. Coll. Toxicol.* 8: 117–125.

Mactutus, C.F. and H.A. Tilson (1985). Evaluation of long-term consequences in behavioral and/or neural function following neonatal chlordecone exposure. *Teratology* 31: 177–186.

Mactutus, C.F., K. Unger, and H.A. Tilson (1982). Neonatal chlordecone exposure impairs early learning and memory in the rat on a multiple measure passive avoidance task. *Neurotoxicology* 3: 27–44.

Mactutus, C.F., K. Unger, and H.A. Tilson (1984). Evaluation of neonatal chlordecone neurotoxicity during early development: initial characterization. *Neurobehav. Toxicol. Teratol.* 6: 67–73.

Marsh, R.R., H.S. Hoffman, and C. Stitt (1978). Reflex inhibition audiometry: a new objective technique. *Acta Oto-Laryngol.* 85: 336–341.

Marshall, J.F. and P. Teitelbaum (1974). Further analysis of sensory inattention following lateral hypothalamic damage in rats. *J. Comp. Physiol. Psychol.* 86: 375–395.

Mattsson, J.L., R.R. Albee, and D. Eisenbrandt (1989). Neurologic approach to neurotoxicologic evaluation in laboratory animals. *J. Am. Coll. Toxicol.* 8: 95–96.

Maurissen, J.P.L., B. Weiss, and H.T. Davis (1983). Somatosensory thresholds in monkeys exposed to acrylamide. *Toxicol. Appl. Pharmacol.* 71: 266–279.

McMillan, D.E. (1981). Effects of chemicals on delayed matching behavior in pigeons. I. Acute effects of drugs. *Neurotoxicology* 2: 485–498.

McMillan, D.E. and A.T. Miller (1974). Interaction between carbon monoxide and d-amphetamine or pentobarbital on schedule-controlled behavior. *Environ. Res.* 8: 53–63.

Merigan, W.H. (1979). Effects of toxicants on visual systems. *Neurobehav. Toxicol.* 1: 15–22.

Meyer, O.A., H.A. Tilson, W.C. Byrd, and M.T. Riley (1979). A method for the routine assessment of fore- and hindlimb grip strength of rats and mice. *Neurobehav. Toxicol. Teratol.* 1: 233–236.

Miczek, K.A. (1974). Intraspecies aggression in rats: effects of d-amphetamine and chlordiazepoxide. *Psychopharmacologia* 39: 275–301.

Miller, D.B. and D.A. Eckerman (1986). Learning and memory measures. In *Neurobehavioral Toxicology.* Annau, Z., Ed., Johns Hopkins Univeristy Press, Baltimore, pp. 94–149.

Mitchell, C.L. and H.A. Tilson (1982). Behavioral toxicology in risk assessment: problems and research needs. *Crit. Rev. Toxicol.* 10: 265–274.

Molinergo, L. and M. Orsetti (1976). Drug action on the "grasping reflex" and on swimming endurance: an attempt to characterize experimentally anti-depressant drugs. *Neuropharmacology* 15: 247–260.

Moser, V.C. (1989). Screening approaches to neurotoxicity: a functional observational battery. *J. Am. Coll. Toxicol.* 8: 85–94.

Nelson, B.K. (1978). Behavioral assessment in the developmental toxicology of energy-related industrial pollutants. In *Developmental Toxicity of Energy-Related Pollutants.* D.D. Mahlum, M.R. Sikov, P.L. Hackett, and F.D. Andrew, Eds., U.S. Government Printing Office, Department of Energy Publication N. Conf-771017, Washington, D.C., pp. 410–424.

Newland, M.C. (1988). Quantification of motor function in toxicology. *Toxicology Lett.* 43: 295–319.

Norton, S. (1982). Methods in behavioral toxicology. In *Principles and Methods of Toxicology*. A. Wallace Hayes, Ed., Raven Press, New York, pp. 353–373.

O'Donoghue, J.L. (1989). Screening for neurotoxicity using a neurologically based examination and neuropathology. *J. Am. Coll. Toxicol.* 9: 97–116.

O'Keefe, J. and L. Nadel (1978). *The Hippocampus as a Cognitive Map*. Oxford University Press, London.

Office of Technology Assessment (1990). Neurotoxicology: identifying and controlling poisons of the nervous system. U.S. Congress Office of Technology Assessment, U.S. Government Printing Office, Washington, D.C.

Olton, D.S., J.T. Becker, and G.E. Handelman (1979). Hippocampus, space and memory. *Behav. Brain Sci.* 2: 313–365.

Overstreet, D.H. (1977). Pharmacological approaches to habituation of the acoustic startle response in rats. *Physiol. Psychol.* 5: 230–238.

Pavlenko, S.M. (1975). Methods for the study of the central nervous system in toxicological tests. In *Methods Used in the USSR for Establishing Biologically Safe Levels for Toxic Substances*. World Health Organization, Geneva, pp. 86–108.

Peele, D.B. (1989). Learning and memory considerations for toxicology. *J. Am. Coll. Toxicol.* 8: 213–223.

Pontecorvo, M.J. (1983). Effects of proactive interference on rats' continuous non-matching-to-sample performance. *Anim. Learn. Behav.* 11: 356–366.

Pryor, G.T., E.T. Uyeno, H.A. Tilson, and C.L. Mitchell (1983). Assessment of chemicals using a battery of neurobehavioral tests: a comparative study. *Neurobehav. Toxicol. Teratol.* 5: 91–117.

Rafales, L.S. (1986). Assessment of locomotor activity. In *Neurobehavioral Toxicology*. Z. Annau, Ed., Johns Hopkins Press, Baltimore, pp. 54–68.

Rafales, L.S., R.L. Bornchein, and V. Caruso (1982). Behavioral and pharmacological responses following acrylamide exposure in rats. *Neurobehav. Toxicol. Teratol.* 4: 355–364.

Ray, O.S. and R.J. Barrett (1975). Behavioral, pharmacological, and biochemical analysis of genetic differences in rats. *Behav. Biol.* 15: 391–417.

Reiter, L.W. (1983). Chemical exposures and animal activity: utility of the figure-eight maze. In *Development in the Science and Practice of Toxicology*. A.W. Hayes, R.C. Schnell, and T.S. Miya, Eds., Elsevier, New York, pp. 73–84.

Reiter, L.W. and R.C. MacPhail (1979). Motor activity: a survey of methods with potential use in toxicity testing. *Neurobehav. Toxicol.* 1: 53–66.

Reiter, L., G. Heavner, P. Ruppert, and K. Kidd (1980). Short-term vs. long-term neurotoxicity: the comparative behavioral toxicity of triethyltin in newborn and adult rats. In *Effects of Food and Drugs on the Development and Function of the Nervous System*. R. Gryder and V. Frankos, Eds., U.S. Government Printing Office, Washington, D.C., pp. 144–154.

Reiter, L.W., R.C. MacPhail, P.H. Ruppert, and D.A. Eckerman (1981). Animal models of toxicity: some comparative data on the sensitivity of behavioral tests. In *Behavioral Consequences of Exposure to Occupational Environments. Proc. 11th Conf. Environ. Toxicol.* 11: 11–23.

Rice, D.C. (1985). Effect of lead on schedule-controlled behavior in monkeys. In *Behavioral Pharmacology: The Current Status*. L.S. Seiden and R.L. Balster, Eds., Alan R. Liss, New York, pp. 473–486.

Rice, D.C. (1988). Quantification of operant behavior. *Toxicology Lett.* 43: 361–379.

Riley, A.L. and D.L. Tuck (1985). Conditioined taste aversions: a behavioral index of toxicity. *Ann. N.Y. Acad. Sci.* 443: 272–292.

Roberts, W.A. (1976). Studies of short-term memory in the pigeon using the delayed matching to sample procedure. In *Processes of Animal Memory.* D.L. Medin, W.A. Roberts, and R.T. David, Eds., Wiley, New York, pp. 79–112.

Rodier, P.M. (1978). Behavioral teratology. In *Handbook of Teratology,* Vol. 4. J.G. Wilson and F.C. Fraser, Eds., Plenum Press, New York, pp. 397–428.

Rodier, P.M. (1979). Critical periods for behavioral anomalies in mice. *Environ. Health Perspect.* 18: 79–83.

Ruppert, P.H., T.J. Walsh, L.W. Reiter, and R.S. Dyer (1982). Trimethyl-induced hyperactivity: time course and pattern. *Neurobehav. Toxicol. Teratol.* 4: 135–139.

Samson, H.H. and J.L. Falk (1974). Ethanol and discriminative motor control: effects on normal and dependent animals. *Pharmacol. Biochem. Behav.* 2: 791–801.

Schallert, T., I.Q. Whishaw, V.D. Ramirez, and P. Teitelbaum (1978). Compulsive, abnormal walking caused by anticholinergics in akinetic 6-hydroxydopamine-treated rats. *Science* 199: 1461.

Schrot, J. and J.R. Thomas (1983). Alteration of response patterning by d-amphetamine on repeated acquisition in rats. *Pharmacol. Biochem. Behav.* 18: 529–534.

Schrot, J., J.R. Thomas, and R.A. Banvard (1980). Modification of the repeated acquisition of response sequences in rats by low-level microwave exposure. *Bioelectromagnetics* 1: 89–99.

Schrot, J., J.R. Thomas, and R.F. Robertson (1984). Temporal changes in repeated acquisition behavior after carbon monoxide exposure. *Neurobehav. Toxicol. Teratol.* 6: 23–28.

Shcard, M.H. (1977). Animal models of aggressive behavior. In *Animal Models in Psychiatry and Neurology.* I. Hanin and E. Usdin, Eds., Pergamon Press, New York, pp. 247–258.

Silverman, A.P., P.B. Banham, K. Extance, and H. Williams (1981). Early change and adaptation in the social behaviour of rats given methylmercury in the diet. *Neurotoxicology* 2: 269–281.

Sobotka, T.J., R.E. Brodie, and M. Cook (1975). Psychophysiologic effects of early lead exposure. *Toxicology* 5: 175–191.

Spyker, J. (1975). Assessing the impact of low level chemicals on development: behavioral and latent effects. *Fed. Proc. Fed. Am. Soc. Exp. Biol.* 34: 1835–1844.

Stebbins, W.C. and D.B. Moody (1979). Comparative behavioral toxicology. *Neurobehav. Toxicol.* 1: 33–44.

Stein, R.B. and R.G. Lee (1981). Tremor and clonus. In *Handbook of Physiology,* Vol. 2, Section 1, Motor Control. V.B. Brooks, Ed., Williams & Wilkins, Baltimore, pp. 325–343.

Swartzwelder, H.S., R.S. Dyer, W. Holahan, and R.D.Myers (1981). Activity changes in rats following acute trimethyltin exposure. *Neurotoxicology* 2: 589–593.

Swartzwelder, H.S., J. Hepler, W. Holahan, S.E. King, H.A. Leverenz, P.A. Miller, and R.D. Myers (1982). Impaired maze performance in the rat caused by trimethyltin treatment: problem-solving deficits and perseveration. *Neurobehav. Toxicol. Teratol.* 4: 169–176.

Tesh, J.M. (1977). An approach to the assessment of postnatal development in laboratory animals. In *Methods in Prenatal Toxicology*. D. Neubert, H.J. Merker, and T.E. Kwasigroch, Eds., Georg Theime, Stuttgart, pp. 186–195.

Thompson, D.M. (1973). Repeated acquisition as a behavioral baseline for studying drug effects. *J. Pharmacol. Exp. Ther.* 184: 506–514.

Thompson, D.M. (1978). Stimulus control and drug effects. In *Contemporary Research in Behavioral Pharmacology*. D.E. Blackman and D.J. Sanger, Eds., Plenum Press, New York, pp. 159–207.

Thompson, R.F. and W.A. Spencer (1966). Habituation: a model phenomenon for the study of neuronal substrates of behavior. *Psychol. Rev.* 173: 16–43.

Tilson, H.A., G.J. Davis, J.A. McLachlan, and G.W. Lucier (1979). The effects of polychlorinated biphenyls given prenatally on the neurobehavioral development of mice. *Environ. Res.* 18: 466–474.

Tilson, H.A., P.A. Cabe, and T.A. Burne (1980). Behavioral procedures for the assessment of neurotoxicity. In *Experimental and Clinical Neurotoxicology*. P.S. Spencer and H.H. Schaumburg, Eds., Williams & Wilkins, Baltimore, pp. 758–766.

Tilson, H.A. and T.A. Burne (1981). Effects of triethyltin on pain reactivity and neuromotor function of rats. *J. Toxicol. Environ. Health* 8: 317–324.

Tilson, H.A. and G.J. Harry (1982). Behavioral principles for use in behavioral toxicology and pharmacology. In *Nervous System Toxicology*. C.L. Mitchell, Ed., Raven Press, New York, pp. 1–27.

Tilson, H.A. and R.E. Squibb (1982). The effects of acrylamide on the behavioral suppression produced by psychoactive agents. *Neurotoxicology* 3: 113–120.

Tilson, H.A., C.F. Mactutus, R.L. McLamb, and T.A. Burne (1982). Characterization of triethyl lead chloride neurotoxicity in adult rats. *Neurobehav. Toxicol. Teratol.* 4: 671–682.

Tilson, H.A. and C.L. Mitchell (1984). Neurobehavioral techniques to assess the effects of chemicals on the nervous system. *Annu. Rev. Pharmacol. Toxicol.* 24: 425–450.

Tilson, H.A., J.S. Hong, and C.F. Mactutus (1985). Effects of 5,5-diphenylhydantoin (phenytoin) or neurobehavioral toxicity of organochlorine insecticies and permethrin. *J. Pharmacol. Exp. Ther.* 233: 285–289.

Tilson, H.A. (1987). Behavioral indices of neurotoxicity: what can be measured? *Neurotoxicol. Teratol.* 9: 427–433.

Tilson, H.A. (1990a). Neurotoxicology in the 1990s. *Neurotoxicol. Teratol.* 12: 293–300.

Tilson, H.A. (1990b). Behavioral indices of neurotoxicity. *Toxicol. Pathol.* 18: 96–104.

Vorhees, C.V. (1974). Some behavioral effects of maternal hypervitaminosis A in rats. *Teratology* 10: 269–274.

Vorhees, C.V. (1983). Influence of early testing on postweaning performance in untreated F344 rats, with comparisons to Sprague-Dawley rats, using a standardized battery of tests for behavioral teratogenesis. *Neurobehav. Toxicol. Teratol.* 5: 587–591.

Wallman, J. (1975). A simple technique using an optomotor response for visual psychophysical measurements in animals. *Vision Res.* 15: 3–8.

Walsh, T.J., M. Gallagher, E. Bostock, and R.S. Dyer (1982a). Trimethyltin impairs retention of a passive avoidance task. *Neurobehav. Toxicol. Teratol.* 4: 163–167.

Walsh, T.J., D.B. Miller, and R.S. Dyer (1982b). Trimethyltin, a selective limbic system neurotoxicant, impairs radial-arm maze performance. *Neurobehav. Toxicol. Teratol.* 4: 177–183.

Walsh, T.J., R.L. McLamb, and H.A. Tilson (1984a). Organometal-induced antinociception: a time- and dose-response comparison of triethyl and trimethyl lead and tin. *Toxicol. Appl. Pharmacol.* 73: 295–299.

Walsh, T.J., H.A. Tilson, D.L. DeHaven, R.B. Mailman, A. Fisher, and I. Hanin (1984b). AF64A, a cholinergic neurotoxin, selectively depletes acetylcholine in hippocampus and cortex, and produces long-term passive avoidance and radial-arm maze deficits in the rat. *Brain Res.* 321: 91–102.

Wasserman, E.A. (1976). Successive matching-to-sample in the pigeon: variations on a theme by Konorski. *Behav. Res. Methods Instrum.* 8: 278–282.

Wecker, J.R. and J.R. Ison (1984). Acute exposure to methyl or ethyl alcohol alters auditory function in the rat. *Toxicol. Appl. Pharmacol.* 74: 258–266.

Weiss, B. and V. Laties (1961). Changes in pain tolerance and other behaviors produced by salicylates. *J. Pharmacol. Exp. Ther.* 131: 120–129.

Weiss, B., R.W. Wood, and W.H. Merigan (1983). Toxicity evaluation needs the intact animal. *Ann. N.Y. Acad. Sci.* 406: 83–91.

Weiss, B. (1985). Experimental implications of behavior as a criterion of toxicity. In *Behavioral Pharmacology: The Current Status.* L.S. Seiden and R.L. Balster, Eds., Alan R. Liss, New York, pp. 467–472.

Wenger, G.R., D.E. McMillan, and L.W. Chang (1984). Behavioral effects of trimethyltin in two strains of mice. II. Multiple fixed ratio, fixed interval. *Toxicol. Appl. Pharmacol.* 73: 89–96.

Wilson, J.G. (1973). *Environment and Birth Defects.* Academic Press, New York, p. 180.

Winneke, G., A. Brockhaus, and R. Baltissen (1977). Neurobehavioral and systemic effects of longterm blood lead-elevation in rats. I. Discrimination learning and open-field behavior. *Arch. Toxicol.* 37: 247–263.

Wood, R.W. (1979). Behavioral evaluation of sensory irritation evoked by ammonia. *Toxicol. Appl. Pharmacol.* 50: 157–162.

Wood, R.W. (1981). Determinants of irritant termination behavior. *Toxicol. Appl. Pharmacol.* 61: 260–268.

Yokel, R.A. (1983). Repeated systemic aluminum exposure effects on classical conditioning of the rabbit. *Neurobehav. Toxicol. Teratol.* 5: 41–46.

Young, J.S. and L.D. Fechter (1983). Reflex inhibition procedures for animal audiometry: a technique for assessing ototoxicity. *J. Acoust. Soc. Am.* 73: 1686–1693.

Zbinden, G. (1963). Experimental and clinical aspects of drug toxicity. *Adv. Pharmacol.* 2: 1–99.

Zbinden, G. (1981). Experimental methods in behavioral teratology. *Arch. Toxicol.* 48: 69–88.

Zenick, H., R. Padick, T. Tokarek, and P. Aragon (1978). Influence of prenatal and postnatal lead exposure on discrimination learning in rats. *Pharmacol. Biochem. Behav.* 8: 347–350.

The Neurological Examination

Cynthia S. Payne
Department of Radiology
Duke University Medical Center
Durham, North Carolina

INTRODUCTION

The clinical neurological examination is an elegant and informative tool in the evaluation of patients with disorders of the nervous system. Despite significant advances in diagnostic technologies such as computed tomography (CT scanning), magnetic resonance imaging (MRI), and positron emission tomography (PET), the clinical examination remains the first and most necessary step. It allows for the intelligent and judicious application of these sophisticated technologies based upon a working differential diagnosis formulated during the examination.

The main goal of this chapter is to provide the reader with an understanding of the general intent and method of a routine neurological examination. The emphasis will be on introducing the basic principles involved in the sometimes baffling activities of clinical neurologists.

It will be assumed that clinically trained readers will have been exposed to the techniques of the neurological exam; thus the text will be primarily directed to nonclinicians. A second purpose will be to provide both clinicians and researchers with an appreciation of how the neurological examination is utilized in the evaluation of patients with toxicologic problems. This is intended to complement the more in-depth discussions of neurotoxicological problems in other chapters of this text. Thirdly, the reader will be directed to comprehensive resources for further study of the subject.

The neurological examination is divided into seven areas:

1. History
2. Mental Status
3. Cranial Nerves
4. Motor System
5. Sensory System
6. Coordination
7. Reflexes

HISTORY

The history is often said to be the most important part of the neurological exam. The nervous system can manifest dysfunction in a limited number of observable ways, although the potential etiologies are numerous. Identical physical findings may have entirely different implications based upon the history. The history often enables the clinician to have an accurate differential diagnosis before actually examining the patient. The subsequent physical examination optimally corroborates the initial impression.

The history is based upon the subjective experiences reported by the patient or other observers. These are symptoms. The observable abnormalities noted on physical examination are signs. The clinician explores the following while reviewing a patient's symptoms:

1. characterization and history of the chief complaint(s)
2. chief complaint in context of past and present medical problems
3. developmental history in children and, when relevant, older individuals
4. social and occupational history
5. family history
6. medications and drug allergies

The nervous system is wonderfully logical even if it seems overwhelmingly complex. Symptoms and, particularly, signs of dysfunction usually implicate the anatomical system subserving that function. The examiner must think anatomically as the patient relates symptoms in lay terms. The clinician tries to first localize the lesion to the area of involvement, e.g., brain, spinal cord, peripheral nerve, or muscle. Questions about the nature of the complaint may be revealing. For example, if a patient complaining of weakness relates that it is more difficult to pull a sweater off over the head than to open a door knob, he will probably be found to have proximal rather than distal motor weakness on exam. Muscle diseases typically cause proximal weakness whereas motor neuron disease typically causes distal weakness. The examiner must remember that all the details of a patient's problems are important to him or her. The skillful interviewer will guide the patient, eliminating extraneous information while asking open-ended questions and showing genuine interest and patience.

The interview begins by recording the patient's age, sex and handedness (the latter is important in assessing cortical lesions involving localized functions such as language). The chief complaint is fully characterized with respect to its nature, onset and progression, and influencing factors. The examiner attempts to localize the patient's symptoms to the appropriate system subserving that function. Past medical problems are documented. The social and occupational history is taken, a particularly important part of the evaluation of neurotoxicological problems. Family history is another important part of the history and one that is often overlooked. This will become even more important as we further understand genetic predispositions to environmentally-induced diseases such as cancer. Medications and drug allergies are noted. The patient's use of cigarettes, alcohol, and other drugs is obviously important although patients are rarely completely forthcoming about the latter two and further information may be gleaned from family members, job history, and previous medical problems which may suggest sustance abuse like infectious hepatitis or pancreatitis.

In children and adolescents, and sometimes in young adults, a developmental history is necessary. This includes any complications of pregnancy, labor, and delivery of the patient, Apgar scores at birth, any peri- or postnatal complications, history of prior and subsequent pregnancies in the patient's mother, and determining if the patient has accomplished appropriate developmental milestones. This is usually divided into language, social skills, and gross and fine motor skills. The Denver Developmental Screen is a standardized reference frequently used to assess these parameters in children.

As in a general medical exam, the examiner will inquire about prior illnesses and any past or current symptoms referable to other areas of the nervous system, as well as other body systems such as cardiovascular, endocrine, etc. This is called a Review of Systems. The history is now concluded. The following is an example case.

A 32-year-old male presents with a chief complaint of a hand tremor and feeling "slowed down." His movements feel like he is "in slow motion." He was entirely well until three weeks prior to presentation when he developed a mild tremor of his hands, which is worse at rest than with action. A family member also adds that his speech has become noticeably softer and more difficult to hear. The slowness in his walking and movements progressed to the point that the day prior he "just froze up" while in the grocery store and had to be almost carried to the car.

The patient has no sensory symptoms, no confusion or memory problems, no weakness, and no family history of neurological disorders, such as Wilson's disease.

At this point in the history, the clinical symptoms, and observable tremor, facial expression, and slowness of movements (bradykinesia), strongly suggest basal ganglion dysfunction. These symptoms are typical of this category of disorders and the examiner has excluded significant involvement of other systems such as sensory, motor, and cognitive functioning. This will be further confirmed during the examination. This would be a typical presentation in an older individual for idiopathic Parkinson's disease, except that the onset of symptoms is unusually rapid. The examiner excludes exposure to neuroleptic drugs and family history. In the review of systems, it is revealed that the patient had infectious hepatitis six years earlier suggesting the possibility of intravenous drug abuse. In a nonjudgmental manner, the examiner elicits confirmation of drug abuse and the further intriguing fact that over the period of several weeks prior to the onset of his symptoms, the patient had tried a new drug "made" by a "home chemist" who told him it was like heroin. In our example case we see how the facts of age, time of onset, and social history of drug exposure have been used to establish a probable history of toxin-induced Parkinson's syndrome before the physical exam has begun. This fictional example was based upon actual case reports of the first patients seen with 1-methyl-4-phenyl-1,2,3,6-tetrahydropyridine (MPTP)-induced Parkinson's. One can appreciate how baffling this clinical history was to the first clinicians encountering these patients. The subsequent unfolding of this fascinating and tragic phenomenon is due in part to the expertise in history-taking and toxicological epidemiology work on the part of the workers involved.

PHYSICAL EXAM

Observation

Observation begins as soon as the examiner encounters the patient. Whether the patient is ambulatory or even comatose, much information is gained by observation. There are a number of neurological diseases that have such distinguishing features that they are virtually diagnosable onsight, so-called "waiting room diagnoses." In the comatose patient, posture, pattern of respirations, and response to stimuli may help determine the level of brain involvement before the clinician actually examines the patient. In ambulatory patients, mode of dress, mannerisms, gait, affect, speech, thought content, and motor function are all features that may have apparent abnormalities observed during the history interview. Traits such as mannerisms and dress must obviously be put into context of the patient's cultural and socioeconomic circumstances.

Mental Status

Mental status testing assesses functions such as orientation, language, intellect, memory, and attention, which are often observed by discreet neuroanatomical structures, just as is the motor system. "Altered mental status" is one of the more common reasons for obtaining neurological consultation. Assessment of higher cortical functions is becoming increasingly important in the field of neurotoxicology with the expanding awareness of the neurobehavioral consequences of environmental agents. Technologies such as PET have enhanced this capability with the ability to image functional metabolism in *in vivo* states.

Level of Consciousness

The classification of a patient's level of consciousness is considered a continuum between normal state of consciousness (alertness) and coma. The intervening stages in succession from alertness are usually labeled lethargy, obtundation, stupor, and coma. Plum and Posner have written the standard monograph on the evaluation of altered mental status. There are only three basic mechanisms which produce coma: (1) bilateral cerebral dysfunction, (2) brainstem dysfunction affecting the reticular activating system, and (3) a combination of both. Coma can also be broken down into structural involvement based upon a rostral to caudal progression. Parameters such as pattern of respirations, pupillary size and response to light, motor response to stimuli, and cranial nerve evaluation (particularly corneal and oculovestibular responses) cannot only be helpful in establishing the level of anatomical involvement, but also have been shown to have prognostic value. Delirium is considered somewhat separately since it involves altered sensorium in usually alert individuals. Hallucinations and distorted perceptions are often seen in delirium whereas the above mentioned stages primarily involve progressive loss of responsiveness to environmental stimuli. Delirium is most often due to metabolic or toxic abnormalities. It is frequently not enough to record a single word to describe level of consciousness. A descriptive sentence of the patient's spontaneous activity and responses to stimulation provides a universally understandable assessment and reference point to gauge progression. For example, in recording the level of consciousness in a drug overdose, the recorded statement might be, "a 17-year-old female who was difficult to arouse by verbal stimulus but would answer questions appropriately but sluggishly with vigorous shaking."

Orientation

Orientation is tested by asking the person to state his or her name, the correct date including year, and the present location. While the inability to correctly recall one's name is unusual, inability to correctly state the year is frequently seen in early dementia when few other signs may be present.

Language

Language is a complex function involving interactions of discreet anatomical structures primarily located in the left cerebral hemisphere in almost all right- and the majority of lefthanded individuals. Testing language includes not only the ability to understand and to execute spoken commands, but grammatically and linguistically correct output of language (within the context of the person's educational attainment and cultural patterns), the ability to read, write, and repeat language, and assessing the motor components of speech such as facial muscles and tongue movements. Abnormalities of language and speech may, when indicated, be more fully assessed by formal testing methods, usually by a Speech and Hearing pathologist.

Attention and Concentration

A patient's inability to concentrate during other components of a mental status test are usually apparent. In addition to these observations, the patient may be asked to do serial subtractions or additions (obviously overlapping with calculation ability), and to recall a series of digits (overlapping with memory function).

Memory

Memory testing is divided into immediate recall (such as 4–7 digit repetition), short term memory (the ability to recall three objects at 5 and 10 min intervals), recent (what the patient had for the previous meal) and remote (recall recent Presidents in order from current through each predecessor up to approximately six).

Calculations

Calculation is primarily a function of the dominant hemisphere, and is tested by giving the patient a series of mental calculations appropriate to his/her level of education.

Abstract Thinking

The patient is given several well-known proverbs and asked for an interpretation.

Constructional Abilities

Constructional abilities are parietal lobe functions and may be tested in a number of ways such as having the patient copy line drawings, number the face of a clock, and perform simple constructions with three-dimensional objects such as blocks.

Cranial Nerves

Cranial nerve evaluation tests the function of the twelve cranial nerves which innervate the muscles of the head and neck, supply sensory innervation to the face, and subserve the special senses of smell, vision, hearing, and taste. The nuclei of most of the cranial nerves lie within the compact brainstem. More than any other area of the neurological exam, a knowledge of the anatomy and function of the cranial nerves allows for precise localization of dysfunction. The following chart lists the twelve cranial nerves, the function subserved, a brief description of the clinically relevant anatomy, how the nerve is tested, the result of lesions to that nerve, and, when applicable, an example of a toxin that affects that nerve.

Several sophisticated technologies exist to test cranial nerve functioning. For the visual system, field perimetry, visual evoked potentials, computed tomography (CT) and magnetic resonance (MR) imaging of the orbits and visual pathways are the most frequently utilized. Hearing and vestibular functioning can be tested with brainstem auditory evoked responses and formal audiometry, in addition to visualization of the temporal bone and brainstem by CT and MR imaging.

Motor System

The motor exam assesses functioning of the central and peripheral motor components, including the peripheral motor nerve, neuromuscular junction, and muscle. The central component is primarily composed of the pyramidal and extrapyramidal systems. Although these are anatomically imprecise terms, the pyramidal tract refers to the upper motor neurons originating in the precentral gyrus and descending through the internal capsule and brainstem to decussate in the medullary pyramids and descend as the corticospinal tracts synapsing on the alpha motor neurons. The alpha motor neuron is the lower motor neuron which exits the spinal cord to innervate specific muscles. The extrapyramidal tract refers to the striatopallidonigral and, separately, the cerebellar pathways. These tracts are primarily concerned with the maintenance of posture, balance, and smooth execution of motor movements. The reader is referred to texts in the reference section which discuss the physiological anatomical differences of these two systems in greater detail. Lesions of each of these systems, imprecise as they are, produce specific and usually readily recognizable patterns. These are listed in Appendix 1, page 587.

The first part of the motor exam is observation. The presence or absence of muscle atrophy, fasciculations, and adventitious movements are noted as well as general muscle bulk. Asymmetry of skeletal structure is noted. Gait is observed, including strength and balance such as the ability to walk on the toes, heels and to tandem walk (heel-to-toe). Some neurologic diseases have distinctive gaits such as the stooped, slow shuffling gait of Parkinson's, the waddling lordotic gait of some muscular dystrophies, and the unsteady jerking gait of Huntington's disease. Abnormal movements fall on a continuum from the rapid tremor seen with mercury poisoning and other diseases, to the quick jerking motions of chorea, the writhing movements of athetosis, and the sustained and other distorting muscle contractions of dystonia.

Muscle tone is evaluated by passively moving the extremities within their range of motion. As noted in Table 1, the type of abnormality in tone may indicate the system involved.

Direct strength testing is done by testing specific muscle groups in their range of action with the examiner providing active resistance. Muscle strength is usually graded on a scale of 0–5 to provide interexaminer reliability. The lowest grade, 0, represents no movement of the limb observed with the patient actively trying and with no resistance from the examiner. Normal strength is graded as 5. Normal strength is obviously individually variable but the examiner should have to exert resonable force to overcome the major motor groups such as the quadriceps even in elderly patients.

Sensory System

The sensory exam is the least objective part of the neurological examination since sensory dysfunction usually cannot be visually observed. A patient may complain of paresthesias (pins-and-needles), but documenting pathologic changes on exam often depends upon the subjective interpretation of the patient to stimuli applied by the examiner. The sensory exam is divided into the following components:

A. Simple modalities
   1. Light touch
   2. Pain
   3. Temperature
B. Complex modalities
   1. Position sense
   2. Vibratory sensation
   3. Two-point discrimination
   4. Stereognosis
   5. Double simultaneous stimulation
   6. Graphesthesia

The simple modalities of pain, light touch, and temperature primarily test the integrity of the spinothalamic tracts. The patient is asked to rate as sharp or dull (or no perception) stimulation from a sharp object. Light touch is tested by lightly stroking the skin with a cotton wisp, and temperature is tested by placing cool and warm objects against the skin. A sensory abnormality may be found unilaterally over an entire extremity (usually indicating a spinal cord or intracranial lesion), over a dermatome on the trunk or extremity indicating dysfunction of a spinal nerve, over the distribution of a peripheral nerve after it has branched from the segmental spinal nerve, or a diffuse peripheral process such as that seen with many toxic and metabolic peripheral neuropathies reported as a "stocking/glove" pattern indicating the

TABLE 1.

Upper Motor Neuron	Lower Motor Neuron[a]	Extrapyramidal
Spasticity	Hypotonia or flaccidity	Cogwheeling (basal ganglia) Hypotonia (cerebellum)
No atrophy (except disuse)	Atrophy	No atrophy
No fasciculations or fibrillations	Fasciculations and fibrillations	None
Babinski response present	No Babinski	No Babinski
Affects groups of muscles	May affect individual muscles	Groups
No adventitious movements	None	Present (chorea, dystonia, tremor)
Deep tendon reflexes hyperactive	Decreased or absent	Usually normal

[a] Ratchet like sensation in the limb on passive range of motion.

stimulus is appreciated more proximally than distally, e.g., more over the midcalf than on the foot.

Simple modalities may even be tested in comatose patients by measuring the threshold of stimulus necessary to evoke a motor response. The motor response is also important in itself. Purposeful movements such as reaching for the painful stimulus implies cortical functioning and an intact brainstem. Nonpurposeful flexion of the upper extremities and extension of the legs implies forebrain involvement to the level of the midbrain red nucleus and is called a decorticate response. Upper extremity extension to a painful stimulus implies involvement of the brainstem caudal to the red nucleus, called decerebrate posturing, and is generally associated with more extensive damage and a worse prognosis.

Joint position sense and vibratory sensation test the integrity of the joint receptors which relay to the thalamus through the dorsal columns of the spinal cord, synapsing in the medullary nuclei gracilis and cuneatus and ascending via secondary afferent fibers in the medial lemniscus. Testing involves applying a vibrating tuning fork to a bony prominence over a distal extremity and recording the minimal threshold. Joint position sense is tested by passively moving a distal joint of the hands or the great toe and asking the patient, with eyes closed, to indicate the first perception and direction of movement. This should be within a few millimeters. These functions may also be tested by having the patient stand with the eyes open. This utilizes both visual input into the vestibular system as well as the dorsal columns. The patient is then asked to close the eyes. The ability to maintain posture with no visual input is now dependent on the dorsal columns. Significant unsteadiness with the eyes closed is called a Romberg sign and indicates dorsal column dysfunction.

Other more complex modalities tested in the sensory exam primarily assess parietal cortex function. This includes the ability to correctly identify numbers traced on the patient's palms with the eyes closed (graphesthesia); the ability to identify an object (e.g., a coin) placed in the patients hand (stereognosis); the ability to appreciate two sharp stimuli presented within a few millimeters of each other over a distal extremity (also requiring intact peripheral pain fibers); and the ability to distinguish two stimuli presented simultaneously. This may be strikingly impaired in lesions of the nondominant (usually right) parietal cortex with the patient entirely denying the existence of the left side of his or her body and environment.

The subjective nature of the sensory examination often necessitates further testing for more quantitative assessment. The use of nerve conduction studies is sensitive in peripheral neuropathies and can distinguish between demyelinating and axonal lesions. Somatosensory evoked responses are utilized for sensitive but nonspecific detection of sensory dysfunction in the spinal cord, brainstem and cortex.

Coordination

Coordination is primarily a function of the cerebellum but coordination testing also involves elements of the vestibular and motor systems. Gait testing, particularly tandem walking, tests the cerebellum as well as the motor system. Lesions of midline cerebellar structures result in a broadbased oscillating, unsteady, ataxic gait. Lesions of lateral cerebellar structures affect the extremities. Dysmetria is the coarse tremor of an extremity demonstrated by having the patient rapidly touch his index finger to his nose and then to the examiner's finger held in front of the patient. Lower extremity dysmetria is elicited by having the patient run his heel down the surface of the contralateral shin. The ability to smoothly and rapidly alternate movements of the extremities is another cerebellar as well as motor function and is tested by having the patient rapidly tap the thumb to each opposing finger or pat the hand on the thigh rapidly alternating back to palm. Clumsiness with this maneuver, when not secondary to motor weakness, is called dysdiadochokinesis. The ability to perform fine motor tasks may also be observed during this part of the examination such as the ability to button a garment, and open and close a safety pin. Other observable aspects of cerebellar dysfunction, while not entirely specific to this structure, include bobbing of the head (titubation), jerking eye movements (nystagmus), characteristic speech patterns, and hypotonia of the muscles.

Reflexes

Reflexes are grouped into three categories:

1. Deep tendon
2. Superficial
3. Pathological

The deep tendon reflexes assess the integrity of the monosynaptic arc from the sensory afferent branch arising in the muscle spindle and synapsing on the alpha motor neuron in the anterior horn of the spinal cord which then by the efferent motor branch causes the muscle to contract. It is frequently forgotten that the reflex requires an intact sensory afferent to be elicited. Tapping on the muscle tendon stretches the muscle spindle activating the afferent fibers. This is strictly a spinal segmental response and does not require the interaction of higher structures. Deep tendon reflexes most often tested are the jaw jerk, biceps, triceps, brachioradialis, patellar, and ankle. Reflexes are graded on a scale which may vary slightly between examiners and institutions. This usually ranges from 0, meaning nonelicitable, to 5, meaning sustained clonus (rhythmic contractions seen in spasticity). "2 +" is usually the recorded grade for normal.

Superficial reflexes are those primarily elicited by stroking an area of the skin which causes a segmental motor response. Already mentioned above is the corneal response obtained by gently stroking the cornea to elicit eye blinking. Other useful superficial reflexes include abdominal reflexes innervated from the lower thoracic levels and elicited by stroking each abdominal quadrant and watching for contraction of the umbilicus towards the stimulus. The anal reflex is demonstrated by the same technique in the perianal area. In males, the cremasteric reflex is elevation of the unilateral testicle when the thigh is stroked with a sharp object. These last three are extremely useful in assessing lower spinal cord lesions.

Pathological reflexes are those which indicate dysfunction of the corticospinal system and are useful in evaluating dementia and other disorders affecting higher cortical functions. The plantar response is the downward flexion of the toes when a noxious stimulus is applied to the plantar surface of the foot, usually by running the tip of a reflex hammer along the sole of the foot. The normal response is flexion of the toes. When the toes extend, this is an indication of dysfunction of the corticospinal tract and is called a Babinski sign or extensor toe sign. It is one of the most useful signs in clinical neurology because of its specificity in indicating organic disease. The extensor toe sign may be elicited by other maneuvers, each designated by its own eponym.

Cortical release signs are useful in evaluating patients with frontal lobe dysfunction, such as dementia, although the anatomy of these responses is not clearly defined. Most of them represent primitive or infantile responses which serve a purpose in neonatal life. These include suck, snout and rooting reflexes of the mouth to direct stimulation or stroking. The grasp response is elicited by stroking the patients palm causing a reflexive grasping of the examiner's hand. Palmomental responses are contractions of the mentalis chin muscle to stroking of the unilateral thenar eminence of the palm. The glabellar response, or Meyerson's sign, often seen in Parkinson's, is the inability to suppress eye blinking to repetitive tapping on the forehead.

Other

A thorough neurological examination must include aspects of a general medical examination. The usefulness of a general review of systems has already been mentioned as well as documenting the patients past medical history. During the examination, the following observations and evaluations may be made based upon the age and chief complaint of the patient: recording of height, weight and head circumference; measurements of vital signs (especially blood pressure); inspection of the skin (informative in a number of neurological disease such as neurofibromatosis, and in toxicology such as palm and sole desquamation in acrylamide exposure); hair and nails (e.g.,

Mees lines in arsenic poisoning and alopecia with thallium); eyes and mouth (lead lines of the gingiva); cardiovascular system; respiratory; and abdominal examination (e.g., hepatomegaly in alcoholism). The autonomic nervous system may also be assessed, particularly when involvement may be suspected such as with organophosphate poisoning. This is a brief list. The reader is directed to Bates' text in the reference section for further information on the general medical examination.

Diagnostic Resources

Once the neurological examination is completed, the clinician may wish to pursue further diagnostic work-up with one of the many technological tools available. This may involve simple blood and urine tests, examination of the cerebrospinal fluid, electromyography and nerve conduction studies as previously discussed, evoked responses, diagnostic imaging such as computed tomography, magnetic resonance, and positron emission tomography, quantitative clinical assessment of intelligence and higher cognitive functioning, and electroencephalography (EEG). Advances in immunology and molecular genetics promise to have significant impact upon clinical neurology. DNA polymorphisms linked to the Huntington's disease, Duchenne's muscular dystrophy and myotonic muscular dystrophy genes and several other disorders have been developed and will allow for the accurate identification of gene carriers in asymptomatic patients and *in utero*. The tools of molecular biology will undoubtedly impact upon the field of clinical neurotoxicology. Already the acetylcholine receptor has been sequenced and will allow further elucidation of toxin interactions at the neuromuscular junction. The ability to understand the effects of toxins on gene expression, as well as genetic predisposition to certain toxins, will be enhanced by these techniques. This is an exciting era in neurobiological research that promises to greatly expand our knowledge of the interactions of the nervous system with the environment and to increase our capacity to care for patients afflicted with neurological disorders.

REFERENCES

Adams, R. and M. Victor (1989). *Principles of Neurology,* 4th ed. McGraw-Hill, New York.

Ballard, P., J.W. Tetrud, and J.W. Langston (1985). Permanent human Parkinsonism due to 1-methyl-4-phenyl-1,2,3,6-tetrahydropyridine (MPTP): seven cases. *Neurology* 35: 949–956.

Bates, B. (1991). *A Guide to Physical Examination,* 5th ed. J.B. Lippincott, Philadelphia.

Berg, B., Ed. (1984). *Child Neurology. A Clinical Manual*. Jones Medical Publications, Greenbrae, CA.

Burns, R.S., P.A. LeWitt, M.H. Ebert, H. Pakkenberg, and I.J. Koplin (1985). The clinical syndrome of striatal dopamine deficiency. Parkinsonisms induced by 1-methyl-4-phenyl-1,2,3,6-tetrahydropyridine (MPTP). *N. Engl. J. Med.*, May 30, 1985, pp. 1418–1421.

deGroot, J. and Chusid, J.G., Eds. (1988). *Correlative Neuroanatomy and Functional Neurology*, 20th ed. Appleton and Lange Medical Publications, Norwalk, CT.

DeJong, R.N. (1979). *The Neurologic Examination*. Harper & Row, Philadelphia.

Gilman, S. and S. Winans-Newman, Eds. (1992). *Manter and Gatz's Essentials of Clinical Neuroanatomy and Neurophysiology*, 8th ed. F.A. Davis, Philadelphia.

Mancall, E.L. (1981). *Alpers and Mancall's Essentials of the Neurologic Examination*, 2nd ed. F.A. Davis, Philadelphia.

Mayo Clinic and Mayo Foundation (1981). *Clinical Examinations in Neurology*, 5th ed. W.B. Saunders, Philadelphia.

Plum, F. and J. Posner (1980). *The Diagnosis of Stupor and Coma*, 3rd ed. F.A. Davis, Philadelphia.

Rowland, L.P., Ed. (1989). *Merritt's Textbook of Neurology*, 8th ed. Lea & Febiger, Philadelphia.

APPENDIX I

Key: A, function; B, anatomy; C, method of testing; D, effect of lesions; E, example of toxin(s)

I. Olfactory
   A. Smell
   B. Olfactory nerve endings in nasal cavity terminate in olfactory bulb Olfactory bulb projects to primary olfactory cortex
   C. Ability to perceive nonirritating odors (e.g., coffee)
   D. Hyposmia or anosmia
   E. Nasally inhaled drugs
II. Optic
   A. Vision
   B. Ganglion cells of retina to optic nerve. Nerve partially decussates at optic chiasm. Optic tract posteriorly to optic cortex through lateral geniculate nucleus of thalamus
   C. Examinations of optic nerve head and retina by opthalmoscopy. Visual acuity testing. Perception of object presented in different quadrants of fields of vision. Pupillary response to light (direct and consensual)
   D. Impairment or loss of vision. Pattern of loss depends upon site of lesion
   E. Methanol
III. Oculomotor
IV. Trochlear
V. Trigeminal
   A. Sensory division innervates face and anterior two thirds of tongue. Includes conjuctivae and mucosa of mouth and nasal cavity. Motor division supplies muscles of mastication
   B. Sensory cell bodies in Gasserion ganglion. Synapse in the pontine principal sensory and mesencephalic nuclei. Also descend in spinal trigeminal tract to upper cervical levels. Motor nucleus in midpons
   C. Light touch and pain sensation over face. Corneal reflex-stroking cornea with cotton wisp and observing lid blinking. Jaw clenching, opening against resistance, and lateral movement. Jaw jerk reflex-contraction of masseters and temporalis when chin is tapped downward
   D. Sensory impairment. Absent corneal reflex. Weakness of jaw muscles. Brisk jaw jerk
   E. Clostridium tetani
VI. Abducens
   A. Ocular motility. Parasympathetic innervation to pupil and lid (III)

B. Nucleus of III in midline midbrain level of superior colliculi (includes parasymphathetic Edinger-Westphal nucleus). Innervates medial, superior and inferior recti, inferior oblique, levator of lid and pupillary sphincter and ciliary muscles. Nucleus of IV caudal medial midbrain. Innervates superior oblique. Nucleus of VI dorsal tegmentum of pons. Innervates lateral recti

C. Observation of conjugate eye movements in each axis of vision. Lid position. Pupillary size and ability to accommodate

D. Impairment of ocular movement. Lid droop (ptosis). Inability to constrict pupil to accommodation. Presence of pupillary dilatation (mydriasis)

E. Botulinum toxin and narcotics (pupil). Neuromuscular blockers

VII. Facial

A. Sensory branch supplies taste to anterior two thirds of tongue. Motor supplies muscles of facial expression and eye closure

B. Nucleus in caudal pons. Supranuclear fibers arise in contralateral precentral motor strip and bilaterally innervate frontalis and obicularis oculi muscles

C. Taste testing on anterior tongue. Ability to move facial muscles (smile, grimace, and wrinkle forehead) and resist forced eye opening

D. Loss of taste on anterior tongue. Motor impairment of speech (dysarthria). Weakness of facial muscles. Supranuclear lesions produce only lower facial weakness whereas nuclear or, more commonly seen, peripheral lesions produce complete unilateral weakness (Bell's palsy). In lower motor neuron lesions, may lose motor component of corneal reflex

E. Neuromuscular blockers

VIII. Acoustic

A. *Cochlear division:* hearing. *Vestibular division:* maintenance of posture and balance

B. *Cochlear:* Organ of Corti in inner ear. Dorsal and ventral cochlear nuclei pontomedullary junction. Bilaterally project to auditory cortex via lateral lemniscus and medial geniculate nuclei of thalamus. *Vestibular:* Semicircular canals. Terminate in four vestibular nuclei of medulla. Diffuse connections via medial longitudinal fasciculus, vestibulocerebellar and vestibulospinal pathways.

C. *Cochlear:* Perception of sounds. Localization of vibrating tuning fork held to midline of forehead (Weber test). Comparison of bone and air condution with tuning fork (Rinne test). *Vestibular:* Observation of gait and balance, and eye movements for nystagmus. Ability to maintain midline gaze with forced passive lateral head movements in comatose patients (oculocephalic or Doll's head phenomenon). Injection of cold water against tympanic membranes in unresponsive patients, observing for eye movements (oculovestibular response or cold water calorics)

    D. *Cochlear:* Hearing impairment. Rarely results from cortical lesion because bilateral innervation. *Vestibular:* Unsteady gait. Nystagmus. Absent Doll's eyes and cold water responses (see above). May be unilateral indicating focal process.

    E. *Cochlear:* aminoglycoside antibiotics, salicylates. *Vestibular:* Ethanol and many other CNS active drugs

IX. Glossopharyngeal

X. Vagus

    A. Muscles of soft palate, pharnyx, and larynx. IX taste to posterior tongue. X parasympathetic innervation to heart and viscera

    B. Medullary nucleus ambiguus, salivatory nuclei, dorsal motor nucleus of X, and fasciculus solitarius.

    C. Ability to raise soft palate (saying "aaah"). Gag reflex (stimulus applied to posterior pharynx). Quality of voice

    D. Raspy voice. Asymmetric palate or gag reflex. Dysphagia (difficulty swallowing)

    E. Rabies. Botulinum toxin

XI. Accessory

    A. Innervation to trapezius and sternocleidomastoid muscles

    B. Spinal and cranial roots of medulla and upper cervical cord

    C. Ability to shrug shoulders and turn head against resistance

    D. Weakness of shoulder elevation or head turning

    E. Idiosyncratic response to phenothiazines may cause sustained contractions (dystonia)

XII. Hypoglossal

    A. Motor innervation of tongue

    B. Nuclei extend length of medulla

    C. Observation of tongue. Direct strength testing against tongue depressor or examiner's hand against cheek. Ability to protrude in midline

    D. Wasting and fasciculations. Weakness. Deviation on protrusion

    E. Clostridia botulinum and tetani. Dyskinesias may result from chronic neuroleptic drugs

INDEX

A

Abducens nerve
  evaluation of, 587–588
  function of, 11
Abstract thinking, 578
Acaricides, organophosphorus compounds
    in, 447
Accessory nerve
  evaluation of, 589
  function of, 11
Acetaldehyde, formation of, 339
Acetyl CoA, 327
  acetylcholine synthesis from, 192
Acetylcholine
  accumulation of, at neuromuscular
    junction, 201
  activating ion channels, 168
  blocking of, 194
  chemical inhibition of, 19–20
  dissociation of, from receptor sites, 193
  effect of, on hippocampus, 8
  functions of, 14–15
  in insect nervous system, 442
  organophosphorus compounds and, 450
  in parasympathetic system, 441
  release of, from synaptic cleft, 193
  release of, into synapse, 191
  respiratory paralysis with interference
    with, 17
  synthesis and release of, 192
  toxins mimicking and antagonizing
    action of, 501–502
Acetylcholine-activated ion channels,
    blocking of, 170–171
Acetylcholine esterase inhibitors, 17
Acetylcholine receptors
  naturally occurring toxins acting on,
    501–502
  organophosphorus compounds and,
    450–451
Acetylcholinesterase, 14
  action of, 451
  aging of phosphorylated, 452
  in basal lamina, 192
  blocking of, 194
  inhibition of, 14–15, 123
    by organophosphorus insecticides, 453

inhibitors of, 456
interaction of, with acetylcholine and
    organophosphorus esters, 449
reactivation of phosphorylated, 452
Acetylenes, see Alkynes
3-Acetylpyridine, neurologic dysfunction
    with, 539
ACh, see Acetylcholine
AChE, see Acetylcholinesterase
Acoustic nerve, evaluation, 588–589
Acoustic startle reflex, 541, 545
Acrylamide, 473–474
  catabolism of, 140–141
  effect of, on motor coordination, 537
  effect of, on smooth membrane axonal
    transport, 102–104, 106–107
  effects of, on axonal transport, 96, 98
  effects of, on high affinity binding
    interactions, 138–143
  increased dopamine receptor binding
    with, 558
  interference of, with neurofilaments, 52
  neurologic dysfunction with, 539
  neurotoxicant effect of, 109
  neurotoxicity of, 201–202
  somatosensory tresholds and, 543
  use of, in schedule-controlled operant
    responding, 555
Actin
  alteration of transport rate of, 56
  association of, with myosin, 193
  IDPN effect on transport of, 51
  inhibition of growth rate of, 55
Actin filaments, stabilization of, 56
Action potential
  changes in muscle compound, 197
  mechanism of generation of, 156–157
  motor unit, duration and amplitude of,
    199–200
  of presynaptic nerve terminal, 175
  range of, 195
Activated charcoal, for benzodiazepine
    intoxication, 497
Active avoidance, 550–552
Acyl CoA, 327
Additive effect, 519
β-Adrenergic receptors, 123–124, 127
Adrenocorticotropic hormone (ACTH)
    release, 9